OPTIMAL CONTROL OF DISCRETE SYSTEMS

V. G. Boltyanskii

Translated from Russian by Ron Hardin

A HALSTED PRESS BOOK

JOHN WILEY & SONS, New York · Toronto

ISRAEL PROGRAM FOR SCIENTIFIC TRANSLATIONS, Jerusalem

Sole distributors for the Western Hemisphere
HALSTED PRESS, a division of
JOHN WILEY & SONS, INC., NEW YORK
ISBN: 0-470-26530-2

Library of Congress Catalog Card No.:
78-67814

Distributors for the U.K., Europe, Africa and the Middle East
JOHN WILEY & SONS, LTD., CHICHESTER

Distributors for Japan, Southeast Asia and India
TOPPAN COMPANY LTD., TOKYO AND SINGAPORE

Distributed in the rest of the world by
KETER PUBLISHING HOUSE JERUSALEM LTD.
ISBN 0 7065 1578 1
IPST cat. no. 222070

This book is a translation from Russian of
OPTIMAL'NOE UPRAVLENIE DISKRETNYMI SISTEMAMI
Izdatel'stvo "Nauka"
Moscow, 1973

Printed and bound by Keterpress Enterprises, Jerusalem
Printed in Israel

CONTENTS

PREFACE v

Chapter I. STATEMENT OF PROBLEM AND NATURE OF RESULTS

§ 1. Optimization of Discrete Processes 1
1. Maximum of product 1
2. Some applied problems 4
3. Optimal control of discrete objects 9
4. Other statements of discrete-control problems 12
5. Maximization of several functionals 20

§ 2. Connection between Discrete-Optimization Problems and Other Extremum Problems 26
6. Extremum of function 26
7. Problem of mathematical programming 30
8. Controlled processes with continuous time 36

§ 3. Methods of Solving Discrete-Optimization Problems 43
9. Dynamic programming 43
10. Discrete maximum principle 50
11. Ideas of mathematical programming 63

Chapter II. BASIC CONCEPTS OF MULTIDIMENSIONAL GEOMETRY

§ 4. Vector Space 76
12. Definition of vector space 76
13. Dimension and basis 81
14. Subspace 88
15. Homomorphisms of vector spaces 95
16. Euclidean vector space 103

§ 5. Euclidean Geometry 112
17. Definition of affine space 112
18. Planes in affine space 120
19. Affine mappings 129
20. Affine functions 136
21. Euclidean space 142
22. Topology of Euclidean space 150
23. Coordinates 154

Chapter III. ELEMENTS OF THEORY OF CONVEX SETS

§ 6. Convex Sets 162
24. Definition of convex set 162
25. Convex hull 166
26. Boundary of convex field 173
27. Convex polyhedron 179

§ 7. Supporting Properties of Convex Sets 187
28. Supporting cone 187
29. Affine function on a convex set 197

§ 8. Theorems on Separability of Convex Cones 204
30. Separation of convex sets 204
31. Dual cone 214
32. Separability of system of convex cones 218

Chapter IV. EXTREMA OF FUNCTIONS
§ 9. Existence Theorems 226
33. Tangent mapping 226
34. Covering of set 240
35. Theorem on intersection 249
§10. Extremum Criteria 265
36. Necessary condition for extremum of function 265
37. Sufficient condition for extremum of function 287
38. Maximum principle 300
39. Method of dynamic programming 308

Chapter V. OPTIMALITY CRITERIA FOR DISCRETE PROCESSES
§11. Dynamic Programming 317
40. Description of method 317
41. Connection with theory of extrema of functions 326
§12. Necessary Conditions for Optimality 329
42. Fundamental problem 330
43. Problem with phase constraints 342
44. Existence theorem 351
45. Discrete objects with variable control domain 354
46. Discrete maximum principle (method of local sections) 358
§13. Sufficient Conditions for Optimality 375
47. Objects with constant control domain 375
48. Objects with variable control domain 384

AUTHOR INDEX 388

SUBJECT INDEX 389

Chapter III contains the necessary information from the theory of convex sets, presented in terms of complete proofs. The theory of convex fields finds wide application in problems of optimal control (including discrete problems), and a knowledge of the concepts and facts associated with this theory is now a necessary element in the education of engineers as well as mathematicians. At the end of the chapter, theorems on the separability of convex cones, which generalize the method of Milyutin and Dubovitskii, are presented.

This chapter can also be regarded as a separate complete entity. For the engineer not wishing to enter into the subtleties of the mathematical proofs, this chapter can be regarded as a handy reference. Taken together, Chapters II and III contain all the necessary geometrical information.

Chapter IV is also a completely self-contained work. It deals mainly with problems of mathematical programming. Results like the Kuhn-Tucker theorem are presented, including some results which are probably new. This chapter serves as a basis for formulating (in the next chapter) the theory of discrete optimal control.

Chapter V, which discusses optimality criteria for discrete control, is the main chapter of the book, with regard to content. Here the necessary and sufficient conditions for optimality are presented in as general a form as possible. In addition, a new version of the discrete maximum principle is given, obtained using the "method of local sections," a technique applied earlier by the author to the theory of continuous optimal control.

It should be noted that problems involving a lag are not considered in detail here (but a method for reducing lag problems to problems without a lag is indicated in Chapter I). The corresponding optimality criteria are given in the works of Fam Huishak; they can also be obtained using the methods described in this book. The works of R. Gabasov and others have also gone beyond the material given in Chapter V, in that they give optimality criteria of orders higher than the first, which can be applied to detect special (degenerate) optimal processes.

These, then, are the specifics of the book: five essentially self-contained, small books which together constitute a single entity. The success of this format for the book is left to the judgment of the reader.

In conclusion, I would like to express my appreciation to all those colleagues and friends who helped make this book possible by offering their attention, remarks, and advice. In particular, the following should be singled out: L.S. Pontryagin, my teacher, N.N. Krasovskii, B.N. Petrov, E.F. Mishchenko, N.Kh. Rozov, and R.M. Mukurdumov.

V. Boltyanskii

And now a few words about the methods on which the theory is based. Two methods were used. The first involves problems on the separability of a system of convex cones. It has been described elsewhere by Milyutin and Dubovitskii. This method is presented here (Subsection 32) in a form which is considerably more general. It differs from the method of Milyutin and Dubovitskii only in that they assume all the cones except one to be solid, whereas we consider the general case.

The second method employed in the book (Subsection 35) involves concepts of topology (and in particular, the theory of intersections dating back to the works of the American mathematician S. Lefschetz). Apropos of this method, the following should be mentioned. In the preface to the book by him referred to above, Pshenichnyi called the initial proof of the maximum principle "somewhat sensational." It is my opinion that the unusual element in this proof was the use of ideas of topology, against which the majority of applied mathematicians are unfortunately prejudiced. I have tried to show here that topological methods are very useful in the mathematical theory of control, and also that they lead to more profound and detailed results. Moreover, a reader to whom topological concepts still seem strange will get an idea of how to do without classical methods of analysis.

There are five chapters in the book. Chapter I contains a preliminary discussion of the optimal control of discrete objects. For a certain circle of readers this discussion will be quite sufficient. In this chapter the problem is posed, some possible methods for solving it are presented, some preliminary formulations of the results are suggested, and the theory of discrete optimal control is compared briefly with other "extremal" theories in mathematics.

This introductory chapter can essentially be considered as a separate small book designed to acquaint the reader with the subject. It is to be hoped that, once familiar with the mathematical statement of the problem and having become reconciled to the type of formulation used (and in particular to the presence of auxiliary variables in the theorems), an engineer using the book will then consider the last chapter as a handy reference, for the use of which he has already been psychologically prepared. If a more profound study of the subject is necessary, the reader is referred to the intermediate chapters.

Chapter II presents the basic facts associated with the theory of vector (finite-dimensional) spaces, affine geometry, and Euclidean geometry. The presentation of these subjects in linear-algebra courses is algebraic, so that it does not ordinarily provide an introduction to the geometrical concepts. Consequently, engineers are often unable, for example, to understand the notation of a hyperplane equation or, in the best case, they write such an equation by analogy with equations known from a course in solid analytic geometry.

This chapter presents a complete formulation of Euclidean (multidimensional) geometry, on the basis of the axiomatics of Weyl. It actually also constitutes an individual small book. The reader can (if he chooses not to go into the geometry so deeply or if he considers himself already familiar with the general points of the subject) use this chapter as a reference for the study of the subsequent chapters, where the geometrical concepts are applied.

PREFACE

A number of examples can be cited from the history of mathematics in which the discrete variant of a theory appeared before the continuous variant, and thus paved the way for the development of the latter. In the theory of optimal control, however, the situation has been otherwise. Here the continuous theory, which was worked out during the past 15 to 20 years, was the first to appear. L. S. Pontryagin's maximum principle proved to be a central, supporting result of the theory. Due to the great significance and popularity of this result, in discrete problems of optimization as well (which became of wide interest somewhat more recently), attempts have been made to find a discrete analog of the maximum principle.

It should be noted that not a few of the many works devoted to this subject have contained mistakes. For example, the book of Fan and Wang (Fan, L.T., and C.S. Wang, *The Discrete Maximum Principle* – John Wiley, New York, 1964) is mathematically incorrect. Consequently, we see that the development of the discrete variant of the theory of optimal control has involved a number of difficulties.

Of the works which have helped create a theory of discrete optimal processes, we should cite the papers of Soviet investigators like N.N. Krasovskii, L.I. Rozonoer, F.M. Kirillova, R. Gabasov, A.G. Butkovskii, and A.I. Propoi, and those of non-Soviet authors like S.S.L. Chang, R. Bellman, E.S. Lee, H. Halkin, J.M. Holtzman, B.W. Jordan, E. Polak, and Fam Hui-shak.

In addition to the papers by the above authors, we should mention as well the small, excellently written book by B.N. Pshenichnyi entitled *The Necessary Conditions for an Extremum* ("Nauka," Moskva, 1969). In particular, this work briefly outlines the theorems which are generalized under the name: "the discrete maximum principle." Finally, there are a number of papers and books which have been especially devoted to describing the method of dynamic programming devised by the American mathematician R. Bellman, a method which differs completely from the discrete maximum principle. The cited studies are essentially all the existing publications concerning the theory of discrete optimal control.

Therefore, this book, which we now offer to the reader's attention, constitutes a first attempt at presenting systematically the theory of discrete optimal control. The presentation is purely mathematical, and no applied problems are considered. On the other hand, even with respect to the mathematical theory of discrete optimal processes, this work does not pretend to be exhaustive. For instance, the following are not considered at all: discrete processes with an infinite number of steps, the approximation of continuous optimal processes by discrete ones, and calculating methods in discrete-control problems. Some of these subjects have been dealt with in the recent book by N.N. Moiseev, *Numerical Methods in the Theory of Optimal Systems* ("Nauka," Moskva, 1971).

FROM THE SOVIET PUBLISHER

The development of the theory of optimal control must be counted as one of the greatest achievements of modern mathematics. This theory has two aspects: continuous and discrete. In its continuous variant, the theory deals with controlled objects that can be described by differential equations; this has been dealt with in a number of detailed monographs. On the other hand, the discrete variant of the theory, which is no less important with regard to both theory and applications, has nowhere been given a comprehensive treatment.

This book fills the referred-to gap in Soviet and non-Soviet literature on mathematics and engineering. The mathematical theory of optimal control for objects with discrete time is presented in a form accessible to an engineer with an undergraduate training in mathematics. Since some new methods and results are described, this book will also be of interest to the mathematician. For the convenience of the reader, the work is divided into five chapters, each of which may be considered to be a separate study. The contents of the chapters are outlined in some detail in the Preface.

FROM THE TRANSLATOR

In the original Russian this book was very carefully organized and written, and the editing and proofreading were almost flawless. All important theorems (followed by their proofs), examples, notes, corollaries, definitions, and problems have been numbered consecutively (according to subsection) throughout the book by the author, so that they can be readily referred to and easily found. Moreover, the number of cited references is happily kept to a minimum.

A work like this is evidence that mathematical analysis, like the entire body of science, is truly an international effort, a human effort. Soviet results and techniques, like Pontryagin's maximum principle and the Dubovitskii-Milyutin theorem, are presented here along with non-Soviet results and techniques, like Bellman's method of dynamic programming.

The language used, like the symbolic notation, can be said to be international as well. In most cases the English cognate of the Russian word was the correct term. For instance, the Russian term "diskretnyi ob"ekt," which occurs often in the book, has been translated literally as "discrete object." This extremely general term, not used so much in English, refers to a quantity, system, or device characterized by discreteness.

An author index and a subject index have been compiled for the convenience of the reader. If properly used, the subject index constitutes a glossary of the terminology of the mathematical theory of optimization, since most terms emphasized and defined in the text have been included in this index.

Remarks or additions made by the translator are either labeled as such or enclosed in square brackets. Note that the sections (1 through 13) and the subsections (1 through 48) are numbered consecutively throughout the book, regardless of which chapter they are in.

Chapter I. STATEMENT OF PROBLEM AND NATURE OF RESULTS

§ 1. Optimization of Discrete Processes

1. **Maximum of product.** In this and the following subsections, several statements of the problem will be considered. Their very different natures notwithstanding, these statements lead, after some mathematical interpretation, to similar posings of a mathematical problem: the problem of optimal control of discrete systems.

EXAMPLE 1.1. Let us consider the following well-known mathematical problem: *find N nonnegative numbers whose sum is not greater than a given number* $a > 0$ *and which have a maximum product.*

We will not solve this problem now, but rather will only reformulate it. To do this, let us perform the following gedanken (thought) experiment. Assume that someone wishes to choose N nonnegative numbers having a sum not greater than a and then, after multiplying them together, to see how much larger the obtained product is. During the first second he chooses one nonnegative number (naturally not exceeding a), during the second second he chooses another (the sum of the two numbers selected not exceeding a), during the third second he chooses a third, . . ., and finally he chooses an Nth number during the Nth second. A number selected at a time t is denoted as $u(t)$. Consequently, the collection of N numbers whose sum does not exceed a is replaced by a collection of values of some function $u(t)$, which is however determined not for all t but only for a *discrete* set of values $t = 1, 2, \ldots, N$.

It is natural to assume that, each time one of the other numbers $u(1), u(2), \ldots, u(N)$ is selected, our someone calculates the sum and product of the already selected numbers. The sum is necessary in order to know the limits of selection for the next number (that is, how much still remains up to a). It is convenient to find the product of the already selected numbers each time, so as to have ready the product of all the numbers as the end of the process is approached. Let $x^1(t)$ be the sum and let $x^2(t)$ be the product of all the numbers which have been selected during t seconds:

$$x^1(t) = u(1) + u(2) + \ldots + u(t);$$
$$x^2(t) = u(1)\,u(2) \ldots u(t).$$

It is clear that, having selected a number $u(t)$ at a time t, our someone will, in order to find the sum of all the previous numbers, simply have to add this number $u(t)$ to the sum $x^1(t-1)$ of all the previously selected numbers:

$$x^1(t) = x^1(t-1) + u(t). \qquad (1.1)$$

Similarly, $x^2(t)$ is found by multiplying the product of all the previous numbers, that is, $x^2(t-1)$, by the number $u(t)$ just selected:

$$x^2(t) = x^2(t-1) \cdot u(t). \tag{1.2}$$

Formulas (1.1) and (1.2) apply for $t = 2, \ldots, N$. For $t = 1$, we have instead the relations $x^1(1) = u(1)$ and $x^2(1) = u(1)$. These two relations can be included in the system of formulas (1.1), (1.2) if we assume that

$$x^1(0) = 0, \quad x^2(0) = 1. \tag{1.3}$$

Then relations (1.1), (1.2) will be valid for all $t = 1, 2, \ldots, N$.

Finally, let us determine the limits within which $u(t)$ can be selected at time t. Since at time t the sum of all the selected numbers cannot exceed a, that is,

$$x^1(t-1) + u(t) \leqslant a,$$

it follows that

$$u(t) \leqslant a - x^1(t-1).$$

Moreover, by definition $u(t)$ must be a nonnegative number, so that $u(t)$ must lie in the interval

$$0 \leqslant u(t) \leqslant a - x^1(t-1), \tag{1.4}$$

that is,

$$u(t) \in U(x^1(t-1)), \tag{1.5}$$

where $U(x^1(t-1))$ denotes interval (1.4).

Note that the product of all the selected numbers $u(1), u(2), \ldots, u(N)$ is $x^2(N)$. Thus our gedanken experiment can be described as follows: *one after the other, the numbers* $u(1), u(2), \ldots, u(N)$ *are selected, the numbers* $x^1(t)$ *and* $x^2(t)$ (see (1.1, 1.2, 1.3)) *being determined as well, along with* $u(t)$, *for each* $t = 1, 2, \ldots, N$. *In addition, constraint* (1.5) *is imposed, where* $U(x^1(t-1))$ *is the interval defined by relation* (1.4). *At the final moment* $t = N$, *it is the value* $x^2(N)$ *which is of interest.*

Let us agree to say that relations (1.1), (1.2) specify a *discrete controlled object*, where x^1, x^2 are the *phase coordinates* and u is the *controlling parameter*, which has to vary within the *control domain* specified by relations (1.4). We call the arbitrary sequence of numbers $u(1), u(2), \ldots, u(N)$ the *control*, and we call the sequences

$$x^1(0),\ x^1(1),\ \ldots,\ x^1(N);$$
$$x^2(0),\ x^2(1),\ \ldots,\ x^2(N),$$

defined by relations (1.1) and (1.2) with the aid of *initial conditions* (1.3), the *trajectory* corresponding to this control. If relation (1.5) is satisfied at each moment $t = 1, 2, \ldots, N$, then the control $u(1), \ldots, u(N)$ is called *admissible*.

Now it is clear that the problem of the maximum of the product can be reformulated as follows: *for discrete controlled object* (1.1), (1.2), *we wish to find an admissible control, and the corresponding trajectory with initial conditions* (1.3), *such that constraint* (1.5) *is satisfied, while the quantity* $x^2(N)$ *has a maximum value.*

EXAMPLE 1.2. Let us consider another means of reducing the problem of the product maximum to a discrete controlled object. To do this, we carry out the gedanken experiment somewhat differently. Assume that the someone in Example 1.1 wishes to take into account only the case where all the multipliers are *positive* (if one of the multipliers were zero, then the product would be zero as well, which naturally does not give a maximum of the product), and that, in order to simplify the multiplication, he calculates not the product itself but rather its logarithm, that is, the *sum* of the logarithms of the multipliers.

Using this approach, the control $u(t)$, $t = 1, 2, \ldots, N$, and the first phase coordinate $x^1(t)$, $t = 0, 1, \ldots, N$, retain their original meaning; relation (1.1) and the first of equations (1.3) are also retained. The second phase coordinate and the relations associated with it become unnecessary; moreover, our someone, after selecting successively the numbers $u(1), u(2), \ldots, u(N)$, calculates the sum of their logarithms and imposes the condition that this sum

$$J = \ln u(1) + \ln u(2) + \ldots + \ln u(N) \tag{1.6}$$

be as large as possible (note that here natural logarithms are used, that is, logarithms to base e).

Next we impose constraints (1.4) in the following manner. Let M_1 be the interval $[0, a]$ and let the following inclusion be satisfied:

$$x^1(N) \in M_1. \tag{1.7}$$

Since, in view of (1.1), we have $x^1(N) = u(1) + u(2) + \ldots + u(N)$, inclusion (1.7) signifies that the sum of all the selected numbers $u(1), u(2), \ldots, u(N)$ does not exceed a. Consequently, it remains just to see to it that all these numbers are positive, that is, constraints (1.4) are replaced by the simpler expressions

$$u(t) > 0, \qquad t = 1, 2, \ldots, N. \tag{1.8}$$

Thus the control domain $U(x^1(t-1))$ (cf. (1.5)) is now the *ray* $U = (0, \infty)$, being independent of $x^1(t-1)$, so that relation (1.8) has a simpler form than (1.5):

$$u(t) \in U. \tag{1.9}$$

The discrete controlled object is thus now specified by relation (1.1), where x^1 is the (unique) phase coordinate and u is the controlling parameter, which has to vary within the limits of the control domain specified by relation (1.8). As a result, the problem of the product maximum considered in the foregoing example can be formulated as follows: *for discrete controlled object* (1.1), *find a control* $u(1), u(2), \ldots, u(N)$ *satisfying condition* (1.8), *such that, for the corresponding trajectory* $x^1(0)$, $x^1(1), \ldots, x^1(N)$ *with an initial condition* $x^1(0)$, *inclusion* (1.7) *is satisfied and sum* (1.6) *has the highest possible value.*

2. Some applied problems. EXAMPLE 2.1.* A farm has a herd of cattle. Each year some of the herd are marketed for beef and the rest are kept on the farm for breeding. The profit from the sale of the cattle to the meat-packer is expressed as a function $\varphi(x)$, where x is the number of cattle sold (function $\varphi(x)$ may, for instance, have the form shown in Figure 1: meat deliveries greater than the specified amount b command a higher price). The number of cattle remaining on the farm for breeding increases by a factor of a (where $a > 1$) in the following year (until the beginning of marketing). How can the farm earn the maximum profit over a period of N years, if the *minimum* yearly amounts of meat marketed are equal to b?

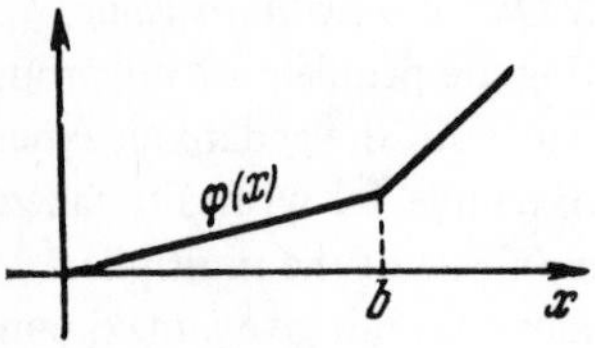

FIGURE 1

Let $x^1(0)$ be the initial number of cattle on the farm, and let $x^1(t)$ be the number of cattle remaining on the farm until the end of year t ($t = 1, 2, \ldots, N$). The number of cattle sold as beef during year t is $u(t)$, $t = 1, 2, \ldots, N$. During year $(t-1)$ a number of cattle equal to $x^1(t-1)$ was kept on the farm for breeding. Consequently, during year t (before marketing) the number of cattle on the farm will be $ax^1(t-1)$. Of this number, $u(t)$ will be marketed for beef, while the rest, that is, $ax^1(t-1) - u(t)$, will stay on the farm to breed until the end of year t. Thus,

$$x^1(t) = ax^1(t-1) - u(t), \qquad t = 1, 2, \ldots, N. \tag{2.1}$$

The farm's profit over N years will be

$$J = \varphi(u(1)) + \varphi(u(2)) + \ldots + \varphi(u(N)) = \sum_{t=1}^{N} \varphi(u(t)). \tag{2.2}$$

Taking into account the contracted deliveries, we impose on the controlling parameter u the following constraints:

$$u(t) \geqslant b, \qquad t = 1, 2, \ldots, N. \tag{2.3}$$

In addition, because of the nature of the problem, phase coordinate x^1 (that is, the number of cattle kept for breeding) is nonnegative:

$$x(t) \geqslant 0, \qquad t = 1, 2, \ldots, N. \tag{2.4}$$

Thus the given problem on the maximum profit of a livestock farm over a period of N years can be formulated as follows: *for discrete controlled object* (2.1), *find a control* $u(1), u(2), \ldots, u(N)$ *satisfying condition* (2.3), *such that, for the corresponding trajectory* $x^1(0), x^1(1), \ldots, x^1(N)$ *with the specified initial condition* $x^1(0) = c$, *relation* (2.4) *is satisfied and sum* (2.2) *is as large as possible.*

It should be noted that in Example 2.1 the number $x^1(N)$, that is, the number of cattle remaining on the farm until the end of year N, is in no way regulated by the

* See: Bellman, R. *Dynamic Programming.* Princeton Univ. Press, Princeton, N.J. 1960.

conditions of the problem. Thus it is clear that, in order to obtain the maximum profit over N years, it would be advisable during year N to market *all* the cattle remaining at the end of that year, the result being that the farm would cease to exist after year N. If, on the other hand, this is not the plan, then the last of conditions (2.4) should be replaced by the condition $x^1(N) \geqslant d$, where d is the scheduled breeding of the cattle at the end of the N-year period.

EXAMPLE 2.2. We wish to design an N-stage space rocket having a specified starting weight G. We also know the weight H of the spacecraft which is to be put on trajectory by the last stage of the rocket vehicle. Each stage of the rocket vehicle has a fuel supply. During the operation of each successive stage (from the moment its engine ignites to detachment of the burned-out stage), the rocket acquires an additional speed Δv, which is a function of the payload P carried by this stage and the weight Q of the stage itself (determining the fuel supply needed):

$$\Delta v = f(P, Q).$$

Let us find the distribution of weight among the stages for which the velocity of the spacecraft (after detachment of the last stages of the rocket vehicle) will be a maximum.*

If $u(t)$ is the weight of stage t, *reckoned from the spacecraft*, then the weight of the final stage (which places the spacecraft on trajectory and detaches after all the other stages) is denoted as $u(1)$, the weight of the penultimate stage as $u(2)$, etc. If, in addition, the weight of the spacecraft plus the t stages attached to it is $x^1(t)$, $t = 0, 1, \ldots, N$ (Figure 2), then it is obvious that

$$x^1(t) = x^1(t-1) + u(t), \qquad (2.5)$$
$$t = 1, 2, \ldots, N,$$

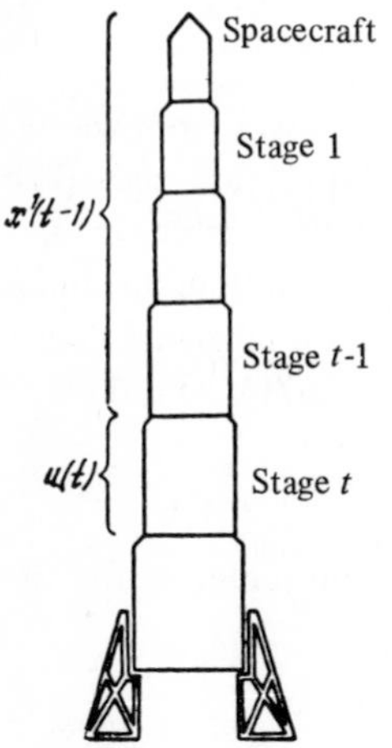

FIGURE 2

the weight of the spacecraft and the initial weight of the rocket being specified by the following conditions:

$$x^1(0) = H, \quad x^1(N) = G.$$

In view of the foregoing, the additional velocity imparted to stage t while it burns will be

$$\Delta v_t = f(x^1(t-1), u(t)),$$

since $u(t)$ is the weight of stage t, and $x^1(t-1)$ is the weight of the load that it carries. The total velocity imparted to the spacecraft by all the stages of the rocket vehicle is thus

$$J = \sum_{t=1}^{N} f(x^1(t-1), u(t)). \qquad (2.6)$$

* This problem was taken from the book by E.S. Venttsel', *Elementy dinamicheskogo programmirovaniya (Elements of Dynamic Programming)*, pp.99–102. – "Nauka," Moskva. 1964.

Consequently, the problem of the maximum spacecraft velocity can be formulated as follows: *for discrete controlled object* (2.5), *find a control* $u(1), u(2), \ldots, u(N)$ *such that, for the corresponding trajectory* $x^1(0), x^1(1), \ldots, x^1(N)$ *with initial condition* $x^1(0) = H$, *the relation* $x^1(N) = G$ *is valid and sum* (2.6) *is as large as possible.* Here controlling parameter u naturally also has constraints of type $u(t) \geqslant b_t$ imposed upon it, although this was not mentioned.

EXAMPLE 2.3.* A certain substance (product of a chemical plant) is in solution and is to be extracted by washing. The washing is carried out in N identical units, through which the solution flows successively (Figure 3), there being no mixing of the solution and the water during the washing process (for instance, solution and washwater can be separated by a thin film with a selective permeability). The amount of substance extracted during a given stage of washing is expressed as a function $\varphi(x, u)$, where x is the amount of substance in solution (at the inflow of the washing unit) and u is the amount of water used. The profit earned by the chemical plant is determined as $\beta_1 X - \beta_2 U$, where X is the amount of substance extracted (all washing stages) and U is the quantity of water used. The problem consists in finding the regime of washing that yields maximum profit.**

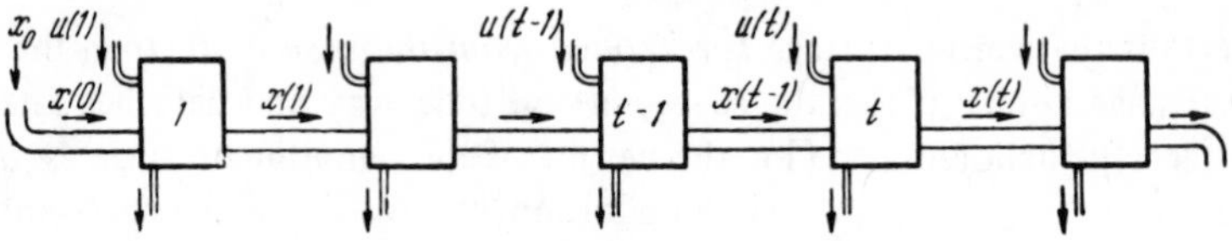

FIGURE 3

Let $x(t)$ be the amount of substance in the solution when it emerges from washing stage t (or, what comes to the same thing, when it enters stage $t+1$), and let $u(t)$ be the amount of washwater supplied to unit t. Then, as a result of the washing in unit t, an amount of substance equal to $\varphi(x(t-1), u(t))$ will be extracted from the solution (which initially contained an amount $x(t-1)$), so that the amount of substance in the solution emerging from stage t drops to $x(t-1) - \varphi(x(t-1), u(t))$, and thus

$$x(t) = x(t-1) - \varphi(x(t-1), u(t)), \qquad t = 1, 2, \ldots, N, \quad (2.7)$$

where $x(0) = x_0$ is the amount of substance in the initial solution (entering the first washing stage). The total profit of the plant is

$$J = \beta_1 (x(0) - x(N)) - \beta_2 \sum_{t=1}^{N} u(t). \quad (2.8)$$

* See: Fan, L.T., and C.S. Wang. *The Discrete Maximum Principle.* – John Wiley, New York. 1964.

** Note that, in reality, the solution would flow *continuously* along a conduit passing through all the washing stages. We can, however, imagine a part of the solution to be separate and assume that it remains in the first washing unit for a certain time, is then in the second unit for this same time, etc. In such a case u will denote the amount of water entering the washing unit during the time of passage of the considered part of the solution, that is, it is the rate of the water supply, while x can be taken to be the *concentration* of the substance in the solution.

Here the controlling parameter $u(t)$ has imposed on it the constraints

$$0 \leqslant u(t) \leqslant b, \qquad t = 1, 2, \ldots, N, \tag{2.9}$$

where b is determined by the structural properties of the washing unit.

Consequently, the problem of the most favorable washing regime can be formulated as follows: *for discrete controlled object* (2.7), (2.9), *find a control* $u(1), u(2), \ldots, u(N)$ *such that, for the corresponding trajectory* $x(0), x(1), \ldots, x(N)$ *and initial condition* $x(0) = x_0$, *sum* (2.8) *is as large as possible.*

EXAMPLE 2.4.* As our last example, we consider the *transport problem:* the delivery of a commodity ("raw material") from the suppliers ("warehouse") to the consumers ("factories").

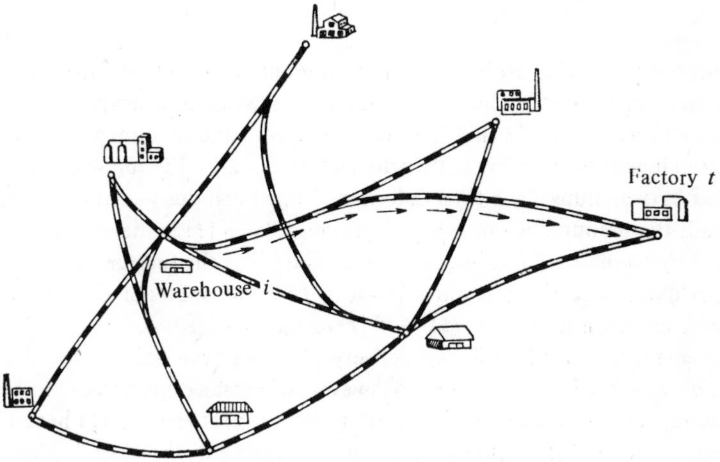

FIGURE 4

For greater simplicity and clarity, we assume that three warehouses are involved, and that there are N factories involved. Moreover, we only consider one kind of raw material and we assume that the supply is equal to the demand; in other words, if a_1, a_2, a_3 are the stocks of raw material at the warehouses, and if $b_1, b_2, \ldots, b_N$ are the requirements of the factories, then we have

$$a_1 + a_2 + a_3 = b_1 + b_2 + \ldots + b_N. \tag{2.10}$$

The cost of delivering u units of raw material from warehouse i to factory t depends on u (that is, on *how much* raw material has to be brought), and also on i and t

* The reduction of linear-programming problems to problems of the control of discrete systems has been considered by N.N. Moiseev, *O primenenii metodov optimal'nogo upravleniya k zadache optimal'nogo planirovaniya (The Application of Optimal-Control Methods to the Problem of Optimal Planning).* – Kibernetika, No. 2. 1966.

(these numbers characterize the delivery *distance*, Figure 4); this cost is called $\varphi_t^i(u)$ (Figure 5). The problem consists in delivering the raw material from the warehouses to the factories at minimal transport cost.

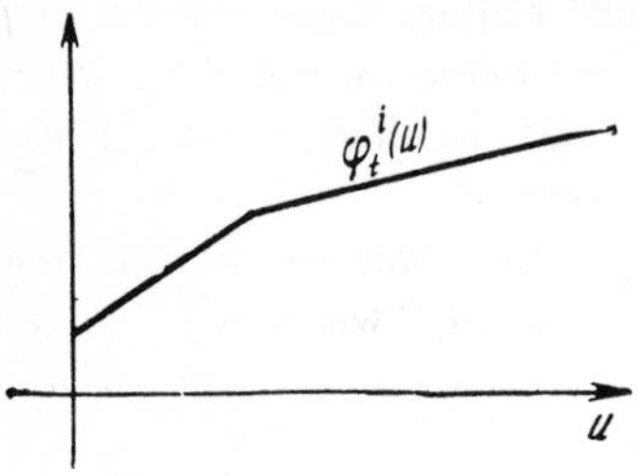

FIGURE 5

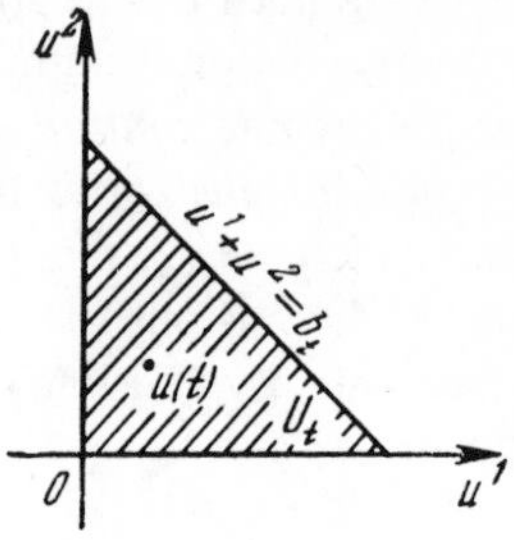

FIGURE 6

The solution of this problem can be represented as follows. In order to work out the delivery plan, we first note the amount of raw material which has to be transported from warehouses 1 and 2 to factory 1, next the amount to be brought from these warehouses to factory 2, then to factory 3, etc. The amount of raw material supplied by warehouse 3 is then determined unambiguously: it must supply each factory with the amount of raw material lacking. If $u^1(t)$ is the amount of raw material supplied by warehouse 1 to factory t, and if $u^2(t)$ is the amount supplied to it by warehouse 2, then, since the demand of factory t for raw material is b_t, warehouse 3 has to deliver an amount $b_t - u^1(t) - u^2(t)$ to factory t.

Consequently, to work out the delivery plan, we have only to specify the numbers $u^1(t), u^2(t)$, $t = 1, 2, \ldots, N$. However, when selecting these numbers, several precautions must be taken. First of all, numbers $u^1(t)$ and $u^2(t)$ have to be non-negative, in view of the nature of the problem, and their sum should not exceed b_t:

$$u^1(t) \geqslant 0, \quad u^2(t) \geqslant 0, \quad u^1(t) + u^2(t) \leqslant b_t, \qquad t = 1, 2, \ldots, N. \quad (2.11)$$

In other words, the "controlling point" $u(t) = (u^1(t), \ u^2(t))$ must lie within the triangle U_t shown in Figure 6:

$$u(t) \in U_t. \quad (2.12)$$

On the other hand, this is not enough. If we do not plan the deliveries (that is, choose numbers $u^1(t)$, $u^2(t)$, $t = 1, 2, \ldots, N$) carefully, then it may be that we will "plan" to take from one of the warehouses more raw material than it holds. To keep this from happening, the amount of raw material destined to be removed from each warehouse has to be taken into account. Thus let $x^1(t)$ and $x^2(t)$ be the amounts of raw material delivered to the first t factories from the first and second warehouses, respectively:

$$x^1(t) = u^1(1) + u^1(2) + \ldots + u^1(t);$$
$$x^2(t) = u^2(1) + u^2(2) + \ldots + u^2(t).$$

Then, when scheduling the deliveries to factory $t+1$, we will have an idea of the amounts of raw material still available at the first two warehouses. Obviously,

$$x^1(t)=x^1(t-1)+u^1(t),\quad x^2(t)=x^2(t-1)+u^2(t),\qquad (2.13)$$
$$t=1,\ 2,\ \ldots,\ N,$$

where, for convenience, it is assumed that

$$x^1(0)=0,\quad x^2(0)=0.\qquad (2.14)$$

Since the supply is equal to the demand, that is, relation (2.10) is valid, it follows that after the deliveries to all N factories the warehouses will be empty, that is,

$$x^1(N)=a_1,\quad x^2(N)=a_2.\qquad (2.15)$$

Clearly, if relations (2.11) and (2.15) hold, then the delivery plan will be satisfactory (the factories will receive the required amounts of raw material from the three warehouses). Then the cost of all the deliveries (according to the plan drawn up) will be

$$J=\sum_{t=1}^{N}\left[\varphi_t^1(u^1(t))+\varphi_t^2(u^2(t))+\varphi_t^3(b_t-u^1(t)-u^2(t))\right].\qquad (2.16)$$

Consequently, the transport problem can be formulated as follows: *for discrete controlled object* (2.13), *find an admissible control* $u^1(t),\ u^2(t),\ t=1,\ 2,\ \ldots,\ N$ *(that is, one which satisfies condition* (2.11)), *such that, for the corresponding trajectory* $x^1(t), x^2(t),\ t=0, 1, \ldots, N$ *and initial condition* (2.14), *final condition* (2.15) *is satisfied and sum* (2.16) *is as small as possible.*

3. Optimal control of discrete objects. By generalizing the examples considered in Subsections 1 and 2, we will now arrive at an overall mathematical description of discrete controlled objects. Assume, as before, that variable t (which will sometimes be referred to as the "time") can take on only a discrete set of values: $t=0,\ 1,\ \ldots,\ N$, where N is some definite natural number.

In all the foregoing examples except the last, there was only a single controlling parameter u, whereas in Example 2.4 there were two controlling parameters, u^1 and u^2. In the general case, it is assumed that on the controlled object there may act a (more or less arbitrary) selection of r *controlling parameters* $u^1,\ \ldots,\ u^r$, or, what comes to the same thing, a point u of the space of variables $u^1,\ \ldots,\ u^r$. Thus at each moment t the controlling point $u(t)$ has r coordinates:

$$u(t)=(u^1(t),\ \ldots,\ u^r(t)).$$

We arbitrarily define the *control* as the sequence of points

$$u(1),\ u(2),\ \ldots,\ u(N)\qquad (3.1)$$

in the space of variables $u^1,\ \ldots,\ u^r$.

In the problems considered above, the state of the object was characterized by one or two phase coordinates. In the general case, however, it can be assumed that at each moment t the state of the object is characterized by n *phase coordinates* $x^1, \ldots, x^n$, that is, by a point x in the space E^n of variables $x^1, \ldots, x^n$. Therefore, at each moment t the phase state $x(t)$ has n coordinates:

$$x(t) = (x^1(t), \ldots, x^n(t)).$$

The sequence

$$x(0),\ x(1),\ \ldots,\ x(N) \tag{3.2}$$

of states of the object at moments $t = 0, 1, \ldots, N$ will be called the *trajectory* of motion of the object.

The *initial state* $x(0)$ must be specified. The subsequent behavior of the object will be determined unambiguously if we select a certain control (3.1), with the aid of the relations

$$x(t) = f_t(x(t-1), u(t)), \qquad t = 1, \ldots, N, \tag{3.3}$$

where $f_t(x, u) = \left(f_t^1(x, u), \ldots, f_t^n(x, u)\right)$ is some vector function with values in space E^n. The subscript t of function $f(x, u)$ indicates that we consider not a single function $f_t(x, u)$ for all moments $t = 1, \ldots, N$, but rather various functions that change from one moment to the next. If, on the other hand, function $f_t(x, u)$ does not actually depend on t, then we will have only a single function $f(x, u)$, and relation (3.3) becomes

$$x(t) = f(x(t-1), u(t)).$$

Such was precisely the case in the problems considered previously; for instance, in Example 1.1 the behavior of the discrete controlled object is described by relations (1.1) and (1.2), the right-hand sides of which do not depend explicitly on t.

Relations 3.3 constitute a *law of motion* of the discrete controlled object. Trajectory (3.2), which satisfies relation (3.3), is said to *correspond* to the initial state $x(0)$ and control (3.1).

Furthermore, for each point $x \in E^n$ and each $t = 1, \ldots, N$, some nonempty set $U_t(x)$ is specified in the space of variables $u^1, \ldots, u^r$; this is the *control domain*, corresponding to phase state x at time t. Let us consider only those controls (3.1) which satisfy the following condition (cf. (1.5), (1.9), (2.12)):

$$u(t) \in U_t(x(t-1)), \qquad t = 1, \ldots, N, \tag{3.4}$$

where trajectory (3.2) starts from initial point $x(0)$ and corresponds to control (3.1). Controls satisfying this condition are called *admissible* (relative to initial state $x(0)$).

Relations (3.3) and (3.4) also define the *discrete controlled object*. The process of control of this object is effected in the following manner. Since the initial phase state $x(0)$ is specified, we know the corresponding control domain $U_1(x(0))$. In view of (3.4), we can select an arbitrary controlling point $u(1) \in U_1(x(0))$, after which phase state $x(1)$ is determined at moment $t = 1$ (see (3.3)). Then, knowing

$x(1)$, we can consider the corresponding control domain $U_2(x(1))$. Next, having selected an arbitrary controlling point $u(2) \in U_2(x(1))$ (see (3.4)), we can find the next phase state $x(2)$ (see (3.3)), etc. It is clear that control (3.1), obtained as a result of such a successive procedure, will be admissible (relative to the initial state $x(0)$) and that the obtained trajectory (3.2) will correspond to this control.

Now let us state the *problem of optimal control* for discrete controlled object (3.3), (3.4). To do this, we assume that certain functions $f_t^0(x, u)$, $t = 1, \ldots, N$, are specified. As an *effectiveness criterion,* that is, a functional indicating how "favorable" the selected process (3.1), (3.2) will be, we take the following:

$$J = f_1^0(x(0), u(1)) + f_2^0(x(1), u(2)) + \ldots + f_N^0(x(N-1), u(N)) = \sum_{t=1}^{N} f_t^0(x(t-1), u(t)). \tag{3.5}$$

The problem of optimal control consists in *selecting, for a known initial state* $x(0)$, *an admissible control* (3.1) *for object* (3.3), (3.4), *such that functional* (3.5) *has a maximum value* (in some cases functional (3.5) will have a *minimum* value; this was the case in Example 2.4).

This problem, which we will refer to as the *fundamental problem,* can be characterized as a problem of optimal control with a *fixed left-hand endpoint and a free right-hand endpoint.* In other words, the initial state $x(0)$ is assumed to be given, while the state at the right-hand endpoint of the time interval, that is, $x(N)$, is in no way restricted (except that the value of functional (3.5) must be a maximum). The problem considered in Example 2.3 is a typical optimal-control problem with a fixed left-hand endpoint and a free right-hand endpoint.

In addition to the fundamental problem, we can also consider a *problem with moving endpoints.* Here we assume that, in phase space E^n, two sets M_0 and M_N are specified, and we state the problem as follows: *find an initial state* $x(0) \in M_0$ *and an admissible (relative to* $x(0)$*) control* (3.1), *such that the relation* $x(N) \in M_N$ *is valid and at the same time functional* (3.5) *is as large as possible.*

Obviously, if M_0 consists of one point, while M_N coincides with the entire phase space E^n, then the problem with moving endpoints becomes the fundamental problem considered earlier.

Moreover, if each of the sets M_0, M_N consists of only a single point (that is, both initial state $x(0)$ and final state $x(N)$ are specified in advance), then we have a *problem with fixed endpoints.*

In Example 1.2 we had a problem with a fixed left-hand endpoint and a moving right-hand endpoint (see (1.7)). In Examples 2.2 and 2.4 we considered problems with fixed endpoints.

Finally, it is also possible that, for *each* $t = 0, 1, \ldots, N$, a certain set M_t is specified in phase space E^n, the problem being stated as follows: *find an initial condition* $x(0)$ *and an admissible (relative to* $x(0)$*) control* (3.1), *such that relations* $x(t) \in M_t$, $t = 0, 1, \ldots, N$, *are satisfied and at the same time functional* (3.5) *is as large as possible.* Let us call this a *problem with constraints on the phase coordinates.* When $M_1 = \ldots = M_{N-1} = E^n$, it becomes a problem with moving endpoints.

Thus the problem with constraints on the phase coordinates is the most general of all the problems considered. This is the problem we had in Example 2.1 (see relation (2.4), which has the form $x(t) \in M_t$, where M_t is the ray $[0, \infty)$).

4. Other statements of discrete-control problems. The problems of optimal control of discrete objects formulated in the previous subsection constitute the basic theme of the studies in this book. It is for problems thus formulated that the main results will be proven. However, these problems (even the most general of them, namely the problem with constraints on the phase coordinates) are not all-embracing. Some other statements of problems are also encountered in the theory of discrete systems: optimal problems for systems with feedback, discrete objects with lag, etc. Here we will consider some of these problems and we will show how they can be reduced to problems formulated in the previous subsection.

Systems with feedback. Let us consider again the problem of the successive extraction of a substance from solution (Example 2.3). If the washing process takes place very slowly, then the solution emerging from washing unit N will still contain a considerable amount of substance, and the pouring off of the solution will lead to sizable losses. In such cases it is advisable to pour off only a small part of the solution emerging from washing stage N, and to send the rest of it back to the intake of the first unit, thereby providing "feedback" (Figure 7).

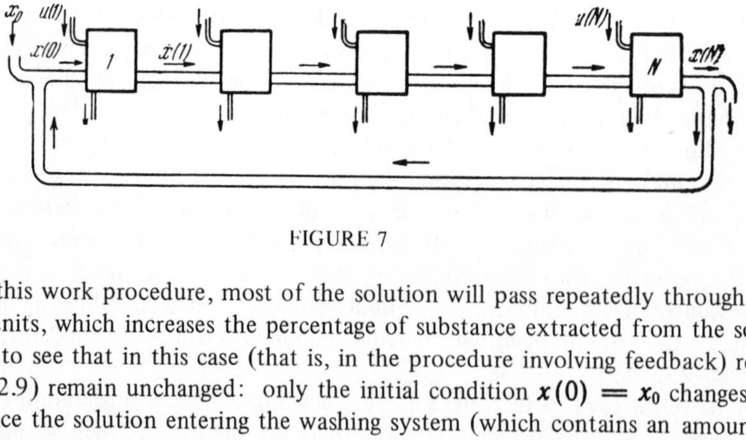

FIGURE 7

Using this work procedure, most of the solution will pass repeatedly through all N washing units, which increases the percentage of substance extracted from the solution. It is easy to see that in this case (that is, in the procedure involving feedback) relations (2.7) to (2.9) remain unchanged: only the initial condition $x(0) = x_0$ changes. Actually, since the solution entering the washing system (which contains an amount x_0 of the product substance) mixes with some of the solution emerging from washing stage N, it follows that $x(0)$(that is, the amount of substance in solution at the intake of the first stage) will already be different from x_0, being some combination of the quantities x_0 and $x(N)$, namely

$$x(0) = \frac{x_0 + \lambda x(N)}{1 + \lambda}.$$

In other words, instead of the equality $x(0) = x_0$, we now have some *relation* between $x(0)$ and $x(N)$), which can in general be written as $\varphi(x(0), x(N)) = 0$.

In general, a discrete system with feedback can be described mathematically by relations (3.3) and (3.4), with additional conditions of the form

$$\varphi_i(x(0), x(N)) = 0, \qquad i = 1, 2, \ldots, l \tag{4.1}$$

(these replace the previously imposed requirements $x(0) \in M_0$, $x(N) \in M_N$). In addition, constraints may be placed on the phase coordinates: $x(t) \in M_t$, $t = 1, 2, \ldots, N-1$. The problem, as previously, consists in maximizing functional (3.5) with all these requirements satisfied.

The problem stated can be reduced to the problems considered in Subsection 3 using the following technique. Let us add to phase variable $x = (x^1, x^2, \ldots, x^n)$ another variable $y = (x^{n+1}, \ldots, x^{2n})$, which obeys the conditions

$$y(t) = y(t-1), \qquad t = 1, \ldots, N; \tag{4.2}$$

$$y(0) = x(0). \tag{4.3}$$

Obviously, in view of these conditions, $y(0) = y(1) = \ldots = y(N) = x(0)$, so that relation (4.1) can be written as

$$\varphi_i(y(N), x(N)) = 0, \qquad i = 1, 2, \ldots, l. \tag{4.4}$$

Thus we arrive at the following statement of the problem. Given $2n$ phase coordinates $x^1, \ldots, x^n, x^{n+1}, \ldots, x^{2n}$ and r controlling parameters $u^1, \ldots, u^r$, as previously, relations (3.3), (3.4), (4.2) define the discrete controlled object. Relation (4.3) indicates that the initial point $(x(0), y(0))$ belongs to the set M_0^* defined in the space of variables $x^1, \ldots, x^{2n}$ by equations $x^i = x^{i+n}$, $i = 1, 2, \ldots, n$. Relation (4.4) can be written as $(x(N), y(N)) \in M_N^*$, where set M_N^* is defined by the equations $\varphi_i(y, x) = 0$, $i = 1, \ldots, l$, while functional (3.5) turns out to be dependent only on coordinates $x^1, \ldots, x^n$.

Consequently, by increasing (doubling) the number of phase coordinates, we can reduce a discrete problem with feedback to a problem with moving endpoints (or to a problem with constraints on the phase coordinates, if constraints $x(t) \in M_t$, $t = 1, 2, \ldots, N-1$) are introduced into the statement of the problem with feedback). Note that doubling the number of phase coordinates does not in practice complicate the problem much, since relations (4.2) are especially simple in form.

Systems with lag. In order to represent problems with lag, let us restate the problem considered in Example 2.1. Now we take into account that there are young stock (yearlings) on the farm each year, who will not produce progeny during the coming year, in addition to adult animals. Assume that, if the number of adult animals remaining on the farm is y, their offspring during the next year will number ky. Also assume that only adult cattle are delivered to the slaughterhouse.

The constraints $x(t)$, $u(t)$ in Example 2.1 are retained in this case. Then, during year t, offspring will be obtained only from those animals born during year $t-2$ or earlier, that is, from animals that were left on the farm until the end of year $t-2$ and were kept there until year t. However, $x(t-2)$ animals were left on the farm during year $t-2$, and $u(t-1)$ of these were shipped for beef during year $t-1$.

Thus in year t there were $x(t-2)-u(t-1)$ *adult* animals on the farm, so that in year t the progeny amounted to $k(x(t-2)-u(t-1))$ animals. From this it follows that, before the meat shipments of year t, there were $x(t-1)+k(x(t-2)-u(t-1))$ animals on the farm. After the shipments this number was reduced by $u(t)$, to the number of animals on the farm at the end of year t. Therefore,

$$x(t)=x(t-1)+k(x(t-2)-u(t-1))-u(t), \qquad t=2, 3, \ldots, N. \tag{4.5}$$

It is seen that, in contrast to (3.3), this relation has the form

$$x(t)=f_t(x(t-1), x(t-2), u(t), u(t-1)), \quad t=2, 3, \ldots, N, \tag{4.6}$$

that is, the phase state $x(t)$ depends not only on $x(t-1)$ and $u(t)$ but also on the values $x(t-2)$ and $u(t-1)$ preceding them. Relation (4.5) is written only for $t=2, \ldots, N$, since if $t=1$ is substituted into it the quantities $x(-1)$ and $u(0)$ will appear on the right-hand side, and these were not defined in the problem. Thus the values of $x(0)$ and $x(1)$ should be specified separately, and the other states $x(2), \ldots, x(N)$ can be obtained from formula (4.5).

On the other hand, if we know the number of animals on the farm per year up to the planning date (that is, $x(-1)$), as well as the meat shipments during the year prior to the first year of planning (that is, $u(0)$), then relation (4.5) will be meaningful for $t=1$ as well.

Continued fractions (or, as they are still called, *chain fractions*) constitute another example of a system with a lag. A continued fraction has the form

$$a_0+\cfrac{1}{a_1+\cfrac{1}{a_2+\cfrac{\cdot}{\cdot\, \cdot + \cfrac{1}{a_N}}}} \tag{4.7}$$

where $a_0, a_1, \ldots, a_N$ are certain numbers. If we break off fraction (4.7) at the number a_t, we obtain the expression

$$a_0+\cfrac{1}{a_1+\cfrac{1}{a_2+\cfrac{\cdot}{\cdot\, \cdot + \cfrac{1}{a_t}}}} \tag{4.8}$$

known as the t-th *convergent* of continuous fraction (4.7). If we eliminate the "multistage" fractions in (4.8), then we can write this expression as an ordinary fraction, which we will denote as $\frac{P_t}{Q_t}$. For instance,

$$a_0 = \frac{a_0}{1} = \frac{P_0}{Q_0}; \quad a_0 + \frac{1}{a_1} = \frac{a_0 a_1 + 1}{a_1} = \frac{P_1}{Q_1};$$

$$a_0 + \cfrac{1}{a_1 + \cfrac{1}{a_2}} = \frac{a_0 a_1 a_2 + a_0 + a_2}{a_1 a_2 + 1} = \frac{P_2}{Q_2}, \text{ etc.}$$

Thus the numerators and denominators of the convergents can be determined as follows:

$$P_0 = a_0, \quad P_1 = a_0 a_1 + 1, \quad P_2 = a_0 a_1 a_2 + a_0 + a_2, \ldots;$$
$$Q_0 = 1, \quad Q_1 = a_1, \quad Q_2 = a_1 a_2 + 1, \ldots$$

It can be shown by induction* that the numerators and denominators of the convergents are defined by the relations

$$P_t = a_t P_{t-1} + P_{t-2}, \quad Q_t = a_t Q_{t-1} + Q_{t-2}. \tag{4.9}$$

Note that these relations are valid for $t = 2, 3, \ldots N$; however, if (purely formally) we set $P_{-1} = 1$ and $Q_{-1} = 0$, then these relations will be valid for $t = 1$ as well.

Now we can assume that someone, on the basis of his own judgment, selects the *control*

$$u(1) = a_1, \quad u(2) = a_2, \ldots, \quad u(N) = a_N$$

and calculates successively the convergents (first, second, etc.), considering their numerators and denominators as the phase coordinates:

$$x^1(1) = P_1, \quad x^2(1) = Q_1; \quad x^1(2) = P_2,$$
$$x^2(2) = Q_2; \ \ldots; \ x^1(N) = P_N, \quad x^2(N) = Q_N.$$

Then relations (4.9) define the discrete controlled object

$$\begin{aligned} x^1(t) &= u(t)\, x^1(t-1) + x^1(t-2), \\ x^2(t) &= u(t)\, x^2(t-1) + x^2(t-2). \end{aligned} \tag{4.10}$$

Here, in contrast to (3.3), the right-hand sides depend on $x(t-2)$ as well as on $u(t)$, $x(t-1)$.

The discrete objects (4.5), (4.10) considered above are called *objects with a lag* (since $x(t-2)$, that is, the information on the state of the object at moment $t-2$, not only participates in the formation of state $x(t-1)$ but also, as if it were delayed or retarded somewhere, participates in the formation of state $x(t)$ at moment t).

* See, for instance, the book by A.Ya. Khinchin, *Tsepnye drobi (Chain Fractions)*, pp.10–12. – Gostekhizdat, Moskva-Leningrad. 1949.

In general, a *discrete object with a lag of one step* can be described as follows. Let $x = (x^1, \ldots, x^n)$ be the vector of the phase state, and let $u = (u^1, \ldots, u^r)$ be the controlling vector. The *prehistory* of the process is known, that is, states $x(-1)$ and $x(0)$, and the further behavior of the object is determined by the relations

$$x(t) = f_t(x(t-1), \quad x(t-2), \ u(t)), \qquad t = 1, 2, \ldots, N. \tag{4.11}$$

The problem consists in finding a control $u(1), u(2), \ldots, u(N)$, having determined according to which trajectory $x(-1), x(0), x(1), \ldots, x(N)$ we can maximize the functional

$$J = \sum_{t=1}^{N} f_t^0(x(t-1), x(t-2), u(t))$$

(it may be that here constraints of type $u(t) \in U_t(x(t-1), \ x(t-2))$ are placed on the controls, while constraints of type $x(t) \in M_t$ are placed on the phase coordinates).

In order to reduce this problem to the problems considered in Subsection 3, we introduce an additional phase variable $y = (x^{n+1}, \ldots, x^{2n})$, subject to the conditions:

$$y(t) = x(t-1), \qquad t = 0, 1, \ldots, N. \tag{4.12}$$

Then $y(0) = x(-1)$, while relation (4.11) can be written in the form

$$x(t) = f_t(x(t-1), y(t-1), u(t)), \qquad t = 1, 2, \ldots, N, \tag{4.13}$$

while restraint $u(t) \in U_t(x(t-1), x(t-2))$ takes the form

$$u(t) \in U_t(x(t-1), y(t-1)). \tag{4.14}$$

Consequently, we arrive at the following statement of the problem. There are now $2n$ phase coordinates $x^1, \ldots, x^n, x^{n+1}, \ldots, x^{2n}$, and the same r controlling parameters $u^1, \ldots, u^r$ as before. Relations (4.12)–(4.14) define the discrete controlled object, for which the initial point $(x(0), y(0))$ is specified; we have to maximize the functional

$$J = \sum_{t=1}^{N} f_t^0(x(t-1), y(t-1), u(t)).$$

Accordingly, by increasing (doubling) the number of phase coordinates, we reduce a discrete problem with a lag (of one step) to a problem with a fixed left-hand endpoint.

If the term $u(t-1)$ also enters into the right-hand side of relation (4.11) (cf. (4.6)), then additional phase coordinates again must be introduced. By introducing (along with x and y) the phase coordinate $z = (x^{2n+1}, \ldots, x^{2n+r})$, and imposing upon it the conditions

$$z(t) = u(t),$$

we are able to write a relation of type (4.6) in the form

$$x(t) = f_t(x(t-1), y(t-1), u(t), z(t-1)),$$

which is in complete agreement with the description of the controlled object in form (3.3).

Finally, it should be noted that discrete controlled objects with lags of θ steps (where θ is a natural number lower than N) may be encountered. These objects are described by relations of the type

$$x(t) = f_t(x(t-1), x(t-2), \ldots, x(t-\theta-1), u(t)), \qquad (4.15)$$
$$t = 1, 2, \ldots, N,$$

the *prehistory* $x(-\theta)$, $x(-\theta+1)$, $\ldots$, $x(-1)$, $x(0)$ being specified, and constraints may be placed on the controlling parameters and on the phase coordinates. For such objects the introduction of an additional phase variable $y = (x^{n+1}, \ldots, x^{2n})$ satisfying condition (4.12) reduces θ by unity, that is, it reduces the considered object to another object having a lag of $\theta - 1$ steps. Thus, by introducing enough additional phase coordinates, we can reduce object (4.15) to an object without a lag.

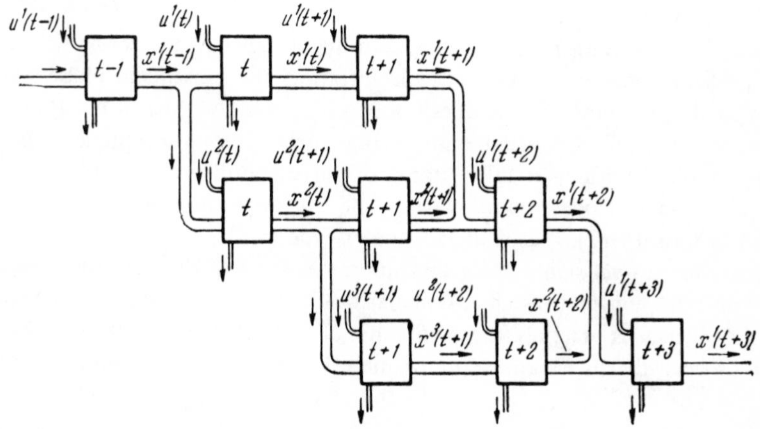

FIGURE 8

Systems with variable number of phase coordinates. In some cases, for the processes considered, the number of phase coordinates describing the behavior of the object varies along with the discrete time t. Assume, for example, that in a washing system (see Example 2.3) after stage $t-1$ the flow divides, only to come together again later, as shown in Figure 8. Then, in stage $t-1$ we have only one phase coordinate x^1, in stages t and $t+2$ two coordinates x^1, x^2, and in stage $t+1$ three phase coordinates x^1, x^2, x^3. The same will be the case for the number of controlling parameters. In general, an object with a variable number of phase coordinates and controlling parameters is described just as in Subsection 3 (see (3.3)), but here $x(t)$ is a vector, the number of components of which depends on t:

$$x(t) = (x^1(t), x^2(t), \dots, x^{n_t}(t)), \qquad t = 0, 1, \dots, N;$$

the same will be true for the control:

$$u(t) = (u^1(t), u^2(t), \dots, u^{r_t}(t)), \qquad t = 1, 2, \dots, N.$$

It is easy to reduce this object to the objects considered in Subsection 3. Let n be the *largest* of the numbers $n_0, n_1, \dots, n_N$. Then each vector $x(t)$ will have *no more* than n components. However, we can assume that vector $x(t)$ has *exactly* n components, after putting zeros in the missing places:

$$x(t) = (x^1(t), x^2(t), \dots, x^{n_t}(t), 0, 0, \dots, 0) \tag{4.16}$$

(here $n - n_t$ zeros are inserted) and rewriting relation (3.3) as follows in coordinate form:

$$x^i(t) = \begin{cases} f_t^i(x(t-1), u(t)) & \text{for} \quad i = 1, \dots, n_t; \\ 0 & \text{for} \quad i = n_t + 1, \dots, n. \end{cases} \tag{4.17}$$

Naturally, in (4.16) the last $n - n_t$ coordinates (that is, the zeros) will not depend on the selected control process and will not affect the subsequent phase states. However, this already amounts to a question of the specific writing of equations (4.17). For us here, the only important thing is that such a formal technique succeeds in reducing an object with a variable number of phase coordinates to a notation with a *constant* number of phase coordinates, while the law of motion of this object is written in form (4.17), that is, it falls under type (3.3).

Functions of final state. Note that, in expression (3.5) for the functional, the final state $x(N)$ did not enter at all. At the same time, problems are sometimes encountered in which it is precisely the final state $x(N)$ that characterizes the effectiveness of the process, that is, as a measure of the effectiveness we take some function

$$J = \varphi(x(N)) \tag{4.18}$$

of the final state $x(N)$. This is the case we had in Example 1.1.

A more general case is also possible, in which the effectiveness criterion contains sum (3.5) and an expression of type (4.18), that is, it has the form

$$J = \varphi(x(N)) + \sum_{t=1}^{n} f_t^0(x(t-1), u(t)). \tag{4.19}$$

This is the case we had in Example 2.3.

Assume that for controlled object (3.3), (3.4) the effectiveness criterion is given in form (4.18). Then, taking into account the relation $x(N) = f_N(x(N-1), u(N))$

(see (3.3)) and setting

$$f_1^0 = f_2^0 = \dots = f_{N-1}^0 = 0, \quad f_N^0(x(N-1), u(N)) = \\ = \varphi(f_N(x(N-1), u(N))) = \varphi(x(N)),$$

we obtain

$$\varphi(x(N)) = f_N^0(x(N-1), u(N)) = \sum_{t=1}^{N} f_t^0(x(t-1), u(t)).$$

Thus a functional of form (4.18) is readily reduced to form (3.5). Similarly, a functional of form (4.19) can be reduced to form (3.5).

It is useful to keep in mind, too, that a functional of form (3.5) can in turn be reduced to form (4.18). Let us assume that discrete controlled object (3.3), (3.4) is given and that the problem of maximizing functional (3.5) for it has been stated. Assuming that $x = (x^1, \dots, x^n)$ and $u = (u^1, \dots, u^r)$ have their previous meanings, we introduce an additional phase coordinate x^{n+1}, which is subject to the conditions:

$$x^{n+1}(0) = 0;$$
$$x^{n+1}(t) = x^{n+1}(t-1) + f_t^0(x(t-1), u(t)).$$

Then it is clear that

$$x^{n+1}(\theta) = \sum_{t=1}^{\theta} f_t^0(x(t-1), u(t)), \qquad \theta = 1, 2, \dots, N,$$

and, in particular, $x^{n+1}(N) = J$ (see (3.5)). Consequently, functional (3.5) takes the form $x^{n+1}(N)$, that is, it reduces to form (4.18) (for this, only one additional phase coordinate has to be introduced).

Isoperimetric problem. The following problem is called isoperimetric (relative to discrete objects). For discrete controlled object (3.3), (3.4), let us consider the functionals of the same form:

$$\left.\begin{aligned} J_1 &= \sum_{t=1}^{N} g_t^1(x(t-1), u(t)), \\ &\dots\dots\dots\dots \\ J_k &= \sum_{t=1}^{N} g_t^k(x(t-1), u(t)). \end{aligned}\right\} \tag{4.20}$$

Finally, k real numbers $c_1, c_2, \dots, c_k$ are specified. The problem consists in considering only those processes (3.1), (3.2) for which functionals (4.20) assume the *specified* values:

$$J_1 = c_1, \dots, J_k = c_k,$$

and finding among all these processes the one for which functional (3.5) has the largest possible value (it may be that in this case end conditions or constraints are imposed on the phase coordinates).

This problem is also readily reduced to the problems considered in Subsection 3. Let us introduce the additional phase coordinates $x^{n+1}, \ldots, x^{n+k}$, which are subject to the conditions

$$x^{n+i}(0) = 0, \qquad i = 1, 2, \ldots, k; \tag{4.21}$$

$$x^{n+i}(t) = x^{n+i}(t-1) + g_t^i(x(t-1), u(t)), \quad i = 1, 2, \ldots, k, \tag{4.22}$$

where $x = (x^1, \ldots, x^n)$ and $u = (u^1, \ldots, u^r)$ have their previous meanings. Then

$$x^{n+i}(\theta) = \sum_{t=1}^{\theta} g_t^i(x(t-1), u(t)); \;\; \theta = 1, 2, \ldots, N; \; i = 1, 2, \ldots, k;$$

in particular,

$$x^{n+i}(N) = J_i, \qquad i = 1, 2, \ldots, k. \tag{4.23}$$

The conditions $J_1 = c_1, \ldots, J_k = c_k$ imposed during the statement of the problem can be written in the form

$$x^{n+i}(N) = c_i, \qquad i = 1, 2, \ldots, k,$$

that is, they become conditions at the right-hand endpoint. Consequently, because of the increase in the number of phase coordinates, the isoperimetric problem reduces to the problems discussed in Subsection 3.

We have considered several different modifications of the statement of the problem involving controlled objects. Naturally, combinations of these may also be encountered (for instance, an object with both feedback and lag). In addition, other, more specialized, modifications not considered here may be encountered. However, in view of the foregoing, we are sure that, by introducing additional variables, these statements of the problems can also be reduced to the problems examined in Subsection 3.

5. Maximization of several functionals. Engineers frequently present mathematicians with the following problem: how must an object be controlled in order to maximize not just one functional (3.5), but rather *several* specified functionals of this type. It is easy to see, however, that such a statement of the problem is incorrect: the desiderata have to be matched properly to what is possible.

Actually, if for discrete controlled object (3.3), (3.4) we pose the problem of the maximum of functional (3.5), then, *as a rule,* we obtain one completely determined optimal process making it possible to achieve this. Now, if for the same object (3.3), (3.4) we consider the problem of the maximum of some *other* functional, then we also obtain, as a rule, one completely determined optimal process solving this problem. The two optimal processes thus obtained by no means have to be identical; on the contrary, they are usually *different,* that is, a control providing a maximum of the first functional will not maximize the second (and vice versa).

Thus it is in general futile to seek a control which *simultaneously* maximizes two functionals; this will be feasible only in certain very special cases. The analogy with the problem of the maximum of a *function* provides a good illustration of the foregoing. If $f(x)$ and $g(x)$ are two functions specified over a certain interval, the maximum* of function $f(x)$ is as a rule attained at a *single* point (Figure 9), as is the case for the maximum of function $g(x)$. These functions usually go through their maxima at *different* points, and only in very exceptional cases does it turn out that both functions have their maxima at the same point. The problem of "finding the point at which *both* functions f and g go through a maximum" is thus inappropriate (that is, it is, as a rule, insoluble).

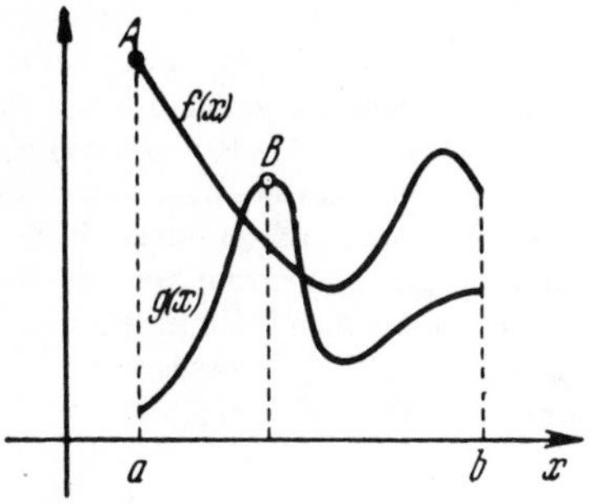

FIGURE 9

Nevertheless, we are constantly confronted by the problem (or, better said, by the desire) to maximize simultaneously *several* functions (or several functionals, in the case of a discrete controlled object). In Example 2.1 this means that we wish to *maximize* the meat shipments and at the same time to *maximize* the number of cattle on the farm at the end of the planning period. Clearly, these desires are contradictory (the greater the meat shipments, the fewer animals will remain on the farm), another illustration of the unfeasibility of attaining simultaneously the maxima of two (or more) functionals.

Various ways out of this situation can be found. One of them (see Example 2.1) consists in fixing the number of cattle on the farm at the end of the planning period (and thus rejecting the second functional) and under these conditions finding the maximum of the *one* remaining functional (the profit from the meat shipments).

Consequently, if it is desired to attain simultaneously values which are as large as possible for two or more functionals, the problem has to be restated somewhat. For instance, in this same problem concerning the livestock farm, the annual meat shipments can be fixed at some allowable level that ensures a maximum number of cattle on the farm at the end of the planning period. Only after such a stipulation is made will it be possible to obtain a clear mathematical statement of the problem. Mathematics is not all-powerful; as a minimum, we must know which of the desired alternatives is more important and which is less important.

On the other hand, there are mathematical *techniques* for clarifying the situation with regard to the simultaneous maximization of several functionals. Some of these will be described in the present subsection. It should be stressed that the question of *which* of these techniques to choose, and in what form, does not lie within the sphere of mathematics, but rather has to be decided by the designers, planners, contractors, and other engineering workers.

Thus, let us assume that discrete controlled object (3.3), (3.4) is considered and that for this object we wish to attain values which are as large as possible for several functionals of type (3.5), namely the functionals

$$\left.\begin{aligned} J_1 &= \sum_{t=1}^{N} g_t^1\,(x\,(t-1),\ u\,(t)),\\ &\cdots\cdots\cdots\cdots\cdots\\ J_k &= \sum_{t=1}^{N} g_t^k\,(x\,(t-1),\ u\,(t)). \end{aligned}\right\} \tag{5.1}$$

Of course, there may also be end conditions or constraints on the phase coordinates, but this will not be discussed here, since it is not related to the problem at hand.

* We refer throughout only to the *absolute* maximum, that is, to the largest value of the function; relative (local) maxima are not taken into account.

One of the simplest means of modifying this desire (and reducing it to a mathematically intelligible problem of the optimal control of a discrete object) is to *fix* the minimum acceptable (from the point of view of the statement of the problem) values of all the functionals (5.1) except one, and under these conditions to seek the maximum of the remaining functional.

Let us illustrate this technique by assuming that we wish to attain simultaneously the maxima of two functions. Since this is unfeasible (see Figure 9), we stipulate that function $g(x)$ must have a value *not less* than some number c, and it is on this condition that we seek a maximum of function $f(x)$. The condition $g(x) \geqslant c$ sets off a subset (shown by a heavy line in Figure 10) on the interval $[a, b]$, within which functions $f(x)$ and $g(x)$ are considered. In this subset we also have to seek the maximum of function $f(x)$; point x_0 in Figure 11 represents the solution of this problem. It should be noted that we had to affect the value of function $f(x)$ somewhat (that is, if we put aside the condition $g(x) \geqslant c$, then the value of function f would be greater; cf. points A and C in Figure 11), but such a restatement of the problem (that is, imposing the condition $g(x) \geqslant c$) succeeds to some extent in "reconciling" the mutually exclusive aims of maximizing functions $f(x)$ and $g(x)$.

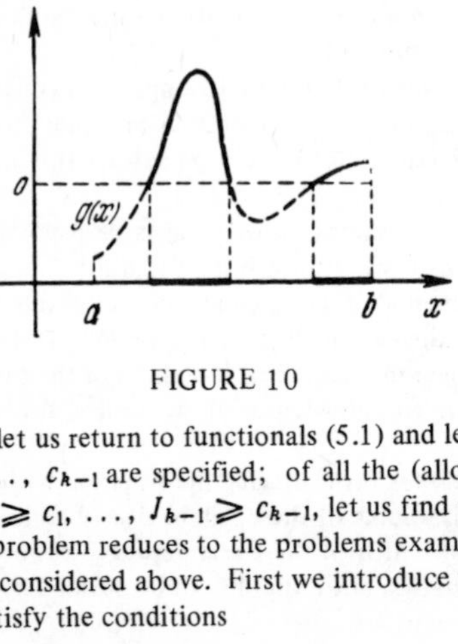

FIGURE 10

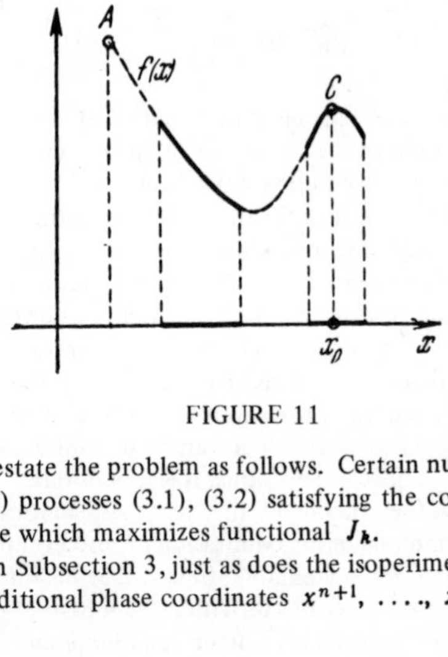

FIGURE 11

Now let us return to functionals (5.1) and let us restate the problem as follows. Certain numbers $c_1, c_2, \ldots, c_{k-1}$ are specified; of all the (allowable) processes (3.1), (3.2) satisfying the conditions $J_1 \geqslant c_1, \ldots, J_{k-1} \geqslant c_{k-1}$, let us find the one which maximizes functional J_k.

This problem reduces to the problems examined in Subsection 3, just as does the isoperimetric problem considered above. First we introduce the additional phase coordinates $x^{n+1}, \ldots, x^{n+k-1}$, which satisfy the conditions

$$x^{n+i}(0) = 0, \qquad i = 1, 2, \ldots, k-1;$$
$$x^{n+i}(t) = x^{n+i}(t-1) + g_t^i(x(t-1), u(t)), \qquad i = 1, 2, \ldots, k-1$$

(cf. (4.21), (4.22)). Then (cf. (4.23))

$$x^{n+i}(N) = J_i, \qquad i = 1, 2, \ldots, k-1,$$

so that the imposed conditions $J_1 \geqslant c_1, \ldots, J_{k-1} \geqslant c_{k-1}$ can be written as

$$x^{n+i}(N) \geqslant c_i, \qquad i = 1, 2, \ldots, k-1,$$

that is, they become conditions on the right end. Now only a single functional J_k remains to be maximized.

Let us next consider another technique which is widely used. We again consider discrete controlled object (3.3), (3.4), and we wish to attain values of the k functionals (5.1) which are as large as possible. However, these functionals are disparate; some of them are, from our point of view, more important, and some are less important.

Accordingly, it is advisable to specify certain *weights*, that is, some positive numbers $\beta_1, \beta_2, \ldots, \beta_k$ which characterize, in our opinion, the importance of functionals $J_1, J_2, \ldots, J_k$ (a more important functional being accorded a greater weight). Now, instead of the k functionals (5.1), we will have only one functional J, in which functionals $J_1, J_2, \ldots, J_k$ have the appropriate weights:

$$J = \beta_1 J_1 + \ldots + \beta_k J_k = \sum_{t=1}^{N} \left(\beta_1 g_t^1 (x(t-1), u(t)) + \ldots \right.$$
$$\left. \ldots + \beta_k g_t^k (x(t-1), u(t))\right), \qquad (5.2)$$

and we arrive at the problem of maximizing a *single* functional J, defined by formula (5.2).

This method was applied back in Example 2.3. There we had two functionals which we wished to make as large as possible: functional $J_1 = x(0) - x(N)$, that is, the amount of substance extracted from the solution, and functional $J_2 = -\sum_{t=1}^{N} u(t)$, expressing the negative of the quantity of water used (a maximum of this functional means a *minimum* water consumption). Clearly, it is impossible to attain a maximum of both these functionals *simultaneously*: the maximum of functional J_2 corresponds to *zero* water consumption, in which case no substance is extracted (that is, functional J_1 is a minimum rather than a maximum).

In order to reconcile these contradictory aims, let us consider functional $J = \beta_1 J_1 + \beta_2 J_2$ (see (2.8)), where "weights" β_1, and β_2, can, for instance, be selected as follows: β_1 is the value of the product substance, and β_2 is the cost of water. In this case J can be interpreted as the profit of the enterprise.

It should be noted that replacement of the several functionals (5.1) by a linear combination (5.2) of these is, naturally, not the sole possibility. In general the problem can be restated by taking the functional

$$J = F(J_1, J_2, \ldots, J_k), \qquad (5.3)$$

where F is some function of the k independent variables. Exactly which function F is selected for this purpose (or even, in the simpler case (5.2), which "weights" $\beta_1, \beta_2, \ldots, \beta_k$ are used) does not lie within the sphere of mathematics. However, if function (5.3) (or, in the particular case, (5.2)) is selected on the basis of certain "applied" considerations, then we obtain a mathematical problem of the type considered in Subsection 3.

There is still another approach to the maximization of several functionals, albeit one which has a very limited sphere of application. Let us first consider it, using as an example functions of two variables. Let $f(x, y)$ and $g(x, y)$ be the two functions of variables x and y, specified in some set M of the x, y plane. We wish to maximize (as much as possible) both of these functions, but the most important of the two is function $f(x, y)$. Now we can restate the problem as follows: find the point (x_0, y_0) at which function $f(x, y)$ has its largest value, but if this largest value is reached at more than one point then the point is chosen for which function $g(x, y)$ has the largest value.

We consider, for instance, the function $f(x, y) = 2 - (x + y)^2$ (defined over the entire xy plane). The largest value, 2, of this function is reached not at a single point in the xy plane, but rather at all points lying on the line $x + y = 0$. Thus, of all the points on this line, it is reasonable to seek the one at which the other function $g(x, y)$ is as large as possible.

The problem can be stated similarly for discrete controlled objects. Assume that functionals (5.1) are ordered according to their importance: $J_1, J_2, \ldots, J_k$, the most important being J_1. If there are many processes maximizing functional J_1, then, *of these,* the most important one will be the process maximizing functional J_2. Correspondingly, if there are many such processes, then functional J_3 is to be maximized, and so on. This problem can, at any rate formally, be reduced to the isoperimetric problem considered in Subsection 4.

If c_1 is the maximum possible value of functional J_1 (with constraints (3.4) placed on the controls and with constraints on the phase coordinates, if they are imposed), then the equality $J_1 = c_1$ signifies maximization of functional J_1. It should be noted that finding the value of c_1 represents the solution of a problem of this type, which was considered in Subsection 3, namely to find the maximum of functional J_1 for object (3.3), (3.4). Next, let c_2 be the maximum possible value of functional J_2, reached when the condition $J_1 = c_1$ is satisfied. Finding the value of c_2 is then equivalent to solving the isoperimetric problem, namely to find the maximum of functional J_2 for object (3.3), (3.4), on condition that $J_1 = c_1$. Then we go on to find the maximum of functional J_3 on condition that $J_1 = c_1$ and $J_2 = c_2$, and so on. Ultimately we arrive at the last isoperimetric problem, namely to find the maximum of functional J_k on condition that $J_1 = c_1, \ldots, J_{k-1} = c_{k-1}$.

In conclusion, we mention one more statement of the problem of attaining the maximum possible values of several functionals. Again we consider discrete controlled object (3.3), (3.4) and functionals (5.1). For greater clarity, let us first examine in more detail the case $k = 2$, that is, the case of only *two* functionals J_1 and J_2.

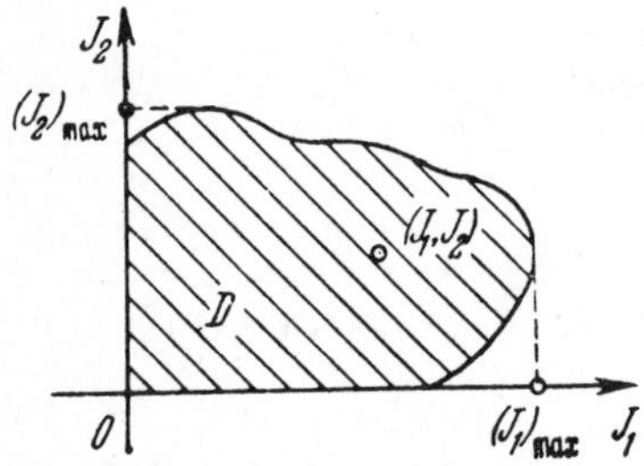

FIGURE 12

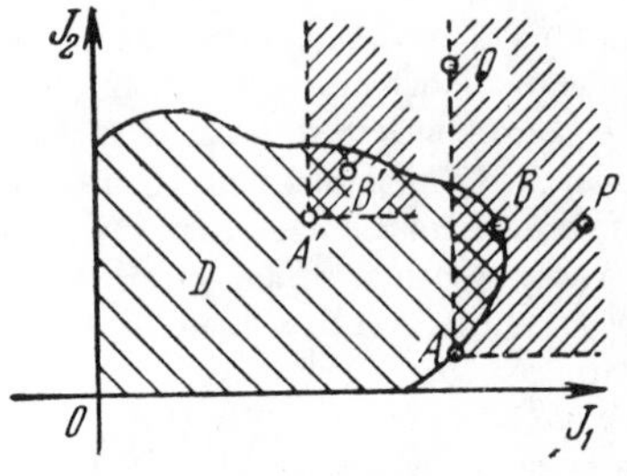

FIGURE 13

For some admissible process (3.1), (3.2) in the specified discrete controlled object (that is, a process satisfying conditions (3.3), (3.4) and the constraints on the phase coordinates, if they are imposed), functionals J_1, J_2 take on certain values which determine a *point* in the plane of variables J_1, J_2 (Figure 12). If we consider *all the possible* allowed processes in the given discrete controlled object, and if for each such process we cite the corresponding point (J_1, J_2), then we obtain in the plane of variables J_1, J_2 a certain set D (see Figure 12), which provides us with a complete representation of the possible values of the functional pair J_1, J_2. For example, inspection of Figure 12 shows which *maximum* values each of functionals J_1 and J_2 can take, and we see that functionals J_1 and J_2 cannot *simultaneously* reach a maximum.

Now let us take some point $A = \left(\overset{*}{J_1}, \overset{*}{J_2}\right)$ of set D (Figure 13). Since this point belongs to set D, therefore there *exists* some admissible process (3.1), (3.3) which gives to functional J_1 the value $\overset{*}{J_1}$, and to functional J_2 the value $\overset{*}{J_2}$. Rays drawn from point A parallel to the positive J_1 and J_2 semi-axes form a right angle with its vertex at A (shaded region in Figure 13). Each interior point $P = (J_1, J_2)$ of this angle possesses the property that *both* of its coordinates J_1 and J_2 are greater than the coordinates of point A, that is $J_1 > \overset{*}{J_1}$, $J_2 > \overset{*}{J_2}$. However, if a point $Q = \left(J_1', J_2'\right)$ lies on one of the sides of this angle but does not coincide with point A, then one of its coordinates will be the same as the corresponding coordinate of A, while the second will be larger; for example, in Figure 13 we have: $J_1' = \overset{*}{J_1}$, $J_2' > \overset{*}{J_2}$.

Hence it is clear that, if the given right angle (with its vertex at A) has in common with set D some point B different from A, then the process leading to the point $A = \left(\overset{*}{J_1}, \overset{*}{J_2}\right)$ will not be optimal *in any sense of the word* (it should be stressed that we are referring here to *maximization* of functionals J_1 and J_2). Actually, the process leading to point B (which exists, since point B belongs to set D) increases, relative to the point $A = \left(\overset{*}{J_1}, \overset{*}{J_2}\right)$, the values of both functionals J_1 and J_2 (or, at any rate, it increases the value of one of these functionals, without affecting the value of the other).

Thus, in order for a point $A = (J_1, J_2)$ belonging to set D to correspond to an *optimal* process (at least in some sense of the word), it is *necessary* for the right angle with its vertex at point A and its sides parallel to the positive coordinate semiaxes not to contain any points of set D (except A). In particular, it is clear that an *interior* point of set D (for instance, point A' in Figure 13) does not satisfy this necessary condition and thus cannot correspond to an optimal process, no matter how we modify the meaning of the term "optimality" for functionals J_1 and J_2. In other words, a point corresponding to an optimal process has to lie on the boundary of set D. On the other hand, not every boundary point of set D satisfies this necessary condition (for example, point A in Figure 13). Figure 14 shows three points, C_1, C_2, C_3, which satisfy the formulated necessary condition.

The idea now is to take the above-formulated necessary condition for optimality as the *definition* of optimality (for the case of two functionals J_1 and J_2). In other words, let us now consider the following restatement of the problem of maximizing two functionals J_1 and J_2. If (3.1) and (3.2) represent some admissible process for the specified discrete controlled object, a process yielding the values J_1^* and J_2^* for the functionals considered, then we call this process *optimal,* provided no admissible process exists for which $J_1 \geqslant J_1^*$ and $J_2 \geqslant J_2^*$, at least one of these two inequalities being strict.

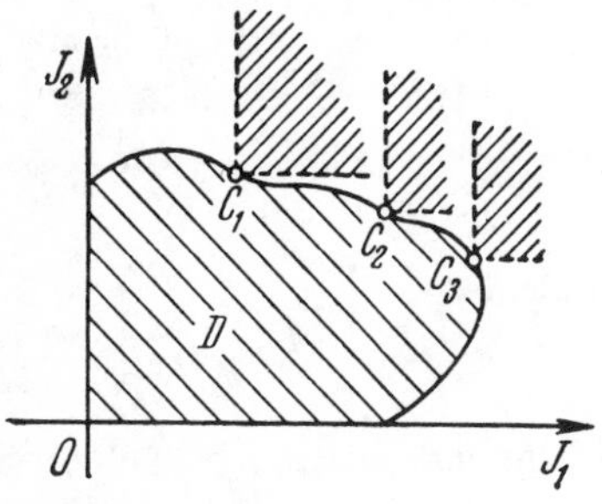

FIGURE 14

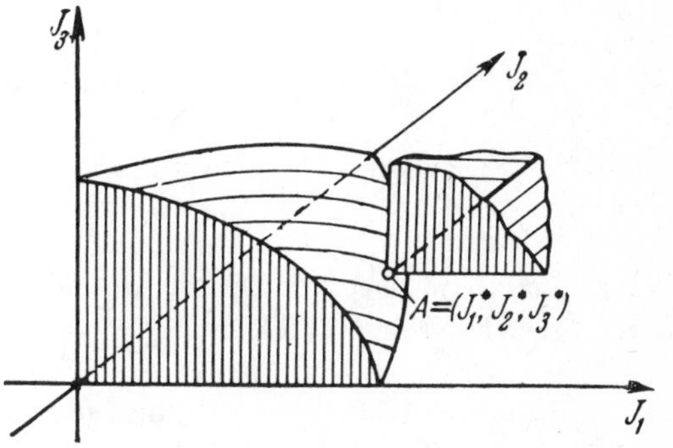

FIGURE 15

Geometrically this means that a right angle with a vertex at point $C = (J_1^*, J_2^*)$ and sides parallel to the positive coordinate semiaxes does not contain any points in set D other than C. As shown by Figure 14, according to this definition of optimality (in contrast to the previous ones), we will have, as a rule, an *infinite number* of optimal processes for the given discrete controlled object.

A similar definition can be formulated for the case of the k functionals (5.1). Thus, an admissible process (3.1), (3.2) giving to functionals (5.1) the values $J_1^*, J_2^*, \ldots, J_k^*$ is called *optimal* if there is no admissible process for which $J_1 \geqslant J_1^*$, $J_2 \geqslant J_2^*$, $\ldots$, $J_k \geqslant J_k^*$, at least one of these inequalities being strict. In other words, the process is optimal if it is impossible to increase the value of one of functionals (5.1) without thereby reducing the values of the other functionals.

This definition of optimality can also be interpreted geometrically, as in the case of $k = 2$. However, here, instead of a right angle with a vertex at the point (J_1^*, J_2^*), let us consider a solid angle in k-dimensional space, consisting of all the points $(J_1, J_2, \ldots, J_k)$ satisfying the conditions $J_1 \geqslant J_1^*$, $J_2 \geqslant J_2^*$, $\ldots$, $J_k \geqslant J_k^*$. Figure 15 illustrates the case $k = 3$.

Next let us show that, for this definition of optimality as well, the search for the optimal processes can be reduced (somehow) to the problems considered in Subsection 3. For simplicity, we again limit ourselves to the case of $k = 2$ and we assume that set D is *convex* (Figure 16). Let (3.1), (3.2) be some optimal process corresponding to a point $A = (J_1^*, J_2^*)$ lying on the boundary of set D; a right angle with its vertex at A and its sides parallel to the positive coordinate semiaxes will not have any common points with set D except A.

Call this right angle K; there will be some straight line Γ through point A which *separates* set D from angle K. Let $n = (\beta_1, \beta_2)$ be a vector orthogonal to line Γ and directed toward angle K. Then the function

$$F(J_1, J_2) = \beta_1 J_1 + \beta_2 J_2, \tag{5.4}$$

considered in the plane of variables J_1, J_2, has a constant value on line Γ, equal to the value of this function at point A, while at all points of set D this function has a *lower* value. Consequently, function (5.4), considered in set D, reaches a maximum at point A. In other words, the specified admissible process (3.1), (3.2) (to which point A corresponds) maximizes functional (5.4). The numbers β_1 and β_2 are *nonnegative,* since vector n lies inside angle K (or goes along one side of it).

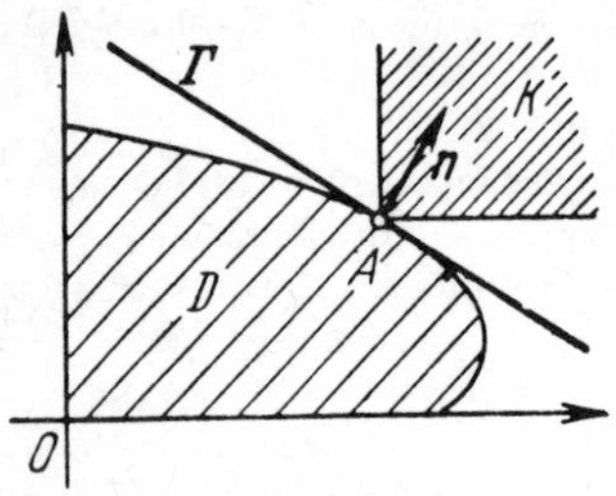

FIGURE 16

Therefore, the above definition of optimality is, in the final analysis, equivalent to optimality in the sense of the maximum of a single functional (5.4) for *several* nonnegative numbers β_1, β_2. A similar statement can also be made for a larger number of functionals.

To sum up, we note that the problems considered in Subsection 3 are fundamental. Many of the problems of optimal control of discrete objects considered in Subsections 4 and 5 can be reduced to these.

§ 2. Connection between Discrete-Optimization Problems and Other Extremum Problems

6. Extremum of function. In this subsection we consider the relationship between the problem of optimal control of discrete objects, posed in § 1, and the well-known problem of the extremum (maximum or minimum) of a function. If some function f is defined on a set M (or, as is still said, has a set M as its *domain of definition*) and takes on real values, then this means that to each point $a \in M$ there corresponds some number $f(a)$, which is called the *value* of function f at point a.

A point $x_0 \in M$ is called a *minimum point* of function f if for each point $x \in M$ the inequality $f(x) \geqslant f(x_0)$ is valid. If x_0 is the minimum point of function f, then the corresponding value $f(x_0)$ is called the *least value* (or *minimum*) of function f. The least value of function f is written as

$$\min_{x \in M} f(x). \tag{6.1}$$

Therefore, symbol (6.1) is defined by the following two conditions:

1) $f(x) \geqslant \min_{x \in M} f(x)$ for any $x \in M$;

2) there exists a point $x_0 \in M$ such that $f(x_0) = \min_{x \in M} f(x)$.

The *maximum* of a function is defined similarly, and the problem of finding the maximum of function f is equivalent to finding the minimum of the function $-f$.

Since the problem of determining the extrema of functions is too general when stated this way, certain constraints are as a rule placed on function f itself and on its domain of definition. Many branches, and even whole fields of study, of modern mathematics are entirely devoted to problems involving the extrema of functions for certain constraints. Examples are the numerous branches of analysis dealing with different extremum problems, the calculus of variations, linear programming, and the theory of games.

The mathematical theory of optimal control of discrete objects is also intimately connected with the consideration of a special class of problems on extrema of functions. Thus we will demonstrate here that the *problem of optimal control of a discrete object is* **equivalent** *to the problem of the extremum of a function defined in some subset of Euclidean space.* In other words, we will show that each problem of the extremum of a function can be reformulated as a problem of optimal control for some discrete object (albeit of a very special kind). Conversely, each problem of optimal control of a discrete object can be reduced to a problem of the extremum of some function.

Thus, let us assume that we have to solve a problem of the extremum (to be definite, the *maximum*) of a function defined on a subset of Euclidean space; that is, in the n-dimensional space E^n of variables $z^1, \ldots, z^n$ we have a set Ω on which the function $F(z)=F(z^1, \ldots, z^n)$ is defined. The problem is to find the point $z_0= =(z_0^1, \ldots, z_0^n) \in \Omega$ at which function F (considered on set Ω) reaches its greatest value.

Let E^n be the phase space of the discrete controlled object and let $N=1$. In this case (that is, for $N=1$) the problem of the optimal control of a discrete object (with a fixed left-hand endpoint and a free right-hand endpoint) is described as follows. An initial point $x(0)=x_0$ and a set $U=U_0(x(0))$ are given. The trajectory consists of, in addition to $x(0)$, one more point $x(1)$, defined by the relation (cf. (3.3), (3.4))

$$x(1)=f_1(x(0),\ u(1)), \qquad \text{where} \quad u(1) \in U \tag{6.2}$$

In addition, the function $f_1^0(x(0),\ u(1))$ is defined. A control (that is, a point $u(1) \in U$) must be so selected that functional

$$J=f_1^0(x(0),\ u(1)) \tag{6.3}$$

(cf. 3.5)) is as large as possible.

Next let us set

$$U=\Omega, \quad f_1(x,\ u)=u, \quad f_1^0(x,\ u)=F(u).$$

Then the discrete problem of optimal control described by relations (6.2), (6.3) takes the following form: the trajectory of the object is described by the equation $x(1)= u(1)$, and a point $u(1) \in \Omega$ must be so chosen that functional $J=F(u(1))$ is as large as possible. Here the quantities $x(0)$ and $x(1)$ do not enter in, and the problem posed is actually the following one: find a point $u(1) \in \Omega$ such that $J=F(u(1))$ is a maximum. However, this obviously also constitutes the initial problem concerning the maximum of function $F(z)$ on set Ω.

Thus, any problem concerning an extremum of a function defined on a subset of a Euclidean space can be reformulated (albeit in a trivial form) as a problem of optimal control of a discrete object. Naturally, there also exist other methods for writing the problem of the extremum of a function as a problem of optimal control of a discrete object.

For instance, in Examples 1.1 and 1.2 we constructed the discrete controlled object differently, starting from the problem of maximizing the function $F(z^1, z^2, \ldots, z^n) = z^1 z^2 \ldots z^n$ defined on a set Ω described by the inequalities $z^1 \geqslant 0, \ldots, z^n \geqslant 0$, $z^1 + \ldots + z^n \leqslant a$. Actually, for the reduction method indicated there, we obtained a discrete object with a *two-dimensional* (in Example 1.1) or *one-dimensional* (in Example 1.2) phase space, for which $r = 1$ and $N = n$; on the other hand, using the method just described, we obtain from the same problem concerning the maximum of a function a discrete object for which $r = n$ and $N = 1$. Other methods for the reduction of this problem to a problem of optimal control of a discrete object are also possible. The same is true for any other problem of the extremum of a function.

Conversely, let us assume that some problem of optimal control of a discrete object has been specified. We consider the most general of the problems examined in Subsection 3: a controlled object is described by relations (3.3), (3.4), and we wish to select an initial point $x(0) \in M_0$ and a control (3.1) such that the corresponding trajectory (3.2) satisfies the conditions $x(t) \in M_t$, $t = 1, 2, \ldots, N$, while functional (3.5) is as large as possible. We recall that $U_t(x)$ is a subset of the r-dimensional space of variables $u^1, \ldots, u^r$.

Now let us introduce the new variables u_t^j and x_s^i, where $i = 1, \ldots, n$; $j = 1, \ldots, r$; $s = 0, 1, \ldots, N$; $t = 1, 2, \ldots, N$. The number of these variables will thus be $k = n(N+1) + rN$. The space in which all these variables constitute coordinates is called E^k. If $z = (u_t^j, x_s^i)$ is some point in this space, then for each $s = 0, 1, \ldots, N$ we can consider a point in space E^n having the numbers $x_s^1, \ldots, x_s^n$ as its coordinates; this point is called $\xi_s(z)$. Similarly, we can consider (for each $t = 1, \ldots, N$) a point in the space of variables $u^1, \ldots, u^r$ having the numbers $u_t^1, \ldots, u_t^r$ as its coordinates; call this point $\eta_t(z)$. Now we can write the system of relations:

$$\xi_s(z) \in M_s, \qquad s = 0, 1, \ldots, N; \tag{6.4}$$

$$\eta_t(z) \in U_t(\xi_{t-1}(z)), \qquad t = 1, \ldots, N; \tag{6.5}$$

$$\xi_t(z) = f_t(\xi_{t-1}(z), \eta_t(z)), \qquad t = 1, \ldots, N. \tag{6.6}$$

The set of all points $z \in E^k$ satisfying relations (6.4) through (6.6) is called Ω; therefore $\Omega \subset E^k$. Further, we define function $F(z)$ (which will be considered only in set Ω) as

$$F(z) = f_1^0(\xi_0(z), \eta_1(z)) + f_2^0(\xi_1(z), \eta_2(z)) + \ldots \\ \ldots + f_N^0(\xi_{N-1}(z), \eta_N(z)), \tag{6.7}$$

and we consider how to find the maximum of this function in set Ω. We will show that this problem is equivalent to the initial problem concerning optimal control of a discrete object.

Next we define $x(0) \in M_0$ as some initial point, (3.1) as an admissible (relative to this initial point) control, and (3.2) as the corresponding trajectory of the discrete controlled object being examined. In addition, $x_s^1, \ldots, x_s^n$ are the coordinates of point $x(s)$ $(s = 0, 1, \ldots, N)$ and $u_t^1, \ldots, u_t^r$ are the coordinates of point $u(t)$ $(t = 1, \ldots, N)$. Then we obtain a collection of numbers x_s^i, u_t^j for all the indexes $i = 1, \ldots, n$; $j = 1, \ldots, r$; $s = 0, 1, \ldots, N$; $t = 1, \ldots, N$, that is, we obtain some point $z = (u_t^j, x_s^i)$ in space E^k. This point z will obviously satisfy the relations

$$\left.\begin{aligned} \xi_s(z) &= x(s), \qquad s = 0, 1, \ldots, N; \\ \eta_t(z) &= u(t), \qquad t = 1, \ldots, N \end{aligned}\right\} \tag{6.8}$$

(and it will be defined unambiguously by these relations). From (3.3), (3.4) and the relations $x(t) \in M_t$, $t = 0, 1, \ldots, N$, it follows that point z satisfies relations (6.4) through (6.6).

On the other hand, if z is an arbitrary point in set Ω, then after constructing points (3.2) and (3.1) according to formulas (6.8) we find (with the aid of relations (6.4)–(6.6)) that (3.2) and (3.1) represent the trajectory and control satisfying relations (3.3) and (3.4) and inclusions $x(t) \in M_t$, $t = 0, 1, \ldots, N$. Consequently, relations (6.8) establish a one-to-one correspondence between the points of set Ω and the admissible processes in the given discrete object (that is, between the processes (3.1), (3.2) which satisfy relations (3.3), (3.4) and inclusions $x(t) \in M_t$, $t = 0, 1, \ldots, N$). Moreover, in view of (6.8), the value of function F at a point $z \in \Omega$ (see (6.7)) is *equal* to the value of functional J (see (3.5)) for the process corresponding to point z. Hence it follows that, in particular, the maximum points of the function F being considered in set Ω correspond, in view of (6.8), to the optimal processes of the discrete object being considered.

Therefore, the problem of finding the optimal processes in the discrete system being considered is *equivalent** to the problem of finding the maximum points of the constructed function $F(z)$.

The equivalence of the problems considered naturally raises the question of whether it is really necessary to construct separately a theory of optimal processes in discrete systems. Wouldn't it be advisable just to stop here and say that the problem of finding optimal processes in discrete systems is thus reduced to a problem of the extremum of a function defined on some subset of Euclidean space, and that this problem is considered, for instance, in the theory of mathematical programming? This point of view

* Of course, it may be easier to reduce the problem of optimizing the discrete processes to a problem of the minimum of some function. It would be sufficient to designate as Ω^* the set of all *admissible* processes (3.1), (3.2) (that is, processes satisfying relations (3.3), (3.4)), and in set Ω^* to consider the function J defined by equation (3.5); then finding the maximum points of the function J defined on set Ω^* would mean, by definition, finding the optimal processes in the discrete object being considered. However, the reduction given above is useful in that it reduces the problem of optimizing discrete processes to a problem of the maximum of a function defined on some *subset of Euclidean space* rather than to finding the maximum of a function defined on *some* abstract set.

is quite justified; in any case, the theory of optimal processes in discrete systems can be taken to be a *chapter* in the theory of the extrema of functions.

However, the question is not at all whether one of the theories is formally inferior to the other. Although the fundamental problem of the mathematical theory of optimal processes in discrete systems *reduces* to a problem of the extremum of a function defined on some subset of Euclidean space, still the problem of optimizing discrete processes has its own idiosyncrasies. Actually, the system of relations (6.4) through (6.7), defining the set $\Omega \subset E^k$ and the function $F(z)$ in it, is very specific (for instance, each of the variables u_t^j enters into no more than *two** of the $3N + 1$ relations (6.4)–(6.6) defining the set Ω on which function F is specified, although the number N may also be very large, while each of variables x enters into no more than *four* of them). Accordingly, it is to be expected that the necessary or sufficient conditions for optimality will also be formulated more specifically than the *general* theorems expressing the criteria for the extrema of functions. These considerations justify making the optimal control of discrete systems a separate *object of study* in mathematical theory.

Naturally, in view of the reduction carried out in this subsection, all of the specific results obtained in this way could be formulated in terms of the extrema of functions (that is, if the domain of definition and the function itself have a certain form, then certain precise extremum conditions are satisfied). However, this is not advisable either, since many practical applications (cf. Subsection 2) require that the problem be stated as a problem of a controlled object passing over from one state to another. Moreover, discrete controlled processes also show up in mathematical problems, for example, as a means of obtaining a discrete simplified variant of the problem of optimal control of *continuous* processes (which will be discussed below). All this indicates that the problem of discrete optimal control not only differs from the problem of the extremum of a function in the *form of its notation* but also differs with regard to its very characteristics, making it an independent branch of mathematics.

7. **Problem of mathematical programming**. As we saw in Subsection 6, the problem of optimal control of discrete objects reduces to a problem of the extremum of a function defined on some subset Ω of Euclidean space. This narrows considerably the class of extremum problems which can be considered. However, even in this form, the problem is extremely general, and in order to obtain meaningful results certain constraints have to be placed on set Ω and the function considered in it.

In the case encountered most often, set Ω is defined in Euclidean space by some *system of equalities and inequalities,* while the function considered in it (the extremum of which is being sought) is *smooth.* This can be explained in greater detail as follows.

* If relations (6.4)–(6.6) are depicted in coordinate form, then it is seen that each variable u_t^j, x_s^i enters into more of the relations, but in this case as well the *ratio* of the number of relations containing the given variable to the total number of relations will be approximately the same. For instance, if (6.6) is depicted in coordinate form, then we obtain nN relations, of which no more than $n + 1$ contain the given variable x_s^i.

In the Euclidean space E^n of variables $z^1, \ldots, z^n$ the following functions are defined:

$$F^0(z) = F^0(z^1, \ldots, z^n),$$
$$F^1(z) = F^1(z^1, \ldots, z^n), \ \ldots, \ F^k(z) = F^k(z^1, \ldots, z^n),$$
$$g^1(z) = g^1(z^1, \ldots, z^n), \ \ldots, \ g^s(z) = g^s(z^1, \ldots, z^n),$$

and assumed to be smooth, that is, they have continuous derivatives with respect to all their independent variables. Moreover, Ω is defined as the set of all points $z \in E^n$ satisfying the system of relations

$$F^1(z) = 0, \ \ldots, \ F^k(z) = 0, \tag{7.1}$$
$$g^1(z) \leqslant 0, \ \ldots, \ g^s(z) \leqslant 0. \tag{7.2}$$

In this set Ω let us consider function $F^0(z)$ and let us state the problem of finding the maximum (or minimum) of this function in set Ω. The formulated problem is called a *problem of mathematical programming.*

For the problem thus posed, a large number of extremum conditions have been worked out, necessary as well as sufficient. This makes it possible (in accordance with what was said in Subsection 6) to obtain a number of optimality criteria for discrete processes.

Now let us say a few words about the meaning of relations (7.1). The equation $F^1(z) = 0$ defines an $(n-1)$*-dimensional surface* (or, alternatively, a *hypersurface*) in the n-dimensional space E^n. For instance, for $n = 2$ the equation $F^1(z^1, z^2) = 0$ defines a *line* in the plane E^2 of variables z^1, z^2. For $n = 3$ the equation $F^1(z^1, z^2, z^3) = 0$ defines a *surface* (two-dimensional) in the three-dimensional space E^3. The same will be true of each of equations (7.1). The set defined by *system* of equations (7.1) thus consists of points belonging to *each* of the hypersurfaces $F^1(z) = 0, \ldots, F^k(z) = 0$, that is, this set represents the *intersection* of all the indicated hypersurfaces. For example, for $n = 3$ the *two* equations $F^1(z) = 0$ and $F^2(z) = 0$ determine the intersection of two surfaces, that is, a *line,* in the three-dimensional space E^3 (Figure 17). In general it should be assumed that equations (7.1) define, in an n-dimensional space E^n, an intersection between k hypersurfaces, that is, they define an $(n-k)$*-dimensional surface* in E^n.

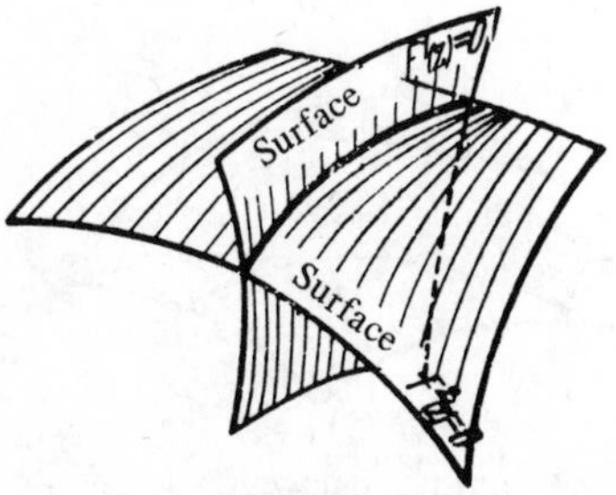

FIGURE 17

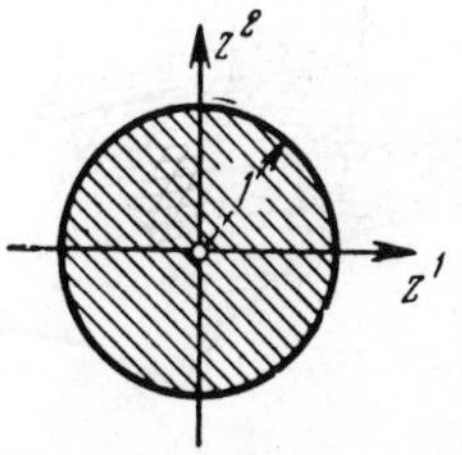

FIGURE 18

Let us discuss further the meaning of inequalities (7.2). The *equation* $g^1(z) = 0$ defines, as we noted, some hypersurface in E^n. This hypersurface divides space E^n into two regions: in one the inequality $g^1(z) > 0$ is valid, while in the other the inequality $g^1(z) < 0$ is valid. Therefore, the inequality $g^1(z) \leqslant 0$ defines a *closed domain* in E^n, which includes all the points of the region $g^1(z) < 0$ and the points on the hypersurface $g^1(z) = 0$ bounding it.

As an example, we consider the closed domain defined in the plane E^2 of variables z^1, z^2 by the inequality $g(z) \leqslant 0$, where $g(z) = (z^1)^2 + (z^2)^2 - 1$. This inequality defines a *closed unit circle* (Figure 18); for points *inside* this circle the inequality $g(z) < 0$ is valid, on the bounding circumference the equality $g(z) = 0$ is valid, and for external points $g(z) > 0$ is valid.

Thus each of inequalities (7.2) defines a closed domain in E^n. However, the set defined by *all* the inequalities (7.2) consists of the points belonging to *each* of the closed domains $g^1(z) \leqslant 0, \ldots, g^s(z) \leqslant 0$, that is, this set represents the *intersection* of all the indicated closed domains.

Finally, the set defined by the system of all the relations (7.1), (7.2) represents the intersection between surface (7.1) and the set defined by inequalities (7.2). This set (defined by system of relations (7.1), (7.2)) can be represented as an $(n-k)$-dimensional "curvilinear polyhedron."

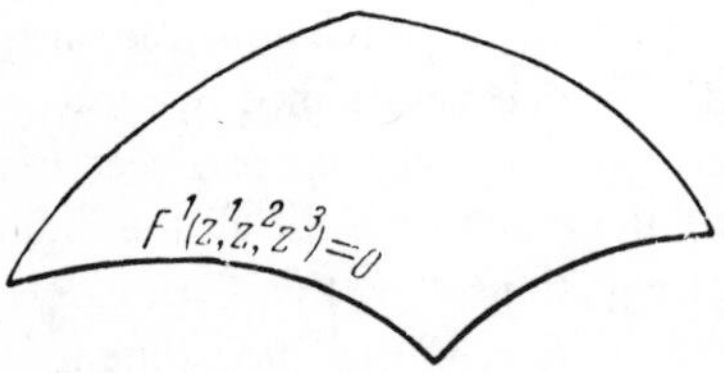

FIGURE 19

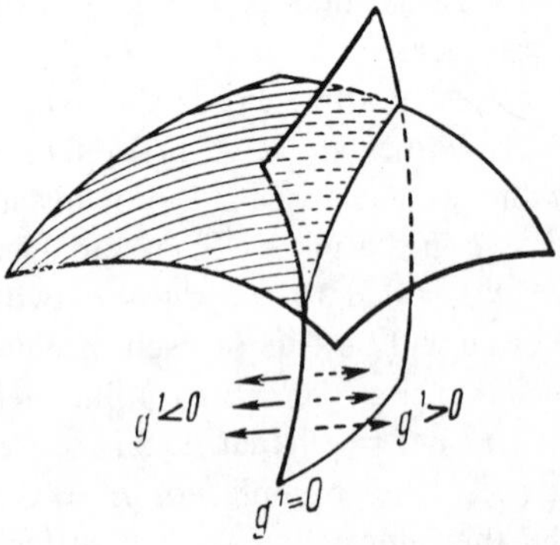

FIGURE 20

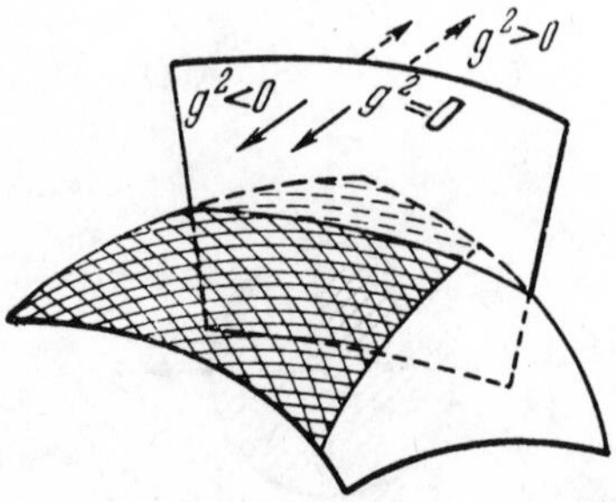

FIGURE 21

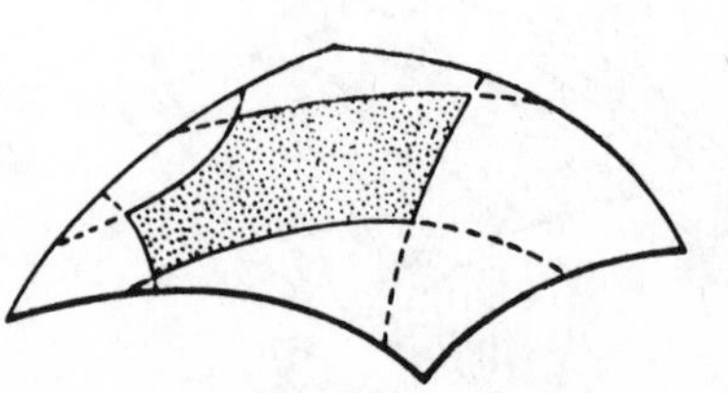

FIGURE 22

Assume, for instance, that $n = 3$ and $k = 1$. Then we have only a single equation

$$F^1(z^1, z^2, z^3) = 0 \tag{7.3}$$

(cf. (7.1)), which defines a *surface* in E^3 (Figure 19). By adding to (7.3) one inequality $g^1(z^1, z^2, z^3) \leqslant 0$, we obtain a closed domain on this surface (Figure 20), of which the second inequality $g^2(z^1, z^2, z^3) \leqslant 0$ represents a "fragment" (Figure 21), etc. Finally, the entire system of inequalities

$$g^1(z^1, z^2, z^3) \leqslant 0, \ \ldots, \ g^s(z^1, z^2, z^3) \leqslant 0$$

defines on surface (7.3) some "curvilinear polyhedron" (that is, a "curvilinear polyhedron" of two dimensions; Figure 22).

The point of minimum (or maximum) of function $F^0(z)$, defined on the "curvilinear polyhedron" (7.1), (7.2), can be found using classical methods. Let us consider a simple example.

EXAMPLE 7.1. *Find the minimum of the function*

$$F = x^2y^2 - 4xy^2 - 2x^2y + 8xy + \frac{2}{3}x^3 + x^2 - 4x + 3, \tag{7.4}$$

considered on the rectangle $-1 \leqslant x \leqslant 3, \ 0 \leqslant y \leqslant 2$ (Figure 23).

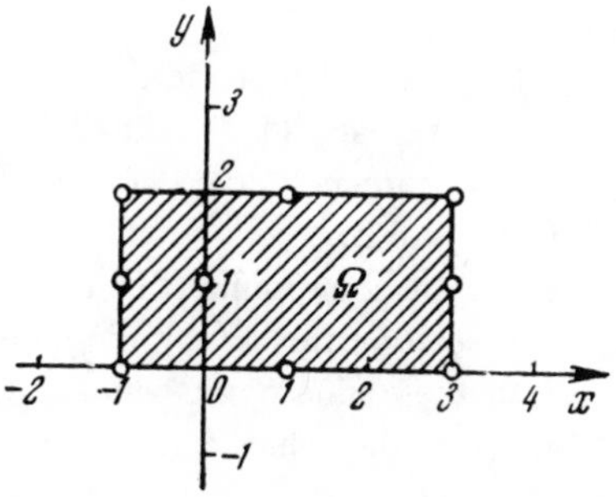

FIGURE 23

Solution. Here the set Ω on which function F is defined is specified on the plane E^2 of variables $z^1 = x$, $z^2 = y$ by the inequalities

$$-z^1 - 1 \leqslant 0, \quad z^1 - 3 \leqslant 0, \quad -z^2 \leqslant 0, \quad z^2 - 2 \leqslant 0, \tag{7.5}$$

that is, equations (7.1) are absent, while inequality (7.2) has form (7.5).

We assume first that the desired point of minimum lies *inside* the rectangle being considered. In this case the following necessary condition for an extremum, known from analysis, has to be satisfied: the partial derivatives of function F with respect to x and y must go to zero at this point. After taking the partial derivatives and setting them equal to zero, we get the system:

$$\left.\begin{aligned} F_x &= 2xy^2 - 4xy - 4y^2 + 8y + 2x^2 + 2x - 4 = 0, \\ F_y &= 2x^2y - 2x^2 - 8xy + 8x = 2x(x-4)(y-1) = 0. \end{aligned}\right\} \tag{7.6}$$

The second of relations (7.6) may be valid, inside the given rectangle, for either $y = 1$ or $x = 0$. In the former case ($y = 1$) we obtain $x = 0$ from the first equation (7.6), and in the latter case the first equation (7.6) gives $y = 1$. Thus, inside the rectangle there is only one point at which the two derivatives F_x and F_y go to zero simultaneously, namely the point (0,1). In other words, of all the *interior* points of the rectangle, only point (0,1) *can* represent a point of minimum.

However, it may be that function F reaches a minimum at the boundary of the rectangle. Let us first consider the side of the rectangle defined by the relations $x = -1$ and $0 \leqslant y \leqslant 2$. By substituting the value $x \equiv -1$ into (7.4), we find that on this side function F has the following value:

$$F_{(x \equiv -1)} = 5y^2 - 10y + \frac{22}{3}. \tag{7.7}$$

We assume that function F reaches a minimum at some point M_0 of the side in question. In this case, for any point M of the rectangle, the inequality $F(M) \geqslant F(M_0)$ should be satisfied. In particular, this inequality should be satisfied for any point M lying on the given side; in other words, function F, considered *only* on the segment $x = -1,\ 0 \leqslant y \leqslant 2$ (that is, function (7.7)), must reach a minimum at point M_0. If the minimum is reached at some *interior* point on this segment, then at this point the derivative of function (7.7) has to go to zero, that is, the relation $10y - 10 = 0$ is satisfied for $y = 1$. Thus, $(-1, 1)$ is the only interior point on the side $x = -1,\ 0 \leqslant y \leqslant 2$ at which it is possible for function F to reach a minimum.

Similar considerations for the other sides of the rectangle give us three more possible points for the minimum: (3,1), (1,0), and (1,2). Finally, it is *possible* for function F to reach a minimum at one of the corners of the considered rectangle, that is, at one of the points (–1,0), (3,0), (–1,2), (3,2).

Consequently, there are a total of nine points (Figure 23) at which function F *may* reach a minimum. In order to ascertain the actual minimum point, it remains only to find the values of function F at these nine points and to take the lowest one. A direct calculation shows that the lowest of the values at these points are $F(1, 0) = 2/3$ and $F(1, 2) = 2/3$. Thus function F goes through minima at the two points (1,0) and (1,2), the minimum value being 2/3.

The method employed to solve Example 7.1 suggests a generalization to include all cases. Assume that we have to find the minimum of a function $F^0(z)$, defined on a set Ω which is described by relations (7.1), (7.2).

We take any *interior* point $z_0 = (z_0^1, z_0^2, \ldots, z_0^n)$ of the "polyhedron" Ω, that is, some point satisfying all of relations (7.1), (7.2), the inequalities in (7.2) being *strict* (Figure 24). In order for function F^0 to be a minimum at point z_0, the partial derivatives of F^0 with respect to all directions tangent to "polyhedron" Ω have to go to zero (see Figure 24). This gives $n - k$ equations (since "polyhedron" Ω has dimension $n - k$). In addition, we have another k equations (7.1) (all valid at point z_0, since $z_0 \in \Omega$). Thus we obtain a total of n equations establishing the necessary condition for function $F^0(z)$ to reach a minimum at an *interior* point of "polyhedron" Ω.

This necessary condition constitutes a system of n equations in n unknowns z_0^1, $z_0^2, \ldots, z_0^n$, so that (as a rule) inside Ω there are only a finite number of points at which function $F^0(z)$ *may* have a minimum.

Now let us assume that at point $z_0 \in \Omega$ one of relations (7.2) turns into an equality, while the other inequalities remain strict. For instance, $g^1(z_0) = 0$, $g^2(z_0) < 0$, $\ldots$, $g^s(z_0) < 0$. Then point z_0 satisfies the relations

$$F^1(z) = 0, \ \ldots, \ F^k(z) = 0, \quad g^1(z) = 0; \qquad (7.8)$$
$$g^2(z) \leqslant 0, \ \ldots, \ g^s(z) \leqslant 0.$$

These relations define a "curvilinear polyhedron" Ω' of dimension $n-(k+1) = (n-k)-1$ (that is, with a dimension one less than Ω), which is actually a *face* of "curvilinear polyhedron" Ω (Figure 25). Obviously, z_0 is an *interior* point of polyhedron Ω' (since relations (7.8) are *strict* inequalities at point z_0). Therefore, we can write out similarly the necessary condition for reaching a minimum at point z_0, a condition comprising n equations in the coordinates $z_0^1, \ z_0^2, \ldots, \ z_0^n$ of point z_0. Hence it follows that, on face Ω' as well, there are only a finite number of points at which a minimum *may* be reached.

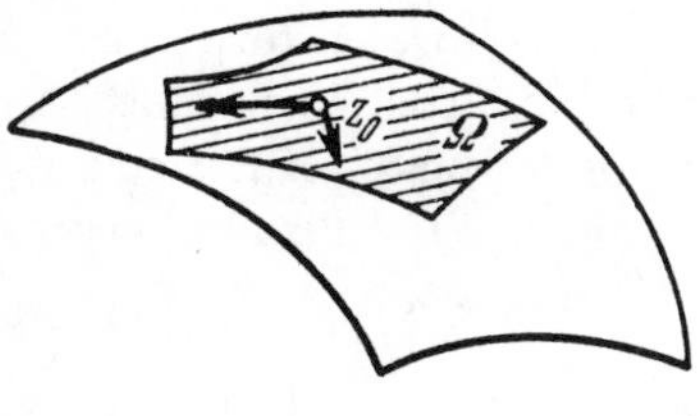

FIGURE 24

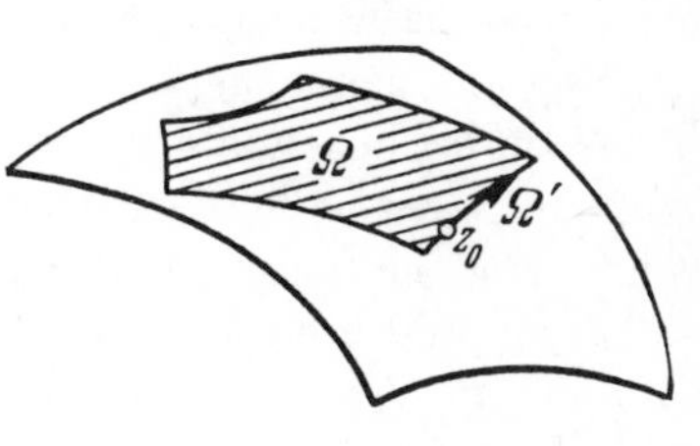

FIGURE 25

The situation is similar at faces having fewer dimensions. Thus, in all we obtain (inside the "polyhedron" and on all of its faces) a finite number of points at which a minimum *may* be reached. Then the values of function F at these points are compared and the lowest one is selected.*

In general, the described method becomes very time-consuming if the "polyhedron" considered has a large number of faces. Since with an increase in n (that is, the number of variables $z^1, \ z^2, \ \ldots, \ z^n$) the number of faces of the polyhedron must increase as well, the method becomes increasingly cumbersome. For instance, it is easy to calculate that an *n-dimensional parallelepiped,* that is, a polyhedron defined in space E^n by the inequalities

$$a^1 \leqslant z^1 \leqslant b^1, \quad a^2 \leqslant z^2 \leqslant b^2, \ \ldots, \ a^n \leqslant z^n \leqslant b^n,$$

* Strictly speaking, the described method guarantees finding a minimum only provided the *existence* of a minimum is established beforehand. However, if, as tacitly assumed above, function F is differentiable (and thus continuous), and if, as was also tacitly assumed above, the "polyhedron" being considered is a bounded closed set, then function F will definitely reach a minimum on this "polyhedron," so that the described method is applicable.

will have $3^n - 1$ faces (of all dimensions $0, 1, \ldots, n-1$). For $n = 4$ we already have 80 faces, and each of these requires a separate calculation to find the points at which there may be a minimum. Therefore, the above-described method (calling for consideration of all the faces of a "curvilinear polyhedron" and being somewhat haphazard) is rarely applied, other methods being used instead of it in mathematical programming (see §3).

8. Controlled processes with continuous time. Mathematical programming is one means of obtaining optimality criteria for discrete controlled objects. Here we consider another source of theorems for discrete controlled objects, by stating the problem of optimal control for *nondiscrete* processes, that is, for processes with continuous time. This problem is of interest for two reasons. First, the results obtained in the theory of optimal control of continuous objects make it possible to formulate *by analogy* "similar theorems" relevant to discrete objects (see below, Subsection 10). Therefore, the theory of optimal control of continuous objects is for the theory of discrete controlled objects an important heuristic technique indicating certain directions of study and at the same time suggesting the nature of the results. Second, the relationship between discrete and continuous controlled objects, whereby continuous objects can be taken as the "limiting case" of discrete objects, is important in itself, for example, as a means of describing approximately continuous controlled processes (which are as a rule more complex) in terms of discrete processes (or else as a means of describing approximately discrete processes in terms of continuous processes).

EXAMPLE 8.1. Let us consider the discrete controlled object of Example 1.2 and let us assume in it the transition at the limit to a continuous controlled object. Thus we consider the discrete object described by the relations

$$x^1(t) = x^1(t-1) + u(t), \tag{8.1}$$

$$u(t) \in U, \tag{8.2}$$

$$x^1(0) = 0, \quad x^1(N) \in M_1, \tag{8.3}$$

where $U = (0, \infty)$ and $M_1 = [0, a]$. For this object we wish to find the process which imparts the largest value to the functional

$$J = \sum_{t=1}^{N} \ln u(t). \tag{8.4}$$

Assume that N is "large" ($N = 50$ or more, say) and that we want a *graphical* representation of the behavior of the control

$$u(1),\ u(2),\ \ldots,\ u(N) \tag{8.5}$$

and the corresponding trajectory

$$x^1(0),\ x^1(1),\ \ldots,\ x^1(N). \tag{8.6}$$

Since N is large, a graphical representation to the ordinary time scale is unsuitable. Thus it is natural to select some "step" h and instead of t to introduce a new independent variable τ, defined as

$$\tau = th, \qquad t = 0,\ 1,\ \ldots,\ N. \tag{8.7}$$

Hence τ will have the values $0,\ h,\ 2h,\ \ldots,\ Nh$. The choice of "step" h is characterized as follows:

$$h = \frac{a}{N} \tag{8.8}$$

(so that for large N "step" h is small).

If we introduce the new variables y^1 and v, depending on τ, according to the formulas:

$$y^1(\tau) = x^1\left(\frac{\tau}{h}\right) = x^1(t), \quad v(\tau) = \frac{1}{h}u\left(\frac{\tau}{h}\right) = \frac{1}{h}u(t), \tag{8.9}$$

then relation (8.1) becomes $\left(\text{taking into account that } y^1(\tau - h) = x^1\left(\frac{\tau - h}{h}\right) = x^1\left(\frac{\tau}{h} - 1\right) = x^1(t-1)\right)$:

$$y^1(\tau) = y^1(\tau - h) + hv(\tau),$$

or

$$\frac{y^1(\tau) - y^1(\tau - h)}{h} = v(\tau). \tag{8.10}$$

The left-hand side of this equation is an expression having as its limit for $h \to 0$ the *derivative* $dy^1/d\tau$ (it was to achieve this that the factor $1/h$) was introduced on the right-hand side of the second formula of (8.9)). Further, in terms of the new variables, relations (8.2) and (8.3) become

$$v(\tau) \in U, \tag{8.11}$$

$$y^1(0) = 0, \quad y^1(a) \in M_1 \tag{8.12}$$

(since in view of (8.8) and (8.9) we have $y^1(a) = x^1(a/h) = x^1(N)$), where as before $U = (0, \infty)$ and $M_1 = [0, a]$. Finally, functional (8.4) takes the form

$$J = \sum_{t=1}^{N} \ln u(t) = \sum \ln hv(\tau) =$$

$$= N \ln h + \sum \ln v(\tau) = N \ln h + \frac{1}{h}\sum h \ln v(\tau),$$

where the summation on the right-hand side is over $\tau = h,\ 2h,\ \ldots,\ Nh$. Note that the term $N \ln h$ and the positive factor $1/h$ are both constants, so that the problem of maximizing functional J is equivalent to the problem of the maximum of the functional

$$I = \sum h \ln v(\tau). \tag{8.13}$$

Everything done so far has constituted a simple change of variables. The problem of maximizing function (8.13) for controlled object (8.10)–(8.12) with a discrete time $\tau = 0,\ h,\ 2h,\ \ldots,\ Nh$ differs only in its notation from the problem of maximizing functional (8.4) for controlled object (8.1)–(8.3) with a discrete time $t = 0,\ 1,\ \ldots,\ N$.

However, formulas (8.10)–(8.13) can be reinterpreted. Relation (8.10) is an approximation of the equation

$$\frac{dy^1(\tau)}{d\tau} = v(\tau), \tag{8.14}$$

while the expression on the right-hand side of (8.13) (where, it should be recalled, the summation is over $\tau = h,\ 2h,\ \ldots,\ Nh$, with $Nh = a$) is an *integral sum* for the definite integral

$$I^* = \int_0^a \ln v(\tau)\, d\tau. \tag{8.15}$$

Consequently, it can be assumed that the initial *discrete* problem of optimal control (see (8.1) through (8.4)) corresponds approximately to the following problem for a *nondiscrete* ("continuous") variable τ which changes over the interval $0 \leqslant \tau \leqslant a$.

Find a **control** $v(\tau)$, *that is, a function defined in the interval* $0 \leqslant \tau \leqslant a$ *and assuming values on set* U (see 8.11)), *such that the solution* $y^1(\tau)$ *of differential equation* (8.14) *with initial condition* $y^1(0) = 0$ *satisfies the final condition* $y^1(a) \in M_1$ (see (8.12)), *integral* I^* *being as large as possible.*

This is also a "continuous" problem of optimal control; instead of a "discrete time" t, here we consider a "continuous" variable τ (time). In what sense, then, do these two problems of optimal control (discrete and continuous) "correspond" approximately to one another? Apparently, we can expect to find for each admissible process $u(t),\ x^1(t)$ in discrete object (8.1)–(8.3) an admissible process of continuous controlled object (8.14), (8.11), (8.12) which is close to it (in the sense of conversion (8.9)), and vice versa.

Consequently, if the continuous problem turns out to be *simpler,* then by solving it (that is, by finding the optimal process $v(\tau),\ y^1(\tau)$) we obtain using formulas (8.9) an *approximate* solution of the problem of discrete optimal control (8.1)–(8.4). Of course, so that we will have a reliable method of simplifying the problem, rather than just good intentions, the error introduced by replacing one problem by the other should be known; however, this will not be discussed here.

The "continuous" problem (8.14), (8.11), (8.12), (8.15) in question is easy to solve. First we note that, in view of (8.14) and the first relation (8.12), we have

$$y^1(a) = y^1(0) + \int_0^a \frac{dy^1(\tau)}{d\tau}\, d\tau = \int_0^a v(\tau)\, d\tau,$$

so that, from the second relation (8.12),

$$\int_0^a v(\tau)\,d\tau \leqslant a. \tag{8.16}$$

Now let us make use of the fact that $\ln v \leqslant v - 1$ on the ray $(0, \infty)$, equality being reached only at the point $v = 1$.† Thus $\ln v(\tau) \leqslant v(\tau) - 1$ (we recall that $v(\tau) > 0$ in view of (8.11)), and so

$$\int_0^a \ln v(\tau)\,d\tau \leqslant \int_0^a (v(\tau) - 1)\,d\tau = \int_0^a v(\tau)\,d\tau - a \leqslant 0$$

(see (8.16)), equality being reached only for $v(\tau) \equiv 1$.

This also gives the solution of the above-formulated problem for a continuous controlled object: the sought optimal control has the form $v(\tau) \equiv 1$ and it imparts to functional (8.15) a value $I^* = 0$; for any other admissible control, functional I^* has a lower (that is, negative) value.

Once we have an *accurate* solution $v(\tau) \equiv 1$ for the problem of optimal control of the given *continuous* object, we can use formulas (8.9) to obtain an *approximate* solution for the initial problem of optimal control of a discrete object:

$$u(t) = hv(\tau) \equiv h = \frac{a}{N} \qquad (t = 1, 2, \ldots, N).$$

Thus (without raising the question of the accuracy of the approximation), for the discrete problem (8.1)–(8.3) (in the sense of the maximum of functional (8.4)), that is, for the discrete object considered in Example 1.2, the following control is optimal:

$$u(1) = u(2) = \ldots = u(N) = \frac{a}{N},$$

this optimal control being *unique.* Returning to the problem of the maximum of the product (Example 1.1), we find that the *product of N nonnegative numbers, the sum of which does not exceed a, reaches a maximum if each of the numbers is equal to a/N; in any other case the product will be lower.*

We obtained this result as an *approximation,* starting from considerations associated with continuous controlled processes. Actually, as we saw in §3, the obtained result is *accurate.*

The reduction to a continuous problem of optimal control carried out in Example 8.1 can be applied similarly for many discrete problems. However, the opposite transition is the more frequent one, namely from a continuous problem to a discrete problem. As a rule, a continuous problem of optimal control is not amenable to accurate solution,

† Actually, function $\ln v - v + 1$ goes to zero for $v = 1$, decreases for $v > 1$, and increases for $0 < v < 1$.

so that some approximate solution has to be sought. Replacing the continuous problem by the simpler discrete problem is one possible means of obtaining such an approximate solution.

Let us consider, in general terms, the reduction of a continuous problem to a discrete one, using the following "continuous" problem of optimal control as an example.

We start with the differential equation

$$\frac{dy}{d\tau} = g(y, v) \tag{8.17}$$

(cf. (8.14)), *in which* $y = (y^1, y^2, \dots, y^n)$ *is the vector of the phase state, and* $v = (v^1, \dots, v^r)$ *is the control vector. Moreover, a set* V, *called the control domain, and a positive number* T *are specified (in the space of variables* $v^1, \dots, v^r$*). Any (piecewise-continuous) function* $v(\tau)$, *defined in the interval* $0 \leqslant \tau \leqslant T$ *and taking the following values on set* V:

$$v(\tau) \in V \tag{8.18}$$

(cf. (8.11)), *is called an admissible control. In addition, a point* y_0 *and a set* M_1 *are defined in the space of variables* $y^1, \dots, y^n$. *Finally, we have a function* $g^0(y, v)$ *which can be used, provided functions* $y(\tau)$ *and* $v(\tau)$ *are known, to find the definite integral*

$$I^* = \int_0^T g^0(y(\tau), v(\tau))\, d\tau \tag{8.19}$$

(cf. (8.15)). *An admissible control* $v(\tau)$ *is sought, such that the corresponding solution* $y(\tau)$ *of differential equation* (8.17) *with initial condition*

$$y(0) = y_0 \tag{8.20}$$

also satisfies the final condition

$$y(T) \in M_1 \tag{8.21}$$

(cf. (8.12)), *functional* (8.19) *being as large as possible.*

In order to reduce this "continuous" problem of optimal control to a discrete problem, we specify some natural number N and we set

$$h = \frac{T}{N} \tag{8.22}$$

(cf. (8.8)). Next we give independent variable τ only the values $0, h, 2h, \dots, Nh = T$, and we introduce a new (discrete) variable t according to the formula

$$t = \frac{\tau}{h} \tag{8.23}$$

(cf. (8.7)), so that t takes only the values $0, 1, \ldots, N$. Further, instead of variables y and v let us introduce the new variables x and u, defined as

$$x(t) = y(th) = y(\tau), \quad u(t) = v(th) = v(\tau) \tag{8.24}$$

(this substitution is somewhat different from substitution (8.9); in connection with this, see Note 8.3, below).

Now let us assume that $v(\tau)$ is an admissible control (in the sense of the "continuous" problem being considered), and that $y(\tau)$ is the corresponding trajectory, that is, the solution of differential equation (8.17) for $v = v(\tau)$ with initial condition (8.20). Consequently,

$$\frac{dy(\tau)}{d\tau} = g(y(\tau), v(\tau)).$$

This relation can be replaced by the approximation

$$\frac{y(\tau) - y(\tau - h)}{h} \approx g(y(\tau), v(\tau)) \approx g(y(\tau - h), v(\tau)),$$

which we will consider only for the values $\tau = h, 2h, \ldots, Nh$. This gives

$$y(\tau) \approx y(\tau - h) + hg(y(\tau - h), v(\tau)),$$

or, in view of (8.24),

$$x(t) \approx x(t-1) + hg(x(t-1), u(t)), \qquad t = 1, 2, \ldots, N. \tag{8.25}$$

In addition, integral (8.19) can be replaced approximately by the integral sum:

$$I^* \approx \sum_{t=1}^{N} hg^0(y(th), v(th)) \approx \sum_{t=1}^{N} hg^0(y(th - h), v(th)) =$$

$$= h \sum_{t=1}^{N} g^0(x(t-1), u(t)). \tag{8.26}$$

Finally, relations (8.18), (8.20), and (8.21), after substitution (8.24), become

$$u(t) \in V, \tag{8.27}$$

$$x(0) = y_0, \quad x(N) \in M_1. \tag{8.28}$$

Next let f be the function on the right-hand side of (8.25):

$$f(x, u) = x + hg(x, u)$$

and let us replace (introducing some error) the approximate equation (8.25) by the precise equation

$$x(t) = f(x(t-1), u(t)). \tag{8.29}$$

Further, we replace (again introducing some error) integral I^* by the expression on the right-hand side of (8.26), that is, we go from the problem of the maximum of functional I^* to the problem of the maximum of functional

$$J = h \sum_{t=1}^{N} g^0 (x(t-1),\ u(t)). \tag{8.30}$$

As a result, we also obtain the discrete problem of optimal control (8.29), (8.27), (8.30) with a fixed left-hand endpoint and a moving right-hand endpoint (see (8.28)). During the transition to this problem from the initial continuous problem, we introduced errors: first, of the entire interval $0 \leqslant \tau \leqslant T$, only the finite number of values $\tau = 0,\ h,\ 2h,\ \ldots,\ Nh$ remained, and second, the approximate equations (8.25), (8.26) were replaced by relations (8.29), (8.30). However, under certain conditions (say, if function $g(y, v)$ is "sufficiently smooth," and if the number h is "sufficiently small," that is, if the derivative $\partial g(y(\tau),\ v(\tau))/\partial y$ does not vary much over a segment of length h), it can be expected that for each admissible process $u(t),\ \ x(t)$ in discrete object (8.27)–(8.30) there will be an admissible process "close" to it in continuous controlled object (8.17)–(8.21), and vice versa (Figure 26).

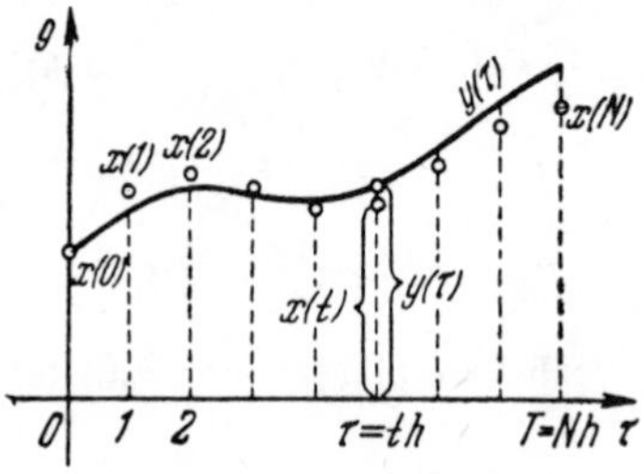

FIGURE 26

Therefore, if we pose a continuous problem of optimal control $I^* = \max$ for conditions (8.17)–(8.21) and if we find that the solution of this problem involves serious difficulty, then, by choosing some sufficiently large number N and passing with the aid of formulas (8.22)–(8.24) to the discrete problem (8.27)–(8.30), we can attempt to solve this discrete problem. The solution, if successfully found, can then be transformed using formulas (8.24), giving an approximate solution to the initial continuous problem. Of course, strictly speaking, to apply this method, it is necessary to make *estimates* of the error introduced by substituting the discrete problem for the continuous one.

NOTE 8.2. Above we spoke of optimality in the sense of reaching a *maximum* of integral functional I^* (see (8.19)). Accordingly, passing to the discrete case, we also had a problem of a *maximum* of functional J (see (8.30)). Naturally, in both cases we could have considered the problem of a *minimum* of the corresponding functional, rather than a maximum (this is the situation encountered most frequently in the theory of continuous controlled processes: minimum fuel consumption, minimum cost of electrical energy, minimum use of washwater, etc.).

NOTE 8.3. The above technique of replacing a continuous problem with a discrete problem (see (8.24)) is the simplest method, but it is by no means the only one possible. Often the form of equation (8.17) may suggest another reduction method, which for the case at hand may well be more suitable. For instance, in Example 8.1 the transition from discrete problem (8.1)–(8.4) to the continuous problem (which naturally could also have been in the opposite direction, from continuous to discrete) was effected somewhat differently: in the second formula (8.9) it turned out to be convenient to introduce the factor $1/h$. As another example, let us take the case in which equation (8.17) has the form

$$\frac{dy}{d\tau} = y^2 + v$$

(y and v are scalars). Here, instead of substitution (8.24), it is more convenient to make the substitution

$$x(t) = hy(th) = hy(\tau), \quad u(t) = h^2 v(th) = h^2 v(\tau). \tag{8.31}$$

Actually, using this substitution (by the same means as that employed to derive relation (8.25)), we obtain the discrete process

$$x(t) = x(t-1) + (x(t-1))^2 + u(t),$$

not containing parameter h on the right-hand side. Here, however, relation (8.18) is replaced (in view of the second formula (8.31)) by the inclusion $u(t) \in U$, where U is the set of all points of form $h^2 v$ $(v \in V)$.

NOTE 8.4. In the theory of optimal control of continuous processes† problems with a *nonfixed* time are fundamental, that is, problems in which the time T (cf. (8.19)) is not specified in advance. The case of a *prespecified* interval $0 \leqslant \tau \leqslant T$ within which the desired control is considered is called (in the theory of optimal processes with continuous time) a *problem with fixed time.* Since in the discrete problems being considered the set of values which t may take are specified in advance (and are the same for all the considered admissible controls), it follows that the analog of these problems for a *continuous* time t should be a problem with a *fixed* time. The transition examined in this subsection had just this character: from a continuous problem with a *fixed time* to a discrete problem of optimal control.

§3. Methods of Solving Discrete-Optimization Problems

9. Dynamic programming. In this and the following subsections we will describe the basic *methods* for finding the optimal regimes in discrete controlled objects. One of these is the method of *dynamic programming,* developed by the American mathematician R. Bellman.

The method of dynamic programming essentially consists in the following: for optimality of the entire process as a whole, during each intermediate stage the subsequent part of the process must also possess the property of optimality. In more detail, we

† See, for instance, Pontryagin, L.S., V.G. Boltyanskii, R.V. Gamkrelidze, and E.F. Mishchenko. *Matematicheskaya teoriya optimal'nykh protsessov (The Mathematical Theory of Optimal Processes).* – "Nauka," Moskva. 1969 [see the earlier edition published in English, under the same title, by Interscience (Wiley), New York. 1962]; Boltyanskii, V.G. *Matematicheskie metody optimal'nogo upravleniya (Mathematical Methods of Optimal Control).* – "Nauka," Moskva. 1969.

assume that, in controlling discrete object (3.3)–(3.5), we have already *somehow* selected control $u(1), u(2), \ldots, u(k)$ and trajectory $x(0), x(1), \ldots, x(k)$ from the start of the process until moment $t = k$. We wish to complete the process, that is, to select $u(k+1), \ldots, u(N)$ and $x(k+1), \ldots, x(N)$. Then, if the completed part of the process (from $t = k$ until $t = N$) is not optimal, in the sense of maximizing the functional

$$J_k = \sum_{t=k+1}^{N} f_t^0(x(t-1), u(t)), \tag{9.1}$$

it follows that the process as a whole will not be optimal either. Actually, if we could facilitate the "tail end" of the process, that is, if we could so vary $x(k+1), \ldots, x(N)$ and $u(k+1), \ldots, u(N)$ that the sum of the corresponding terms in (3.5) is greater, then this would facilitate the process as a whole.

This obvious idea can, according to Bellman, be realized as follows (for simplicity, we consider only the fundamental problem stated back in Subsection 3). Assume that at time $t = k$ we are already at point $x = x(k)$ (without raising the question of how we reached this point from the initial state). Using all possible methods, we carry out the concluding part of the process:

$$u(k+1), \ldots, u(N); \quad x(k+1), \ldots, x(N) \tag{9.2}$$

(so that relations (3.3) and (3.4) are satisfied). For each such completed part of the process, we calculate sum (9.1), denoting the *largest* of all these sums as $\omega_k(x) = \omega_k(x(k))$.

Thus, starting from point $x = x(k)$, we *may* choose the concluding part of process (9.2) in such a way that sum (9.1) has a value $\omega_k(x)$, no *larger* value of this sum (starting from point x) being possible. By setting $k = 0, 1, \ldots, N$, we obtain $N+1$ functions $\omega_0(x), \omega_1(x), \ldots, \omega_N(x)$. Clearly, $\omega_N(x) \equiv 0$, since in sum (9.1) there will be *no* terms for $k = N$, that is, this sum equals zero. In addition, it is obvious that $\omega_0(x_0)$ is the maximum value of sum (3.5), that is, of functional J, which we are able to obtain if the motion begins from point $x = x_0$ and if all possible admissible processes (3.1), (3.2) are employed. In other words, $\omega_0(x_0)$ is the value of functional J corresponding to an *optimal* process with an initial state $x(0) = x_0$, so that the problem of discrete optimal control just consists in finding the value of $\omega_0(x_0)$ and a means of reaching it.

Consequently, if we could successively calculate functions $\omega_N(x), \omega_{N-1}(x), \ldots, \omega_1(x), \omega_0(x)$, beginning from $\omega_N(x) \equiv 0$ and continuing to the desired function $\omega_0(x)$, then this would mean that we have a method for solving the discrete problem of optimal control. And it is easy to obtain such a method.

Assuming that we start from a point $x = x(k-1)$ at a time $t = k-1$, let us choose the best conclusion of the process, that is, we so select $u(t), x(t)$ for $t = k, k+1, \ldots, N$ that the sum

$$J_{k-1} = \sum_{t=k}^{N} f_t^0(x(t-1), u(t))$$

has its maximum value $\omega_{k-1}(x)$. This sum can be written as

$$J_{k-1} = f_k^0(x(k-1),\ u(k)) + \sum_{t=k+1}^{N} f_t^0(x(t-1),\ u(t)),$$

the second term on the right being equal to J_k. Obviously, this term is $\omega_k(x(k))$, that is, it is equal to the largest possible value of sum J_k for motion from point $x(k)$ (improving the "tail end" of the process, that is, increasing the sum J_k, would, according to the foregoing, increase the sum J_{k-1}, which is impossible, since $J_{k-1} = \omega_{k-1}(x)$). Accordingly,

$$\omega_{k-1}(x) = f_k^0(x(k-1),\ u(k)) + \omega_k(x(k)). \tag{9.3}$$

Next we note that this relation is valid if $x(k)$ and $u(k)$ enter into the *best* conclusion of the process starting from point $x = x(k-1)$. However, if we *arbitrarily* (while observing relations (3.3) and (3.4)) choose $x(k)$ and $u(k)$, then relation (9.3) has to be replaced by the inequality

$$\omega_{k-1}(x) \geqslant f_k^0(x(k-1),\ u(k)) + \omega_k(x(k)), \tag{9.4}$$

since now there will be *some* (possibly not maximum) sum J_{k-1} on the right-hand side of (9.4). Consequently, $\omega_{k-1}(x)$ is the *largest* of all the values of sum

$$f_k^0(x(k-1),\ u(k)) + \omega_k(x(k)),$$

which can be obtained by choosing $u(k)$ and $x(k)$ in the allowed manner, that is,

$$\omega_{k-1}(x) = \max\left[f_k^0(x(k-1),\ u(k)) + \omega_k(x(k))\right].$$

Replacing $x(k)$ by $f_k(x(k-1),\ u(k))$ on the right-hand side (see (3.3)) and recalling that $x(k-1) = x$, we obtain this relation in its final form:

$$\omega_{k-1}(x) = \max_{u \in U_k(x)} \left[f_k^0(x,\ u) + \omega_k(f_k(x,\ u))\right], \quad k = 1,\ 2,\ \ldots,\ N. \tag{9.5}$$

Here the maximum is taken over all $u = u(k)$ (since $x(k)$ does not enter into the right-hand side at all), that is, over all points $u \in U_k(x)$.

Relation (9.5), known as *Bellman's equation*, makes it possible to calculate successively the functions $\omega_N, \omega_{N-1}, \ldots, \omega_1, \omega_0$, beginning from $\omega_N \equiv 0$ and continuing to the desired function $\omega_0(x)$. The technique used to find the desired function $\omega_0(x)$ with the aid of relations (9.5) is called *dynamic programming.*

The practical application of this technique usually calls for the use of computers with large memory capacities. The thing is that, for a successive calculation of functions $\omega_k(x)$ using relation (9.5), we will not know until the last moment (that is, until the number $\omega_0(x_0)$ is found) which states $x(t)$ give the optimal trajectory starting from point $x(0) = x_0$ (this will be illustrated below, in Example 9.2). Therefore, the values of functions $\omega_k(x)$ have to be found (and stored), in theory, for all (in practice, for a dense enough network) values of x. This is equivalent to finding *all at once* the optimal trajectories (corresponding to the different initial states $x(0)$).

For the above reason, the method of dynamic programming is not very effective, in spite of the unusually wide range of applications which are in principle possible. In some (albeit very rare) instances the introduction of additional arguments enables a final solution of the problem.

As an illustration, let us solve the problem stated in Example 1.1 using dynamic programming. For the solution we will need the following lemma, which, so as not to interrupt the description later, we present now:

LEMMA 9.1. *A function* $u\left(\frac{c-u}{k}\right)^k$, *considered in the interval* $0 \leqslant u \leqslant c$, *reaches a maximum at a single point* $u=\frac{c}{k+1}$, *this maximum being equal to* $\left(\frac{c}{k+1}\right)^{k+1}$.

This is readiiy *proven* by differentiation.

EXAMPLE 9.2. Consider the discrete problem of optimal control posed in Example 1.1. In this case there are only two phase coordinates x^1, x^2 and a single controlling parameter u. Thus equation (3.3) in coordinate form becomes

$$x^1(t)=f_t^1(x^1(t-1),\ x^2(t-1),\ u(t)),$$
$$x^2(t)=f_t^2(x^1(t-1),\ x^2(t-1),\ u(t)),$$

where as f_t^1 and f_t^2 we take the functions (see (1.1), (1.2))

$$f_t^1(x^1,\ x^2,\ u)=x^1+u, \quad f_t^2(x^1,\ x^2,\ u)=x^2u. \tag{9.6}$$

Moreover, the domain $U(x)=U_t(x)=U_t(x^1, x^2)$ is defined by the inequalities (see (1.4))

$$0 \leqslant u \leqslant a-x^1. \tag{9.7}$$

Now let us introduce the function $f_t^0(x^1,\ x^2,\ u)$ defining functional (3.5). Since the entire sum (3.5) should be (according to the statement of the problem) equal to $x^2(N)$, that is, it should equal $x^2(N-1)u(N)$ (see (1.2)):

$$f_1^0(x^1(0), x^2(0), u(1))+\ \dots\ +f_N^0(x^1(N-1), x^2(N-1), u(N))=$$
$$=x^2(N-1)\,u(N),$$

it is natural to assume that functions f_t^0 have the following form:

$$f_1^0=\ \dots\ =f_{N-1}^0\equiv 0, \quad f_N^0(x^1,\ x^2,\ u)=x^2u. \tag{9.8}$$

Let us now consider an application of the method of dynamic programming. From (9.5) we have (in view of the relation $\omega_N\equiv 0$):

$$\omega_{N-1}(x^1,\ x^2)=\max_{u\in U(x)} f_N^0(x^1,\ x^2,\ u)=\max_{u\in U(x)} x^2u=x^2(a-x^1).$$

Here we made use of the fact that $x^2 \geqslant 0$, so that the expression x^2u has a maximum for $u = a - x^1$ (see (9.7)). Moreover, from this it follows that the optimal value $u(N)$, that is, the value for which J_{N-1} has its highest value $\omega_{N-1}(x(N-1))$, is

$$u(N) = a - x^1(N-1).$$

Clearly, carrying out a single step (that is, going from ω_N to ω_{N-1}) by no means gives us the values of $u(N)$; all we have done is to ascertain how it depends on the previous state $x(N-1)$ (which is as yet unknown).

Let us take the process a little further. To do this, we note that, in view of (9.8), relation (9.5) for $k < N$ becomes

$$\omega_{k-1}(x^1, x^2) = \max_{u \in U(x)} \omega_k\left(f_k^1(x^1, x^2, u), f_k^2(x^1, x^2, u)\right) = \\ = \max_{u \in U(x)} \omega_k(x^1 + u, x^2 u).$$

For $k = N - 1$ we obtain

$$\omega_{N-2}(x^1, x^2) = \max_{u \in U(x)} x^2 u(a - x^1 - u) = x^2 \max_{u \in U(x)} u(a - x^1 - u).$$

According to Lemma 9.1, expression $u(a - x^1 - u)$ reaches a maximum at $u = \frac{a - x^1}{2}$, the value being $\left(\frac{a - x^1}{2}\right)^2$. Consequently,

$$\omega_{N-2}(x^1, x^2) = x^2\left(\frac{a - x^1}{2}\right)^2.$$

Since the maximum is reached for $u = (a - x^1)/2$, therefore the optimal control $u(N-1)$ has the value:

$$u(N-1) = \frac{a - x^1(N-2)}{2}.$$

In the next stage:

$$\omega_{N-3}(x^1, x^2) = \max_{u \in U(x)} x^2 u\left(\frac{a - x^1 - u}{2}\right)^2 = x^2 \max_{u \in U(x)} u\left(\frac{a - x^1 - u}{2}\right)^2.$$

According to Lemma 9.1, the expression $u\left(\frac{a - x^1 - u}{2}\right)^2$ has a maximum for $u = (a - x^1)/3$, the value being $\left(\frac{a - x^1}{3}\right)^3$. Therefore

$$\omega_{N-3}(x^1, x^2) = x^2\left(\frac{a - x^1}{3}\right)^3.$$

In addition, optimal control $u(N-2)$ has the form

$$u(N-2)=\frac{a-x^1(N-3)}{3}.$$

Continuing this process, we find by induction:

$$\omega_{N-k}(x^1,\ x^2)=x^2\left(\frac{a-x^1}{k}\right)^k,$$
$$u(N-k+1)=\frac{a-x^1(N-k)}{k};\qquad k=1,\ 2,\ \ldots,\ N. \tag{9.9}$$

So far we have carried out the process "blindly," without knowing where it was going. Thus we found only the general relation (9.9) applicable for *all* optimal processes. Now let us use the initial values $x^1(0)=0$ and $x^2(0)=1$ (see (1.3)). For $k=N$ we get from (9.9): $u(1)=a/N$, and according to (1.1) this gives $x^1(1)=$ $=a/N$. Now, for $k=N-1$, we obtain from (9.9): $u(2)=a/N$, so that (see (1.1)) $x^1(2)=2a/N$. Next, for $k=N-2$, (9.9) gives $u(3)=a/N$, and so forth.

Accordingly, using the method of dynamic programming, we found the *unique* optimal process:

$$u(1)=u(2)=\ \ldots\ =u(N)=\frac{a}{N}$$

which is in complete agreement with the result of Example 8.1 (so that actually the result of Example 8.1 is *precise* rather than approximate).

The example considered leads us to an interesting fact. Back on p. 28 we noted that the same problem of the minimum of a function can be reduced in different ways to a problem of finding optimal processes in discrete systems. For instance, the problem considered in Example 1.1 can be described by the discrete object referred to in Examples 1.1 and 1.2, or it can be reduced to a discrete object using the method of pp. 27–28. In the former case, as we saw in Example 9.2, dynamic programming provides an easy solution of the problem. However, in the latter case (that is, for reduction to a discrete object using the method of pp. 27–28), we have $N=1$, so that it is advisable to use dynamic programming only to find the "function" (that is, in the given case, the *number*)

$$\omega_0=\max_{u\in\Omega} f_1^0(x,\ u)=\max_{u\in\Omega} F(u).$$

But this simply amounts to a formulation of the stated problem and nothing more. Consequently, for this reduction to a discrete controlled object, the method of dynamic programming contributes *nothing whatever* to the solution of the original problem.

It is clear that the "effectiveness" of the method of dynamic programming (if it is applied to the problem of the extremum of a function) depends considerably on how skillfully we select the means of reducing the given problem of the extremum of a function to some problem of optimal control of a discrete object.

In conclusion, let us consider the application of dynamic programming to the problem of optimal control of *continuous* objects. We begin with the "continuous" problem of optimal control formulated in Subsection 8 (see formulas (8.17–(8.21)), assuming that set M_1 (see (8.21)) coincides with the entire space of variables $y^1, \ldots, y^n$, and we carry out the reduction to the discrete problem (8.27)–(8.30). The method of dynamic programming is applied to this discrete problem.

Bellman's equation (see (9.5)) in this case takes the form

$$\omega_{k-1}(x) = \max_{u \in V} [hg^0(x, u) + \omega_k(x + hg(x, u))], \qquad k = 1, 2, \ldots, N$$

(cf. relations (8.29) and (8.25)). Here we have

$$\omega_k(x) = \max \sum_{t=k+1}^{N} hg^0(x(t-1), u(t)),$$

where the maximum pertains to all possible completed parts of the process (see (9.2)), starting from the point $x(k) = x$. If we make substitution (8.23), (8.24) in these relations and introduce the notation $\tilde{\omega}_\tau(y) = \tilde{\omega}_{th}(y) = \omega_t(x)$, then we get the relations

$$\left.\begin{aligned} \tilde{\omega}_{\tau-h}(y) &= \max_{v \in V} [hg^0(y, v) + \tilde{\omega}_\tau(y + hg(y, v))], \quad \tau = h, 2h, \ldots Nh; \\ \tilde{\omega}_\tau(y) &= \max \sum_{\theta=\tau+h}^{Nh} hg^0(y(\theta - h), v(\theta)). \end{aligned}\right\} \tag{9.10}$$

In the second relation (9.10) the summation is over successive values of θ, differing from one another by h (that is, $\theta = \tau + h, \tau + 2h, \ldots, Nh$), so that this relation constitutes an approximate expression of the equation

$$\tilde{\omega}_\tau(y) = \max \int_\tau^T g^0(y(\theta), v(\theta))\, d\theta. \tag{9.11}$$

Here the maximum pertains to all admissible controls $v(\tau)$ and to the corresponding solutions of differential equation (8.17), with the initial condition $y(\tau) = y$.

In view of the equation

$$\tilde{\omega}_\tau(y + hg(y, v)) \approx \tilde{\omega}_\tau(y) + h \sum_{i=1}^{n} \frac{\partial \tilde{\omega}_\tau(y)}{\partial y^i} g^i(y, v)$$

(following from Taylor's formula), the first of relations (9.10) can be rewritten as

$$\frac{\tilde{\omega}_\tau(y) - \tilde{\omega}_{\tau-h}(y)}{h} + \max_{v \in V} \left[g^0(y, v) + \sum_{i=1}^{n} \frac{\partial \tilde{\omega}_\tau(y)}{\partial y^i} g^i(y, v) \right] \approx 0.$$

This relation is an approximate expression of the equation

$$\frac{\partial \tilde{\omega}_\tau (y)}{\partial \tau} + \max_{v \in V} \left[g^0 (y, v) + \sum_{i=1}^{n} \frac{\partial \tilde{\omega}_\tau (y)}{\partial y^i} g^i (y, v) \right] = 0. \tag{9.12}$$

Relations (9.11) and (9.12) also represent the principle of dynamic programming for the "continuous" problem of optimal control being considered. This principle can be formulated as follows.

Let τ *be some point in the interval* $[0, T]$ *and let* y *be some point in the phase space (that is, the space of variables* $y^1, \ldots, y^n$*). We will consider all possible admissible controls* $v(\theta)$, $\tau \leqslant \theta \leqslant T$ *(see* (8.18)) *and the corresponding solutions* $y(\theta)$ *of differential equation* (8.17), *with initial condition* $y(\tau) = y$. *Taking the maximum for all such processes* $v(\theta)$ *and* $y(\theta)$, *we can determine* $\tilde{\omega}_\tau (y)$ *with formula* (9.11). *It is found that function* $\tilde{\omega}_\tau (y)$ *satisfies relation* (9.12), *which is known as Bellman's equation for the "continuous" problem being considered.*

It is clear from the foregoing that the number $\tilde{\omega}_\tau (y)$ for $\tau = 0$ and $y = y_0$ gives the largest possible value of functional (8.19) satisfying conditions (8.17), (8.18), and (8.20), and also that the optimal process $v(\tau)$, $y(\tau)$ (for which functional (8.19) is a maximum) satisfies Bellman's equation, that is, it turns it into the identity

$$\frac{\partial \tilde{\omega}_\tau (y(\tau))}{\partial \tau} + g^0 (y(\tau), v(\tau)) + \sum_{i=1}^{n} \frac{\partial \tilde{\omega}_\tau (y(\tau))}{\partial y^i} g^i (y(\tau), v(\tau)) \equiv 0.$$

Consequently, the method of dynamic programming provides certain information about optimal processes in the "continuous" problem being considered, so that it can be used to find the optimal processes.

We will not discuss here the difficulties encountered in applying this method (to the "continuous" case), that is, the difficulties involved in solving partial differential equation (9.12), which is further complicated by the presence of the maximum. Let us note only that the above considerations naturally do not constitute a proof of the correctness of the method of dynamic programming (that is, of the satisfaction of relation (9.12) for function (9.11)). It is enough to state that the above transition from the "continuous" problem to the discrete problem, and vice versa, did not have a correct basis.

For a correct substantiation of the method of dynamic programming (as applied to the "continuous" problem being discussed), certain conditions have to be imposed on the statement of the problem. Unfortunately, however, the function $\tilde{\omega}_\tau (y)$ enters into these conditions (finding this function also means the problem is solved), so that *prior to* solution of the problem these conditions are practically unenforceable. All these things tend to make the method of dynamic programming ineffective (more details on the problems involved in substantiating the method of dynamic programming, as applied, to be sure, to a "continuous" problem with *nonfixed* time, are given on page 29 of: Boltyanskii, V.G. *Matematicheskie metody optimal'nogo upravleniya (Mathematical Methods of Optimal Control)*. – "Nauka," Moskva. 1969).

It should be emphasized again that, as applied to *discrete* problems, which are the ones of main interest in this book, the substantiation of the method of dynamic programming is quite correct. Below (in Subsection 40) we will describe the most general version of the method of dynamic programming, as applied to a discrete problem of optimal control.

10. Discrete maximum principle. The method of dynamic programming considered in the preceding subsection is (despite its shortcomings when applied to "continuous" problems) one of the two main techniques used in the theory of optimal control of continuous processes. The other method is based on the application of L.S. Pontryagin's *maximum principle.* This method is most effective, and at present it is the main means of solving "continuous" problems of optimal control. It is not surprising that, right after the discovery and proof of the maximum principle, attempts were made to find a "discrete" version of this principle. The present subsection will deal with this subject.

First of all, let us give the formulation of the maximum principle, keeping our basic goal (the discrete problem) in mind all the while, even though we consider the "continuous" problem of optimal control with a *fixed* time. Thus we return to problem (8.17)–(8.21) formulated in Subsection 8, but for simplicity we take only the case of a free right-hand endpoint, that is, we assume that set M_1 in relation (8.21) is identical to the entire space of variables $y^1, \ldots, y^n$. In this case the maximum principle is formulated as follows (see page 375 of: Boltyanskii, V.G., *Matematicheskie metody optimal'nogo upravleniya (Mathematical Methods of Optimal Control).* – "Nauka," Moskva. 1969).

Auxiliary variables $\varphi_0, \varphi_1, \ldots, \varphi_n$ *are introduced (here* n *is the number of phase variables* $y^1, \ldots, y^n$ *) and used to write the auxiliary formula*

$$\tilde{H}(\varphi, y, v) = \sum_{i=0}^{n} \varphi_i g^i(y, v), \tag{10.1}$$

where g^0 *is the function entering into the definition of functional* (8.19), *and* $g^1, \ldots, g^n$ *are the components of the vector function* $g(v, y)$ *on the right-hand side of equation* (8.17). *In addition, we write the following system of differential equations for the auxiliary unknowns:*

$$\frac{d\varphi_j}{d\tau} = -\frac{\partial \tilde{H}}{\partial y^j} = -\sum_{i=0}^{n} \varphi_i \frac{\partial g^i(y, v)}{\partial y^j}, \qquad j = 1, \ldots, n, \tag{10.2}$$

where we set

$$\varphi_0 = 1. \tag{10.3}$$

Next we assume that $v(\tau)$, $0 \leqslant \tau \leqslant T$, *is some admissible control (that is, it satisfies condition* (8.18)*), while* $y(\tau)$ *is the corresponding solution of differential equation* (8.17) *with initial condition* (8.20). *We then substitute functions* $y = y(\tau)$ *and* $v = v(\tau)$ *into the right-hand sides of equations* (10.2) *and we call* $\varphi(\tau)$ *the solution of this system with the initial condition*

$$\varphi_1(T) = \varphi_2(T) = \ldots = \varphi_n(T) = 0. \tag{10.4}$$

If the process $v(\tau)$, $y(\tau)$ *is optimal, that is, if it imparts a maximum to functional* (8.19), *then for each moment* τ $(0 \leqslant \tau \leqslant T)$ *the following maximum relation will be valid:*

$$\tilde{H}(\varphi(\tau), y(\tau), v(\tau)) = \max_{v \in V} \tilde{H}(\varphi(\tau), y(\tau), v), \tag{10.5}$$

that is, $\tilde{H}(\varphi(\tau), y(\tau), v(\tau)) \geqslant \tilde{H}(\varphi(\tau), y(\tau), v)$ for any $v \in V$

If we pose the problem of the minimum (rather than the maximum) of functional (8.19), *then the formulation remains the same, except that relation* (10.3) *becomes*

$$\varphi_0 = -1. \tag{10.3'}$$

The formulated theorem (maximum principle) gives the *necessary* condition for optimality (10.5), in many cases making it possible to select, out of all the processes with initial condition (8.20), a *single* process $v(\tau)$, $y(\tau)$ "suspected" of being optimal (or else making it possible to select a finite number of such processes, from which then the truly optimal process is chosen).

Now let us try to formulate a "discrete" analog of the maximum principle. To do this, the same technique of transition from the "continuous" problem to the discrete problem which was applied in Subsection 8 is used.

First, substitution (8.24) is made, but with the additional introduction of another variable ψ, defined by the formula

$$\psi(t) = \varphi(th) = \varphi(\tau). \tag{10.6}$$

Then $\tilde{H}$ turns out to be a function of the independent variables ψ, x, u:

$$\tilde{H} = \sum_{i=0}^{n} \psi_i(t)\, g^i(x(t-1),\ u(t))$$

(where $\psi_0 = \varphi_0$), while maximum relation (10.5) (which we will consider only for the values $\tau = h, 2h, \ldots, Nh$) becomes, in terms of the new variables,

$$\tilde{H}(\psi(t),\ x(t-1),\ u(t)) \geqslant \tilde{H}(\psi(t),\ x(t-1),\ u) \quad \text{for any} \quad u \in V; \quad t = 1, \ldots, N,$$

that is,

$$\sum_{i=0}^{n} \psi_i(t)\, g^i(x(t-1),\ u(t)) \geqslant \sum_{i=0}^{n} \psi_i(t)\, g^i(x(t-1),\ u) \quad \text{for any} \quad u \in V; \quad t = 1, \ldots, N. \tag{10.7}$$

In addition, relation (10.2) can be replaced by the approximate equation

$$\frac{\varphi_j(\tau) - \varphi_j(\tau - h)}{h} \approx -\sum_{i=0}^{n} \varphi_i(\tau) \frac{\partial g^i(y(\tau),\ v(\tau))}{\partial y^j} \approx -\sum_{i=0}^{n} \varphi_i(\tau) \frac{\partial g^i(y(\tau - h),\ v(\tau))}{\partial y^j},$$

which we will consider only for the values $\tau = h, 2h, \ldots, Nh$. In terms of the new variables, this approximate equation is written as

$$\psi_j(t-1) \approx \psi_j(t) + h \sum_{i=0}^{n} \psi_i(t) \frac{\partial g^i(x(t-1), u(t))}{\partial x^j}; \qquad (10.8)$$

$$j = 1, \ldots, n; \ t = 1, \ldots, N.$$

Finally, relations (10.3) and (10.4) can be written in the form:

$$\psi_0 = 1, \qquad (10.9)$$

$$\psi_1(N) = \psi_2(N) = \ldots = \psi_n(N) = 0. \qquad (10.10)$$

Now we introduce, as in Subsection 8, the function

$$f(x, u) = x + hg(x, u)$$

and, for uniformity of notation, the function $hg^0(x, u)$ is called $f^0(x, u)$, so that functional (8.30) is written as

$$J = \sum_{t=0}^{N} f^0(x(t-1), u(t)). \qquad (10.11)$$

In addition, we introduce the function

$$H(\psi, x, u) = hg^0(x, u) + \sum_{i=1}^{n} \psi_i (x^i + hg^i(x, u)) = \sum_{i=0}^{n} \psi_i f^i(x, u).$$

Then, after replacing approximate equation (10.8) by the precise equation (and introducing a certain error in the process), we obtain

$$\psi_j(t-1) = \sum_{i=0}^{n} \psi_i(t) \frac{\partial f^i(x(t-1), u(t))}{\partial x^j} = \frac{\partial H(\psi(t), x(t-1), u(t))}{\partial x^j}; \qquad (10.12)$$

$$j = 1, \ldots, n; \quad t = 1, 2, \ldots, N.$$

Maximum relation (10.7) can now be written as

$$H(\psi(t), x(t-1), u(t)) \geqslant H(\psi(t), x(t-1), u) \qquad \text{for any} \quad u \in V,$$

that is,

$$H(\psi(t), x(t-1), u(t)) = \max_{u \in V} H(\psi(t), x(t-1), u); \qquad (10.13)$$

$$t = 1, 2, \ldots, N.$$

Finally, as in Subsection 8, we replace (thereby introducing a certain error) approximate equation (8.25) by the precise equation:

$$x(t) = f(x(t-1), u(t)) \qquad (10.14)$$

(see (8.29)).

The relations obtained also express the so-called *discrete maximum principle.* We formulate this principle (by analogy with the "continuous" maximum principle) as follows.

For a discrete controlled object (10.14) *with a control domain* V (see (8.27)) *and an effectiveness criterion* (10.11), *we consider the fundamental problem (that is, the problem with a free right-hand endpoint and without any constraints on the phase coordinates) with initial condition* $x(0) = y_0$ (see (8.28)), *optimality being understood here to mean the maximum of functional* (10.11).

To solve this problem, we introduce the auxiliary variables $\psi_0, \psi_1, \ldots, \psi_n$ *(here* n *is the number of phase variables* $x^1, \ldots, x^n$ *) and we use them to formulate the following auxiliary function:*

$$H(\psi, x, u) = \sum_{i=0}^{n} \psi_i f^i(x, u), \tag{10.15}$$

where f^0 *is the function entering into the definition of functional* (10.11), *and* $f^1, \ldots, f^n$ *are the components of the vector function* $f(x, u)$ *appearing on the right-hand side of equation* (10.14). *In addition, we write a system of relations* (10.12) *for the auxiliary unknowns and we set* $\psi_0 = 1$ (see (10.9)).

Next let $u(t)$, $t = 1, 2, \ldots, N$ *be some admissible (that is, satisfying inclusion* $u(t) \in V$ *for all* $t = 1, 2, \ldots, N$*) control, and let* $x(t)$, $t = 0, 1, \ldots, N$, *be the corresponding (that is, satisfying relation* (10.14)*) trajectory, starting from point* $x(0) = y_0$. *After substituting functions* $x = x(t)$ *and* $u = u(t)$ *into the right-hand sides of relations* (10.12), *we can use relations* (10.9) *and* (10.10) *to find successively* $\psi(N-1), \psi(N-2), \ldots, \psi(1)$.

If process $u(t)$, $x(t)$ *is optimal, that is, if it makes functional* (10.11) *a maximum, then for each* $t = 1, 2, \ldots, N$ *maximum relation* (10.3) *is satisfied.*

If optimality is understood to mean the minimum (rather than the maximum) of functional (10.11), *then the formulation remains the same, except that relation* (10.9) *becomes*

$$\psi_0 = -1. \tag{10.9'}$$

We have formulated the discrete maximum principle as a necessary condition for optimality (by analogy with the "continuous" maximum principle), and we have limited ourselves just to the case of the fundamental problem formulated in Subsection 3. Naturally, the considerations presented cannot serve as a *proof* of the discrete maximum principle, since the complete transition from a continuous object to a discrete object was not substantiated and introduced some error (when the approximate equations were replaced by the precise ones).

Attempts have been made (beginning with the studies of S. Katz) to prove the discrete maximum principle in general form. However, all of these considerations were mathematically naive and contained obvious errors.*

* The Soviet reader is acquainted with one such attempt, in the translation of: Fan, L.T., and C.S. Wang, *The Discrete Maximum Principle*. – John Wiley, New York. 1964. This book discusses some interesting applied problems, but it is mathematically incorrect (as was noted by its editor); in particular, the considerations on pp.33–35 of the book contain an obvious mistake and no proof of the discrete maximum principle is given.

Thus it is interesting to interpret mathematically the formulation of the discrete maximum principle and to ascertain whether the principle is correct in this form. It is seen that, *in the presented formulation, the maximum principle will be neither a necessary nor a sufficient condition for optimality.*

In order to interpret mathematically the formulation of the discrete maximum principle, let us reformulate the considered discrete problem of optimal control as a problem of the extremum of a function defined on a subset of Euclidean space (see Subsection 6). To do this, we introduce variables u_t^j and x_s^i (as in Subsection 6) and we denote as E_z the space in which all these variables are coordinates ($i = 1, \ldots, n$; $j = 1, \ldots, r$; $s = 1, \ldots, N$; $t = 1, \ldots, N$). For any point $z \in E_z$ we then, as in Subsection 6, determine the points $\xi_s(z)$ and $\eta_t(z)$.

Since the fundamental problem is being considered, we do not have relation (6.4), and relations (6.5) and (6.6) take the form

$$\eta_t(z) \in V, \qquad t = 1, \ldots, N; \tag{10.16}$$

$$\xi_t(z) = f(\xi_{t-1}(z), \eta_t(z)), \qquad t = 1, \ldots, N \tag{10.17}$$

(where it is assumed that $\xi_0(z) = y_0$). The set of all points $z \in E_z$ satisfying all the relations (10.16), (10.17) is called Ω. Then the specified discrete problem of optimal control is equivalent to the problem of finding the maximum of the function

$$F(z) = f^0(\xi_0(z), \eta_1(z)) + f^0(\xi_1(z), \eta_2(z)) + \ldots + f^0(\xi_{N-1}(z), \eta_N(z)), \tag{10.18}$$

considered on set Ω (in view of formulas (6.8)). Relation (10.17) can be written in coordinate form as

$$-\xi_t^i(z) + f^i(\xi_{t-1}(z), \eta_t(z)) = 0; \quad t = 1, \ldots, N; \; i = 1, \ldots, n.$$

Let us call the left-hand side of this equation $F_t^i(z)$:

$$F_t^i(z) = -\xi_t^i(z) + f^i(\xi_{t-1}(z), \eta_t(z)) \tag{10.19}$$

(here, obviously, $\xi_t^i(z) = x_t^i$). From (10.19) it is easy to obtain expressions for the partial derivatives of function $F_t^i(z)$ with respect to variables x_s^j. These have the following form:

$$\frac{\partial F_t^i(z)}{\partial x_t^i} = -1; \quad \frac{\partial F_t^i}{\partial x_{t-1}^j} = \frac{\partial f^i(\xi_{t-1}(z), \eta_t(z))}{\partial x^j}, \quad j = 1, \ldots, n \tag{10.20}$$

(the other partial derivatives are zero, since the rest of the variables x_s^j do not enter into the right-hand side of (10.19)).

Now let us consider the function obtained from $F(z)$ by adding on a linear combination of functions (10.19):

$$G(z) = F(z) + \sum_{s=1}^{N} \sum_{j=1}^{n} \psi_j^s F_s^j(z), \tag{10.21}$$

where ψ_j^s are certain constants. Since all the functions $F_s^j(z)$ go identically to zero on set Ω (see (10.17), (10.19)), it follows that

$$F(z) \equiv G(z) \quad \text{on set} \quad \Omega$$

for any choice of constants ψ_j^s. Consequently, the problem of the maximum of function $F(z)$ on set Ω is equivalent to the problem of the maximum of function $G(z)$ on set Ω. Note also that, in view of (10.20), function $G(z)$ has the following partial derivatives with respect to variables x_{t-1}^i:

$$\frac{\partial G(z)}{\partial x_{t-1}^i} = \frac{\partial F(z)}{\partial x_{t-1}^i} + \sum_{s=1}^{N}\sum_{j=1}^{n} \psi_j^s \frac{\partial F_s^j(z)}{\partial x_{t-1}^i} =$$

$$= \frac{\partial f^0(\xi_{t-1}(z), \eta_t(z))}{\partial x^i} - \psi_i^{t-1} + \sum_{j=1}^{n} \psi_j^t \frac{\partial f^j(\xi_{t-1}(z), \eta_t(z))}{\partial x^i};$$

$$t = 2, 3, \ldots, N;$$

$$\frac{\partial G(z)}{\partial x_N^i} = -\psi_i^N.$$

Therefore, if we set $\psi_0^t = 1$, these partial derivatives will have the form

$$\frac{\partial G(z)}{\partial x_{t-1}^i} = -\psi_i^{t-1} + \sum_{j=0}^{n} \psi_j^t \frac{\partial f^j(\xi_{t-1}(z), \eta_t(z))}{\partial x^i}, \quad \frac{\partial G(z)}{\partial x_N^i} = -\psi_i^N. \tag{10.22}$$

Next we assume that $u(t)$, $x(t)$ is an optimal process for the discrete problem being considered, and that z_0 is the corresponding point of set Ω, that is, $\xi_t(z_0) = x(t)$, $\eta_t(z_0) = u(t)$; $t = 1, \ldots, N$ (see (6.8)). In addition, assume that $\psi(1)$, $\psi(2), \ldots, \psi(N)$ correspond to this optimal process, in view of the discrete maximum principle (see (10.10), (10.12)). The constants ψ_j^s entering into the definition of function $G(z)$ (see (10.21)) are selected as follows:

$$\psi_j^s = \psi_j(s); \qquad j = 1, \ldots, n; \ s = 1, \ldots, N.$$

Then, from (10.22) we have

$$\frac{\partial G(z_0)}{\partial x_{t-1}^i} = -\psi_i(t-1) + \sum_{j=0}^{n} \psi_j(t) \frac{\partial f^j(x(t-1), u(t))}{\partial x^i}, \quad t = 2, \ldots, N;$$

$$\frac{\partial G(z_0)}{\partial x_N^i} = -\psi_i(N),$$

so that, according to (10.10) and (10.12),

$$\frac{\partial G(z_0)}{\partial x_t^i} = 0; \qquad t = 1, \ldots, N; \quad t = 1, \ldots, n.$$

Thus, at point $z = z_0$, function $G(z)$ has zero partial derivatives with respect to all the variables x_t^i.

Finally, let us bring in maximum condition (10.13). By writing it in the form

$$\sum_{i=0}^{n} \psi_i(t) f^i(x(t-1), u(t)) = \max_{u \in V} \sum_{i=0}^{n} \psi_i(t) f^i(x(t-1), u)$$

and changing to variables x_t^i, u_t^j, ψ_i^t, we obtain

$$\sum_{i=0}^{n} \psi_i^t f^i(\xi_{t-1}(z_0), \eta_t(z_0)) = \max_{u_t \in V} \sum_{i=0}^{n} \psi_i^t f^i(\xi_{t-1}(z_0), u_t), \quad t = 1, \ldots, N.$$

If the term $-\sum_{i=1}^{n} \psi_i^t \xi_t^i(z_0)$, which does not depend on u_t, is added to both sides of this equation, then we get (taking into account that $\psi_0^t = 1$):

$$f^0(\xi_{t-1}(z_0), \eta_t(z_0)) + \sum_{i=1}^{n} \psi_i^t(-\xi_t^i(z_0) + f^i(\xi_{t-1}(z_0), \eta_t(z_0))) =$$

$$= \max_{u_t \in V} \left[f^0(\xi_{t-1}(z_0), u_t) + \sum_{i=1}^{n} \psi_i^t(-\xi_t^i(z_0) + f^i(\xi_{t-1}(z_0), u_t)) \right].$$

By summing these relations over $t = 1, \ldots, N$ and taking (10.18), (10.19), (10.21) into account, we now obtain

$$G(z_0) = \max_{\tilde{z}_0} G(\tilde{z}_0), \tag{10.23}$$

where $\tilde{z}_0$ is a point having the same coordinates x_t^i as point z_0 (that is $\xi_t^i(\tilde{z}_0) = \xi_t^i(z_0)$ for all i, t) and arbitrary coordinates u_s^t satisfying conditions (10.16). In other words, at point z_0 function $G(z)$ reaches a maximum for all variables u_s^j (with variables x_t^i fixed). Conversely, if we carry out the above operations in reverse order, we obtain from relation (10.23) the maximum relation (10.13) for all $t = 1, \ldots, N$. Consequently, equation (10.23) is equivalent to maximum condition (10.13), in terms of variables u_s^j and x_t^i.

Let us next interpret the formulas obtained. To do this, we call E_u the space of all the variables u_s^j ($j = 1, \ldots, r$, $s = 1, \ldots, N$) and for any point $u^* \in E_u$ we set $\eta_t(u^*) = (u_t^1, \ldots, u_t^r)$. In addition, let Ξ be the set of all points u^* of space E_u, the coordinates of which satisfy conditions $\eta_t(u^*) \in V$, $t = 1, \ldots, N$ (cf. 10.16)). Therefore, specifying the point $u^* \in \Xi$ is equivalent to specifying the admissible control $u(1), \ldots, u(N)$. It should be mentioned here that specifying point $u^* \in \Xi$ (that is, specifying the values of all the variables u_s^j) provides an unambiguous determination of the values of variables x_t^i, in view of relations (10.17) (this is equivalent to saying that specifying control $u(1), \ldots, u(N)$ defines uniquely the trajectory $x(0), x(1), \ldots, x(N)$, in view of relations (10.14), since the initial point $x(0) = y_0$ is given).

Let $\zeta(u^*)$ be the point $x^* = \{x_t^i\}$ (in the space E_x of all the variables x_t^i), defined by relations (10.17), starting from point $u^* \in \Xi$. Consequently, for any point $u^* \in \Xi$, the point $z = (u^*, \zeta(u^*))$ satisfies all of relations (10.16), (10.17), that is, it belongs to set Ω, and obviously any point $z \in \Omega$ is of this type.

If all the variables u_s^j are represented arbitrarily on a graph by a single axis u^*, while all the variables x_t^i are represented by a single axis x^*, then we get a function $x^* = \zeta(u^*)$, defined in set Ξ (lying on the u^* "axis"). The set of all points $z = (u^*, x^*)$ for which $x^* = \zeta(u^*)$, that is, the *graph* of function ζ, will also represent set Ω (Figure 27).

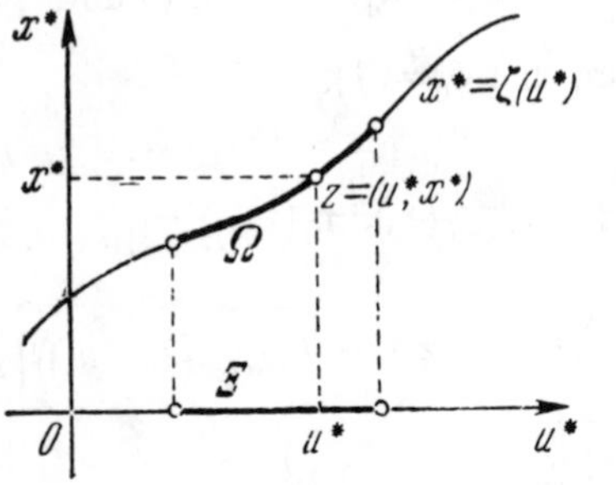

FIGURE 27

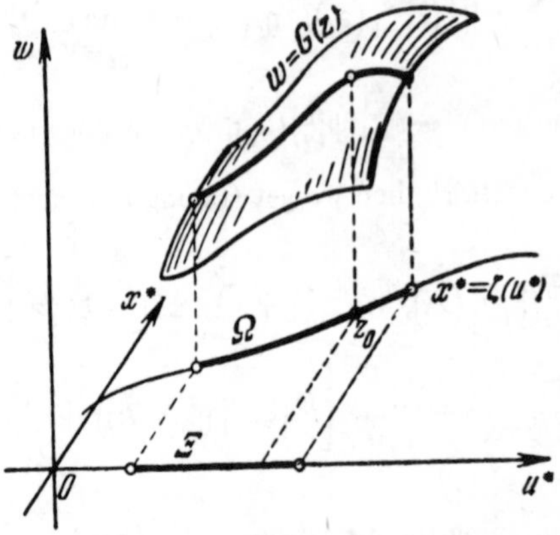

FIGURE 28

Moreover, function $G(z)$ is defined in space E_z and the maximum of this function on set Ω is sought (Figure 28); this is equivalent to looking for the maximum of function $F(z)$ on set Ω. Here function $G(z)$ is selected in such a way that all of its derivatives with respect to the coordinates of point x^* (that is, with respect to x_t^i) go to zero at some point $z_0 \in \Omega$. It is definite that, under these circumstances, fulfillment of the maximum condition, that is, validity of equation (10.23), constitutes a necessary condition for function $G(z)$, considered on set Ω, to reach a maximum at point z_0.

This also constitutes a formulation of the discrete maximum principle in terms of coordinates u_s^j and x_t^i. Consequently, without going into the specific forms of function $G(z)$, function $\zeta(u^*)$, and set Ξ, we can sum up the discrete maximum principle as follows:

A set Ξ is given in space E_u, and in addition some mapping ζ: $\Xi \to E_x$ of this set Ξ is defined in space E_x. If Ω is the set of all points $z = (u^*, \zeta(u^*))$, where $u^* \in \Xi$, that is, the graph of mapping ζ, then $\Omega \subset E_z$, where $E_z = E_u \times E_x$ is *the direct product of spaces E_u and E_x. In space E_z some function $G(z)$ is defined, and in set Ω some point $z_0 = (u_0^*, x_0^*)$ is defined, where at z_0 all the partial derivatives of function $G(z)$ with respect to the directions lying in E_x are zero. In order for (under these circumstances) function $G(z)$, considered only on set Ω, to reach a maximum at point z_0, this function $G(z)$ at point z_0 has to reach a maximum with respect to variable u^*, that is, $G(u^*, x_0^*) \leqslant G(u_0^*, x_0^*)$ for any $u^* \in \Xi$.*

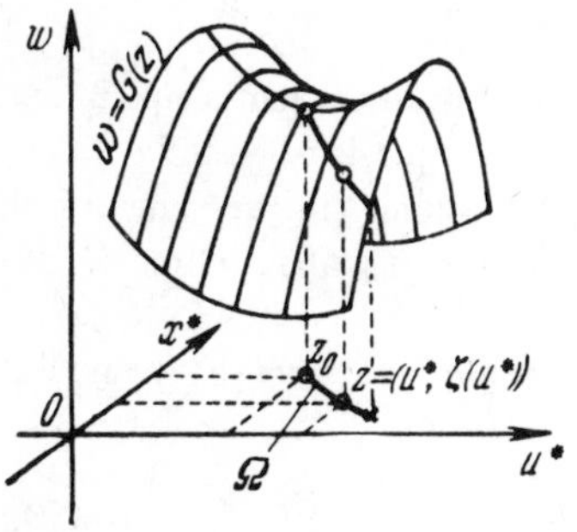

FIGURE 29

Since we are discussing functions in Euclidean spaces, this statement (the validity of which we will now assess) is known as the *Euclidean maximum principle.* Simple examples indicate that, in its above general formulation, the Euclidean maximum principle is not valid. Figure 29 illustrates this fact: function $G(z)$ is expressed graphically as a surface having a *saddle* over point z_0 (so that *all* the derivatives of function $G(z)$ at poing z_0 go to zero). In the direction of the u^* axis the surface *rises* (and thus function $G(z)$ *does not reach a maximum* with respect to variable z_0 at point u^*), and in the direction of the x^* axis it falls. Now, if the line Ω, that is, the plot of function $x^* = \zeta(u^*)$, is sufficiently inclined with respect to the u^* axis, then for motion along this line away from point z_0 function $G(z)$ *decreases.* Thus, function $G(z)$, considered on set Ω, reaches a maximum at point z_0, but the necessary condition specified in the Euclidean maximum principle (that is, reaching a maximum with respect to variable u^*) is not satisfied. In the following example, this idea will be corroborated by an accurate calculation.

EXAMPLE 10.1. Let E_u be the space of variables $u_1, u_2, \ldots, u_N$, and let E_x be the space of variables $x_1, x_2, \ldots, x_N$. For the set Ξ we take all of space E_u, while the mapping $\zeta : \Xi \to E_x$ is defined by the relation

$$\zeta(u_1, u_2, \ldots, u_N) = (u_1, u_2, \ldots, u_N),$$

that is, point $\zeta(u^*)$ has in space E_x the same coordinates as point u^* in space E_u. Moreover, function $G(z)$ is given by the formula

$$G(z) = G(u_1, \ldots, u_N, x_1, \ldots, x_N) = \\ = u_1^2 + \ldots + u_N^2 - 2(x_1^2 + \ldots + x_N^2), \tag{10.24}$$

and z_0 is taken to be the coordinate origin of space E_z. Then at point z_0 *all* the derivatives of function $G(z)$ go to zero, including the derivatives with respect to variables $x_1, x_2, \ldots, x_N$.

Further, in view of the definition of mapping ζ, Ω is defined in space E_z by the equations

$$x_1 = u_1, \quad x_2 = u_2, \quad \ldots, \quad x_N = u_N,$$

so that on this set function $G(z)$ has the form

$$-(x_1^2 + x_2^2 + \dots + x_N^2),$$

that is, this function, considered on set Ω, reaches a maximum at point z_0. However, the necessary condition contained in the Euclidean maximum principle is not satisfied, since $G(z)$, taken to be a function just of variables $u_1, \dots, u_N$, has the form

$$u_1^2 + u_2^2 + \dots + u_N^2,$$

so that at point z_0 it reaches a minimum (with respect to u^*) rather than a maximum. Consequently, in the given case the Euclidean maximum principle is not valid.

EXAMPLE 10.2. The Euclidean maximum principle also cannot (in the general case) be considered as a *sufficient* condition for a maximum at point z_0 of function $G(z)$, considered only on set Ω.

Let us retain the same notation as in the previous example, but as $G(z)$ we now take the same function with the opposite sign:

$$G(z) = -u_1^2 - \dots - u_N^2 + 2(x_1^2 + \dots + x_N^2). \tag{10.25}$$

As previously, all the derivatives of function $G(z)$ go to zero at point z_0 but now the condition contained in the Euclidean maximum principle is satisfied, that is, $G(z)$, taken to be a function just of variables $u_1, \dots, u_N$, reaches a maximum at z_0. On the other hand, function G, considered on set Ω, does not reach a maximum at point z_0.

Thus, in the general case (in the form in which it was formulated above), the Euclidean maximum principle is neither necessary nor sufficient as a condition for reaching a maximum on set Ω.

The example considered can easily be reformulated in terms of discrete controlled objects. As a result, we obtain examples indicating that in general the discrete maximum principle is neither a necessary nor a sufficient condition for optimality.† To arrive at these examples, we set $n = r = 1$ (that is, we assume that there is only one phase coordinate and only one controlling parameter), and we apply formulas (6.8), so that u_t and x_t are replaced by $u(t)$ and $x(t)$. In addition, in (10.24) and (10.25), we discard the term x_N^2 (since the independent variable $x(N)$ does not enter into functional (10.11)).

EXAMPLE 10.3. Let us consider the discrete controlled object

$$x(t) = u(t), \qquad t = 1, 2, \dots, N \tag{10.26}$$

(that is, $f(x(t-1), u(t)) = u(t)$, see (10.14)), with the initial condition $x(0) = 0$, where $x(t)$ and $u(t)$ are *scalars* (that is, $n = r = 1$). As a control domain V (see (8.27)) we take the interval $[-1, 1]$. For this discrete object we consider the fundamental

† The first to propose an example of this sort was A.G. Butkovskii, in the article: *O neobkhodimykh i dostatochnykh usloviyakh optimal'nosti dlya impul'snykh sistem upravleniya (On the Necessary and Sufficient Conditions for Optimality in Pulse Systems of Control)*. – Avtomatika i Telemekhanika, **24**, No.8. 1963.

problem (the problem with a free right-hand endpoint and with no constraints on the phase coordinates), in the sense of a maximum of functional (10.11), where

$$f^0(x(t-1),\ u(t)) = (u(t))^2 - 2(x(t-1))^2. \qquad (10.27)$$

According to the discrete maximum principle, we introduce the auxiliary variables ψ_0, ψ_1, where $\psi_0 = 1$. Function H (see (10.15)) here has the form

$$\begin{aligned} H(\psi(t),\ x(t-1),\ u(t)) &= \\ &= \psi_0 f^0(x(t-1),\ u(t)) + \psi_1(t) f^1(x(t-1),\ u(t)) = \\ &= (u(t))^2 - 2(x(t-1))^2 + \psi_1(t)\, u(t), \end{aligned} \qquad (10.28)$$

so that system of relations (10.12) becomes

$$\psi_1(t-1) = \frac{\partial H(\psi(t),\ x(t-1),\ u(t))}{\partial x} = -4x(t-1). \qquad (10.29)$$

Now let us consider for this discrete object the process

$$\begin{aligned} &u(1) = u(2) = \ldots = u(N-1) = 0, \quad u(N) = 1, \\ &x(0) = x(1) = \ldots = x(N-1) = 0, \quad x(N) = 1 \end{aligned} \qquad (10.30)$$

(obviously, this process satisfies relation (10.26)). For this process, in view of (10.10) and (10.29), it follows that

$$\psi_1(N) = \psi_1(N-1) = \ldots = \psi_1(2) = \psi_1(1) = 0. \qquad (10.31)$$

From (10.26) and (10.27) we obtain (for any admissible process)

$$\begin{aligned} J &= (u(1))^2 + (u(2))^2 + \ldots + (u(N))^2 - 2(x(1))^2 - 2(x(2))^2 - \ldots \\ &\ldots - 2(x(N-1))^2 = (u(N))^2 - (u(1))^2 - (u(2))^2 - \ldots \\ &\ldots - (u(N-1))^2 \leqslant (u(N))^2 \leqslant 1. \end{aligned}$$

At the same time, for process (10.30) we obviously have $J = 1$, and thus this process is optimal. However, maximum condition (10.13) for this process, with $t = 1, \ldots, N-1$, is not satisfied (see (10.28), (10.30), (10.31)):

$$\begin{aligned} &H(\psi(t),\ x(t-1),\ u(t)) = 0, \qquad t = 1, 2, \ldots, N-1; \\ &\max_{u \in V} H(\psi(t),\ x(t-1),\ u) = \max_{u \in V} u^2 = 1. \end{aligned}$$

Therefore, for this optimal process the necessary condition specified in the discrete maximum principle is not satisfied. This example shows that, in the formulation given above, the discrete maximum principle is not valid.

EXAMPLE 10.4. Let us consider the same discrete object as in the previous example, but with the opposite sign of functional J:

$$f^0(x(t-1),\ u(t)) = 2(x(t-1))^2 - u(t))^2. \qquad (10.32)$$

Then relations (10.28) and (10.29) become

$$H(\psi(t),\ x(t-1),\ u(t)) = 2(x(t-1))^2 - (u(t))^2 + \psi_1(t)\,u(t),$$
$$\psi_1(t-1) = 4x(t-1).$$

Thus for the process

$$u(1) = \ldots = u(N) = 0,\quad x(0) = x(1) = \ldots = x(N) = 0 \qquad (10.33)$$

relation (10.31) is valid. The following maximum relation (for all $t = 1, \ldots, N$) is also satisfied for process (10.33):

$$\max_{u \in V} H(\psi(t),\ x(t-1),\ u) = \max_{u \in V}(-u^2) = 0 = H(\psi(t),\ x(t-1), u(t)).$$

However, process (10.33) is not optimal: for it functional J (see (10.32)) is equal to zero, whereas, for instance for the process

$$u(1) = 1,\ u(2) = \ldots = u(N) = 0;\quad x(1) = 1,$$
$$x(0) = x(2) = \ldots = x(N) = 0,$$

we have $J = 1$. Consequently, the discrete maximum principle is not a sufficient condition for optimality either.

Thus, in contrast to the "continuous" case, the discrete maximum principle in general gives neither a necessary nor a sufficient condition for optimality. However, the great popularity of Pontryagin's maximum principle, which is a convenient, widely applied necessary condition for the optimality of continuous controlled processes, directed the efforts of investigators toward finding optimality conditions in the form of a maximum principle for discrete processes as well.

First, a demonstration was given of the discrete maximum principle, for a limited class of discrete controlled objects. The first work in this direction was that of Rozonoer,* who demonstrated that, for the fundamental problem as applied to discrete objects that are linear with respect to the phase coordinates, the discrete maximum principle is valid, as both a necessary and a sufficient condition for optimality.

Second, for discrete controlled objects of general form, theorems were established which represented refinements or modifications of the above-formulated discrete maximum principle. A formulation of the maximum principle for discrete systems possessing some convex structure was presented by Propoi,** but his proof has a flaw in

* Rozonoer, L.I. *Printsip maksimuma L.S. Pontryagina v teorii optimal'nykh sistem (L.S. Pontryagin's Maximum Principle in the Theory of Optimal Systems)*, III. – Avtomatika i Telemekhanika, **20**, No.12. 1959.

** Propoi, A.I. *O printsipe maksimuma dlya diskretnykh sistem upravleniya (A Maximum Principle for Discrete Control Systems)*. – Avtomatika i Telemekhanika, **26**, No.7. 1965.

it. Correct proofs have been offered by Halkin, Holtzman, and others.* Gabasov and Kirillova** introduced a quasimaximum principle, establishing a relationship between the "continuous" and "discrete" maximum principles.

Consequently, the discrete maximum principle, in spite of its originally incorrect formulation, constitutes one of the basic directions of study in the theory of optimal control of discrete systems.

11. Ideas of mathematical programming. In this subsection, we will characterize briefly another group of results, partly associated with the discrete maximum principle and also making it possible to gain information about optimal processes in discrete objects. Methods for solving the problem of mathematical programming formulated in Subsection 7 will be discussed.

Let us begin by recalling some classical concepts of mathematical analysis. Let $f(z) = f(z^1, \ldots, z^n)$ be some function defined in space E^n and having the property of *smoothness,* that is, having continuous derivatives

$$\frac{\partial f}{\partial z^1}, \quad \frac{\partial f}{\partial z^2}, \quad \ldots, \quad \frac{\partial f}{\partial z^n}. \tag{11.1}$$

Therefore, at each point z_0, n numbers (11.1) are specified; the vector having these numbers as its coordinates is called the *gradient* of function f (at point z_0) and is designated as $\operatorname{grad} f(z_0)$:

$$\operatorname{grad} f(z_0) = \left\{ \frac{\partial f(z_0)}{\partial z^1}, \frac{\partial f(z_0)}{\partial z^2}, \ldots, \frac{\partial f(z_0)}{\partial z^n} \right\}.$$

If δz is an arbitrary vector originating at point z_0, and if the end of this vector is called z, then the coordinates of vector δz are $z^1 - z_0^1,\ z^2 - z_0^2,\ \ldots,\ z^n - z_0^n$. If the system of coordinates $z^1, z^2, \ldots, z^n$ in space E^n is rectangular, then the scalar product of vectors δz and $\operatorname{grad} f(z_0)$ is calculated as follows:

$$\delta z \operatorname{grad} f(z_0) = \\ = \frac{\partial f(z_0)}{\partial z^1}(z^1 - z_0^1) + \frac{\partial f(z_0)}{\partial z^2}(z^2 - z_0^2) + \ldots + \frac{\partial f(z_0)}{\partial z^n}(z^n - z_0^n).$$

* Jordan, B.W., and E. Polak. *Theory of a Class of Discrete Optimal Control Systems.* – J. Electron. and Control, **17**, 6. 1964; Pearson, J. *The Discrete Maximum Principle.* – Int. J. of Control, **11**, 2. 1965; Halkin, H. *A Maximum Principle of the Pontryagin Type for Systems Described by Nonlinear Difference Equations.* – J. SIAM on Control, **4**, 1. 1966; Holtzman, J.M., and H. Halkin. *Directional Convexity and the Maximum Principle for Discrete Systems.* – J. SIAM on Control, **4**, 2. 1966; Cannon, M., C. Cullum, and E. Polak. *Constrained Minimization Problem in Finite-Dimensional Spaces.* – J. SIAM on Control, **4**, 3. 1966.

** Gabasov, R., and F.M. Kirillova. *K voprosu o rasprostranenii printsipa maksimuma L.S. Pontryagina na diskretnye sistemy (On the Extension of L.S. Pontryagin's Maximum Principle to Discrete Systems).* – Avtomatika i Telemekhanika, **27**, No.11. 1966.

According to Taylor's formula, we have

$$f(z)=f(z_0)+\frac{\partial f(z_0)}{\partial z^1}(z^1-z_0^1)+\ \dots\ +\frac{\partial f(z_0)}{\partial z^n}(z^n-z_0^n)+\ \dots,$$

where the dots at the end indicate a quantity with a higher order of smallness than the distance between points z and z_0 (that is, than the length $|\boldsymbol{\delta z}|$ of vector $\boldsymbol{\delta z}$). Therefore,

$$f(z)-f(z_0)\approx \boldsymbol{\delta z}\,\mathrm{grad}\, f(z_0), \tag{11.2}$$

where $\boldsymbol{\delta z}$ is the vector from point z_0 to point z, and equation (11.2) is approximate (the error being a quantity of higher order of smallness than $|\boldsymbol{\delta z}|$).

Now, if Λ is a curve drawn from point z_0 and $\boldsymbol{a}$ is a unit vector tangent to this curve at z_0 (Figure 30), then for a point z different from z_0 on curve Λ we can write

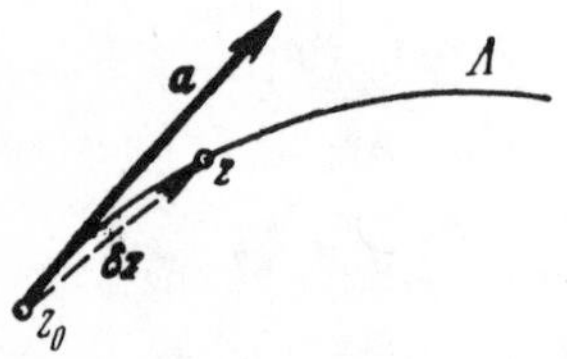

FIGURE 30

$$\boldsymbol{\delta z}\approx|\boldsymbol{\delta z}|\boldsymbol{a}, \tag{11.3}$$

where $\boldsymbol{\delta z}$ is the vector from point z_0 to point z and the equation is approximate (error with higher order of smallness than $|\boldsymbol{\delta z}|$). From (11.2) and (11.3), we see that the following approximate equation is valid for point z on curve Λ:

$$f(z)-f(z_0)\approx|\boldsymbol{\delta z}|\cdot(\boldsymbol{a}\,\mathrm{grad}\, f(z_0)). \tag{11.4}$$

If the scalar product $\boldsymbol{a}\,\mathrm{grad}\, f(z_0)$ is different from zero, then the right-hand side of relation (11.4) will have the same sign as the product $\boldsymbol{a}\,\mathrm{grad}\, f(z_0)$ (since $|\boldsymbol{\delta z}|>0$), so that for sufficiently small $|\boldsymbol{\delta z}|$ the left-hand side will have the same sign as well (Figure 31), since for small enough $|\boldsymbol{\delta z}|$ the quantity dropped, having a higher order of smallness, will be less than the right-hand side of relation (11.4). Thus the following assumption is valid:

THEOREM 11.1. *If Λ is a curve originating at point z_0 and having a tangent vector $\boldsymbol{a}$ at z_0, the scalar product $\boldsymbol{a}\,\mathrm{grad}\, f(z_0)$ being different from zero, then for any point z on curve Λ different from z_0 and close enough to z_0, the difference $f(z)-f(z_0)$ has the same sign as the scalar product $\boldsymbol{a}\,\mathrm{grad}\, f(z_0)$.*

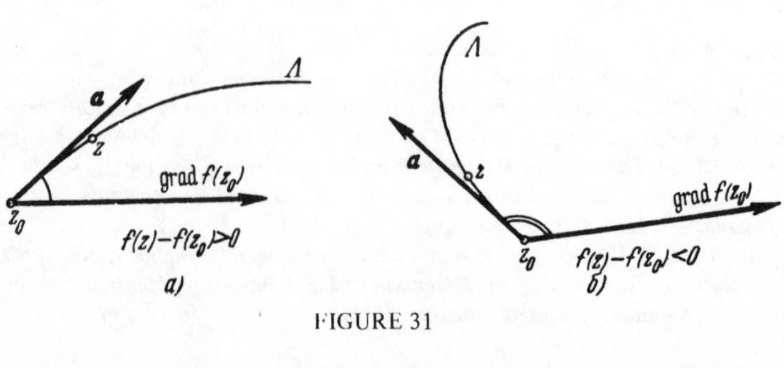

FIGURE 31

Now assume that at point z_0 function f goes to zero, that is, point z_0 lies on the hypersurface $f(z) = 0$ (or, alternatively, on the boundary of the closed domain $f(z) \leqslant 0$). We consider some curve Λ starting from point z_0 and lying on the hypersurface $f(z)=0$ (Figure 32). If $\boldsymbol{a}$ is the tangent vector to this curve at point z_0, then the relation $\boldsymbol{a}\,\mathrm{grad}\,f(z_0)=0$ is satisfied (since $f(z)-f(z_0)=0$ for any point z on curve Λ so that the scalar product $\boldsymbol{a}\,\mathrm{grad}\,f(z_0)$ cannot be different from zero; see Theorem 11.1).

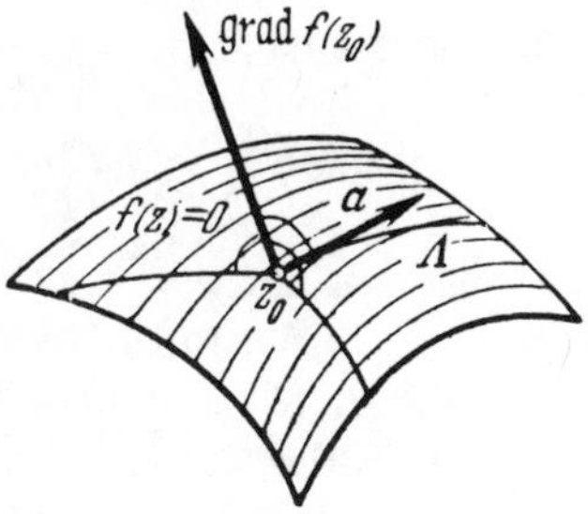

FIGURE 32

In other words, vector $\mathrm{grad}\,f(z_0)$ is "orthogonal" (at point z_0) to any curve lying on hypersurface $f(z)=0$, that is, it is "orthogonal" to hypersurface $f(z)=0$ at point z_0. Hence the vector $\mathrm{grad}\,f(z_0)$ is called the *normal vector* of hypersurface $f(z)=0$ at point z_0.

Next we consider the position of curve Λ with respect to the closed domain $f(z) \leqslant 0$. The following statement is a direct consequence of Theorem 11.1.

THEOREM 11.2. *Let $f(z)$ be a smooth function going to zero at point z_0, and let Λ be a curve originating at point z_0 and having at z_0 a tangent vector $\boldsymbol{a}$. If $\boldsymbol{a}\,\mathrm{grad}\,f(z_0)<0$, then there exists a point $z_1 \in \Lambda$ such that the arc of curve Λ from point z_0 to point z_1 lies wholly in the region $f(z) \leqslant 0$, whereas this arc does not have any points in common with hypersurface $f(z)=0$ other than z_0* (Figure 33). *If, on the other hand, $\boldsymbol{a}\,\mathrm{grad}\,f(z_0)>0$, then there will exist a point $z_1 \in \Lambda$ such that the arc of curve Λ from z_0 to z_1 does not have any points in common with region $f(z) \leqslant 0$ other than z_0* (Figure 34).

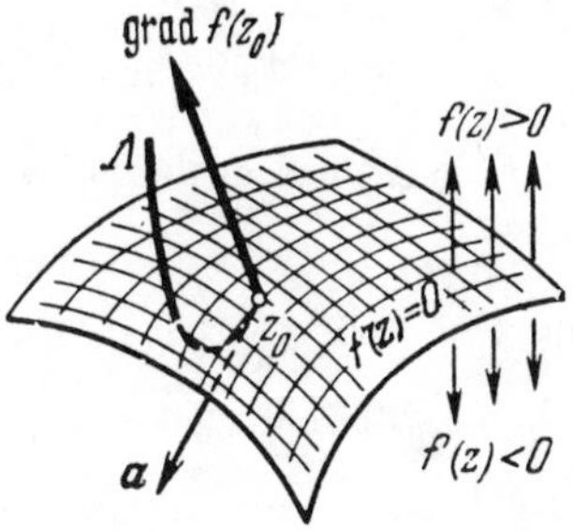

FIGURE 33

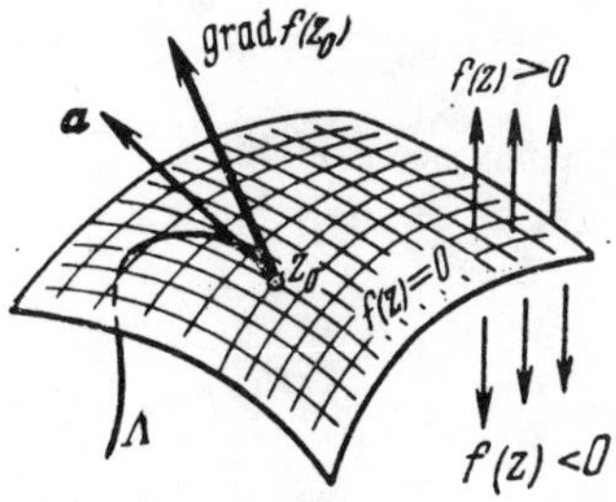

FIGURE 34

Let us apply these considerations to the problem of mathematical programming. Assume that function $F^0(z)$, considered on set Ω (see (7.1) and (7.2)), reaches a maximum at point $z_0 \in \Omega$. If we draw an arbitrary curve Λ starting from point z_0 and located in set Ω, then for any point z on Λ the inequality $F^0(z) \leqslant F^0(z_0)$ will be satisfied (since z_0 is the point of maximum).

Consequently, for a vector $\boldsymbol{a}$ tangent to curve Λ at point z_0, the inequality $\boldsymbol{a}\,\mathrm{grad}\,F^0(z_0) \leqslant 0$ is valid (otherwise, that is, if $\boldsymbol{a}\,\mathrm{grad}\,F^0(z_0) > 0$, according to Theorem 11.1 we would have $F^0(z) - F^0(z_0) > 0$ for points on curve Λ quite close to z_0, which is impossible). Thus we are led to the following theorem:

THEOREM 11.3. *In order for a smooth function* $F^0(z)$ *considered on set* Ω *to reach a maximum at a point* $z_0 \in \Omega$, *the inequality* $\boldsymbol{a}\,\mathrm{grad}\,F^0(z_0) \leqslant 0$ *has to be satisfied for any vector* $\boldsymbol{a}$ *tangent to the curve* $\Lambda \subset \Omega$ *at* z_0.

However, this theorem is unsuitable for applications, since it is not yet very clear how the vectors $\boldsymbol{a}$ possessing the indicated properties (that is, tangent to curves in set Ω at point z_0) are to be found. Actually, the specifics of set Ω (as expressed by relations (7.1) and (7.2)) are not taken into account at all in this theorem. In order to improve this theorem and to arrive at a more suitable necessary extremum condition, let us consider the simplest, most obvious case, $n = 2$, and then attempt to reformulate the results obtained for any n.

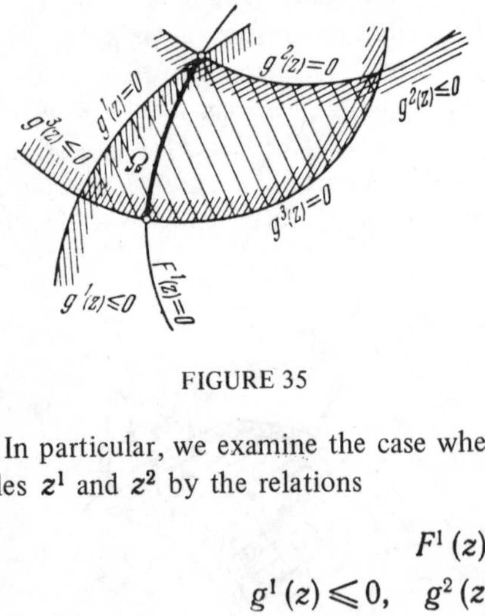

FIGURE 35

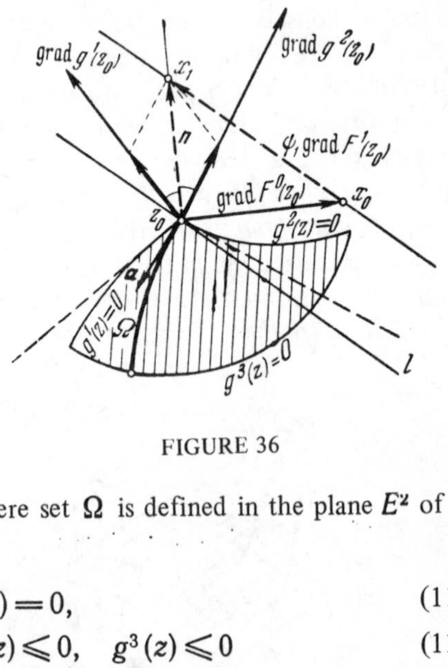

FIGURE 36

In particular, we examine the case where set Ω is defined in the plane E^2 of variables z^1 and z^2 by the relations

$$F^1(z) = 0, \tag{11.5}$$

$$g^1(z) \leqslant 0, \quad g^2(z) \leqslant 0, \quad g^3(z) \leqslant 0 \tag{11.6}$$

(Figure 35), and we pose the problem of finding the maximum of function $F^0(z)$ on set Ω. First we assume the maximum to be reached at a point $z_0 \in \Omega$ at which the first two constraints (11.6) are *active* (that is, the equations $g^2(z_0) = 0$ and $g^1(z_0) = 0$ are valid) and the third constraint is *inactive* (that is, $g^3(z_0) < 0$, see Figure 36). The vectors

$$\mathrm{grad}\,g^1(z_0), \quad \mathrm{grad}\,g^2(z_0) \tag{11.7}$$

form in Figure 36 an *acute* angle, while the tangents to the curves $g^1(z) = 0$ and $g^2(z) = 0$ at point z_0 (dashed in Figure 36) form an obtuse angle, the sides of which are perpendicular, respectively, to vectors (11.7). Set Ω is also located inside this obtuse angle, the set being represented in Figure 36 by a *curve* starting from point z_0.

Let $\boldsymbol{a}$ be the tangent vector to this curve Ω at point z_0 and let l be the line perpendicular to vector $\boldsymbol{a}$ and passing through point z_0. The vector $\operatorname{grad} F^1(z_0)$ (orthogonal to curve Ω at point z_0) is directed along line l. Since, for any point z on curve Ω we have $F^0(z) \leqslant F^0(z_0)$ (z_0 being the maximum point), according to Theorem 11.1 the scalar product $\boldsymbol{a} \operatorname{grad} F^0(z_0)$ is *nonpositive.* For clarity we assume that this product is *negative,* that is, the vector $\operatorname{grad} F^0(z_0)$ is on the side of line l *opposite to* vector $\boldsymbol{a}$ (see Figure 36). In other words, the vector $\operatorname{grad} F^0(z_0)$ is located in the same half-plane determined by line l as are vectors (11.7). Consequently, the line parallel to l and passing through the end x_0 of the vector $\operatorname{grad} F^0(z_0)$ intersects the bisector of the angle between vectors (11.7) at some point x_1.

The vector going from point x_0 to point x_1 is parallel to line l, so that it must have the form $\psi_1 \operatorname{grad} F^1(z_0)$, where ψ_1 is some number. Thus the vector

$$\boldsymbol{n} = \operatorname{grad} F^0(z_0) + \psi_1 \operatorname{grad} F^1(z_0)$$

(that is, the vector from point z_0 to point z_1) is directed along the bisector of the angle between vectors (11.7). Consequently, vector $\boldsymbol{n}$ makes *acute* angles with vectors (11.7) and so

$$\boldsymbol{n} = \sigma_1 \operatorname{grad} g^1(z_0) + \sigma_2 \operatorname{grad} g^2(z_0),$$

where σ_1 and σ_2 are *positive* numbers (see Figure 36). By equating to one another the two expressions found for vector $\boldsymbol{n}$, we get a relation that can be written as follows:

$$\psi_0 \operatorname{grad} F^0(z_0) + \psi_1 \operatorname{grad} F^1(z_0) = \\ = \sigma_1 \operatorname{grad} g^1(z_0) + \sigma_2 \operatorname{grad} g^2(z_0) + \sigma_3 \operatorname{grad} g^3(z_0), \quad (11.8)$$

where $\psi_0 = 1$, $\sigma_1 > 0$, $\sigma_2 > 0$, $\sigma_3 = 0$.

Accordingly, relation (11.8), in which all the coefficients σ_i are *nonnegative,* is valid. As for ψ_0, instead of $\psi_0 = 1$ it is sufficient to write $\psi_0 > 0$, since multiplication of all the coefficients ψ_0, ψ_1, σ_1, σ_2, σ_3 by the same positive multiplier does not alter the form of relation (11.8) or the signs of the coefficients. We note too that the relations

$$\sigma_1 g^1(z_0) = 0, \quad \sigma_2 g^2(z_0) = 0, \quad \sigma_3 g^3(z_0) = 0 \quad (11.9)$$

are valid, since $g^1(z_0) = 0$, $g^2(z_0) = 0$, $\sigma_3 = 0$.

An analogous statement also holds true for the case of not just two, but some other number of active constraints at point z_0. Let us assume, for example, that function $F^0(z)$ reaches a maximum on set Ω at point z_0, at which only the three constraints (11.6) are active, that is, $g^1(z_0) < 0$, $g^2(z_0) < 0$, $g^3(z_0) = 0$ (Figure 37). Set Ω is represented in Figure 37 by the heavy part of the curve proceeding from point z_0. Let $\boldsymbol{a}$ be the vector tangent to this curve at point z_0, and let l be the line perpendicular to vector $\boldsymbol{a}$ and passing through z_0. The vector grad $F^1(z_0)$ (orthogonal to curve Ω at point z_0) is directed along line l.

As in the previous case, the scalar product $\boldsymbol{a}$ grad $F^0(z_0)$ is *nonpositive.* The scalar product $\boldsymbol{a}$ grad $g^3(z_0)$ is, in view of Theorem 11.2, also nonpositive (since curve Ω, proceeding from point z_0, lies completely in the domain $g^3(z) \leqslant 0$). We assume here, for clarity, that these products are *negative,* that is, the vectors grad $F^0(z_0)$ and grad $g^3(z_0)$ are on the side of line l *opposite to* vector $\boldsymbol{a}$ (see Figure 37). Consequently, there exists a *positive* number σ_3 such that the ends of the vectors grad $F^0(z_0)$ and σ_3 grad $g^3(z_0)$ lie along a single line parallel to l; hence the vector σ_3 grad $g^3(z_0)$ — grad $F^0(z_0)$ is parallel to l, that is, proportional to the vector grad $F^1(z_0)$:

$$\sigma_3 \operatorname{grad} g^3(z_0) - \operatorname{grad} F^0(z_0) = \psi_1 \operatorname{grad} F^1(z_0).$$

However, this means that, in this case as well, relations (11.8) and (11.9) are valid (where now $\psi_0 = 1$, $\sigma_3 > 0$, and $\sigma_1 = \sigma_2 = 0$).

Thus, *if function* $F^0(z)$, *considered on set* Ω (see (11.5) and (11.6)), *reaches a maximum at a point* $z_0 \in \Omega$, *then there exist numbers* ψ_0, ψ_1, *at least one of which is different from zero, and nonnegative numbers* σ_1, σ_2, *and* σ_3, *all such that* $\psi_0 \geqslant 0$ *and relations* (11.8) *and* (11.9) *are valid.*

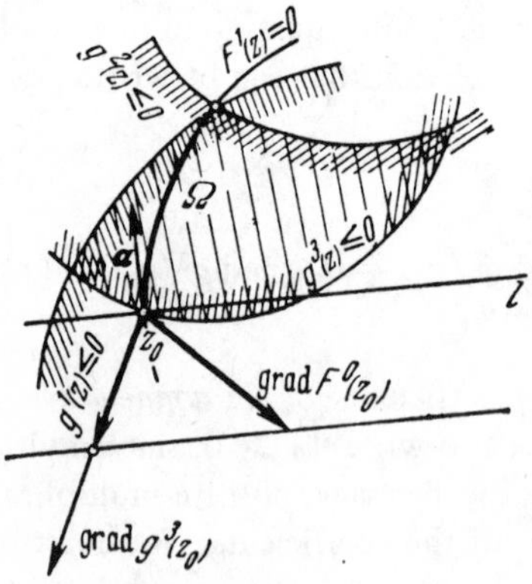

FIGURE 37

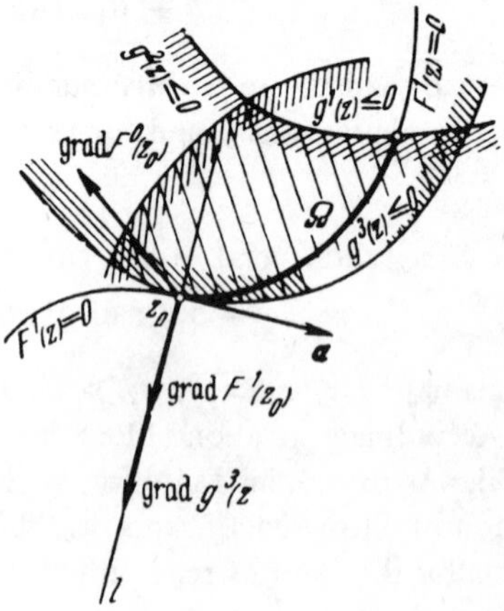

FIGURE 38

Here we have written $\psi_0 \geqslant 0$ rather than $\psi_0 > 0$, since certain "extreme" possibilities were not taken into account; for instance, if $\boldsymbol{a}$ grad $g^3(z_0) = 0$, then Figure 37 takes on the aspect shown in Figure 38, and relation (11.8) has the form ψ_1 grad $F^1(z_0) = \sigma_3$ grad $g^3(z_0)$, that is, here $\psi_0 = 0$ and $\psi_1 \neq 0$.

We have dealt with (albeit not completely) only the case where $n=2$ and $k=1$ (cf. (7.1) and (11.5)). However, the nature of the result obtained makes the following theorem very likely:

THEOREM 11.4. *Let $F^0(z)$ be a smooth function defined on the specified set Ω by system of relations* (7.1), (7.2). *For function $F^0(z)$, considered on set Ω, to reach a maximum at a point $z_0 \in \Omega$, there must exist numbers $\psi_0, \psi_1, \ldots, \psi_k$ and $\sigma_1, \ldots, \sigma_s$, at least one of which is different from zero, such that the numbers $\psi_0, \sigma_1, \ldots, \sigma_s$ are nonnegative and satisfy the relations*

$$\sum_{i=0}^{k} \psi_i \operatorname{grad} F^i(z_0) = \sum_{j=1}^{s} \sigma_j \operatorname{grad} g^j(z_0), \tag{11.10}$$

$$\sigma_1 g^1(z_0) = \sigma_2 g^2(z_0) = \ldots = \sigma_s g^s(z_0) = 0. \tag{11.11}$$

This theorem is actually valid, and its proof will be presented in Chapter IV; here we limit ourselves just to stating it.

NOTE 11.5. In the particular case where constraints (7.2) (and hence numbers σ_i as well) are completely absent, relation (11.10) becomes

$$\sum_{i=0}^{k} \psi_i \operatorname{grad} F^i(z_0) = 0. \tag{11.12}$$

Consequently, *in order for function $F^0(z)$, considered on the set Ω specified by the equations $F^1(z) = 0, \ldots, F^k(z) = 0$, to reach a maximum (or an extremum in general) at a point $z_0 \in \Omega$, there must exist numbers $\psi_0, \psi_1, \ldots, \psi_k$, at least one of which is different from zero, such that relation* (11.12) *is valid.*

This is the classical Lagrange theorem of the conditional extremum. Theorem 11.4 is its generalization.

EXAMPLE 11.6. As an example, let us solve the problem of the maximum of the product considered in Example 1.1. We wish to maximize the function

$$F^0(z) = z^1 z^2 \ldots z^N, \tag{11.13}$$

considered on the set Ω specified by the inequalities

$$z^1 \geqslant 0, \quad z^2 \geqslant 0, \ldots, z^N \geqslant 0, \quad z^1 + z^2 + \ldots + z^N \leqslant a. \tag{11.14}$$

Thus *equalities* (7.1) are completely absent in the definition of set Ω, the only existing relations being the inequalities, which can be written as

$$-z^1 \leqslant 0, \ldots, -z^N \leqslant 0, \quad z^1 + \ldots + z^N - a \leqslant 0,$$

that is, in form (7.2), where

$$g^1(z) = -z^1, \ldots, g^N(z) = -z^N, \quad g^{N+1}(z) = z^1 + \ldots + z^N - a$$

(so that in the given case $s = N + 1$).

Let function (11.13) reach a maximum at a point $z_0 \in \Omega$. Then, according to Theorem 11.4, there will exist *nonnegative* numbers $\psi_0, \sigma_1, \ldots, \sigma_{N+1}$, not all equal to zero, such that the following relations are valid:

$$\psi_0 \operatorname{grad} F^0(z_0) = \sum_{j=1}^{N+1} \sigma_j \operatorname{grad} g^j(z_0), \tag{11.15}$$

$$\sigma_j g^j(z_0) = 0, \qquad j = 1, 2, \ldots, N + 1. \tag{11.16}$$

Note that at point $z_0=(z_0^1, z_0^2, \ldots, z_0^N)$ not a single one of coordinates z_0^1, $z_0^2, \ldots, z_0^N$ goes to zero, since otherwise we would have $F^0(z_0)=0$ (see (11.13)), which obviously does not give a *maximum* of $F^0(z)$. Hence, $z_0^1>0, \ldots, z_0^N>0$, so that $g^1(z_0)<0, \ldots, g^N(z_0)<0$.

From (11.16) it now follows that $\sigma_1=\sigma_2= \ldots =\sigma_N=0$, so that (11.15) becomes

$$\psi_0 \operatorname{grad} F^0(z_0)=\sigma_{N+1} \operatorname{grad} g^{N+1}(z_0); \quad \psi_0 \geqslant 0, \quad \sigma_{N+1} \geqslant 0,$$

that is,

$$\psi_0 \frac{\partial F^0(z_0)}{\partial z^i}=\sigma_{N+1} \frac{\partial g^{N+1}(z_0)}{\partial z^i}=\sigma_{N+1}, \qquad i=1, \ldots, N.$$

However, the derivative $\frac{\partial F(z_0)}{\partial z^i}$ is obviously equal to $\frac{z_0^1 \cdot z_0^2 \cdot \ldots \cdot z_0^N}{z_0^i}$ (since $z_0^i \neq 0$). Consequently,

$$\frac{\psi_0 \cdot z_0^1 \cdot z_0^2 \cdot \ldots \cdot z_0^N}{z_0^i}=\sigma_{N+1}, \qquad i=1, \ldots, N.$$

From this it follows that $\sigma_{N+1} \neq 0$ (since all the numbers $z_0^1, z_0^2, \ldots, z_0^N$ are different from zero) and also that all the numbers $z_0^1, \ldots, z_0^N$ are equal:

$$z_0^1=z_0^2= \ldots =z_0^N\left(=\frac{\psi_0 \cdot z_0^1 \cdot z_0^2 \cdot \ldots \cdot z_0^N}{\sigma_{N+1}}\right).$$

But since $\sigma_{N+1} \neq 0$, we see from (11.16) that

$$g^{N+1}(z_0)=0, \quad \text{that is,} \quad z_0^1+z_0^2+ \ldots +z_0^N-a=0.$$

Therefore,

$$z_0^1=z_0^2= \ldots =z_0^N=\frac{a}{N}. \tag{11.17}$$

Thus the maximum of function (11.13), subject to constraints (11.14), can be reached *only* at point (11.17). On the other hand, set Ω, specified in space E^N by inequalities (11.14), is a *closed, bounded* set. Accordingly, the continuous function (11.13), defined on this set, has to reach a maximum at at least one point in set Ω. This means that point (11.17) is actually a maximum point of function $F^0(z)$, albeit the *only* maximum point.

Theorems 11.3 and 11.4 formulated above make it possible to obtain the necessary optimality conditions for discrete controlled objects. Actually, as we saw in Subsection 6, a problem of optimal control of discrete systems reduces to a problem of the extremum of a function considered in some subset of Euclidean space, and the necessary extremum conditions for such functions were also considered in this subsection.

However, we will not now formulate all the optimality conditions obtained in this way, rather we will limit ourselves just to a single example: the application of Theorem 11.4 to the fundamental problem considered in Subsection 10 (p.54). We assume that the control domain V is defined in the space of variables $u^1, \ldots, u^r$ by the system of inequalities

$$g^1(u^1, \ldots, u^r) \leqslant 0, \quad g^2(u^1, \ldots, u^r) \leqslant 0, \ldots, g^l(u^1, \ldots, u^r) \leqslant 0. \tag{11.18}$$

Passing to variables u_t^l and x_s^l as in Subsection 10, we now reformulate this problem as a problem of finding the maximum of function (10.18) on set Ω, the latter being once again determined by system of relations (10.16), (10.17). According to (11.18), relations (10.16) can in the given case be written as

$$g^j(u_t^1, u_t^2, \ldots, u_t^r) \leqslant 0; \qquad j = 1, \ldots, l; \quad t = 1, \ldots, N. \tag{11.19}$$

Thus, the problem is to maximize function (10.18) on the set Ω determined by the system of equalities (10.17) and inequalities (11.19).

Let this function (considered on set Ω) have a maximum at a point $z_0 \in \Omega$. Then, according to Theorem 11.4, there will exist numbers ψ_0, ψ_i^t (where $i = 1, \ldots, n$; $t = 1, \ldots, N$) and nonnegative numbers σ_j^t (where $j = 1, \ldots, l$; $t = 1, \ldots, N$), all such that

$$\psi_0 \operatorname{grad} F(z_0) + \sum_{s=1}^{N} \sum_{j=1}^{n} \psi_j^s \operatorname{grad} F_s^j(z_0) = \sum_{s=1}^{N} \sum_{j=1}^{l} \sigma_j^s \operatorname{grad} g_s^j(z_0), \tag{11.20}$$

$$\sigma_j^t g_t^j(z_0) = 0; \qquad j = 1, \ldots, l; \quad t = 1, \ldots, N, \tag{11.21}$$

where $g_t^j(z)$ is a function $g^j(u^1, \ldots, u^r)$, having independent variables $u_t^1, \ldots, u_t^r$, that is, $g_t^j(z) = g^j(\eta_t(z))$. According to (11.20), it then follows that the function

$$\psi_0 F(z) + \sum_{s=1}^{N} \sum_{j=1}^{n} \psi_j^s F_s^j(z) - \sum_{s=1}^{N} \sum_{j=1}^{l} \sigma_j^s g_s^j(z) \tag{11.22}$$

has zero derivatives with respect to all variables (that is, u_t^j, x_t^i) at point z_0.

First of all, let us equate to zero the partial derivatives with respect to variables x_t^i. Since functions $g_t^j(z)$ do not depend on these variables, by taking (10.20) into account, we can write the result of differentiating function (11.22) with respect to x_{t-1}^i as

$$\psi_0 \frac{\partial f^0(\xi_{t-1}(z_0), \eta_t(z_0))}{\partial x^i} + \sum_{j=1}^{n} \psi_j^t \frac{\partial f^j(\xi_{t-1}(z_0), \eta_t(z_0))}{\partial x^i} - \psi_i^{t-1} = 0;$$

$$i = 1, \ldots, n; \quad t = 2, \ldots, N, \tag{11.23}$$

and the result of differentiating with respect to x_N^i as

$$-\psi_i^N = 0, \qquad i = 1, \ldots, n. \tag{11.24}$$

Now let us consider the derivatives with respect to variables u_t^i. By equating to zero the derivative of function (11.22) with respect to u_t^i at point z_0, we obtain (in view of (10.18), (10.19), and the definition of functions $g_s^j(z)$) the following relations:

$$\psi_0 \frac{\partial f^0(\xi_{t-1}(z_0), \eta_t(z_0))}{\partial u^i} + \sum_{j=1}^{n} \psi_j^t \frac{\partial f^j(\xi_{t-1}(z_0), \eta_t(z_0))}{\partial u^i} =$$

$$= \sum_{j=1}^{l} \sigma_j^t \frac{\partial g^j(\eta_t(z_0))}{\partial u^i}; \qquad i = 1, \ldots, r;\ t = 1, \ldots, N,$$

that is,

$$\operatorname{grad}_u \left(\sum_{j=0}^{n} \psi_j^t f^j(\xi_{t-1}(z_0), \eta_t(z_0)) \right) = \sum_{j=1}^{l} \sigma_j^t \operatorname{grad} g^j(\eta_t(z_0)), \qquad (11.25)$$
$$t = 1, \ldots, N.$$

Next we specify a certain number t $(= 1, \ldots, N)$. The point $\eta_t(z_0)$ belongs to set V (see (8.27)). To be more definite, we assume that at point $\eta_t(z_0)$ the first k inequalities (11.18) are active (that is, they become equalities), while the rest are inactive:

$$g^1(\eta_t(z_0)) = 0, \ldots, g^k(\eta_t(z_0)) = 0, \ g^{k+1}(\eta_t(z_0)) < 0, \ldots, g^l(\eta_t(z_0)) < 0.$$

In other words,

$$g_t^1(z_0) = 0, \ldots, g_t^k(z_0) = 0, \ g_t^{k+1}(z_0) < 0, \ldots, g_t^l(z_0) < 0.$$

From these relations it follows, in view of (11.21), that

$$\sigma_{k+1}^t = \ldots = \sigma_l^t = 0. \qquad (11.26)$$

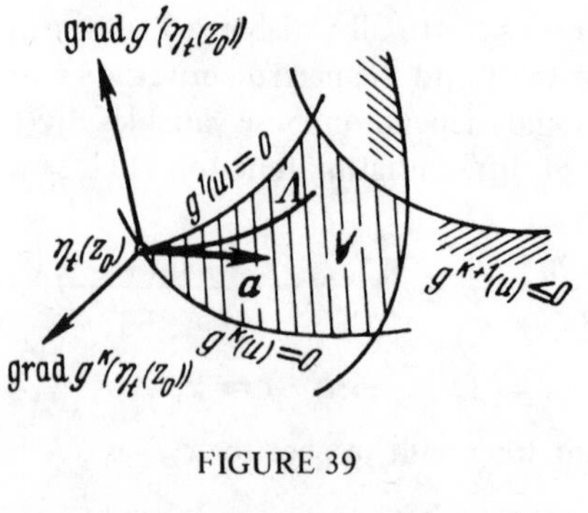

FIGURE 39

Now let $\boldsymbol{a}$ be a vector proceeding (in the space of variables $u^1, \ldots, u^r$) from point $\eta_t(z_0)$ and lying tangent to some curve in set V (Figure 39), that is, forming a nonpositive scalar product with each of the vectors

$$\text{grad } g^1(\eta_t(z_0)), \ \ldots, \ \text{grad } g^k(\eta_t(z_0)),$$

which we now assume to be linearly independent. Taking relations (11.26) and inequalities $\sigma_j^t \geqslant 0$ into account, we find that the scalar product of vector $\boldsymbol{a}$ times the vector on the right-hand side of (11.25) is *nonpositive.* Consequently,

$$\boldsymbol{a} \, \text{grad}_u \times \left(\sum_{j=0}^{n} \psi_j^t f^j(\xi_{t-1}(z_0), \ \eta_t(z_0)) \right) \leqslant 0.$$

However, this means that the derivative of function

$$\sum_{j=0}^{n} \psi_j^t f^j(\xi_{t-1}(z), \ \eta_t(z)) \tag{11.27}$$

at point z_0 in the direction specified by vector $\boldsymbol{a}$ is nonpositive.

Thus, if function (10.18), considered on set Ω, reaches a maximum at point z_0, there must exist constants $\psi_0 \geqslant 0$, ψ_i^t ($i = 0, 1, \ldots, n$; $t = 1, \ldots, N$), such that relations (11.23) and (11.24) are satisfied, and for each t function (11.27) has at point z_0 a nonpositive derivative in any direction included in set V (at point $\eta_t(z_0)$).

Finally, we note that $\psi_0 \neq 0$; if the opposite were true, then from (10.23) and (10.24) we would obtain successively $\psi_i^{N-1} = 0, \psi_i^{N-2} = 0, \ldots, \psi_i^1 = 0$ ($i = 1, \ldots, n$), that is, all the constants ψ_0, ψ_i^t would be zero, which is impossible (see (11.20)). Thus $\psi_0 > 0$. Since constants ψ_0, ψ_i^t enter into all the relations (11.23), (11.24), (11.27) *uniformly* (so that multiplication of all these constants by the same factor $k > 0$ does not make the conditions considered invalid), we can assume that $\psi_0 = 1$.

Next let us reformulate the result obtained in terms of discrete controlled processes, that is, we replace $\xi_t(z)$ by $x(t)$, $\eta_t(z)$ by $u(t)$, and ψ_i^t by $\psi_i(t)$. Then, obviously, relations (11.23) and (11.24) become (10.12) and (10.10), while function (11.27) becomes $H(\psi(t), x(t-1), u(t))$. Hence we are led to the following theorem:*

THEOREM 11.7. *We consider object* (10.14) *with control domain* V (see (8.27)). *At each point* $u \in V$ *the gradients of the active inequalities (11.18) are assumed to be linearly independent. The fundamental problem is considered (with a free right-hand endpoint and with no constraints on the phase coordinates), with initial condition* $x(0) = y_0$ (see (8.28)), *optimality being understood to mean a maximum of functional* (10.11).

To solve this problem, we introduce the auxiliary variables $\psi_0, \psi_1, \ldots, \psi_n$ *(here* n *is the number of phase coordinates* $x^1, \ldots, x^n$*) and we use them to formulate the following auxiliary function:*

$$H(\psi, x, u) = \sum_{i=0}^{n} \psi_i f^i(x, u),$$

* Optimality conditions of this type (consisting in nonpositivity of the derivative of function $H(\psi, x, u)$ in any permissible direction) have been presented by Propoi and others (see reference** on p. 62 and reference * on p. 63).

where f^0 *is the function entering into the definition of functional* (10.11), *and* $f^1, \ldots, f^n$ *are the components of vector-function* $f(x, u)$, *appearing on the right-hand side of* (10.14). *In addition, we write the system of relations* (10.12) *for the auxiliary unknowns and we set* $\psi_0 = 1$ (see (10.9)).

Now let $u(t)$, $t = 1, \ldots, N$, *be some admissible (that is, satisfying inclusion* $u(t) \in V$ *for all* $t = 1, \ldots, N$*) control, and let* $x(t)$, $t = 0, 1, \ldots, N$, *be the corresponding (that is, satisfying relation* (10.14)*) trajectory, starting from point* $x(0) = y_0$. *If functions* $x = x(t)$, $u = u(t)$ *are substituted into the right-hand sides of* (10.12), *then we can use relations* (10.9) *and* (10.10) *to find successively* $\psi(N-1), \psi(N-2), \ldots, \psi(1)$.

If process $u(t)$, $x(t)$ *is optimal, meaning that it maximizes functional* (10.11), *then for each* $t = 1, \ldots, N$ *function* $H(\psi(t), x(t-1), u)$ *will have at a point* $u = u(t)$ *a nonpositive derivative in any direction entering into set* V *(at point* $u(t)$*).*

If optimality is understood to mean a minimum (rather than a maximum) of functional (10.11), *then the formulation remains the same, except that relation* (10.9) *becomes*

$$\psi_0 = -1.$$

Theorem 11.7 constitutes a correctly proven (albeit with reference to Theorem 11.4) necessary condition for optimality. It is of interest to compare this theorem with the discrete maximum principle (p. 54). If the derivative of function $H(\psi(t), x(t-1), u)$ at point $u = u(t)$, in any direction entering into set V, is *negative* in addition to being nonpositive (as assumed in Theorem 11.7), then this would mean that in a certain neighborhood around point $u(t)$ the function $H(\psi(t), x(t-1), u)$ *decreases* for a shift away from point $u(t)$ in any direction (provided we stay in set V). In other words, function $H(\psi(t), x(t-1), u)$ of variable $u \in V$ reaches at point $u(t)$ a *local* maximum (Figure 40).

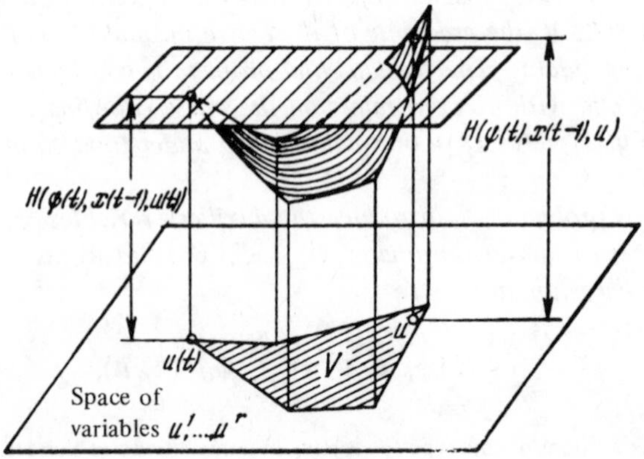

FIGURE 40

Therefore, even if the condition of nonpositivity of the derivatives, which was specified in Theorem 11.7, is fortified somewhat (by replacing it with the condition of negativeness of the derivatives), we obtain a condition of *local* maximum of function $\bar{H}(\psi(t), x(t-1), u)$ of variable $u \in V$ at point $u(t)$ that is considerably weaker than the condition of *absolute* maximum of this function contained in the discrete maximum principle.

It is obvious that the necessary condition contained in Theorem 11.7 is considerably weaker than the necessary condition specified in the discrete maximum principle. At the same time, Theorem 11.7 turns out to be general, while the discrete maximum principle is applicable only to a narrow class of discrete controlled objects.

It should be noted that in Example 10.3 the necessary condition specified in Theorem 11.7 is, as is easily proven, satisfied for process (10.30), whereas the necessary condition contained in the discrete maximum principle is not satisfied.

Chapter II. BASIC CONCEPTS OF MULTIDIMENSIONAL GEOMETRY

§4. Vector Space

12. Definition of vector space. A set R is called a *vector space* if in it two operations obeying the two groups of axioms listed below are defined. The elements of set R are known as *vectors;* they will be denoted by boldface letters. In order to show that a vector $\boldsymbol{a}$ belongs to a space R (that is, it is an element of this space), we write $\boldsymbol{a} \in R$.

The first of the two operations specified in a vector space is *vector addition.* With every two vectors $\boldsymbol{a} \in R$, $\boldsymbol{b} \in R$, this operation associates some vector (of the same space) known as the *sum* of vectors $\boldsymbol{a}$ and $\boldsymbol{b}$ and denoted as $\boldsymbol{a} + \boldsymbol{b}$. The second operation is *multiplication of a vector by a number.* This operation associates with each vector $\boldsymbol{a} \in R$ and each real number λ some vector (of the same space) which is called the product of vector $\boldsymbol{a}$ times number λ and is designated as $\lambda\boldsymbol{a}$. In addition, we note that *equality* of two vectors always implies their *coincidence.* Finally, the axioms (fulfillment of which entitles a set R with the above two operations defined in it to be called a vector space) are formulated as follows:

GROUP I: *axioms of vector addition.*

I_1. *The addition of vectors is commutative, that is, for any two vectors* $\boldsymbol{a} \in R$, $\boldsymbol{b} \in R$ *the following is true:*

$$\boldsymbol{a} + \boldsymbol{b} = \boldsymbol{b} + \boldsymbol{a}.$$

I_2. *The addition of vectors is associative, that is, for any three vectors* $\boldsymbol{a} \in R$, $\boldsymbol{b} \in R$, $\boldsymbol{c} \in R$ *the following is true:*

$$(\boldsymbol{a} + \boldsymbol{b}) + \boldsymbol{c} = \boldsymbol{a} + (\boldsymbol{b} + \boldsymbol{c}).$$

I_3. *In set* R *there exists an element called the zero vector (or simply the zero), designated as* $\mathbf{0}$, *for which the following is true with regard to any vector* $\boldsymbol{a} \in R$:

$$\boldsymbol{a} + \mathbf{0} = \boldsymbol{a}.$$

I_4. *For any element* $\boldsymbol{a} \in R$ *there exists in* R *an element known as the negative of vector* $\boldsymbol{a}$: *this is denoted as* $-\boldsymbol{a}$, *where* $\boldsymbol{a} + (-\boldsymbol{a}) = \mathbf{0}$.

GROUP II: *axioms of multiplication of a vector by a number.*

II_1. *The multiplication of vectors by numbers is associative, that is, for any vector* $\boldsymbol{a} \in R$ *and any two real numbers* λ, μ, *the following is true:*

$$\lambda(\mu\boldsymbol{a}) = (\lambda\mu)\boldsymbol{a}.$$

II_2. *The multiplication of vectors by numbers is distributive with regard to numbers, that is, for any vector* $\boldsymbol{a} \in R$ *and any two real numbers* λ, μ, *the following is true:*

$$(\lambda + \mu)\,\boldsymbol{a} = \lambda\boldsymbol{a} + \mu\boldsymbol{a}.$$

II_3. *The multiplication of vectors by numbers is distributive with regard to vectors, that is, for any vectors* $\boldsymbol{a} \in R$, $\boldsymbol{b} \in R$ *and any real number* λ, *the following is true:*

$$\lambda\,(\boldsymbol{a} + \boldsymbol{b}) = \lambda\boldsymbol{a} + \lambda\boldsymbol{b}.$$

II_4. *For any vector* $\boldsymbol{a} \in R$ *the following is true:*

$$1\boldsymbol{a} = \boldsymbol{a}.$$

Let us consider two examples of vector spaces.

EXAMPLE 12.1. On a plane Π we fix some point O and we define as a *vector* an arbitrary point of plane Π (different from point O or coinciding with it). If A and B are two arbitrary vectors (that is, two points of plane Π), then their *sum* $A + B$ is defined as a point C for which the middle of segment OC coincides with the middle of segment AB (Figure 41). Hence, provided that points O, A, and B do not lie along a straight line, $OACB$ will be a parallelogram. If A is an arbitrary vector and λ is a real number, then, provided that point A does not coincide with O and that $\lambda > 0$, the *product* λA is defined as the point D on radial line OA for which $\frac{OD}{OA} = \lambda$; if point A does not coincide with O and if $\lambda < 0$, the the product λA is defined as the point D on the extension of line OA past point O for which $\frac{OD}{OA} =$ $= -\lambda$ (Figure 42); in the remaining cases (if A coincides with O or if $\lambda = 0$) we will say that $\lambda A = 0$.

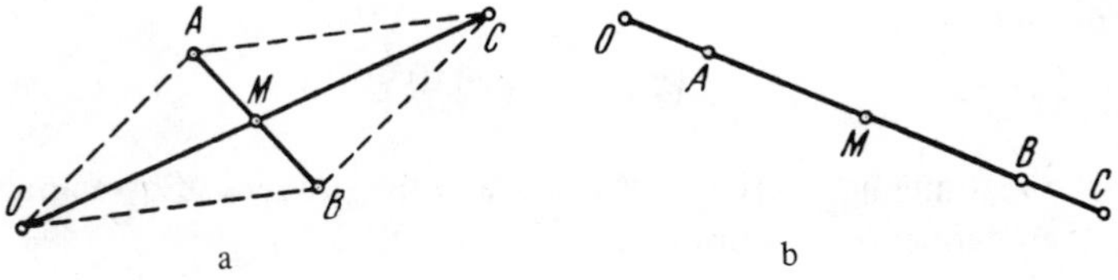

FIGURE 41

For this definition of vectors and operations performed on them, the set of all points of plane Π becomes a vector space. The zero element of this vector space is point O; moreover, if A is an arbitrary vector, then the negative vector $-A$ is the point *symmetrical* to point A relative to O (Figure 43). In practice, it is customary when considering such a vector space to portray the vectors not as points A, B, C, ...

of the plane Π, but rather as directed line segments ("radius vectors") $\overrightarrow{OA}$, $\overrightarrow{OB}$, $\overrightarrow{OC}$, ..., from point O to the corresponding points (Figure 44).

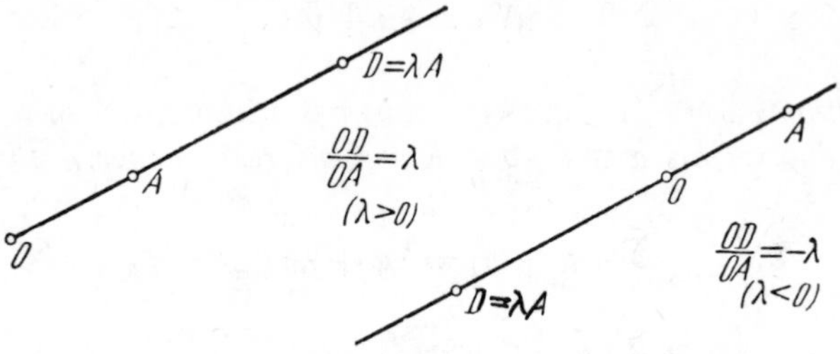

FIGURE 42

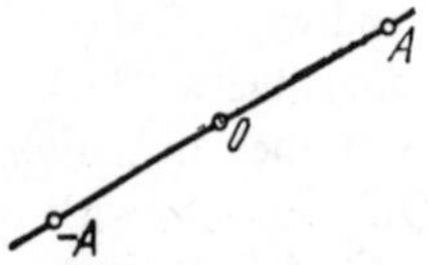

FIGURE 43

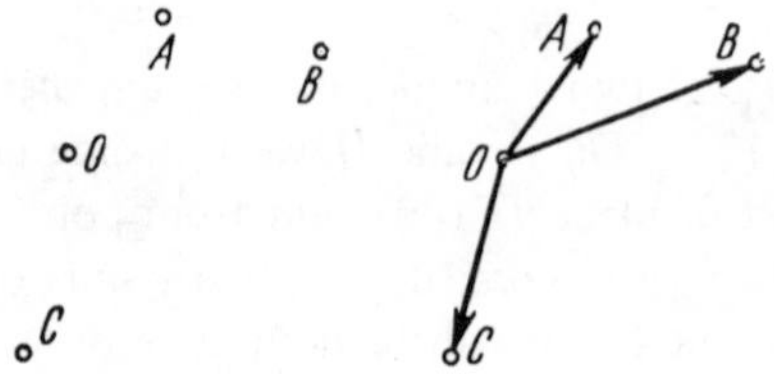

FIGURE 44

Similarly, we can obtain a vector space if we fix some point O in ordinary three-dimensional space Σ (known from elementary geometry) and define as a vector an arbitrary point A (or the directed line segment $\overrightarrow{OA}$) in the space, the operations being defined as previously.

This "geometrical model" of a vector space serves as a basic source of geometrical intuition, and it provides a "visual" geometrical interpretation of the facts corresponding to different vector spaces.

EXAMPLE 12.2. Let us select some natural number n and define as a *vector* the arbitrary sequence

$$\{x^1, \ldots, x^n\},$$

consisting of n real numbers. The vector sum and the product of a vector times a number are then defined by the formulas:

$$\{x^1, \ldots, x^n\} + \{y^1, \ldots, y^n\} = \{x^1 + y^1, \ldots, x^n + y^n\};$$
$$\lambda\{x^1, \ldots, x^n\} = \{\lambda x^1, \ldots, \lambda x^n\}.$$

It is easy to show that these vectors and the indicated operations on them obey all the axioms of a vector space. This vector space is called an *n-dimensional arithmetical vector space.*

Let us now consider the most elementary facts following from the axioms of a vector space.

THEOREM 12.3. *For any two vectors* $\boldsymbol{a} \in R$, $\boldsymbol{b} \in R$ *the equation* $\boldsymbol{a} + \boldsymbol{x} = \boldsymbol{b}$ *always has one, and only one, solution.*

Proof. In order to demonstrate the *existence* of a solution, let us verify that the vector $\boldsymbol{x} = (-\boldsymbol{a}) + \boldsymbol{b}$ is a solution of this equation. Actually, substituting this quantity into the left-hand side, we can successively apply axioms I_2, I_4, I_1, I_3 to obtain:

$$\boldsymbol{a} + \boldsymbol{x} = \boldsymbol{a} + ((-\boldsymbol{a}) + \boldsymbol{b}) = (\boldsymbol{a} + (-\boldsymbol{a})) + \boldsymbol{b} = \boldsymbol{0} + \boldsymbol{b} = \boldsymbol{b} + \boldsymbol{0} = \boldsymbol{b}.$$

The existence of a solution is thereby proven.

Next let us establish the *uniqueness* of the solution. If $\boldsymbol{x}_1$ and $\boldsymbol{x}_2$ are two solutions of the given equation, that is, if the equations

$$\boldsymbol{a} + \boldsymbol{x}_1 = \boldsymbol{b}, \quad \boldsymbol{a} + \boldsymbol{x}_2 = \boldsymbol{b}$$

are valid, then it follows that

$$\boldsymbol{a} + \boldsymbol{x}_1 = \boldsymbol{a} + \boldsymbol{x}_2.$$

Here (given the meaning of the word "equation") the *very same* vector is on both sides of the equals sign. By adding the negative vector $(-\boldsymbol{a})$ to this vector, we obtain

$$(-\boldsymbol{a}) + (\boldsymbol{a} + \boldsymbol{x}_1) = (-\boldsymbol{a}) + (\boldsymbol{a} + \boldsymbol{x}_2),$$

or, according to Axiom I_2,

$$((-\boldsymbol{a}) + \boldsymbol{a}) + \boldsymbol{x}_1 = ((-\boldsymbol{a}) + \boldsymbol{a}) + \boldsymbol{x}_2;$$

moreover, according to Axiom I_1,

$$(\boldsymbol{a} + (-\boldsymbol{a})) + \boldsymbol{x}_1 = (\boldsymbol{a} + (-\boldsymbol{a})) + \boldsymbol{x}_2,$$

and thus (Axiom I_4)

$$\boldsymbol{0} + \boldsymbol{x}_1 = \boldsymbol{0} + \boldsymbol{x}_2,$$

or, in view of Axiom I_1,

$$\boldsymbol{x}_1 + \boldsymbol{0} = \boldsymbol{x}_2 + \boldsymbol{0};$$

finally, from Axiom I_3 it follows that $\boldsymbol{x}_1 = \boldsymbol{x}_2$.

A particular corollary following from Theorem 12.3 is the *uniqueness of the zero.* Let us assume that in vector space R we have, in addition to the zero element $\boldsymbol{0}$, another element $\boldsymbol{0}'$ possessing similar properties (and, in particular, having the property cited in Axiom I_3, that is, $\boldsymbol{a} + \boldsymbol{0}' = \boldsymbol{a}$ for any $\boldsymbol{a} \in R$). The two vectors $\boldsymbol{0}$ and $\boldsymbol{0}'$ will then be solutions of the equation $\boldsymbol{a} + \boldsymbol{x} = \boldsymbol{a}$, so that in view of Theorem 12.3 they must be identical: $\boldsymbol{0} = \boldsymbol{0}'$.

Theorem 12.3 also implies validity of the equations

$$0\boldsymbol{a}=\boldsymbol{0}, \quad (-1)\boldsymbol{a}=-\boldsymbol{a} \tag{12.1}$$

for any vector $\boldsymbol{a}$. In fact, by applying Axioms II_4, II_2, II_4 successively, we obtain

$$\boldsymbol{a}=1\boldsymbol{a}=(1+0)\boldsymbol{a}=1\boldsymbol{a}+0\boldsymbol{a}=\boldsymbol{a}+0\boldsymbol{a}.$$

Consequently, $\boldsymbol{a}+0\boldsymbol{a}=\boldsymbol{a}$, that is, the vector $\boldsymbol{x}=0\boldsymbol{a}$ is a solution of the equation $\boldsymbol{a}+\boldsymbol{x}=\boldsymbol{a}$. Since vector $\boldsymbol{0}$ is also a solution of this equation (Axiom I_3), therefore in view of the uniqueness (Theorem 12.3) it follows that $0\boldsymbol{a}=\boldsymbol{0}$.

Furthermore, by applying successively Axioms II_4 and II_2, as well as the first of equations (12.1) proven above, we get

$$\boldsymbol{a}+(-1)\boldsymbol{a}=1\boldsymbol{a}+(-1)\boldsymbol{a}=(1+(-1))\boldsymbol{a}=0\boldsymbol{a}=\boldsymbol{0}.$$

Consequently, vector $\boldsymbol{x}=(-1)\boldsymbol{a}$ is a solution of the equation $\boldsymbol{a}+\boldsymbol{x}=\boldsymbol{0}$. Since the vector $-\boldsymbol{a}$ is also a solution of this equation (Axiom I_4), therefore in view of the uniqueness (Theorem 12.3) we have $(-1)\boldsymbol{a}=-\boldsymbol{a}$.

We have presented such detailed considerations here in order to demonstrate how the axioms work. In the following, the discussion will be somewhat more sketchy.

By definition, the subtraction operation is specified in a vector space by the equation

$$\boldsymbol{a}-\boldsymbol{b}=\boldsymbol{a}+(-\boldsymbol{b}). \tag{12.2}$$

With the aid of this operation, Theorem 12.3 (taking its proof into account) can be reformulated as follows: *for any vectors* $\boldsymbol{a}$, $\boldsymbol{b}$ *the equation* $\boldsymbol{a}+\boldsymbol{x}=\boldsymbol{b}$ *has the unique solution* $\boldsymbol{x}=\boldsymbol{b}-\boldsymbol{a}$. In other words, relations $\boldsymbol{a}+\boldsymbol{c}=\boldsymbol{b}$ and $\boldsymbol{c}=\boldsymbol{b}-\boldsymbol{a}$ are *one and the same.*

Several vectors joined in succession by + and - signs form an *algebraic vector sum.* By definition, the operations carried out in an algebraic sum are successive. For instance,

$$\boldsymbol{a}-\boldsymbol{b}-\boldsymbol{c}+\boldsymbol{d}=((\boldsymbol{a}-\boldsymbol{b})-\boldsymbol{c})+\boldsymbol{d};$$
$$-\boldsymbol{a}+\boldsymbol{b}-\boldsymbol{c}=((-\boldsymbol{a})+\boldsymbol{b})-\boldsymbol{c}\text{, etc.}$$

The definition of the subtraction operation and Axioms I_1 and I_2 imply that an algebraic sum can be calculated in any order (with respect to the signs of the vectors involved); for example,

$$\boldsymbol{a}-\boldsymbol{b}+\boldsymbol{c}-\boldsymbol{d}+\boldsymbol{e}=\boldsymbol{a}+\boldsymbol{c}+\boldsymbol{e}-\boldsymbol{b}-\boldsymbol{d}=$$
$$=-\boldsymbol{b}+\boldsymbol{e}-\boldsymbol{d}+\boldsymbol{a}+\boldsymbol{c}\text{, etc.}$$

Moreover, from Axioms II_2 and II_3 and the second equation of (12.1), we obtain the ordinary rules for such operations:

$$\alpha(\boldsymbol{a}-\boldsymbol{b}+\boldsymbol{c})=\alpha\boldsymbol{a}-\alpha\boldsymbol{b}+\alpha\boldsymbol{c};$$

$$(\alpha+\beta-\gamma)\boldsymbol{a}=\alpha\boldsymbol{a}+\beta\boldsymbol{a}-\gamma\boldsymbol{a};$$
$$-(\boldsymbol{a}-\boldsymbol{b}+\boldsymbol{c})=-\boldsymbol{a}+\boldsymbol{b}-\boldsymbol{c};$$
$$(\alpha-\beta)(\boldsymbol{a}-\boldsymbol{b}+\boldsymbol{c})=\alpha\boldsymbol{a}-\beta\boldsymbol{a}-\alpha\boldsymbol{b}+\beta\boldsymbol{b}+\alpha\boldsymbol{c}-\beta\boldsymbol{c}$$

(where α, β, γ are arbitrary real numbers, and $\boldsymbol{a}$, $\boldsymbol{b}$, $\boldsymbol{c}$ are arbitrary vectors), etc.

Let us next cite some other rules for working with vector equations. As noted, relations $\boldsymbol{a}+\boldsymbol{c}=\boldsymbol{b}$ and $\boldsymbol{c}=\boldsymbol{b}-\boldsymbol{a}$ are the same. In other words, a term can be shifted into another part of a vector equation provided the sign in front of the term is changed. This rule also applies if there are algebraic vector sums on both sides of the equation. For example, the equation

$$\boldsymbol{a}-\boldsymbol{b}+\boldsymbol{c}=\boldsymbol{d}-\boldsymbol{e}$$

is valid if, and only if, the following relation is valid:

$$\boldsymbol{a}+\boldsymbol{c}+\boldsymbol{e}=\boldsymbol{b}+\boldsymbol{d}.$$

Moreover, both sides of the vector equation can be multiplied by any real number and can be divided by the same (nonzero) real number (division by a number $\alpha \neq 0$ is equivalent to multiplication by $\frac{1}{\alpha}$). For instance, for $\alpha \neq 0$ the equation

$$\boldsymbol{a}+\boldsymbol{b}=\boldsymbol{c}-\boldsymbol{d}+\boldsymbol{e}$$

is valid if, and only if,

$$\alpha\boldsymbol{a}+\alpha\boldsymbol{b}=\alpha\boldsymbol{c}-\alpha\boldsymbol{d}+\alpha\boldsymbol{e}.$$

From the foregoing it follows, in particular, that the vector equation $\alpha\boldsymbol{x}=\boldsymbol{a}$ has for $\alpha \neq 0$ the unique solution $\boldsymbol{x}=\frac{1}{\alpha}\boldsymbol{a}$.

In other words, *with regard to the addition and subtraction of vectors and the multiplication of a vector by a number, as well as with regard to the rules about operations with equations, all the properties characteristic of real numbers are conserved in vector spaces.*

13. Dimension and basis. DEFINITION 13.1. If $\boldsymbol{a}_1, \ldots, \boldsymbol{a}_k$ are vectors of vector space R and if $\alpha_1, \ldots, \alpha_k$ are real numbers, then a vector

$$\alpha_1\boldsymbol{a}_1+\ldots+\alpha_k\boldsymbol{a}_k \tag{13.1}$$

is called a *linear combination* of vectors $\boldsymbol{a}_1, \ldots, \boldsymbol{a}_k$ with coefficients $\alpha_1, \ldots, \alpha_k$.

DEFINITION 13.2. Vectors $\boldsymbol{a}_1, \ldots, \boldsymbol{a}_k$ of vector space R are called *linearly dependent* if real numbers $\alpha_1, \ldots, \alpha_k$ exist, at least one of which is different from zero, such that $\alpha_1\boldsymbol{a}_1+\ldots+\alpha_k\boldsymbol{a}_k=\boldsymbol{0}$.

This definition is often stated in a different form. First of all, linear combination (13.1) is called *nontrivial* if at least one of coefficients $\alpha_1, \ldots, \alpha_k$ is different from zero. Definition 13.2 can then be reformulated as follows: vectors $\boldsymbol{a}_1, \ldots, \boldsymbol{a}_k$ are called linearly dependent if there exists a nontrivial linear combination of these vectors equal to the zero vector.

Furthermore, vectors $\boldsymbol{a}_1, \ldots, \boldsymbol{a}_k$ are called *linearly independent* if they are not linear dependent, that is, if *no* nontrivial linear combination of them goes to zero. Frequently the expression "linearly independent vectors" is replaced by the equivalent expression "linearly independent *system* of vectors."

It follows from this definition that, if vectors $\boldsymbol{a}_1, \ldots, \boldsymbol{a}_k$ are linearly independent, then validity of the equation

$$\alpha_1 \boldsymbol{a}_1 + \ldots + \alpha_k \boldsymbol{a}_k = \boldsymbol{0}$$

implies that $\alpha_1 = \ldots = \alpha_k = 0$.

THEOREM 13.3. *If $\boldsymbol{a}_1, \ldots, \boldsymbol{a}_k$ are linearly independent vectors, while vectors $\boldsymbol{b}, \boldsymbol{a}_1, \ldots, \boldsymbol{a}_k$ are linearly dependent, then vector $\boldsymbol{b}$ can be represented as a linear combination of vectors $\boldsymbol{a}_1, \ldots, \boldsymbol{a}_k$, that is, there exist real numbers $\lambda_1, \ldots, \lambda_k$ such that*

$$\boldsymbol{b} = \lambda_1 \boldsymbol{a}_1 + \ldots + \lambda_k \boldsymbol{a}_k.$$

Proof. Since vectors $\boldsymbol{b}, \boldsymbol{a}_1, \ldots, \boldsymbol{a}_k$ are linearly dependent, there must exist numbers $\beta, \alpha_1, \ldots, \alpha_k$, at least one of which is different from zero, such that

$$\beta \boldsymbol{b} + \alpha_1 \boldsymbol{a}_1 + \ldots + \alpha_k \boldsymbol{a}_k = \boldsymbol{0}. \tag{13.2}$$

If we set $\beta = 0$, then relation (13.2) becomes

$$\alpha_1 \boldsymbol{a}_1 + \ldots + \alpha_k \boldsymbol{a}_k = \boldsymbol{0},$$

and thus the linear independence of vectors $\boldsymbol{a}_1, \ldots, \boldsymbol{a}_k$ implies that $\alpha_1 = \ldots = \alpha_k = 0$. However, this is contrary to our assumption that at least one of the numbers $\beta, \alpha_1, \ldots, \alpha_k$ is different from zero. Consequently, $\beta \neq 0$.

Now from (13.2) we obtain

$$\boldsymbol{b} = -\frac{\alpha_1}{\beta} \boldsymbol{a}_1 - \ldots - \frac{\alpha_k}{\beta} \boldsymbol{a}_k,$$

or

$$\boldsymbol{b} = \lambda_1 \boldsymbol{a}_1 + \ldots + \lambda_k \boldsymbol{a}_k,$$

where $\lambda_i = -\alpha_i/\beta$, $i = 1, \ldots, k$.

THEOREM 13.4. *If $\boldsymbol{b}_1, \ldots, \boldsymbol{b}_k, \boldsymbol{c}$ constitute a linearly independent system of vectors, and if $\boldsymbol{a}_{k+1}, \ldots, \boldsymbol{a}_m$ $(m \geqslant k+1)$ are vectors such that the system of vectors*

$$\boldsymbol{b}_1, \ldots, \boldsymbol{b}_k, \boldsymbol{a}_{k+1}, \ldots, \boldsymbol{a}_m \tag{13.3}$$

is also linearly independent, then it is possible to choose some vector $\boldsymbol{a}_s$ of the vectors

$\boldsymbol{a}_{k+1}, \ldots, \boldsymbol{a}_m$ *such that replacement of* $\boldsymbol{a}_s$ *by vector* $\boldsymbol{c}$ *in system* (13.3) *gives a system of vectors which is linearly independent.*

Proof. Let us add vector $\boldsymbol{c}$ to system of vectors (13.3). If the resulting system of vectors

$$\boldsymbol{b}_1, \ldots, \boldsymbol{b}_k, \boldsymbol{a}_{k+1}, \ldots, \boldsymbol{a}_m, \boldsymbol{c} \tag{13.4}$$

turns out to be linearly independent, then in system of vectors (13.3) *any* of vectors $\boldsymbol{a}_{k+1}, \ldots, \boldsymbol{a}_m$ can be replaced by vector $\boldsymbol{c}$ to obtain a linearly independent system.

Now let us assume that system (13.4) is linearly dependent. Then, according to Theorem 13.3, vector $\boldsymbol{c}$ is a linear combination of vectors (13.3):

$$\boldsymbol{c} = \beta_1 \boldsymbol{b}_1 + \ldots + \beta_k \boldsymbol{b}_k + \alpha_{k+1} \boldsymbol{a}_{k+1} + \ldots + \alpha_m \boldsymbol{a}_m. \tag{13.5}$$

If all the coefficients $\alpha_{k+1}, \ldots, \alpha_m$ were equal to zero, then this relation could be rewritten as

$$\boldsymbol{c} - \beta_1 \boldsymbol{b}_1 - \ldots - \beta_k \boldsymbol{b}_k = \boldsymbol{0},$$

which, however, contradicts the linear independence of vectors $\boldsymbol{b}_1, \ldots, \boldsymbol{b}_k, \boldsymbol{c}$. Thus at least one of coefficients $\alpha_{k+1}, \ldots, \alpha_m$ is different from zero. Let us assume, in order to be definite, that $\alpha_{k+1} \neq 0$ (otherwise the numbering order of vectors $\boldsymbol{a}_{k+1}, \ldots, \boldsymbol{a}_m$ would be changed).

Let us show that system of vectors

$$\boldsymbol{b}_1, \ldots, \boldsymbol{b}_k, \boldsymbol{c}, \boldsymbol{a}_{k+2}, \ldots, \boldsymbol{a}_m \tag{13.6}$$

is linearly independent. Assume that

$$\lambda_1 \boldsymbol{b}_1 + \ldots + \lambda_k \boldsymbol{b}_k + \lambda_{k+1} \boldsymbol{c} + \lambda_{k+2} \boldsymbol{a}_{k+2} + \ldots + \lambda_m \boldsymbol{a}_m = \boldsymbol{0}. \tag{13.7}$$

Substitution of expression (13.5) for $\boldsymbol{c}$ then gives

$$\lambda_1 \boldsymbol{b}_1 + \ldots + \lambda_k \boldsymbol{b}_k + \lambda_{k+1} (\beta_1 \boldsymbol{b}_1 + \ldots + \beta_k \boldsymbol{b}_k + \alpha_{k+1} \boldsymbol{a}_{k+1} + \\ + \alpha_{k+2} \boldsymbol{a}_{k+2} + \ldots + \alpha_m \boldsymbol{a}_m) + \lambda_{k+2} \boldsymbol{a}_{k+2} + \ldots + \lambda_m \boldsymbol{a}_m = \boldsymbol{0},$$

or

$$(\lambda_1 + \lambda_{k+1} \beta_1) \boldsymbol{b}_1 + \ldots + (\lambda_k + \lambda_{k+1} \beta_k) \boldsymbol{b}_k + \lambda_{k+1} \alpha_{k+1} \boldsymbol{a}_{k+1} + \\ + (\lambda_{k+2} + \lambda_{k+1} \alpha_{k+2}) \boldsymbol{a}_{k+2} + \ldots + (\lambda_m + \lambda_{k+1} \alpha_m) \boldsymbol{a}_m = \boldsymbol{0}.$$

Since vectors (13.3) are linearly independent, all the coefficients on the left-hand side are zero, and in particular $\lambda_{k+1} \alpha_{k+1} = 0$. Furthermore, since $\alpha_{k+1} \neq 0$, we know that $\lambda_{k+1} = 0$. However, the fact that the rest of the coefficients go to zero then means that $\lambda_1 = \ldots = \lambda_k = 0$, $\lambda_{k+2} = \ldots = \lambda_m = 0$. Hence only the *trivial* linear combination (13.7) can go to zero, that is, vectors (13.6) are linearly independent.

THEOREM 13.5. *If* $\boldsymbol{b}_1, \ldots, \boldsymbol{b}_{m-1}$ *constitute a linearly independent system of vectors of space* R, *while* $\boldsymbol{a}_1, \ldots, \boldsymbol{a}_m$ *form some other linearly independent system of vectors of this space (containing one more vector), then a vector* $\boldsymbol{a}_s$ *can be so selected from* $\boldsymbol{a}_1, \ldots, \boldsymbol{a}_m$ *that the system* $\boldsymbol{b}_1, \ldots, \boldsymbol{b}_{m-1}, \boldsymbol{a}_s$ *will be linearly independent.*

Proof. For $k = 0$ the preceding theorem takes the following form: if $\boldsymbol{c} \neq \boldsymbol{0}$ and system of vectors $\boldsymbol{a}_1, \ldots, \boldsymbol{a}_m$ is linearly independent, then of vectors $\boldsymbol{a}_1, \ldots, \boldsymbol{a}_m$ there must exist one which, when replaced by vector $\boldsymbol{c}$, will once again give a linearly independent system. Here we can assume (after changing, if necessary, the numbering of vectors $\boldsymbol{a}_1, \ldots, \boldsymbol{a}_m$) that vector $\boldsymbol{a}_1$ is replaced by $\boldsymbol{c}$, that is, we assume the system of vectors $\boldsymbol{c}, \boldsymbol{a}_2, \ldots, \boldsymbol{a}_m$ to be linearly independent. Since $\boldsymbol{b}_1 \neq \boldsymbol{0}$ (system $\boldsymbol{b}_1, \ldots, \boldsymbol{b}_{m-1}$ being linearly independent), we can apply this statement to vector $\boldsymbol{c} = \boldsymbol{b}_1$. Consequently, having changed, if necessary, the order of vectors $\boldsymbol{a}_1, \ldots, \boldsymbol{a}_m$, we find that vectors

$$\boldsymbol{b}_1, \boldsymbol{a}_2, \ldots, \boldsymbol{a}_m \tag{13.8}$$

are linearly independent.

Now let us apply Theorem 13.4 for $k = 1$, assuming that $\boldsymbol{c} = \boldsymbol{b}_2$. In this case we can replace one of vectors $\boldsymbol{a}_2, \ldots, \boldsymbol{a}_m$ in system (13.8) by vector $\boldsymbol{b}_2$ and the system will still be linearly independent. If it is vector $\boldsymbol{a}_2$ which is replaced by vector $\boldsymbol{b}_2$ (after changing, if necessary, the number of vectors $\boldsymbol{a}_2, \ldots, \boldsymbol{a}_m$), then, after changing, if necessary, the numbering of vectors $\boldsymbol{a}_1, \ldots, \boldsymbol{a}_m$, we find that vectors

$$\boldsymbol{b}_1, \boldsymbol{b}_2, \boldsymbol{a}_3, \ldots, \boldsymbol{a}_m$$

are linearly independent.

Next we apply Theorem 13.4 for $k = 2$ with $\boldsymbol{c} = \boldsymbol{b}_3$, etc. After $m - 1$ such operations we find that vectors $\boldsymbol{b}_1, \boldsymbol{b}_2, \ldots, \boldsymbol{b}_{m-1}, \boldsymbol{a}_m$ are linearly independent (where $\boldsymbol{a}_m$ is not necessarily the last vector, but may be *any* one of the initial vectors $\boldsymbol{a}_1, \ldots, \boldsymbol{a}_m$, because of the changes in the numbering).

Making use of the concepts of linear dependence and independence, let us now define the *dimension* of a vector space. To do this, it will be convenient to introduce the following group of axioms:

GROUP III: *axioms of degree.*

III_1. *In a vector space* R *there are* n *linearly independent vectors.*

III_2. *Any* $n + 1$ *vectors of vector space* R *are linearly dependent.*

A vector space satisfying (for some $n \geqslant 0$) the axioms of this group is called an *n-dimensional vector space,* or, alternatively, a *vector space of dimension n.* It should be noted that a zero-dimensional vector space consists only of a single element, $\boldsymbol{0}$. The dimension of a vector space R is designated as dim R. Vector spaces also exist which do not have a finite dimension (that is, spaces in which as many linearly independent vectors as desired can be found). In the present book, however, we will consider only vector spaces of *finite* dimension.

DEFINITION 13.6. An ordered system of vectors of a vector space R is called a *basis* of R if its vectors are linearly independent and if the number of its vectors is equal to the dimension of space R.

Axiom III_1 implies that a basis can be found for any n-dimensional vector space.

THEOREM 13.7. *If* $\boldsymbol{a}_1, \ldots, \boldsymbol{a}_n$ *form a basis of* n*-dimensional vector space* R, *then any vector* $\boldsymbol{a} \in R$ *can be represented uniquely as a linear combination of the vectors of this basis. The coefficients of this linear combination are called the coordinates of vector* $\boldsymbol{a}$ *in the basis* $\boldsymbol{a}_1, \ldots, \boldsymbol{a}_n$.

Proof. Let $\boldsymbol{a} \in R$. Since there are a total of $n+1$ vectors $\boldsymbol{a}, \boldsymbol{a}_1, \ldots, \boldsymbol{a}_n$, Axiom III_2 states that these vectors are linearly dependent, so that in view of Theorem 13.3 vector $\boldsymbol{a}$ can be represented as a linear combination of vectors $\boldsymbol{a}_1, \ldots, \boldsymbol{a}_n$.

Let us prove the uniqueness of this representation. If

$$\boldsymbol{a} = \lambda_1 \boldsymbol{a}_1 + \ldots + \lambda_n \boldsymbol{a}_n,$$
$$\boldsymbol{a} = \mu_1 \boldsymbol{a}_1 + \ldots + \mu_n \boldsymbol{a}_n$$

are two representations of vector $\boldsymbol{a}$ as linear combinations of the vectors of the basis, then

$$\lambda_1 \boldsymbol{a}_1 + \ldots + \lambda_n \boldsymbol{a}_n = \mu_1 \boldsymbol{a}_1 + \ldots + \mu_n \boldsymbol{a}_n,$$

and thus

$$(\lambda_1 - \mu_1)\, \boldsymbol{a}_1 + \ldots + (\lambda_n - \mu_n)\, \boldsymbol{a}_n = \boldsymbol{0}.$$

Since vectors $\boldsymbol{a}_1, \ldots, \boldsymbol{a}_n$ are linearly independent, it follows that

$$\lambda_1 - \mu_1 = \ldots = \lambda_n - \mu_n = 0, \quad \text{that is,} \quad \lambda_1 = \mu_1, \ldots, \lambda_n = \mu_n.$$

THEOREM 13.8. *Any linearly independent system of vectors can be augmented until a basis is reached. In other words, if* $\boldsymbol{b}_1, \ldots, \boldsymbol{b}_k$ *are linearly independent vectors of an* n*-dimensional vector space* R, *when* $k < n$, *then there will exist vectors* $\boldsymbol{b}_{k+1}, \ldots, \boldsymbol{b}_n$ *such that* $\boldsymbol{b}_1, \ldots, \boldsymbol{b}_k, \boldsymbol{b}_{k+1}, \ldots, \boldsymbol{b}_n$ *form a basis.*

Proof. Since $k < n$, we can select $k+1$ linearly independent vectors in R. According to Theorem 13.5, of these $k+1$ vectors a vector $\boldsymbol{b}_{k+1}$ can be found such that $\boldsymbol{b}_1, \ldots, \boldsymbol{b}_k, \boldsymbol{b}_{k+1}$ are linearly independent. If $k+1 = n$, then we arrive at the required basis. If, on the other hand, $k+1 < n$, then the linearly independent system $\boldsymbol{b}_1, \ldots, \boldsymbol{b}_k, \boldsymbol{b}_{k+1}$ can be augmented similarly by one more vector, etc.

THEOREM 13.9. *Vectors* $\boldsymbol{a}_1, \ldots, \boldsymbol{a}_m$ *are defined in a vector space* R, *each of these vectors being represented as a linear combination of vectors* $\boldsymbol{b}_1, \ldots, \boldsymbol{b}_k$ *of this space, where* $m > k$. *Vectors* $\boldsymbol{a}_1, \ldots, \boldsymbol{a}_m$ *are then linearly dependent.*

Proof. Of the system of vectors $\boldsymbol{b}_1, \ldots, \boldsymbol{b}_k$, we select the largest (with respect to the number of vectors) linearly independent system. If the latter is, say, system $\boldsymbol{b}_1, \ldots, \boldsymbol{b}_l$ (where $l \leqslant k$), then for $l < k$ each of vectors $\boldsymbol{b}_{l+1}, \ldots, \boldsymbol{b}_k$ is represented as a linear combination of vectors $\boldsymbol{b}_1, \ldots, \boldsymbol{b}_l$ (according to Theorem 13.3). Thus it follows that each vector $\boldsymbol{a}_1, \ldots, \boldsymbol{a}_m$ is represented as a linear combination of vectors $\boldsymbol{b}_1, \ldots, \boldsymbol{b}_l$ (with $l \leqslant k < m$). Now let us assume that vectors $\boldsymbol{a}_1, \ldots, \boldsymbol{a}_m$ are linearly independent. Then, according to Theorem 13.5, there exists a number $s (= 1, \ldots, m)$ such that vectors $\boldsymbol{b}_1, \ldots, \boldsymbol{b}_l, \boldsymbol{a}_s$ are linearly independent. On the other hand, vector a_s is represented as a linear combination of vectors $b_1, \ldots, b_l$, which contradicts the fact that vectors $b_1, \ldots, b_l, a_s$ are linearly independent.

EXAMPLE 13.10. Let us once again consider an arithmetical vector space R^n (Example 12.2), and let us show that its dimension is equal to n. To do this, we consider the following vectors of space R^n:

$$\begin{aligned} e_1 &= \{1, 0, 0, \ldots, 0, 0\}, \\ e_2 &= \{0, 1, 0, \ldots, 0, 0\}, \\ &\cdots\cdots\cdots\cdots \\ e_n &= \{0, 0, 0, \ldots, 0, 1\}. \end{aligned}$$

Then we have

$$\begin{aligned} \alpha_1 e_1 + \alpha_2 e_2 + \ldots + \alpha_n e_n = \alpha_1 \{1, 0, 0, \ldots, 0, 0\} + \\ + \alpha_2 \{0, 1, 0, \ldots, 0, 0\} + \ldots + \alpha_n \{0, 0, 0, \ldots, 0, 1\} = \\ = \{\alpha_1, 0, 0, \ldots, 0, 0\} + \{0, \alpha_2, 0, \ldots, 0, 0\} + \ldots \\ \ldots + \{0, 0, 0, \ldots, 0, \alpha_n\} = \{\alpha_1, \alpha_2, \ldots, \alpha_n\}. \qquad (13.9) \end{aligned}$$

From this it follows that linear combination $\alpha_1 e_1 + \ldots + \alpha_n e_n$ goes to zero if, and only if, $\alpha_1 = \ldots = \alpha_n = 0$, that is, *vectors* $e_1, \ldots, e_n$ *are linearly independent.*

Relation (13.9) indicates that *any* vector $\{\alpha_1, \ldots, \alpha_n\}$ of space R^n is represented as a linear combination of vectors $e_1, \ldots, e_n$:

$$\{\alpha_1, \ldots, \alpha_n\} = \alpha_1 e_1 + \ldots + \alpha_n e_n.$$

In view of Theorem 13.9, this implies that any $n + 1$ vectors of space R^n are linearly dependent. Accordingly, $\dim R^n = n$, and vectors $e_1, \ldots, e_n$ form a *basis* of space R^n.

THEOREM 13.11. *If* $c_1, c_2, \ldots, c_n$ *form a basis of vector space* R *and if in this basis vector* a *has coordinates* $x^1, \ldots, x^n$, *while vector* b *has coordinates* $y^1, \ldots, y^n$, *that is,*

$$\begin{aligned} a &= x^1 e_1 + \ldots + x^n e_n, \\ b &= y^1 e_1 + \ldots + y^n e_n, \end{aligned}$$

then the vector $a + b$ *will have (in this same basis) coordinates* $x^1 + y^1, \ldots, x^n + y^n$, *while vector* λa *will have coordinates* $\lambda x^1, \ldots, \lambda x^n$.

The *proof* of this theorem follows from an obvious calculation.

The established facts make possible a comparison of all n-dimensional vector spaces. Assume that $\varphi : R \to S$ is some mapping of vector space R into vector space S, that is, a function defined on set R and assuming values on set S. Mapping φ is called an *isomorphism* (or an *isomorphic mapping*) if it is *one-to-one* (that is, if each vector $b \in S$ corresponds to one, and only one, vector $a \in R$), while, in addition, *the operations of vector addition and multiplication of a vector by a number are preserved,* that is,

$$\varphi(\boldsymbol{a}+\boldsymbol{b})=\varphi(\boldsymbol{a})+\varphi(\boldsymbol{b}), \quad \varphi(\lambda\boldsymbol{a})=\lambda\varphi(\boldsymbol{a}) \tag{13.10}$$

for any vectors $\boldsymbol{a}$, $\boldsymbol{b}$ of space R and any real number λ. If an isomorphic mapping $\varphi: R \to S$ exists, then vector spaces R and S are called *isomorphic.* Isomorphic vector spaces essentially do not differ from one another; actually there are no operations in a vector space except vector addition and multiplication of a vector by a number, and these operations are preserved during an isomorphic mapping.

THEOREM 13.12. *Two finite-dimensional vector spaces R and S are isomorphic with regard to each other if, and only if,* $\dim R = \dim S$.

Proof. First of all, it should be noted that for isomorphism $\varphi: R \to S$ the zero vector transforms into the zero vector: $\varphi(\mathbf{0}) = \mathbf{0}$. Actually, if $\boldsymbol{a} \in R$, then $\boldsymbol{a}+\mathbf{0}=\boldsymbol{a}$, so that according to (13.10) we have $\varphi(\boldsymbol{a})+\varphi(\mathbf{0}) = \varphi(\boldsymbol{a})$, from which it follows that $\varphi(\mathbf{0}) = \mathbf{0}$. Moreover, a linear combination of vectors transforms into a linear combination with the same coefficients, that is, if $\boldsymbol{b} = \alpha_1\boldsymbol{a}_1 + \ldots + \alpha_k\boldsymbol{a}_k$, then $\varphi(\boldsymbol{b}) = \alpha_1\varphi(\boldsymbol{a}_1) + \ldots + \alpha_k\varphi(\boldsymbol{a}_k)$ (as implied by (13.10)). Consequently, the equation $\alpha_1\boldsymbol{a}_1 + \ldots + \alpha_k\boldsymbol{a}_k = \mathbf{0}$ is valid if, and only if $\alpha_1\varphi(\boldsymbol{a}_1) + \ldots + \alpha_k\varphi(\boldsymbol{a}_k) = \mathbf{0}$. However, then it is obvious that vectors $\boldsymbol{a}_1, \ldots, \boldsymbol{a}_k$ of space R are linearly independent if, and only if, vectors $\varphi(\boldsymbol{a}_1), \varphi(\boldsymbol{a}_2), \ldots, \varphi(\boldsymbol{a}_k)$ of space S are linearly independent. Thus the *maximum* number of linearly independent vectors in spaces R and S will be the same, that is, $\dim R = \dim S$. This means that isomorphic vector spaces must have the same dimension.

Conversely, let us assume that vector spaces R and S have the same dimension n. We consider some basis $\boldsymbol{a}_1, \ldots, \boldsymbol{a}_n$ in space R, and some basis $\boldsymbol{b}_1, \ldots, \boldsymbol{b}_n$ in space S. Then any vector $\boldsymbol{c} \in R$ can be written unambiguously in the form $\boldsymbol{c} = \lambda^1\boldsymbol{a}_1 + \ldots + \lambda^n\boldsymbol{a}_n$ (where $\lambda^1, \ldots, \lambda^n$ are the coordinates of vector $\boldsymbol{c}$ in the basis $\boldsymbol{a}_1, \ldots, \boldsymbol{a}_n$). Now, if $\varphi(\boldsymbol{c})$ is a vector of space S having the same coordinates in basis $\boldsymbol{b}_1, \ldots, \boldsymbol{b}_n$, that is, if $\varphi(\boldsymbol{c}) = \lambda^1\boldsymbol{b}_1 + \ldots + \lambda^n\boldsymbol{b}_n$, then we obtain the mapping $\varphi: R \to S$. It is easy to see that the plotted mapping is isomorphic.

From the theorem proven it follows that, in particular, *any n-dimensional vector space is isomorphic to a space* R^n (see Example 13.10). Let us consider a basis $\boldsymbol{a}_1, \ldots, \boldsymbol{a}_n$ in n-dimensional vector space R, and in space R^n let us take the basis $\boldsymbol{e}_1, \ldots, \boldsymbol{e}_n$ considered in Example 13.10. Then the isomorphic mapping φ constructed to prove Theorem 13.12 associates the vector $\varphi(\boldsymbol{c}) = \lambda^1\boldsymbol{e}_1 + \ldots + \lambda^n\boldsymbol{e}_n = \{\lambda^1, \ldots, \lambda^n\}$ of space R^n with the vector $\boldsymbol{c} = \lambda^1\boldsymbol{a}_1 + \ldots + \lambda^n\boldsymbol{a}_n$ of space R (see (13.9)). Consequently, *by specifying a basis $\boldsymbol{a}_1, \ldots, \boldsymbol{a}_n$ in n-dimensional vector space R and assigning to each vector $\boldsymbol{c} = \lambda^1\boldsymbol{a}_1 + \ldots + \lambda^n\boldsymbol{a}_n \in R$ a vector $\varphi(\boldsymbol{c}) = \{\lambda^1, \ldots, \lambda^n\} \in R^n$, we obtain the isomorphism $\varphi: R \to R^n$.*

Often spaces R and R^n can be *identified* with each other as a result of this isomorphism, that is, we can simply write $\lambda^1\boldsymbol{a}_1 + \ldots + \lambda^n\boldsymbol{a}_n = \{\lambda^1, \ldots, \lambda^n\}$ instead of $\varphi(\lambda^1\boldsymbol{a}_1 + \ldots + \lambda^n\boldsymbol{a}_n) = \{\lambda^1, \ldots, \lambda^n\}$. Thus the notation $\boldsymbol{c} = \{\lambda^1, \ldots, \lambda^n\}$ signifies that vector $\boldsymbol{c} \in R$ has (in basis $\boldsymbol{a}_1, \ldots, \boldsymbol{a}_n$) coordinates $\lambda^1, \ldots, \lambda^n$.

Naturally, this notation is acceptable only provided that basis $\boldsymbol{a}_1, \ldots, \boldsymbol{a}_n$ has been defined beforehand in R (and does not change during the course of the argument).

14. Subspace. In this subsection the symbol R will everywhere denote an n-dimensional vector space.

DEFINITION 14.1. A set $A \subset R$ is called a *subspace* of space R if, for any vector $\boldsymbol{a} \in A$ and any real number λ, the vector $\lambda \boldsymbol{a}$ belongs to set A and, in addition, for every two vectors $\boldsymbol{a}, \boldsymbol{b} \in A$, the vector $\boldsymbol{a} + \boldsymbol{b}$ also belongs to set A.

It follows directly from this definition that, if vectors $\boldsymbol{a}_1, \ldots, \boldsymbol{a}_s$ belong to subspace A, then any linear combination $\lambda^1 \boldsymbol{a}_1 + \ldots + \lambda^s \boldsymbol{a}_s$ of these will likewise belong to A. In particular, the zero element belongs to any given subspace. Obviously, the set consisting only of the single element $\mathbf{0}$ is a subspace; this subspace (which will be denoted by the symbol $\mathbf{0}$) is called the *trivial* subspace. Every other subspace is *nontrivial.* It is also clear that the set A coinciding with all of space R is a subspace; this subspace is called an *improper* subspace. Every other (that is, not coinciding with R) subspace is called a *proper* subspace.

EXAMPLE 14.2. We consider a plane Π in ordinary three-dimensional space Σ and we select some point O on plane Π. Then Σ and Π can be considered as vector spaces, the vectors and the operations on them being defined as indicated in Example 12.1. It is easy to show that Π is a nontrivial proper subspace of vector space Σ.

EXAMPLE 14.3. We consider an arithmetical n-dimensional space R^n; select some natural number $m < n$ and let A_m be a set of vectors having the form $\{\lambda_1, \ldots, \lambda_m, 0, \ldots, 0\}$, that is, a set in which the last $n - m$ coordinates are zero (the first m coordinates may be either zero or nonzero). It is easy to see that A_m is a nontrivial proper subspace of space R^n.

THEOREM 14.4. *Any subspace A of vector space R is itself a vector space (with regard to the operations on vectors which exist in space R), where* $\dim A \leqslant \dim R$. *The equation* $\dim A = \dim R$ *is valid if, and only if, subspace A is improper.*

Proof. For any two vectors $\boldsymbol{a}, \boldsymbol{b} \in A$, the sum $\boldsymbol{a} + \boldsymbol{b}$ is defined as belonging (according to Definition 14.1) to the *same* set A. Axioms I_1, I_2 are obeyed in R for *all* vectors and, in particular, for the vectors belonging to set A. Moreover, since $\mathbf{0} \in A$, Axiom I_3 is also obeyed in A. Finally, since $-\boldsymbol{a} = (-1)\boldsymbol{a}$, it follows that for any $\boldsymbol{a} \in A$ the vector $-\boldsymbol{a}$ also belongs to A, so that Axiom I_4 is also obeyed in A. Thus all the axioms of group I are obeyed in set A. The axioms of group II can be verified just as simply for set A, indicating that the latter is indeed a vector space.

Moreover, if vectors $\boldsymbol{a}_1, \ldots, \boldsymbol{a}_m$ of subspace A are linearly independent in A. then they will be linearly independent throughout space R, since the operations in A are the same as those throughout space R. However, since in R there cannot be more than n linearly independent vectors, this means that in A as well there cannot be more than n such vectors either. Consequently, $\dim A \leqslant n$, that is, $\dim A \leqslant \dim R$.

If A is an improper subspace, that is, if $A = R$, then obviously $\dim A = \dim R$. Conversely, let us first assume that $\dim A = \dim R$, that is, $\dim A = n$. Then in A we can find a total of n linearly independent vectors $\boldsymbol{e}_1, \ldots, \boldsymbol{e}_n$. These vectors form a basis in R, that is, any vector $\boldsymbol{x} \in R$ can be written as $\boldsymbol{x} = \lambda^1 \boldsymbol{e}_1 + \ldots + \lambda^n \boldsymbol{e}_n$. However, then $\boldsymbol{x} \in A$, since a linear combination of vectors from A will also belong to subspace A. Therefore, any vector $\boldsymbol{x} \in R$ belongs to subspace A, that is, $A = R$ and hence A is an improper subspace.

THEOREM 14.5. *The intersection of two (or, in general, any number of) subspaces is a subspace.*

Proof. Let us assume that A and B are subspaces of vector space R and that $\boldsymbol{a}$ and $\boldsymbol{b}$ are two vectors belonging to the set $C = A \cap B$. Since $\boldsymbol{a} \in C$, vector $\boldsymbol{a}$ belongs to both subspaces A and B. Similarly, vector $\boldsymbol{b}$ will also belong to both subspaces A and B. Since $\boldsymbol{a}, \boldsymbol{b} \in A$ and since A is a subspace, it follows that $\boldsymbol{a} + \boldsymbol{b} \in A$. Similarly, $\boldsymbol{a} + \boldsymbol{b} \in B$. Consequently, vector $\boldsymbol{a} + \boldsymbol{b}$ belongs to both subspaces A and B, that is, $\boldsymbol{a} + \boldsymbol{b} \in C$. Thus it is clear that, if $\boldsymbol{a}, \boldsymbol{b} \in C$, then vector $\boldsymbol{a} + \boldsymbol{b}$ also belongs to set C.

Similarly, it can be demonstrated that if $\boldsymbol{a} \in C$ then $\lambda \boldsymbol{a} \in C$ (for any real λ), so that C is a subspace. The argument is analogous for the intersection of any number of subspaces.

It follows from Theorem 14.5 that for any set $Q \subset R$ there exists a *smallest* subspace containing the set Q, that is, a subspace contained in any other subspace also containing Q. Let us consider *all* the subspaces containing set Q (such subspaces do exist, for instance R itself), and let us denote as L_Q the intersection of all these subspaces. Theorem 14.5 implies that L_Q is a subspace. Obviously, $Q \subset L_Q$ and L_Q is contained in any subspace containing set Q. Therefore L_Q is the desired smallest subspace containing Q. It is called a subspace *generated* by set Q.

THEOREM 14.6. *Let $Q \subset R$. A vector $\boldsymbol{x} \in R$ will belong to a subspace L_Q generated by set Q if, and only if, $\boldsymbol{x}$ can be written as a linear combination of vectors belonging to set Q.*

Proof. Let A be the set of all vectors $\boldsymbol{x} \in R$ which can be written as a linear combination of vectors belonging to set Q. It is easy to see that if $\boldsymbol{a}, \boldsymbol{b} \in A$ then $\boldsymbol{a} + \boldsymbol{b} \in A$, and if $\boldsymbol{a} \in A$ then $\lambda \boldsymbol{a} \in A$ for any real λ. Consequently, A is a subspace of vector space R.

Since obviously $Q \subset A$, it follows that (since L_Q is the *smallest* subspace containing Q) the inclusion $A \supset L_Q$ is satisfied. Conversely, if $\boldsymbol{x} \in A$, that is, $\boldsymbol{x} = \lambda^1 \boldsymbol{q}_1 + \ldots + \lambda^s \boldsymbol{q}_s$, where $\boldsymbol{q}_i \in Q$, $i = 1, \ldots, s$, then (in view of the inclusion $Q \subset L_Q$) we have $\boldsymbol{q}_i \in L_Q$, $i = 1, \ldots, s$, so that (since L_Q is a subspace) the vector $\boldsymbol{x} = \lambda^1 \boldsymbol{q}_1 + \ldots + \lambda^s \boldsymbol{q}_s$ belongs to L_Q. Thus $A \subset L_Q$. The foregoing inclusions $A \supset L_Q$ and $A \subset L_Q$ imply that $A = L_Q$.

It follows from Theorem 14.6 that, *for any integer r satisfying inequalities $0 \leqslant r \leqslant n$, there exists in R a subspace of dimension r.* This is obvious for $r = 0$ and $r = n$. For $0 < r < n$ consider in R any r linearly independent vectors $\boldsymbol{e}_1, \ldots, \boldsymbol{e}_r$ (for instance, the first r vectors of some basis), and let Q be the set of all these vectors. Since the subspace L_Q generated by set Q contains r linearly independent vectors $\boldsymbol{e}_1, \ldots, \boldsymbol{e}_r$, we know that $\dim L_Q \geqslant r$. On the other hand, each vector $\boldsymbol{x} \in L_Q$

can according to Theorem 14.6 be represented as a linear combination of vectors $e_1, \ldots, e_r$, so that in view of Theorem 13.9 every $r+1$ vectors of subspace L_Q are linearly dependent. Consequently, $\dim L_Q \leqslant r$. From the foregoing inequalities it follows that $\dim L_Q = r$.

EXAMPLE 14.7. Let us show that any two r-dimensional subspaces are "equally situated" in R. In other words, if A and B are two r-dimensional subspaces of vector space R, then there exists an isomorphism $\varphi : R \to R$ of space R onto itself whereby subspace A is mapped isomorphically onto subspace B.

If $\boldsymbol{a}_1, \ldots, \boldsymbol{a}_r$ form a basis of subspace A, then vectors $\boldsymbol{a}_1, \ldots, \boldsymbol{a}_r$ are linearly independent, so that according to Theorem 13.8 this system of vectors can be augmented until a basis $\boldsymbol{a}_1, \ldots, \boldsymbol{a}_r, \boldsymbol{a}_{r+1}, \ldots, \boldsymbol{a}_n$ of the entire space R is reached. In addition, if $\boldsymbol{b}_1, \ldots, \boldsymbol{b}_r$ form a basis of subspace B, then, similarly, the system of vectors $\boldsymbol{b}_1, \ldots, \boldsymbol{b}_r$ can be augmented until a basis $\boldsymbol{b}_1, \ldots, \boldsymbol{b}_r, \boldsymbol{b}_{r+1}, \ldots, \boldsymbol{b}_n$ of space R is reached.

Now let φ be an isomorphism of space R onto itself which transforms a vector having coordinates $\lambda^1, \ldots, \lambda^n$ relative to basis $\boldsymbol{a}_1, \ldots, \boldsymbol{a}_n$ into a vector having the same coordinates relative to basis $\boldsymbol{b}_1, \ldots, \boldsymbol{b}_n$ (cf. end of proof of Theorem 13.12):

$$\varphi(\lambda^1\boldsymbol{a}_1 + \ldots + \lambda^n\boldsymbol{a}_n) = \lambda^1\boldsymbol{b}_1 + \ldots + \lambda^n\boldsymbol{b}_n.$$

Then a vector $\boldsymbol{x}$ belonging to subspace A, that is, having the form

$$\boldsymbol{x} = \lambda^1\boldsymbol{a}_1 + \ldots + \lambda^r\boldsymbol{a}_r + 0\boldsymbol{a}_{r+1} + \ldots + 0\boldsymbol{a}_n, \tag{14.1}$$

is transformed by mapping φ into a vector

$$\lambda^1\boldsymbol{b}_1 + \ldots + \lambda^r\boldsymbol{b}_r + 0\boldsymbol{b}_{r+1} + \ldots + 0\boldsymbol{b}_n, \tag{14.2}$$

that is, into a vector belonging to subspace B. Hence *any* vector belonging to subspace B, that is, having form (14.2), constitutes an image of vector (14.1) belonging to subspace A. Consequently, for isomorphism φ subspace A is mapped (isomorphically) onto subspace B, that is, isomorphism φ is the desired mapping.

In contrast to the intersection (see Theorem 14.5), the *union* $A \cup B$ of two subspaces A and B is in general not a subspace. However, we can consider the subspace *generated* by union $A \cup B$. This subspace is called the *sum* of subspaces A and B, being written $A + B$.

THEOREM 14.8. *Let A and B be two subspaces of vector space R. A vector $\boldsymbol{x} \in R$ will belong to subspace $A + B$ if, and only if, it can be represented in the form $\boldsymbol{x} = \boldsymbol{a} + \boldsymbol{b}$, where $\boldsymbol{a} \in A$, $\boldsymbol{b} \in B$.*

Proof. Since $A \subset A + B$, $B \subset A + B$, for any $\boldsymbol{a} \in A$, $\boldsymbol{b} \in B$ the vector $\boldsymbol{a} + \boldsymbol{b}$ belongs to subspace $A + B$.

Conversely, let $\boldsymbol{x} \in A + B$. Then, according to Theorem 14.6, $\boldsymbol{x} = \lambda^1\boldsymbol{q}_1 + \ldots + \lambda^s\boldsymbol{q}_s$, where each of vectors $\boldsymbol{q}_1, \ldots, \boldsymbol{q}_s$ belongs to set $A \cup B$, that is, it belongs to one of the subspaces A, B. If, say, $q_1, \ldots, q_k \in A$, $q_{k+1}, \ldots q_s \in B$, then obviously $x = a + b$, where

$$a = \lambda^1 q_1 + \ldots + \lambda^k q_k \in A, \quad b = \lambda^{k+1} q_{k+1} + \ldots + \lambda^s q_s \in B.$$

THEOREM 14.9. *If A and B are two subspaces of vector space R and $L = A + B$ is their sum, while $C = A \cap B$ is their intersection, then*

$$\dim L = \dim A + \dim B - \dim C. \tag{14.3}$$

Proof. The dimension of subspace C is taken to be r, and $e_1, \ldots, e_r$ form some basis of this subspace. The dimensions of subspaces A and B are p and q, respectively. Since vectors $e_1, \ldots, e_r$ are situated in subspace A and are linearly independent, they can be augmented by vectors $f_{r+1}, \ldots, f_p$ until a basis $e_1, \ldots, e_r, f_{r+1}, \ldots f_p$ of subspace A is reached (in view of Theorem 13.8). Similarly, vectors $e_1, \ldots, e_r$ can be augmented by vectors $g_{r+1}, \ldots, g_q$ until a basis $e_1, \ldots, e_r, g_{r+1}, \ldots, g_q$ of subspace B is reached.

Let us show that vectors

$$e_1, \ldots, e_r, f_{r+1}, \ldots, f_p, g_{r+1}, \ldots, g_q \tag{14.4}$$

form a basis of subspace L. Since the number of these vectors is $p + (q - r) = \dim A + \dim B - \dim C$, the validity of formula (14.3) also follows from this.

If x is an arbitrary vector belonging to subspace L, then according to Theorem 14.8 $x = a + b$, where $a \in A$, $b \in B$. Vector $a \in A$ is represented as a linear combination of vectors $e_1, \ldots, e_r, f_{r+1}, \ldots, f_p$, and thus as a linear combination of vectors (14.4) (with zero coefficients for vectors $g_{r+1}, \ldots, g_q$). Similarly, vector $b \in B$ is represented as a linear combination of vectors (14.4). Consequently, vector $x = a + b$ is also represented as a linear combination of vectors (14.4).

It only remains to show that vectors (14.4) are linearly independent. Assume that

$$\mu_1 e_1 + \ldots + \mu_r e_r + \lambda_{r+1} f_{r+1} + \ldots + \lambda_p f_p + \\ + \nu_{r+1} g_{r+1} + \ldots + \nu_q g_q = 0, \tag{14.5}$$

that is,

$$\mu_1 e_1 + \ldots + \mu_r e_r + \lambda_{r+1} f_{r+1} + \ldots + \lambda_p f_p = \\ = -\nu_{r+1} g_{r+1} - \ldots - \nu_q g_q. \tag{14.6}$$

The vector on the left-hand side of (14.6) belongs to subspace A, while the vector on the right-hand side belongs to subspace B. Since this is one and the same vector, it belongs to subspace $A \cap B = C$. Consequently, vector (14.6) is represented as a linear combination of vectors $e_1, \ldots, e_r$:

$$-\nu_{r+1} g_{r+1} - \ldots - \nu_q g_q = \alpha_1 e_1 + \ldots + \alpha_r e_r,$$

or

$$\alpha_1 e_1 + \ldots + \alpha_r e_r + \nu_{r+1} g_{r+1} + \ldots + \nu_q g_q = \mathbf{0}.$$

The linear independence of the vectors $e_1, \ldots, e_r, g_{r+1}, \ldots, g_q$ (which form a basis of subspace B) implies that all the coefficients in this equation are zero. In particular,

$$\nu_{r+1} = \ldots = \nu_q = 0.$$

Hence it follows, in view of (14.6), that

$$\mu_1 e_1 + \ldots + \mu_r e_r + \lambda_{r+1} f_{r+1} + \ldots + \lambda_p f_p = \mathbf{0},$$

so that

$$\mu_1 = \ldots = \mu_r = \lambda_{r+1} = \ldots = \lambda_p = 0$$

(since the vectors $e_1, \ldots, e_r, f_{r+1}, \ldots, f_p$ forming a basis of subspace A are linearly independent). Thus, all the coefficients in (14.5) are zero, that is, vectors (14.4) are linearly independent. The theorem is thereby proven.

Let A and B be two subspaces of an n-dimensional vector space R. If $A + B = R$ and $A \cap B = \mathbf{0}$, then B is called the *direct complement* of subspace A, being written as $R = A \oplus B$ (in this case A is also the direct complement of subspace B, that is, $R = B \oplus A$, since in the indicated definition the roles of subspaces A and B are identical). Therefore, if $R = A \oplus B$, then $\dim(A + B) = n$ and $\dim(A \cap B) = 0$, so that in view of (14.3) $\dim A + \dim B = n$. Conversely, if $A \cap B = \mathbf{0}$ and $\dim A + \dim B = n$, then $\dim(A + B) = n$, that is, $R = A \oplus B$. This leads us to the following statement:

THEOREM 14.10. *Let A and B be two subspaces of n-dimensional vector space R, the intersection of which is a trivial subspace. In order for A and B to be direct complements of one another (that is, $R = A \oplus B$), it is necessary and sufficient that*

$$\dim A + \dim B = n.$$

Note also that, if $R = A \oplus B$ and if $e_1, \ldots, e_p$ form a basis of subspace A, while $f_1, \ldots, f_q$ form a basis of subspace B, then the vectors

$$e_1, \ldots, e_p, \quad f_1, \ldots, f_q$$

constitute a basis of the entire space R (see (14.4)).

THEOREM 14.11. *Let A and B be two subspaces of n-dimensional vector space R. The relation $R = A \oplus B$ will be valid if, and only if, any vector $x \in R$ can be represented uniquely as $x = a + b$, $a \in A$, $b \in B$.*

Proof. If $R = A \oplus B$, then, since in this case $A + B = R$, Theorem 14.8 implies that any vector $x \in R$ can be represented as $x = a + b$, $a \in A$, $b \in B$.

If, in addition, an analogous representation $\boldsymbol{x}=\boldsymbol{a}'+\boldsymbol{b}'$, $\boldsymbol{a}'\in A$, $\boldsymbol{b}'\in B$, exists, then we have

$$\boldsymbol{a}+\boldsymbol{b}=\boldsymbol{a}'+\boldsymbol{b}',$$

and thus $\boldsymbol{a}-\boldsymbol{a}'=\boldsymbol{b}'-\boldsymbol{b}$.

Vector $\boldsymbol{a}-\boldsymbol{a}'$ belongs to subspace A, and since it is equal to vector $\boldsymbol{b}'-\boldsymbol{b}$ it thus also belongs to subspace B. Consequently, $\boldsymbol{a}-\boldsymbol{a}'\in A\cap B$, so that $\boldsymbol{a}-\boldsymbol{a}'=\boldsymbol{0}$ and $\boldsymbol{b}'-\boldsymbol{b}=\boldsymbol{0}$, that is, $\boldsymbol{a}=\boldsymbol{a}'$ and $\boldsymbol{b}=\boldsymbol{b}'$. Consequently, uniqueness of the representation $\boldsymbol{x}=\boldsymbol{a}+\boldsymbol{b}$, $\boldsymbol{a}\in A$, $\boldsymbol{b}\in B$ is proven.

Conversely, let us assume that any vector $\boldsymbol{x}\in R$ can be expressed uniquely as $\boldsymbol{x}=\boldsymbol{a}+\boldsymbol{b}$, $\boldsymbol{a}\in A$, $\boldsymbol{b}\in B$. Then Theorem 14.8 implies that any vector $\boldsymbol{x}\in R$ belongs to subspace $A+B$, that is, $A+B=R$. If some nonzero vector $\boldsymbol{c}\in A\cap B$ existed, then we would have two *different* representations

$$\begin{aligned} \boldsymbol{c}&=\boldsymbol{c}+\boldsymbol{0}, \qquad \boldsymbol{c}\in A, \quad \boldsymbol{0}\in B,\\ \boldsymbol{c}&=\boldsymbol{0}+\boldsymbol{c}, \qquad \boldsymbol{0}\in A, \quad \boldsymbol{c}\in B,\end{aligned}$$

which contradicts the assumption. Consequently, $A\cap B=\boldsymbol{0}$, so that $R=A\oplus B$.

Finally, let us consider the concept of the *direct sum*. If $A_1, \ldots, A_k$ are subspaces of n-dimensional vector space R, then we say that *space R decomposes into the direct sum of subspaces $A_1, \ldots, A_k$* and we write

$$R=A_1\oplus \ldots \oplus A_k, \tag{14.7}$$

if any vector $\boldsymbol{x}\in R$ can be represented uniquely as

$$\boldsymbol{x}=\boldsymbol{a}_1+\ldots+\boldsymbol{a}_k, \quad \boldsymbol{a}_1\in A_1, \ldots, \boldsymbol{a}_k\in A_k.$$

Note that the above definition does not exclude the case in which several of the subspaces in relation (14.7) are trivial (that is, consist only of the single element $\boldsymbol{0}$).

Theorem 14.11 shows that a space R decomposes into the direct sum of two subspaces A and B if, and only if, A and B are the direct complements of one another, that is, the relation $R=A\oplus B$ has the same meaning as previously.

THEOREM 14.12. *If $R=A\oplus B$, where subspace A (considered as a vector space) decomposes into the direct sum of its subspaces $A_1, \ldots, A_k$:*

$$A=A_1\oplus \ldots \oplus A_k, \tag{14.8}$$

then space R decomposes into the direct sum of subspaces $A_1, \ldots, A_k, B$:

$$R=A_1\oplus \ldots \oplus A_k\oplus B. \tag{14.9}$$

Proof. Let $\boldsymbol{x}\in R$. Then, in view of the relation $R=A\oplus B$, we can write $\boldsymbol{x}=\boldsymbol{a}+\boldsymbol{b}$, where $\boldsymbol{a}\in A$, $\boldsymbol{b}\in B$. In addition, vector $\boldsymbol{a}\in A$ can, in view of

(14.8), be written as

$$\boldsymbol{a}=\boldsymbol{a}_1+\ \ldots\ +\boldsymbol{a}_k, \qquad \boldsymbol{a}_1\in A_1,\ \ldots\ \boldsymbol{a}_k\in A_k. \tag{14.10}$$

Consequently,

$$\boldsymbol{x}=\boldsymbol{a}_1+\ \ldots\ +\boldsymbol{a}_k+\boldsymbol{b}; \quad \boldsymbol{a}_1\in A_1,\ \ldots,\ \boldsymbol{a}_k\in A_k, \quad \boldsymbol{b}\in B. \tag{14.11}$$

It now remains just to prove that expansion (14.11) is unique. Let us assume that, in addition to (14.11), the following expansion is valid:

$$\boldsymbol{x}=\boldsymbol{a}'_1+\ \ldots\ +\boldsymbol{a}'_k+\boldsymbol{b}'; \quad \boldsymbol{a}'_1\in A_1,\ \ldots,\ \boldsymbol{a}'_k\in A_k, \quad \boldsymbol{b}'\in B. \tag{14.12}$$

Since $A_i \subset A$, $i=1,\ \ldots,\ k$, it follows that $\boldsymbol{a}'_1+\ \ldots\ +\boldsymbol{a}'_k\in A$, so that the expansion $\boldsymbol{x}=(\boldsymbol{a}'_1+\ \ldots\ +\boldsymbol{a}'_k)+\boldsymbol{b}'$ must be identical to the expansion $\boldsymbol{x}=\boldsymbol{a}+$ $+\boldsymbol{b}$, that is,

$$\boldsymbol{a}=\boldsymbol{a}'_1+\ \ldots\ +\boldsymbol{a}'_k, \quad \boldsymbol{b}=\boldsymbol{b}'. \tag{14.13}$$

However, from (14.8), the first of expansions (14.13) must be identical to expansion (14.10), that is,

$$\boldsymbol{a}_1=\boldsymbol{a}'_1,\ \ldots,\ \boldsymbol{a}_k=\boldsymbol{a}'_k.$$

Consequently, expansions (14.12) and (14.11) are identical.

THEOREM 14.13. *If* $R=A+B$, *then there exists a subspace* $D\subset B$ *such that* $R=A\oplus D$.

Proof. Let us assume that $C=A\cap B$ and let us select in vector space B some subspace D which is a direct complement of subspace $C\subset B$, so that $B=C\oplus D$. Since $D\subset B$, we have $D=B\cap D$ and

$$A\cap D\subset A\cap(B\cap D)=(A\cap B)\cap D=C\cap D,$$

that is, $A\cap D$ is a trivial subspace. Hence, according to Theorem 14.10, it is sufficient to show that $\dim A+\dim D=n$. From formula (14.3) we know that

$$\begin{aligned} n=\dim R&=\dim A+\dim B-\dim C= \\ &=\dim A+(\dim C+\dim D)-\dim C=\dim A+\dim D, \end{aligned}$$

which completes the proof.

Note also that, if $R=A_1\oplus\ldots\oplus A_k$ and if $\boldsymbol{e}_1^{(1)},\ \ldots,\ \boldsymbol{e}_{r_1}^{(1)}$ form a basis of subspace A_1, while in addition $\boldsymbol{e}_1^{(2)},\ \ldots,\ \boldsymbol{e}_{r_2}^{(2)}$ form a basis of subspace $A_2,\ \ldots,$ and finally, $\boldsymbol{e}_1^{(k)},\ \ldots,\ \boldsymbol{e}_{r_k}^{(k)}$ form a basis of subspace A_k, then the vectors

$$\boldsymbol{e}_1^{(1)},\ \ldots,\ \boldsymbol{e}_{r_1}^{(1)},\ \boldsymbol{e}_1^{(2)},\ \ldots,\ \boldsymbol{e}_{r_2}^{(2)},\ \ldots,\ \boldsymbol{e}_1^{(k)},\ \ldots,\ \boldsymbol{e}_{r_k}^{(k)}$$

constitute a basis of space R. This is easy to demonstrate by induction, using the relation

$$A_1 \oplus A_2 \oplus \ldots \oplus A_k = A_1 \oplus (A_2 \oplus \ldots \oplus A_k).$$

15. Homomorphisms of vector spaces. DEFINITION 15.1. A mapping φ: $R \to S$ of a vector space R into a vector space S is called a *homomorphism* if it preserves the operations of vector addition and multiplication of a vector by a number, that is, if conditions (13.10) are satisfied.

In contrast to an isomorphism, a homomorphism does not have to be a one-to-one mapping and it does not have to be a mapping onto the entire space S.

If φ: $R \to S$ is some homomorphism and if A is a subspace of space R, then $\varphi(A)$ denotes the image of subspace A for mapping φ, that is, the set of all vectors of form $\varphi(\boldsymbol{x})$, where $\boldsymbol{x} \in A$. In particular, we can consider an image $\varphi(R)$ of the entire space R for homomorphism φ; it is called the *image of homomorphism* φ, denoted by the symbol $\operatorname{Im} \varphi$ (that is, $\operatorname{Im} \varphi = \varphi(R)$).

THEOREM 15.1. *The image $\varphi(A)$ of any subspace $A \subset R$ for a homomorphism $\varphi: R \to S$ is a subspace of vector space S.*

Proof. If $\boldsymbol{a}$, $\boldsymbol{b} \in \varphi(A)$, that is, if there exist elements $\boldsymbol{x}$, $\boldsymbol{y} \in A$ such that $\varphi(\boldsymbol{x}) = \boldsymbol{a}$, $\varphi(\boldsymbol{y}) = \boldsymbol{b}$, then

$$\boldsymbol{a} + \boldsymbol{b} = \varphi(\boldsymbol{x}) + \varphi(\boldsymbol{y}) = \varphi(\boldsymbol{x} + \boldsymbol{y}).$$

Since $\boldsymbol{x} + \boldsymbol{y} \in A$ (A being a subspace), it follows that $\boldsymbol{a} + \boldsymbol{b} \in \varphi(A)$. Hence, if $\boldsymbol{a}$, $\boldsymbol{b} \in \varphi(A)$, then $\boldsymbol{a} + \boldsymbol{b} \in \varphi(A)$.

If, in addition, $\boldsymbol{a} \in \varphi(A)$, that is, $\boldsymbol{a} = \varphi(\boldsymbol{x})$, where $\boldsymbol{x} \in A$, and if λ is a real number, then

$$\lambda \boldsymbol{a} = \lambda \varphi(\boldsymbol{x}) = \varphi(\lambda \boldsymbol{x}).$$

Since $\lambda \boldsymbol{x} \in A$ (A being a subspace), it follows that $\lambda \boldsymbol{a} \in \varphi(A)$. Thus, if $\boldsymbol{a} \in \varphi(A)$, we know that $\lambda \boldsymbol{a} \in \varphi(A)$ for any real λ. Consequently, $\varphi(A)$ is a subspace.

Again let φ: $R \to S$ be some homomorphism, and let B be a subspace of space S. The set of all elements $\boldsymbol{x} \in R$ for which $\varphi(\boldsymbol{x}) \in B$ is called the *inverse image* of subspace B for homomorphism φ. This inverse image is called $\varphi^{-1}(B)$. In particular, we can assume the inverse image $\varphi^{-1}(\mathbf{0})$ to be a trivial subspace; it is called the *kernel* of homomorphism φ, being denoted by the symbol $\operatorname{Ker} \varphi$. Thus $\operatorname{Ker} \varphi$ consists of all elements $\boldsymbol{x} \in R$ for which $\varphi(\boldsymbol{x}) = \mathbf{0}$.

THEOREM 15.2. *The inverse image $\varphi^{-1}(B)$ of any subspace $B \subset S$ for a homomorphism $\varphi: R \to S$ is a subspace of vector space R.*

Proof. If $\boldsymbol{a}$, $\boldsymbol{b} \in \varphi^{-1}(B)$, that is, if $\varphi(\boldsymbol{a}) \in B$, $\varphi(\boldsymbol{b}) \in B$, then since B is a subspace it follows that $\varphi(\boldsymbol{a}) + \varphi(\boldsymbol{b}) \in B$, that is, $\varphi(\boldsymbol{a} + \boldsymbol{b}) \in B$. Consequently, $\boldsymbol{a} + \boldsymbol{b} \in \varphi^{-1}(B)$, so that if $\boldsymbol{a}$, $\boldsymbol{b} \in \varphi^{-1}(B)$ then $\boldsymbol{a} + \boldsymbol{b} \in \varphi^{-1}(B)$.

If, in addition, $\boldsymbol{a} \in \varphi^{-1}(B)$, that is, $\varphi(\boldsymbol{a}) \in B$, and if λ is a real number, then since B is a subspace it follows that $\lambda\varphi(\boldsymbol{a}) \in B$, that is, $\varphi(\lambda\boldsymbol{a}) \in B$. Consequently, $\lambda\boldsymbol{a} \in \varphi^{-1}(B)$. Accordingly, if $\boldsymbol{a} \in \varphi^{-1}(B)$, this means that $\lambda\boldsymbol{a} \in \varphi^{-1}(B)$ for any real λ.

THEOREM 15.3. *For any homomorphism* φ: $R \to S$, *the following holds true:*

$$\dim R = \dim(\operatorname{Im}\varphi) + \dim(\operatorname{Ker}\varphi). \tag{15.1}$$

Proof. Let p be the dimension of subspace $\operatorname{Im}\varphi$ and let $\boldsymbol{c}_1, \ldots, \boldsymbol{c}_p$ be some basis of this subspace. Since $\boldsymbol{c}_1, \ldots, \boldsymbol{c}_p \in \operatorname{Im}\varphi$, there will exist elements $\boldsymbol{e}_1, \ldots, \boldsymbol{e}_p \in R$ such that

$$\varphi(\boldsymbol{e}_1) = \boldsymbol{c}_1, \ \ldots, \ \varphi(\boldsymbol{e}_p) = \boldsymbol{c}_p.$$

Vectors $\boldsymbol{e}_1, \ldots, \boldsymbol{e}_p$ are linearly independent, since the equation $\alpha_1\boldsymbol{e}_1 + \ldots + \alpha_p\boldsymbol{e}_p = \boldsymbol{0}$ implies that $\alpha_1\varphi_1(\boldsymbol{e}_1) + \ldots + \alpha_p\varphi(\boldsymbol{e}_p) = \boldsymbol{0}$, that is, $\alpha_1\boldsymbol{c}_1 + \ldots + \alpha_p\boldsymbol{c}_p = \boldsymbol{0}$, so that $\alpha_1 = \ldots = \alpha_p = 0$.

If L is a subspace of space R generated by vectors $\boldsymbol{e}_1, \ldots \boldsymbol{e}_p$, then it follows from Theorem 14.6 that subspace L consists of all vectors having the form $x^1\boldsymbol{e}_1 + \ldots + x^p\boldsymbol{e}_p$. It follows directly from the relation

$$\varphi(x^1\boldsymbol{e}_1 + \ldots + x^p\boldsymbol{e}_p) = x^1\boldsymbol{c}_1 + \ldots + x^p\boldsymbol{c}_p$$

that mapping φ, considered in L, will be an *isomorphism* of space L onto the subspace $\operatorname{Im}\varphi$.

Let $\boldsymbol{x} \in R$. Then $\varphi(\boldsymbol{x}) \in \operatorname{Im}\varphi$ and since $\varphi(L) = \operatorname{Im}\varphi$ there is some element $\boldsymbol{a} \in L$ for which $\varphi(\boldsymbol{a}) = \varphi(\boldsymbol{x})$. Therefore, $\varphi(\boldsymbol{x}) - \varphi(\boldsymbol{a}) = \boldsymbol{0}$, or $\varphi(\boldsymbol{x} - \boldsymbol{a}) = \boldsymbol{0}$, so that element $\boldsymbol{b} = \boldsymbol{x} - \boldsymbol{a}$ belongs to subspace $\operatorname{Ker}\varphi$. Accordingly,

$$\boldsymbol{x} = \boldsymbol{a} + \boldsymbol{b}; \qquad \boldsymbol{a} \in L, \quad \boldsymbol{b} \in \operatorname{Ker}\varphi, \tag{15.2}$$

that is, any vector $\boldsymbol{x} \in R$ can be represented in form (15.2).

Let us show that this representation is unique. Suppose that, in addition to (15.2), we have the representation

$$\boldsymbol{x} = \boldsymbol{a}' + \boldsymbol{b}'; \qquad \boldsymbol{a}' \in L, \quad \boldsymbol{b}' \in \operatorname{Ker}\varphi. \tag{15.3}$$

Then $\boldsymbol{a} + \boldsymbol{b} = \boldsymbol{a}' + \boldsymbol{b}'$, and thus $\boldsymbol{a} - \boldsymbol{a}' = \boldsymbol{b}' - \boldsymbol{b} \in \operatorname{Ker}\varphi$, so that $\varphi(\boldsymbol{a} - \boldsymbol{a}') = \boldsymbol{0}$. Consequently, $\varphi(\boldsymbol{a}) = \varphi(\boldsymbol{a}')$. However, since $\boldsymbol{a}, \boldsymbol{a}' \in L$, while mapping φ, considered in L, is an isomorphism, it follows that $\boldsymbol{a} = \boldsymbol{a}'$. Now it follows from the relation $\boldsymbol{a} + \boldsymbol{b} = \boldsymbol{a}' + \boldsymbol{b}'$ that $\boldsymbol{b} = \boldsymbol{b}'$. Thus representations (15.2) and (15.3) must be identical, proving the uniqueness of the representation.

Theorem 14.11 implies that $R = L \oplus \operatorname{Ker}\varphi$, so that

$$\dim R = \dim L + \dim(\operatorname{Ker}\varphi)$$

(Theorem 14.10). However, since space L and $\operatorname{Im}\varphi$ are isomorphic, we know from Theorem 13.12 that $\dim L = \dim(\operatorname{Im}\varphi)$.

THEOREM 15.4. *If R and S are vector spaces and if $\boldsymbol{e}_1, \ldots, \boldsymbol{e}_n$ form a basis of space R, then for any elements $\boldsymbol{a}_1, \ldots, \boldsymbol{a}_n$ of space S there will exist one and only one homomorphism φ: $R \to S$ satisfying the conditions*

$$\varphi(\boldsymbol{e}_1) = \boldsymbol{a}_1, \ \ldots, \ \varphi(\boldsymbol{e}_n) = \boldsymbol{a}_n. \tag{15.4}$$

Proof. Let $\boldsymbol{x}$ be an arbitrary element of space R and let $x^1, \ldots, x^n$ be its coordinates in basis $\boldsymbol{e}_1, \ldots, \boldsymbol{e}_n$:

$$\boldsymbol{x} = x^1\boldsymbol{e}_1 + \ldots + x^n\boldsymbol{e}_n. \tag{15.5}$$

If we set

$$\varphi(\boldsymbol{x}) = x^1\boldsymbol{a}_1 + \ldots + x^n\boldsymbol{a}_n, \tag{15.6}$$

then for any $\boldsymbol{x} \in R$ the element $\varphi(\boldsymbol{x}) \in S$ is defined. It can be proven directly that a mapping φ: $R \to S$ obtained in this way satisfies conditions (13.10), that is, it is a homomorphism. It is also clear that this homomorphism φ satisfies conditions (15.4). The existence of the required homomorphism is thereby proven.

Let us now prove its uniqueness. Let φ_1: $R \to S$ be some homomorphism possessing the required properties, that is,

$$\varphi_1(\boldsymbol{e}_1) = \boldsymbol{a}_1, \ \ldots, \ \varphi_1(\boldsymbol{e}_n) = \boldsymbol{a}_n.$$

Then for vector (15.5) we have

$$\begin{aligned}\varphi_1(\boldsymbol{x}) = \varphi_1(x^1\boldsymbol{e}_1 + \ldots + x^n\boldsymbol{e}_n) = x^1\varphi(\boldsymbol{e}_1) + \ldots + x^n\varphi(\boldsymbol{e}_n) = \\ = x^1\boldsymbol{a}_1 + \ldots + x^n\boldsymbol{a}_n = \varphi(\boldsymbol{x}).\end{aligned}$$

Consequently, $\varphi_1(\boldsymbol{x}) = \varphi(\boldsymbol{x})$ for any element $\boldsymbol{x} \in R$, that is, homomorphism φ_1 is identical to the previously constructed homomorphism φ.

NOTE 15.5. Let $\boldsymbol{f}_1, \ldots, \boldsymbol{f}_k$ be a basis of space S. We can write each of vectors $\boldsymbol{a}_1, \ldots, \boldsymbol{a}_n$ (see 15.4)) as a linear combination of the basis vectors:

$$\boldsymbol{a}_j = c_j^1\boldsymbol{f}_1 + \ldots + c_j^k\boldsymbol{f}_k, \qquad j = 1, \ldots, n. \tag{15.7}$$

Then relation (15.6) can be written as

$$\varphi(\boldsymbol{x}) = \sum_{j=1}^{n} x^j\boldsymbol{a}_j = \sum_{j=1}^{n}\sum_{i=1}^{k} x^j c_j^i \boldsymbol{f}_i.$$

In other words,

$$\varphi(\boldsymbol{x}) = y^1\boldsymbol{f}_1 + \ldots + y^k\boldsymbol{f}_k, \tag{15.8}$$

where

$$y^i = \sum_{j=1}^{n} c^i_j x^j, \qquad i = 1, \ldots, k. \tag{15.9}$$

Thus, a homomorphism $\varphi : R \to S$ satisfying conditions (15.4) can be described as follows: it transforms a vector $\boldsymbol{x}$ having coordinates $x^1, \ldots, x^n$ (see (15.5)) in basis $\boldsymbol{e}_1, \ldots, \boldsymbol{e}_n$ into a vector having coordinates $y^1, \ldots, y^k$ (see 15.8)) in basis $\boldsymbol{f}_1, \ldots, \boldsymbol{f}_k$, the latter coordinates being defined by formulas (15.9): the numbers c^i_j are in this case defined by equations (15.7).

Note that specifying the numbers c^i_j $(i = 1, \ldots, k;\ j = 1, \ldots, n)$, that is, specifying the *matrices* $C = (c^i_j)$, is equivalent to specifying the vectors $\boldsymbol{a}_1, \ldots, \boldsymbol{a}_n$ (see (15.7)). Thus the following statement is valid:

Let R be a vector space with a basis $\boldsymbol{e}_1, \ldots, \boldsymbol{e}_n$ and let S be a vector space with a basis $\boldsymbol{f}_1, \ldots, \boldsymbol{f}_k$. Any homomorphism $\varphi : R \to S$ is determined uniquely by formulas (15.5), (15.8), (15.9) *by specifying a $k \times n$ matrix $C = (c^i_j)$, that is, homomorphism φ transforms an element $\boldsymbol{x} \in R$ with coordinates $x^1, \ldots, x^n$ into an element with coordinates $y^1, \ldots, y^k$ determined by formulas* (15.9). *A one-to-one relationship is thereby established between the homomorphisms $R \to S$ and the $k \times n$ matrices $C = (c^i_j)$.*

DEFINITION 15.6. Let $R = A \oplus B$. According to Theorem 14.11, each vector $\boldsymbol{x} \in R$ is represented uniquely in the form

$$\boldsymbol{x} = \boldsymbol{a} + \boldsymbol{b}; \qquad \boldsymbol{a} \in A, \quad \boldsymbol{b} \in B.$$

Vector $\boldsymbol{a}$ is called the *projection* of vector $\boldsymbol{x}$ onto subspace A in the direction of subspace B.

THEOREM 15.7. *Let $R = A \oplus B$. For each vector $\boldsymbol{x} \in R$ we let $\pi(\boldsymbol{x})$ be the projection of vector $\boldsymbol{x}$ onto subspace A in the direction of subspace B. The obtained mapping π of space R into itself (known as the* **projection** *onto subspace A in the direction of subspace B) is a homomorphism. Here* $\operatorname{Im} \pi = A$ *and* $\operatorname{Ker} \pi = B$.

Proof. Let $\boldsymbol{x}, \boldsymbol{y} \in R$. If vectors $\boldsymbol{x}, \boldsymbol{y}$ are represented as

$$\boldsymbol{x} = \boldsymbol{a}_1 + \boldsymbol{b}_1, \quad \boldsymbol{y} = \boldsymbol{a}_2 + \boldsymbol{b}_2; \qquad \boldsymbol{a}_1, \boldsymbol{a}_2 \in A, \quad \boldsymbol{b}_1, \boldsymbol{b}_2 \in B,$$

then $\boldsymbol{a}_1 = \pi(\boldsymbol{x})$ and $\boldsymbol{a}_2 = \pi(\boldsymbol{y})$. Thus we have

$$\boldsymbol{x} + \boldsymbol{y} = (\boldsymbol{a}_1 + \boldsymbol{b}_1) + (\boldsymbol{a}_2 + \boldsymbol{b}_2) = (\boldsymbol{a}_1 + \boldsymbol{a}_2) + (\boldsymbol{b}_1 + \boldsymbol{b}_2).$$

Since $\boldsymbol{a}_1 + \boldsymbol{a}_2 \in A$ and $\boldsymbol{b}_1 + \boldsymbol{b}_2 \in B$, it follows that $\boldsymbol{a}_1 + \boldsymbol{a}_2 = \pi(\boldsymbol{x} + \boldsymbol{y})$, that is, $\pi(\boldsymbol{x} + \boldsymbol{y}) = \pi(\boldsymbol{x}) + \pi(\boldsymbol{y})$. Similarly, the relation $\pi(\lambda\boldsymbol{x}) = \lambda\pi(\boldsymbol{x})$ is established. Consequently, π is a homomorphism.

Obviously, $\operatorname{Im}\pi \subset A$. However, for any vector $\boldsymbol{a} \in A$ we have $\pi(\boldsymbol{a}) = \boldsymbol{a}$ (since $\boldsymbol{a} = \boldsymbol{a} + \mathbf{0}$; $\boldsymbol{a} \in A$, $\mathbf{0} \in B$), so that $\operatorname{Im}\pi \supset A$. Thus $\operatorname{Im}\pi = A$.

Finally, the inclusion $\boldsymbol{x} \in \operatorname{Ker}\pi$ (that is, $\pi(\boldsymbol{x}) = \mathbf{0}$) is valid if, and only if, $\boldsymbol{x} = \mathbf{0} + \boldsymbol{b}$, where $\boldsymbol{b} \in B$, that is, if $\boldsymbol{x} \in B$. In other words, $\operatorname{Ker}\pi = B$.

Now let us define some operations with homomorphisms. If φ_1 and φ_2 are two homomorphisms of vector space R into vector space S, then for any $\boldsymbol{x} \in R$ the elements $\varphi_1(\boldsymbol{x})$ and $\varphi_2(\boldsymbol{x})$ are vectors of space S, and we can consider their sum $\varphi_1(\boldsymbol{x}) + \varphi_2(\boldsymbol{x})$. The mapping $\psi : R \to S$ is defined by the formula

$$\psi(\boldsymbol{x}) = \varphi_1(\boldsymbol{x}) + \varphi_2(\boldsymbol{x}), \qquad \boldsymbol{x} \in R.$$

It can be shown directly that mapping ψ is a *homomorphism* of vector space R into vector space S. This homomorphism is called the *sum* of homomorphisms φ_1 and φ_2:

$$\psi = \varphi_1 + \varphi_2.$$

Consequently, by definition, $(\varphi_1 + \varphi_2)(\boldsymbol{x}) = \varphi_1(\boldsymbol{x}) + \varphi_2(\boldsymbol{x})$ for any $\boldsymbol{x} \in R$.

In addition, let $\varphi : R \to S$ be some homomorphism and let λ be a real number. The mapping $\chi : R \to S$ is defined by the formula

$$\chi(\boldsymbol{x}) = \lambda\varphi(\boldsymbol{x}), \qquad \boldsymbol{x} \in R.$$

It is easy to show that mapping χ is a *homomorphism* of space R into S. This homomorphism is known as the *product* of φ times the number λ:

$$\chi = \lambda\varphi.$$

Thus, by definition, $(\lambda\varphi)(\boldsymbol{x}) = \lambda(\varphi(\boldsymbol{x}))$ for any $\boldsymbol{x} \in R$.

Finally, let $\varphi : R \to S$ be a homomorphism of space R into space S, while $\psi : S \to T$ is a homomorphism of space S into space T. We define the mapping $\zeta : R \to T$ as

$$\zeta(\boldsymbol{x}) = \psi(\varphi(\boldsymbol{x})), \qquad \boldsymbol{x} \in R.$$

It can be shown that mapping ζ is a *homomorphism* of space R into T. This homomorphism is called the *composition* of homomorphisms ψ and φ, being designated as $\psi \circ \varphi$:

$$\zeta = \psi \circ \varphi.$$

Consequently, by definition, $(\psi \circ \varphi)(\boldsymbol{x}) = \psi(\varphi(\boldsymbol{x}))$ for any $\boldsymbol{x} \in R$.

THEOREM 15.8. *Let R and S be vector spaces. The set of all homomorphisms $R \to S$ with the above-defined operations of addition and multiplication by real numbers is a vector space of dimension* $\dim R \cdot \dim S$. *Let us call this space* $\operatorname{Hom}(R, S)$.

Proof. If $\varphi_1, \varphi_2 \in \operatorname{Hom}(R, S)$, that is, if φ_1 and φ_2 are certain homomorphisms $R \to S$, then we wish to show that $\varphi_1 + \varphi_2 = \varphi_2 + \varphi_1$. We can write (for any $\boldsymbol{x} \in R$):

$$(\varphi_1 + \varphi_2)(\boldsymbol{x}) = \varphi_1(\boldsymbol{x}) + \varphi_2(\boldsymbol{x}), \quad (\varphi_2 + \varphi_1)(\boldsymbol{x}) = \varphi_2(\boldsymbol{x}) + \varphi_1(\boldsymbol{x}).$$

Since $\varphi_1(\boldsymbol{x}) + \varphi_2(\boldsymbol{x}) = \varphi_2(\boldsymbol{x}) + \varphi_1(\boldsymbol{x})$ (addition in vector space S being commutative), it follows that

$$(\varphi_1 + \varphi_2)(\boldsymbol{x}) = (\varphi_2 + \varphi_1)(\boldsymbol{x}) \quad \text{for any } \boldsymbol{x} \in R.$$

This also indicates that homomorphisms $\varphi_1 + \varphi_2$ and $\varphi_2 + \varphi_1$ are identical, that is $\varphi_1 + \varphi_2 = \varphi_2 + \varphi_1$. Thus we see that Axiom I_1 is obeyed in the set $\mathrm{Hom}(R, S)$. Similarly, all the other axioms of Groups I and II can be verified; we note only that the zero element in $\mathrm{Hom}(R, S)$ is a homomorphism transforming the entire space R into the zero element $\mathbf{0} \in S$, while the element $-\varphi$ (for $\varphi \in \mathrm{Hom}(R, S)$) is defined by the equation $(-\varphi)(\boldsymbol{x}) = -(\varphi(\boldsymbol{x}))$ for any $\boldsymbol{x} \in R$ (that is, $-\varphi = (-1)\varphi$). Hence $\mathrm{Hom}(R, S)$ is a vector space.

If n is the dimension of space R and k is the dimension of space S, let us show that the vector space $\mathrm{Hom}(R, S)$ is *isomorphic* to the vector space of all the $k \times n$ matrices $C = (c^i_j)$ with real elements. To do this, we specify in R some basis $\boldsymbol{e}_1, \ldots, \boldsymbol{e}_n$ and in S some basis $\boldsymbol{f}_1, \ldots, \boldsymbol{f}_k$. Each homomorphism φ: $R \to S$ makes it possible to define in space S the vectors $\boldsymbol{a}_1 = \varphi(\boldsymbol{e}_1), \ldots, \boldsymbol{a}_n = \varphi(\boldsymbol{e}_n)$; moreover, by representing these vectors as a linear combination of the vectors of basis $\boldsymbol{f}_1, \ldots, \boldsymbol{f}_k$ (see (15.7)), we can define the numbers c^i_j, that is, we can associate some $k \times n$ matrix $C = (c^i_j)$ with homomorphism φ. According to Note 15.5, each $k \times n$ matrix C is in turn uniquely defined by homomorphism φ, that is, the correspondence established between homomorphisms φ: $R \to S$ and $k \times n$ matrices $C = (c^i_j)$ is one-to-one. It follows that the sum $\varphi' + \varphi''$ of two homomorphisms $R \to S$ conforms to the sum $C' + C'' = (c'^i_j + c''^i_j)$ of the corresponding matrices, while the product $\lambda\varphi$ of homomorphism φ: $R \to S$ times a number λ conforms to the product $\lambda C = (\lambda c^i_j)$ of the corresponding matrix $C = (c^i_j)$ times λ. This also means that the space $\mathrm{Hom}(R, S)$ is *isomorphic* to the vector space of all the $k \times n$ matrices with real elements.

Since the considered spaces are isomorphic, their dimensions are the same (see Theorem 13.12). However, the dimension of the vector space of all the $k \times n$ matrices (with real elements) is kn, so that

$$\dim(\mathrm{Hom}(R, S)) = nk = \dim R \cdot \dim S.$$

THEOREM 15.9. *Let φ: $R \to S$ and ψ: $S \to T$ be homomorphisms of vector spaces. We specify in space R some basis $\boldsymbol{e}_1, \ldots, \boldsymbol{e}_n$, in space S some basis $\boldsymbol{f}_1, \ldots, \boldsymbol{f}_k$, and in space T some basis $\boldsymbol{g}_1, \ldots, \boldsymbol{g}_l$. If in these bases the $k \times n$ matrix C corresponds to homomorphism φ, while the $l \times k$ matrix D corresponds to homomorphism ψ, then the $l \times n$ matrix DC corresponds (in these same bases) to the homomorphism $\psi \circ \varphi$ of space R into T.*

Proof. The elements c_j^i of matrix C are determined from relations (15.4) and (15.7), while the elements of matrix D are determined from the analogous relations

$$\psi(f_j) = b_j = d_j^1 g_1 + \ldots + d_j^l g_l, \qquad j = 1, \ldots, k.$$

These give

$$(\psi \circ \varphi)(e_j) = \psi(\varphi(e_j)) = \psi(a_j) = \psi\left(\sum_{i=1}^{k} c_j^i f_i\right) = \sum_{i=1}^{k} c_j^i \psi(f_i) =$$

$$= \sum_{i=1}^{k} c_j^i b_i = \sum_{i=1}^{k}\left(c_j^i \sum_{\alpha=1}^{l} d_i^\alpha g_\alpha\right) = \sum_{i=1}^{k}\sum_{\alpha=1}^{l} c_j^i d_i^\alpha g_\alpha = \sum_{\alpha=1}^{l}\left(\sum_{i=1}^{k} d_i^\alpha c_j^i\right) g_\alpha.$$

Consequently,

$$(\psi \circ \varphi)(e_j) = \sum_{\alpha=1}^{l} q_j^\alpha g_\alpha,$$

where the elements of the $l \times n$ matrix $Q = (q_j^\alpha)$, that is, of the matrix corresponding to homomorphism $\psi \circ \varphi$, are given by the formula

$$q_j^\alpha = \sum_{i=1}^{k} d_i^\alpha c_j^i.$$

However, this also means that $Q = DC$.

THEOREM 15.10. *Let R, S, T be vector spaces and let ζ: $R \to S$ be some homomorphism. We consider the vector space* $\mathrm{Hom}(R, T)$ *and* $\mathrm{Hom}(S, T)$, *and for an arbitrary element* $\varphi \in \mathrm{Hom}(S, T)$ *(that is, a homomorphism $S \to T$) we set* $\zeta^*(\varphi) = \varphi \circ \zeta$ *(so that $\zeta^*(\varphi)$ is a homomorphism $R \to T$, that is, an element of the space* $\mathrm{Hom}(R, T)$*). The mapping* ζ^*: $\mathrm{Hom}(S, T) \to \mathrm{Hom}(R, T)$ *thus obtained is a homomorphism.*

The *proof* is obtained directly, just like the proof of the following assumption:

THEOREM 15.11. *Let R, S, T be vector spaces and let χ: $S \to T$ be some homomorphism. We consider the vector spaces* $\mathrm{Hom}(R, S)$ *and* $\mathrm{Hom}(R, T)$, *and for an arbitrary element* $\varphi \in \mathrm{Hom}(R, S)$ *(that is, a homomorphism $R \to S$) we set* $\chi_*(\varphi) = \chi \circ \varphi$ *(so that $\chi_*(\varphi)$ is a homomorphism $R \to T$, that is, an element of the space* $\mathrm{Hom}(R, T)$. *The mapping* χ_*: $\mathrm{Hom}(R, S) \to \mathrm{Hom}(R, T)$ *thus obtained is a homomorphism.*

In conclusion, let us introduce the concepts of a *conjugate space* and a *conjugate homomorphism.* Let R be a vector space, and let D denote a number line, which can be thought of as a *one-dimensional* vector space (coinciding with the one-dimensional *arithmetical* vector space R^1). Each homomorphism f: $R \to D$ is called a *linear functional* in space R. Thus the linear functional f: $R \to D$ juxtaposes some real number $f(x)$ with each element $x \in R$, so that

$$f(x_1 + x_2) = f(x_1) + f(x_2), \quad f(\lambda x) = \lambda f(x).$$

The set of all linear functionals in space R, that is, $\mathrm{Hom}(R, D)$, is, in view of Theorem 15.8, a vector space of dimension $\dim R$. This space is called the *conjugate* of vector space R, being denoted as R^*. Consequently,

$$R^* = \mathrm{Hom}(R, D), \quad \dim R^* = \dim R.$$

Now, if R, S are vector spaces and if $\zeta\colon R \to S$ is some homomorphism, then according to Theorem 15.10 a homomorphism $\zeta^*\colon \mathrm{Hom}(S, D) \to \mathrm{Hom}(R, D)$ is defined, that is, a homomorphism $\zeta^*\colon S^* \to R^*$. This homomorphism ζ^* is called the *conjugate* of the initial homomorphism ζ. According to Theorem 15.10, the conjugate homomorphism ζ^* is constructed as follows: to each linear functional $f\colon S \to D$ (that is, to an element of space S^*) it juxtaposes a linear functional $\zeta^*(f) = f \circ \zeta$ defined in space R (that is, constituting an element of space R^*).

In addition, let us mention another means of writing linear functionals, one which possesses certain advantages. Let R be a vector space, and let R^* be the conjugate to this space, with $\boldsymbol{x} \in R$, $f \in R^*$. Since linear functional f is a homomorphism $f\colon R \to D$, the value of $f(\boldsymbol{x}) \in D$ is defined, it being a real number. Let us agree to write this number as $(\boldsymbol{x}, f)$ and, using this notation, to call it the *scalar product* of element $\boldsymbol{x} \in R$ times the linear functional $f \in R^*$:

$$(\boldsymbol{x}, f) = f(\boldsymbol{x}).$$

The following properties of the scalar product follow directly from its definition:

$$\left.\begin{aligned} &(\boldsymbol{x}_1 + \boldsymbol{x}_2, f) = (\boldsymbol{x}_1, f) + (\boldsymbol{x}_2, f); \\ &(\boldsymbol{x}, f_1 + f_2) = (\boldsymbol{x}, f_1) + (\boldsymbol{x}, f_2); \\ &(\lambda \boldsymbol{x}, f) = \lambda(\boldsymbol{x}, f); \quad (\boldsymbol{x}, \lambda f) = \lambda(\boldsymbol{x}, f). \end{aligned}\right\} \tag{15.10}$$

Finally, note the following statement, which also follows directly from the definition of the scalar product and the conjugate homomorphism:

If R, S are vector spaces, $\zeta\colon R \to S$ is some homomorphism, and $\zeta^*\colon S^* \to R^*$ is its conjugate homomorphism, then for any elements $\boldsymbol{x} \in R$, $f \in S^*$ the following relation is valid:

$$(\zeta(\boldsymbol{x}), f) = (\boldsymbol{x}, \zeta^*(f)). \tag{15.11}$$

THEOREM 15.12. *If R is a vector space, R^* is the conjugate to this space, and $f_1, \ldots, f_n$ form an arbitrary basis of space R^*, then there exists a basis $\boldsymbol{e}_1, \ldots, \boldsymbol{e}_n$ of space R such that*

$$f_i(\boldsymbol{e}_j) = \begin{cases} 0 & \text{for} \quad i \neq j; \\ 1 & \text{for} \quad i = j. \end{cases} \tag{15.12}$$

With this choice of bases, for any elements

$$\boldsymbol{x} = x^1\boldsymbol{e}_1 + \ldots + x^n\boldsymbol{e}_n \in R, \quad f = y^1 f_1 + \ldots + y^n f_n \in R^* \tag{15.13}$$

the following relation is valid:

$$(\boldsymbol{x}, f) = f(\boldsymbol{x}) = x^1 y^1 + \ldots + x^n y^n. \tag{15.14}$$

Proof. Let us define a mapping φ: $R \to R^n$ of vector space R into an arithmetical vector space R^n, assuming that for any vector $\boldsymbol{a} \in R$:

$$\varphi(\boldsymbol{a}) = \{f_1(\boldsymbol{a}), \ldots, f_n(\boldsymbol{a})\}.$$

It is easy to show that mapping φ is a homomorphism.

Let us show that $\mathrm{Ker}\,\varphi$, the kernel of this homomorphism, is a trivial subspace. Let $\boldsymbol{a}_1$ be a nonzero vector of space R. We augment vector $\boldsymbol{a}_1$ with vectors $\boldsymbol{a}_2, \ldots, \boldsymbol{a}_n$ until some basis $\boldsymbol{a}_1, \boldsymbol{a}_2, \ldots, \boldsymbol{a}_n$ of space R is reached, and for any vector

$$\boldsymbol{x} = \lambda^1 \boldsymbol{a}_1 + \lambda^2 \boldsymbol{a}_2 + \ldots + \lambda^n \boldsymbol{a}_n \in R$$

we set $f(\boldsymbol{x}) = \lambda^1$. It is easy to show that the mapping f: $R \to D$ so obtained is a linear functional, with $f(\boldsymbol{a}_1) = 1$. If f is expressed as a linear combination of the vectors of basis $f_1, \ldots, f_n$:

$$f = \mu^1 f_1 + \ldots + \mu^n f_n,$$

then it follows that

$$\mu^1 f_1(\boldsymbol{a}_1) + \ldots + \mu^n f_n(\boldsymbol{a}_1) = f(\boldsymbol{a}_1) = 1.$$

Consequently, at least one of the numbers $f_1(\boldsymbol{a}_1), \ldots, f_n(\boldsymbol{a}_1)$ is different from zero, so that $\varphi(\boldsymbol{a}_1) \neq \boldsymbol{0}$. Accordingly, no nonzero vector $\boldsymbol{a}_1 \in R$ belongs to $\mathrm{Ker}\,\varphi$, that is, $\mathrm{Ker}\,\varphi$ is a trivial subspace.

Now it follows from Theorem 15.3 that $\dim(\mathrm{Im}\,\varphi) = \dim R = n$, that is, $\mathrm{Im}\,\varphi = R^n$. Therefore, φ is an *isomorphism* of space R onto space R^n.

Let $\boldsymbol{e}_1, \boldsymbol{e}_2, \ldots, \boldsymbol{e}_n$ be vectors of space R satisfying the relations

$$\begin{aligned}
\varphi(\boldsymbol{e}_1) &= \{1, 0\ 0, \ldots, 0, 0\},\\
\varphi(\boldsymbol{e}_2) &= \{0, 1, 0, \ldots, 0, 0\},\\
&\cdots\cdots\cdots\cdots\cdots\\
\varphi(\boldsymbol{e}_n) &= \{0, 0, 0, \ldots, 0, 1\}.
\end{aligned}$$

Vectors $\boldsymbol{e}_1, \boldsymbol{e}_2, \ldots, \boldsymbol{e}_n$ constitute a basis of vector space R. The relation

$$\{f_1(\boldsymbol{e}_i), f_2(\boldsymbol{e}_i), \ldots, f_n(\boldsymbol{e}_i)\} = \varphi(\boldsymbol{e}_i) = \{0, 0, \ldots, 1, \ldots, 0\}$$

(unity being in the ith place) also indicates that relations (15.12) are valid. Equation (15.14), moreover, follows directly from relations (15.12) and (15.13).

16. Euclidean vector space. DEFINITION 16.1. A vector space R is called a *Euclidean vector space* if in it we can define an operation of *scalar multiplication* of vectors such that a real number $\boldsymbol{ab}$, called the *scalar product* of vectors $\boldsymbol{a}$ and $\boldsymbol{b}$ and obeying the following axioms, is associated with every two vectors $\boldsymbol{a}, \boldsymbol{b} \in R$.

GROUP IV: *axioms of scalar product.*

IV$_1$. *Scalar multiplication is commutative, that is, for any two vectors* $\boldsymbol{a} \in R$, $\boldsymbol{b} \in R$ *the following relation is valid:*

$$\boldsymbol{ab} = \boldsymbol{ba}.$$

IV$_2$. *Scalar multiplication is distributive, that is, for any three vectors* $\boldsymbol{a} \in R$, $\boldsymbol{b} \in R$, $\boldsymbol{c} \in R$ *the following relation is valid:*

$$\boldsymbol{a}(\boldsymbol{b} + \boldsymbol{c}) = \boldsymbol{ab} + \boldsymbol{ac}.$$

IV$_3$. *For any vectors* $\boldsymbol{a} \in R$, $\boldsymbol{b} \in R$ *and any real number* λ, *the following relation is valid:*

$$(\lambda \boldsymbol{a})\boldsymbol{b} = \lambda(\boldsymbol{ab}).$$

IV$_4$. *For any nonzero vector* $\boldsymbol{a} \in R$, *the scalar product* $\boldsymbol{aa}$ *of this vector times itself (known as the* **scalar square** *of vector* $\boldsymbol{a}$ *and written* $\boldsymbol{a}^2$ *) is positive:*

$$\boldsymbol{a}^2 > 0 \quad \text{for} \quad \boldsymbol{a} \neq 0.$$

A set R satisfying all the axioms of Groups I, II, III, and IV is known as an *n-dimensional Euclidean vector space.*

EXAMPLE 16.2. Let us again consider the vector space Π of Example 12.1 and let us introduce into it, as follows, the scalar multiplication of vectors. Let A and B be two arbitrary vectors (that is, two points on plane Π; instead of points, we can consider the corresponding radius vectors $\overrightarrow{OA}$ and $\overrightarrow{OB}$). If at least one of these vectors is zero (that is, coincides with the point O), their scalar product $A \cdot B$ can also be taken to be zero (that is, $O \cdot O = O \cdot B = A \cdot O = 0$). If, on the other hand, vectors A and B are both nonzero (that is, neither of points A and B coincides with O), then the scalar product of these vectors is given by the formula

$$A \cdot B = (\text{length}\, OA)(\text{length}\, OB)\cos(\angle AOB)$$

(Figure 45); the lengths OA and OB in this formula are the ordinary *lengths* of segments OA and OB. It is known from elementary (or analytic) geometry that scalar multiplication defined in this way satisfies all of axioms IV$_1$–IV$_4$, that is, the indicated introduction of scalar multiplication turns Π into a (two-dimensional) *Euclidean vector space.* A similar introduction of the scalar product turns the three-dimensional space Σ of Example 12.1 into a three-dimensional Euclidean vector space.

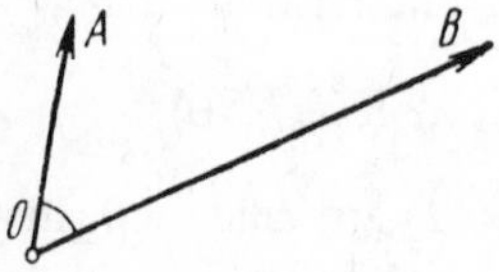

FIGURE 45

EXAMPLE 16.3. Let us consider an n-dimensional arithmetical vector space (see Example 12.2) and let us introduce scalar multiplication in it as follows: for any two vectors

$$\boldsymbol{x} = \{x^1, \ldots, x^n\}, \quad \boldsymbol{y} = \{y^1, \ldots, y^n\}$$

we set

$$\boldsymbol{xy} = x^1y^1 + \ldots + x^ny^n.$$

It can be shown that the scalar product so introduced obeys all of Axioms IV_1–IV_4. Consequently, space R^n with the specified scalar product is an n-dimensional Euclidean vector space.

Now let us derive some corollaries of the axioms of Group IV. First of all, we note that, in view of Axiom IV_1, the following relations are implied by Axioms IV_2 and IV_3:

$$(\boldsymbol{a}+\boldsymbol{b})\boldsymbol{c} = \boldsymbol{ac}+\boldsymbol{bc}, \quad \boldsymbol{a}(\lambda\boldsymbol{b}) = \lambda(\boldsymbol{ab}).$$

From these relations and Axioms IV_2 and IV_3, we get the relation

$$\left(\sum_{i=1}^{p}\alpha_i\boldsymbol{a}_i\right)\left(\sum_{j=1}^{q}\beta_j\boldsymbol{b}_j\right) = \sum_{i=1}^{p}\sum_{j=1}^{q}\alpha_i\beta_j(\boldsymbol{a}_i\boldsymbol{b}_j), \tag{16.1}$$

showing that the ordinary "rule for multiplication of polynomials" can be used for the scalar multiplication of two linear combinations of vectors. Thus (taking into account commutativity Axiom IV_1) the following formulas are valid:

$$(\boldsymbol{a}\pm\boldsymbol{b})^2 = \boldsymbol{a}^2 \pm 2\boldsymbol{ab}+\boldsymbol{b}^2, \quad (\boldsymbol{a}+\boldsymbol{b})(\boldsymbol{a}-\boldsymbol{b}) = \boldsymbol{a}^2-\boldsymbol{b}^2$$

as well as the rule for squaring a polynomial:

$$\left(\sum_{i=1}^{p}\alpha_i\boldsymbol{a}_i\right)^2 = \sum_{i=1}^{p}\alpha_i^2\boldsymbol{a}_i^2 + 2\sum_{i<j}\alpha_i\alpha_j(\boldsymbol{a}_i\boldsymbol{a}_j).$$

It should also be noted that the relations

$$\boldsymbol{0a} = 0, \quad \boldsymbol{a0} = 0$$

(implied, for example by Axiom IV_3) are valid for any vector $\boldsymbol{a}$.

DEFINITION 16.4. The number $\sqrt{\boldsymbol{a}^2}$ is called the *length* of a vector $\boldsymbol{a}$ (belonging to a Euclidean vector space). The length of $\boldsymbol{a}$ is denoted by $|\boldsymbol{a}|$ or else by the same letter as the vector but not written boldface:

$$|\boldsymbol{a}| = a = \sqrt{\boldsymbol{a}^2}.$$

This definition implies that $\boldsymbol{a}^2 = a^2$ (for any vector $\boldsymbol{a}$).

THEOREM 16.5. *For any vectors* $\boldsymbol{a}$, $\boldsymbol{b}$ *of Euclidean vector space* R, *the following relations are valid:*

$$|\boldsymbol{ab}| \leqslant ab, \tag{16.2}$$

$$|\lambda\boldsymbol{a}| = |\lambda|a,$$

$$|\boldsymbol{a}+\boldsymbol{b}| \leqslant |\boldsymbol{a}|+|\boldsymbol{b}| \tag{16.3}$$

(where λ is an arbitrary real number).

Proof. Consider the vector $x\boldsymbol{a} + \boldsymbol{b}$, where x is a real number. According to Axiom IV_4 (and the relation $\boldsymbol{0}^2 = 0$), the scalar square of this vector is a nonnegative number:

$$(x\boldsymbol{a} + \boldsymbol{b})^2 \geqslant 0.$$

Multiplying this out (and taking into account that $\boldsymbol{a}^2 = a^2$ and $\boldsymbol{b}^2 = b^2$), we obtain

$$a^2x^2 + 2(\boldsymbol{ab})\,x + b^2 \geqslant 0.$$

This inequality is valid for any real number x, that is, the quadratic trinomial

$$a^2x^2 + 2(\boldsymbol{ab})\,x + b^2$$

assumes (for all x) nonnegative values only. Thus it follows that the discriminant of this quadratic trinomial is nonpositive:

$$(2\boldsymbol{ab})^2 - 4a^2b^2 \leqslant 0.$$

Consequently,

$$(\boldsymbol{ab})^2 \leqslant a^2b^2, \quad \text{or} \quad |\,\boldsymbol{ab}\,| \leqslant ab.$$

Moreover, we have

$$|\,\lambda\boldsymbol{a}\,| = \sqrt{(\lambda\boldsymbol{a})^2} = \sqrt{\lambda^2\boldsymbol{a}^2} = \sqrt{\lambda^2a^2} = |\,\lambda a\,| = |\,\lambda\,|\,a$$

(since $a \geqslant 0$).

Finally, using inequality (16.2) proven above, we obtain

$$|\,\boldsymbol{a} + \boldsymbol{b}\,| = \sqrt{(\boldsymbol{a} + \boldsymbol{b})^2} = \sqrt{a^2 + 2\boldsymbol{ab} + b^2} \leqslant \sqrt{a^2 + 2ab + b^2} =$$
$$= \sqrt{(a + b)^2} = a + b = |\,\boldsymbol{a}\,| + |\,\boldsymbol{b}\,|.$$

NOTE 16.6. Assume that vectors $\boldsymbol{a}$, $\boldsymbol{b}$ are both different from the zero vector (so that $a > 0$, $b > 0$). Then relation (16.2) can be rewritten as

$$-ab \leqslant \boldsymbol{ab} \leqslant ab, \quad \text{or} \quad -1 \leqslant \frac{\boldsymbol{ab}}{ab} \leqslant 1.$$

It follows from these inequalities that there exists an angle φ (satisfying the inequalities $0 \leqslant \varphi \leqslant \pi$) for which

$$\cos\varphi = \frac{\boldsymbol{ab}}{ab}, \tag{16.4}$$

or

$$\boldsymbol{ab} = ab\cos\varphi. \tag{16.5}$$

By analogy with the usual definition of the scalar product (known from analytic or elementary geometry, cf. Example 16.2), the angle φ defined by equation (16.4) (that is, satisfying relation (16.5)) can be called the *angle between vectors* $\boldsymbol{a}$ and $\boldsymbol{b}$ in the Euclidean vector space. However, as a rule, when considering a Euclidean vector space, the angles between the vectors are not calculated (according to (16.4)), it being preferable to use the scalar products directly. Nevertheless, for reasons of clarity (and by custom), it is often said that "vectors $\boldsymbol{a}$ and $\boldsymbol{b}$ form an acute angle" (or a right angle, or an obtuse angle), depending on whether the value of $\cos\varphi$ calculated using (16.4) is positive (or negative or zero). Hence,

vectors $\boldsymbol{a}$, $\boldsymbol{b}$ *form an acute angle if* $\boldsymbol{ab} > 0$;

vectors $\boldsymbol{a}$, $\boldsymbol{b}$ *form an obtuse angle if* $\boldsymbol{ab} < 0$;

vectors $\boldsymbol{a}$, $\boldsymbol{b}$ *form a right angle if they are both different from the zero vector and if* $\boldsymbol{ab} = 0$.

In the latter case the term "orthogonal" vectors is most often used. Thus, two vectors $\boldsymbol{a}$, $\boldsymbol{b}$ are called *orthogonal* if $\boldsymbol{ab} = 0$ (it not being significant whether the vectors are different from zero). In other words, if vectors $\boldsymbol{a}$ and $\boldsymbol{b}$ are orthogonal, then either they are different from the zero vector and form a right angle, or else at least one of them is zero (so that the angle between them is not defined). This terminology will also be used in the following.

A system of vectors $\boldsymbol{e}_1, \ldots, \boldsymbol{e}_k$ in a Euclidean vector space R is called *orthonormal* if vectors $\boldsymbol{e}_1, \ldots, \boldsymbol{e}_k$ are pairwise orthogonal to each other and if the length of each vector is unity. Thus the condition for orthonormalization of the system of vectors $\boldsymbol{e}_1, \ldots, \boldsymbol{e}_k$ can be written as

$$\boldsymbol{e}_i\boldsymbol{e}_j = \begin{cases} 0 & \text{for} \quad i \neq j; \\ 1 & \text{for} \quad i = j. \end{cases} \tag{16.6}$$

THEOREM 16.7. *Any orthonormal system of vectors is linearly independent.*

Proof. Let $\boldsymbol{e}_1, \ldots, \boldsymbol{e}_k$ be an orthonormal system of vectors. We assume that

$$\alpha^1\boldsymbol{e}_1 + \ldots + \alpha^k\boldsymbol{e}_k = \boldsymbol{0}. \tag{16.7}$$

If we now multiply both sides of this equation (scalar multiplication) by vector $\boldsymbol{e}_i$ (where i is one of the numbers $1, \ldots, k$), then in view of (16.6) we obtain: $\alpha^i = 0$. Since this is valid for any $i = 1, \ldots, k$, all the coefficients on the left-hand side of (16.7) go to zero. Consequently, vectors $\boldsymbol{e}_1, \ldots, \boldsymbol{e}_k$ are linearly independent.

THEOREM 16.8. *Any orthonormal system of vectors* $\boldsymbol{e}_1, \ldots, \boldsymbol{e}_k$ *in an* n*-dimensional Euclidean vector space can (for* $k < n$*) be augmented until an orthonormal basis is reached. In particular, this implies that in any* n*-dimensional Euclidean vector space there exists an orthonormal basis* $\boldsymbol{e}_1, \ldots, \boldsymbol{e}_n$.

Proof. Let $\boldsymbol{e}_1, \ldots, \boldsymbol{e}_k$ be an orthonormal system of vectors in R, with $k < n$. Since vectors $\boldsymbol{e}_1, \ldots, \boldsymbol{e}_k$ are linearly independent (Theorem 16.7), it follows from Theorem 13.8 that this vector system may be augmented by certain vectors $\boldsymbol{f}_{k+1}, \ldots, \boldsymbol{f}_n$ until the basis

$$\boldsymbol{e}_1, \ldots, \boldsymbol{e}_k, \quad \boldsymbol{f}_{k+1}, \ldots, \boldsymbol{f}_n$$

of space R is reached. Naturally, this basis will in general *not be* orthonormal. For vectors $e_1, \ldots, e_k, f_{k+1}$ (which are linearly independent), let us try to select coefficients $\alpha_1, \ldots, \alpha_k$ such that the vector

$$f_{k+1} + \alpha_1 e_1 + \ldots + \alpha_k e_k \tag{16.8}$$

will be orthogonal to all the vectors $e_1, \ldots, e_k$. Then, if we form the scalar product of vector (16.8) times vector e_i (where i is one of the numbers $1, \ldots, k$) and equate this scalar product to zero, we obtain (according to (16.6)):

$$e_i f_{k+1} + \alpha_i = 0,$$

that is,

$$\alpha_i = -e_i f_{k+1} \qquad (i = 1, \ldots, k). \tag{16.9}$$

Thus, if coefficients $\alpha_1, \ldots, \alpha_k$ are defined by relations (16.9), then vector (16.8), which in this case is called e'_{k+1}, will be orthogonal to all the vectors $e_1, \ldots, e_k$. Note that vector e'_{k+1} is different from zero (actually, vector (16.8) is nonzero, since vectors $e_1, \ldots, e_k, f_{k+1}$ are linearly independent). However, it may be that the length of this vector is different from unity. Let us set

$$e_{k+1} = \frac{1}{|e'_{k+1}|} e'_{k+1}$$

and let us assume, as previously, that vector e_{k+1} is orthogonal to all the vectors $e_1, \ldots, e_k$; in addition, its length is unity, as is easily shown. Thus vectors $e_1, \ldots, e_k, e_{k+1}$ form an orthonormal system.

Accordingly, if $k < n$, the orthonormal system $e_1, \ldots, e_k$ can be augmented by a single vector e_{k+1} to produce an orthonormal system consisting of $k+1$ vectors. If $k+1 < n$, then we can similarly add one more vector to get an orthonormal system of $k+2$ vectors. By continuing in this way, we ultimately arrive at an orthonormal system $e_1, \ldots, e_n$, consisting of n vectors, that is (in view of Theorem 16.7), we obtain an orthonormal basis of space R.

THEOREM 16.9. *If $e_1, \ldots, e_n$ form an orthonormal basis of an n-dimensional Euclidean vector space R, then for any vectors*

$$x = x^1 e_1 + \ldots + x^n e_n, \quad y = y^1 e_1 + \ldots + y^n e_n$$

of space R their scalar product is given by the formula

$$xy = x^1 y^1 + \ldots + x^n y^n.$$

In particular,

$$|x| = \sqrt{(x^1)^2 + \ldots + (x^n)^2}.$$

The *proof* follows directly from formulas (16.1) and (16.6).

COROLLARY 16.10. *Any two n-dimensional Euclidean vector spaces are mutually isomorphic. In other words, if R and S are two n-dimensional Euclidean vector spaces, then there exists an isomorphism $\varphi: R \to S$ which preserves the scalar product, that is, $\boldsymbol{xy} = \varphi(\boldsymbol{x})\varphi(\boldsymbol{y})$ for any vectors $\boldsymbol{x}, \boldsymbol{y} \in R$.*

If $\boldsymbol{a}_1, \ldots, \boldsymbol{a}_n$ form an orthonormal basis of space R while $\boldsymbol{b}_1, \ldots, \boldsymbol{b}_n$ form an orthonormal basis of space S, then these bases can be used to construct an isomorphism $\varphi: R \to S$, as in the proof of Theorem 13.12. Since for this isomorphism the coordinates of the vectors are preserved, Theorem 16.9 implies that isomorphism φ preserves the scalar product.

DEFINITION 16.11. Let R be a Euclidean vector space and let A be some subspace of it. A vector $\boldsymbol{x} \in R$ is called *orthogonal* to subspace A if it is orthogonal to each vector belonging to A, that is, $\boldsymbol{xy} = 0$ for $\boldsymbol{y} \in A$. The set B of *all* vectors $\boldsymbol{x} \in R$ orthogonal to subspace A is called the *orthogonal complement* of subspace A (in space R).

THEOREM 16.12. *Let R be a Euclidean vector space, let A be a subspace of it, and let B be the orthogonal complement of A. Then B is the subspace constituting a complement of subspace A, and A is in turn the orthogonal complement of subspace B.*

Proof. Since subspace A is itself a Euclidean vector space (as a subspace of space R), it follows from Theorem 16.8 that there exists in A an orthonormal basis $\boldsymbol{e}_1, \ldots, \boldsymbol{e}_k$, where $k = \dim A$. In view of this same Theorem 16.8, the system of vectors $\boldsymbol{e}_1, \ldots, \boldsymbol{e}_k$ can be augmented by vectors $\boldsymbol{e}_{k+1}, \ldots, \boldsymbol{e}_n$ up to the orthonormal basis $\boldsymbol{e}_1, \ldots, \boldsymbol{e}_k, \boldsymbol{e}_{k+1}, \ldots, \boldsymbol{e}_n$ of space R, where $n = \dim R$. Let C be the subspace generated by vectors $\boldsymbol{e}_{k+1}, \ldots, \boldsymbol{e}_n$, that is, the subspace consisting of all vectors of form $\alpha_{k+1}\boldsymbol{e}_{k+1} + \ldots + \alpha_n\boldsymbol{e}_n$. Since each of vectors $\boldsymbol{e}_{k+1}, \ldots, \boldsymbol{e}_n$ is orthogonal to all of vectors $\boldsymbol{e}_1, \ldots, \boldsymbol{e}_k$, it must also be orthogonal to any linear combination of these vectors, that is, to any vector $\boldsymbol{x} \in A$. However, then in addition any vector $\boldsymbol{y} = \alpha_{k+1}\boldsymbol{e}_{k+1} + \ldots + \alpha_n\boldsymbol{e}_n$ of subspace C is orthogonal to an arbitrary vector $\boldsymbol{x} \in A$, that is $\boldsymbol{y} \in B$.

The inclusion $C \subset B$ can be proven similarly.

Conversely, if a vector

$$\boldsymbol{z} = \alpha_1\boldsymbol{e}_1 + \ldots + \alpha_k\boldsymbol{e}_k + \alpha_{k+1}\boldsymbol{e}_{k+1} + \ldots + \alpha_n\boldsymbol{e}_n$$

belongs to set B, that is, is orthogonal to subspace A, then $\boldsymbol{ze}_1 = 0, \ldots, \boldsymbol{ze}_k = 0$, and hence it follows that $\alpha_1 = \ldots = \alpha_k = 0$, so that

$$\boldsymbol{z} = \alpha_{k+1}\boldsymbol{e}_{k+1} + \ldots + \alpha_n\boldsymbol{e}_n \in C.$$

Consequently, $B \subset C$. We see that $B = C$, meaning that B is the subspace generated by vectors $\boldsymbol{e}_{k+1}, \ldots, \boldsymbol{e}_n$. All the statements of the theorem can be obtained from this result without any difficulty.

COROLLARY 16.13. *Let A be a subspace of Euclidean vector space R. Then each vector $\boldsymbol{x} \in R$ can be represented uniquely in the form $\boldsymbol{x} = \boldsymbol{x}_1 + \boldsymbol{x}_2$, where $\boldsymbol{x}_1 \in A$, while vector $\boldsymbol{x}_2$ is orthogonal to subspace A (that is, belongs to the orthogonal* *complement B of subspace A). Vector $\boldsymbol{x}_1$ is called the* **orthogonal projection** *of vector $\boldsymbol{x}$ onto subspace A.*

The *proof* follows directly from Theorems 16.12 and 14.11.

COROLLARY 16.14. *Let A be a subspace of Euclidean vector space R. For each vector $\boldsymbol{x} \in R$ let $\pi(\boldsymbol{x})$ be the orthogonal projection of vector $\boldsymbol{x}$ onto subspace A. The obtained mapping π of space R into itself (known as the* **orthogonal projecting*** *onto subspace A) is a homomorphism. Here $\operatorname{Im} \pi = A$ and $\operatorname{Ker} \pi = B$, where B is the orthogonal complement of subspace A.*

The *proof* follows directly from Theorem 15.7, Corollary 16.13, and Definition 15.6.

THEOREM 16.15. *Let R and S be Euclidean vector spaces, and let φ: $R \to S$ be some homomorphism. Then there exists a number $M > 0$ such that $|\varphi(\boldsymbol{a})| \leqslant M|\boldsymbol{a}|$ for any $\boldsymbol{a} \in R$.*

Proof. Let $\boldsymbol{e}_1, \ldots, \boldsymbol{e}_n$ be an orthonormal basis of space R. If we set

$$M = |\varphi(\boldsymbol{e}_1)| + \ldots + |\varphi(\boldsymbol{e}_n)|,$$

then for any vector

$$\boldsymbol{a} = x^1\boldsymbol{e}_1 + \ldots + x^n\boldsymbol{e}_n \in R$$

we have

$$|x^i| = \sqrt{(x^i)^2} \leqslant \sqrt{(x^1)^2 + \ldots + (x^n)^2} = |\boldsymbol{a}|, \qquad i = 1, \ldots, n,$$

and thus

$$\begin{aligned} |\varphi(\boldsymbol{a})| &= |\varphi(x^1\boldsymbol{e}_1 + \ldots + x^n\boldsymbol{e}_n)| = |x^1\varphi(\boldsymbol{e}_1) + \ldots + x^n\varphi(\boldsymbol{e}_n)| \leqslant \\ &\leqslant |x^1\varphi(\boldsymbol{e}_1)| + \ldots + |x^n\varphi(\boldsymbol{e}_n)| = |x^1| \cdot |\varphi(\boldsymbol{e}_1)| + \ldots + |x^n| \cdot |\varphi(\boldsymbol{e}_n)| \leqslant \\ &\leqslant |\boldsymbol{a}| \cdot |\varphi(\boldsymbol{e}_1)| + \ldots + |\boldsymbol{a}| \cdot |\varphi(\boldsymbol{e}_n)| = |\boldsymbol{a}| \cdot (|\varphi(\boldsymbol{e}_1)| + \ldots + |\varphi(\boldsymbol{e}_n)|) = \\ &= M|\boldsymbol{a}|. \end{aligned}$$

In conclusion, let us consider how to write linear functionals in Euclidean vector spaces.

THEOREM 16.16. *Let f be an arbitrary linear functional defined in Euclidean vector space R. Then there exists one, and only one, vector $\boldsymbol{b} \in R$ possessing the property that*

$$f(\boldsymbol{x}) = \boldsymbol{b}\boldsymbol{x}$$

(for any $\boldsymbol{x} \in R$). Vector $\boldsymbol{b}$ is called the **gradient** *of linear functional f, being designated as $\operatorname{grad} f$.*

* Here we differentiate between the Russian terms "proektsiya" (projection) and "proektirovanie" (act of projecting) (Translator).

Proof. Let us introduce in space R the orthonormal basis $\boldsymbol{e}_1, \ldots, \boldsymbol{e}_n$ (where $n = \dim R$), and let us set

$$a_1 = f(\boldsymbol{e}_1), \ \ldots, \ a_n = f(\boldsymbol{e}_n).$$

Here $a_1, \ldots, a_n$ are real numbers. In addition, let $\boldsymbol{b}$ be the vector

$$\boldsymbol{b} = a_1\boldsymbol{e}_1 + \ldots + a_n\boldsymbol{e}_n.$$

Now, if

$$\boldsymbol{x} = x^1\boldsymbol{e}_1 + \ldots + x^n\boldsymbol{e}_n$$

is an arbitrary vector of space R, then we obtain

$$f(\boldsymbol{x}) = f(x^1\boldsymbol{e}_1 + \ldots + x^n\boldsymbol{e}_n) = \\ = x^1 f(\boldsymbol{e}_1) + \ldots + x^n f(\boldsymbol{e}_n) = x^1 a_1 + \ldots + x^n a_n = \boldsymbol{x}\boldsymbol{b} = \boldsymbol{b}\boldsymbol{x}$$

(see Theorem 16.9).

Thus the required vector $\boldsymbol{b}$ has been found. Let us show that this vector is unique. Let $\boldsymbol{b}'$ be some other vector possessing the same property:

$$f(\boldsymbol{x}) = \boldsymbol{b}\boldsymbol{x}, \quad f(\boldsymbol{x}) = \boldsymbol{b}'\boldsymbol{x} \quad \text{for any} \quad \boldsymbol{x} \in R.$$

Hence $\boldsymbol{b}\boldsymbol{x} - \boldsymbol{b}'\boldsymbol{x} = 0$, and so

$$(\boldsymbol{b} - \boldsymbol{b}')\boldsymbol{x} = 0.$$

This relation is valid for any $\boldsymbol{x} \in R$. If we substitute, in particular, $\boldsymbol{x} = \boldsymbol{b} - \boldsymbol{b}'$, then we have: $(\boldsymbol{b} - \boldsymbol{b}')^2 = 0$, from which $\boldsymbol{b} - \boldsymbol{b}' = \boldsymbol{0}$, or $\boldsymbol{b} = \boldsymbol{b}'$.

THEOREM 16.17. *Let R be a Euclidean vector space and let R^* be the conjugate of this space. The mapping φ: $R^* \to R$, defined by the formula*

$$\varphi(f) = \operatorname{grad} f, \qquad f \in R^*,$$

is an isomorphism of space R^* onto R.

Proof. Let $f_1, f_2 \in R^*$, that is, f_1 and f_2 are linear functionals in R. If we set

$$\varphi(f_1) = \boldsymbol{b}_1, \ \varphi(f_2) = \boldsymbol{b}_2 \text{ , that is, } \boldsymbol{b}_1 = \operatorname{grad} f_1, \quad \boldsymbol{b}_2 = \operatorname{grad} f_2,$$

then according to Theorem 16.16 we have

$$f_1(\boldsymbol{x}) = \boldsymbol{b}_1\boldsymbol{x}, \ f_2(\boldsymbol{x}) = \boldsymbol{b}_2\boldsymbol{x} \quad \text{for any} \quad \boldsymbol{x} \in R.$$

Consequently,

$$(f_1 + f_2)(\boldsymbol{x}) = f_1(\boldsymbol{x}) + f_2(\boldsymbol{x}) = \boldsymbol{b}_1\boldsymbol{x} + \boldsymbol{b}_2\boldsymbol{x} = (\boldsymbol{b}_1 + \boldsymbol{b}_2)\boldsymbol{x},$$

that is, $\operatorname{grad}(f_1 + f_2) = \boldsymbol{b}_1 + \boldsymbol{b}_2$. In other words, $\varphi(f_1 + f_2) = \varphi(f_1) + \varphi(f_2)$. Similarly, it can be shown that $\varphi(\lambda f) = \lambda\varphi(f)$. Thus mapping φ is a homomorphism.

It remains only to show that mapping φ is one-to-one. Let $\boldsymbol{b} \in R$, and let $f(\boldsymbol{x}) = \boldsymbol{bx}$. According to Axioms IV_1 to IV_3, mapping f possesses properties (13.10), that is, it is a linear functional in R. Then, obviously, $\mathrm{grad}\, f = \boldsymbol{b}$. Therefore, for any vector $\boldsymbol{b} \in R$ there exists a functional f for which $\varphi(f) = \boldsymbol{b}$, that is, φ is a mapping of space R^* onto the whole space R.

Now let $\varphi(f_1) = \varphi(f_2)$. Then (since φ is a homomorphism)

$$\varphi(f_1 - f_2) = \varphi(f_1) - \varphi(f_2) = \mathbf{0}, \qquad \text{that is,} \qquad \mathrm{grad}\,(f_1 - f_2) = \mathbf{0}.$$

This means that $(f_1 - f_2)\boldsymbol{x} = \mathbf{0}\boldsymbol{x} = 0$ for any $\boldsymbol{x} \in R$, so that $f_1(\boldsymbol{x}) - f_2(\boldsymbol{x}) \equiv 0$, or $f_1(\boldsymbol{x}) \equiv f_2(\boldsymbol{x})$. Hence if $\varphi(f_1) = \varphi(f_2)$, it follows that $f_1 = f_2$.

NOTE 16.18. Naturally, the fact that spaces R^* and R are isomorphic is not new: it is valid for *any* n-dimensional vector space (see Theorem 13.12 and the relation $\dim R^* = \dim R$ at the end of Subsection 15). However, the presence of a scalar product in a Euclidean vector space makes it possible to establish an isomorphism between spaces R^* and R, which is not associated with the selection of bases in these spaces; such an isomorphism φ is also constructed in Theorems 16.16 and 16.17.

It should be noted that, on account of this isomorphism, the scalar product considered at the end of Subsection 15 transforms into a scalar product existing in R. In other words, if R is a Euclidean vector space, then for any elements $\boldsymbol{x} \in R$, $f \in R^*$ the following relation is valid:

$$(\boldsymbol{x}, f) = \boldsymbol{x}\varphi(f)$$

(where $\varphi(f) = \mathrm{grad}\, f$); here the left-hand side is the scalar product considered in Subsection 15, and the right-hand side is the scalar product in space R.

§5. Euclidean Geometry

17. Definition of affine space. The properties of the vector spaces considered in the preceding subsection can be geometrically interpreted in the vector space Σ (see Example 12.1). For instance, each line passing through point O in space Σ constitutes a one-dimensional subspace, while any plane passing through point O is a two-dimensional subspace. If the line does not lie in the plane (Figure 46), then it is a direct complement of this plane, whereas if the line is perpendicular to the plane it is an orthogonal complement. Simple geometrical interpretations can also be found for the orthogonal projection (Figure 47) and the other concepts considered.

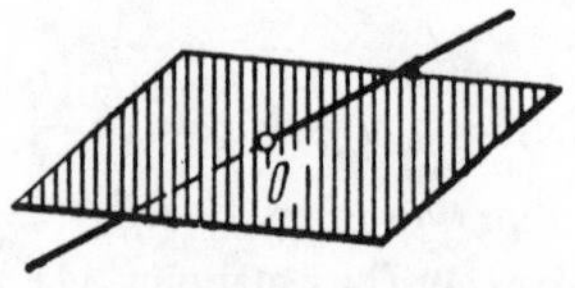

FIGURE 46

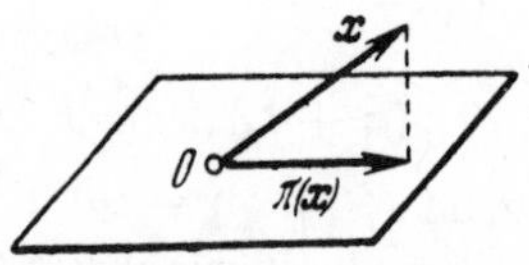

FIGURE 47

Nevertheless, all the concepts introduced and all the theorems proven are more algebraic than geometric in nature. When considering Example 12.1, this manifested itself in the fact that we *fixed* some point O in plane Π (or space Σ) and we examined only vectors (directed line segments) starting from point O (or, as they are called, "bound" vectors), while only lines and planes passing through O were considered as subspaces. All this is by no means dictated by the *geometric* properties of the plane (or space); on the contrary, geometrically all points are equivalent, and it is somewhat artificial to attribute exceptional properties to some point O. Moreover, it is customary in geometry to consider *points* as well as vectors, and to consider also lines and planes not passing through O. In other words, a *vector* space differs from the type of space we are accustomed to deal with in geometry (elementary or analytic).

In the present subsection we will become acquainted with the concept of an affine space, which corresponds exactly to our intuitive ideas about a geometric space with regard to all properties except metric properties. Metric properties will be discussed below, in Subsection 21.

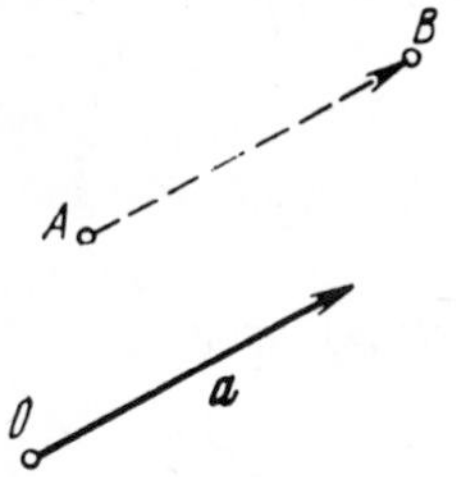

FIGURE 48

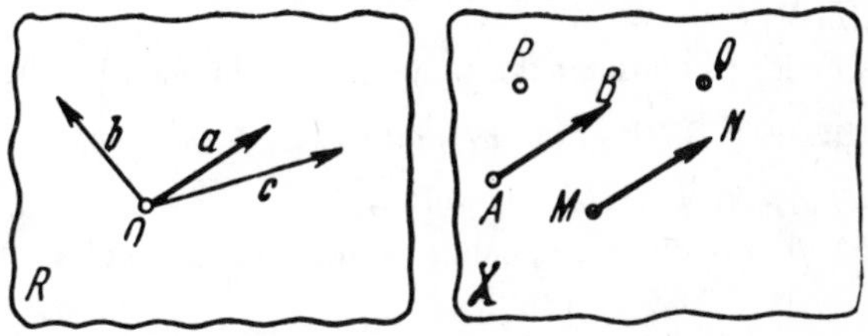

FIGURE 49

In order to make the definition of an affine space (to be given below) clearer and more natural, we recall that in geometry (elementary or analytic) *free* vectors are considered as well as "bound" vectors (starting from point O). In the case of a free vector, a vector $\overrightarrow{AB}$ "drawn" from any point A will be "equal" to a vector $\boldsymbol{a}$ drawn from point O (Figure 48). In other words, if *point* A and *vector* $\boldsymbol{a}$ are specified, then there exists a point B such that $\overrightarrow{AB} = \boldsymbol{a}$. We have been accustomed to consider (and represent) points and vectors in one and the same plane (or in one and the same space), but for greater clarity we can imagine *two* planes located next to one another, in one of which point O is specified and bound vectors are constructed and in the other of which points and free vectors are depicted (Figure 49). Let R be the plane in which bound vectors are represented; it will be a two-dimensional vector space, such as the one in Example 12.1. The second plane (with points and free vectors in it) is called X. There is no fixed point, or origin, in plane X (that is, all points are equivalent); it is also a "geometric" space. For any vector $\boldsymbol{a}$ (that is, an element of vector space R) and any point $A \in X$, there exists in plane X a point B such that the directed line segment $\overrightarrow{AB}$ going from point A to point B in plane X is "equal" to vector $\boldsymbol{a}$:

$$\overrightarrow{AB} = \boldsymbol{a}.$$

The foregoing considerations also form the basis for the axiomatic definition of an *affine* space. This can be presented as follows.

Let R be a vector space of dimension n and let X be some set whose elements are called *points*. The pair (R, X) is known as an *n-dimensional affine space* if to each two points $A, B \in X$ there corresponds some vector $\overrightarrow{AB}$ (also belonging to vector space R), the following axioms being obeyed.

GROUP V: *axioms of vector construction.*

V_1. *For any point $A \in X$ and any vector $\boldsymbol{a} \in R$ there exists a point $B \in X$ such that* $\overrightarrow{AB} = \boldsymbol{a}$ (finding this point B is called *construction* of a vector $\boldsymbol{a}$ from point A).

V_2. *For any three points $A, B, C \in X$ the following relation is valid:*

$$\overrightarrow{AB} + \overrightarrow{BC} + \overrightarrow{CA} = \mathbf{0}.$$

V_3. *If for points $A, B \in X$ the relation $\overrightarrow{AB} = \mathbf{0}$ is valid, then points A and B coincide.*

Now let us consider the simplest facts implied by the axioms of vector construction.

THEOREM 17.1. *For any point $A \in X$ the relation $\overrightarrow{AA} = \mathbf{0}$ is valid. For any two points $A, B \in X$ the relation $\overrightarrow{AB} = -\overrightarrow{BA}$ is valid.*

Proof. For three coinciding points $A = B = C$ we have, in view of Axiom V_2,

$$\overrightarrow{AA} + \overrightarrow{AA} + \overrightarrow{AA} = \mathbf{0},$$

or $3\overrightarrow{AA} = \mathbf{0}$. Multiplication of this equation by 1/3 gives $\overrightarrow{AA} = \mathbf{0}$.

Moreover, if we replace point C in Axiom V_2 by point A, we get

$$\overrightarrow{AB} + \overrightarrow{BA} + \overrightarrow{AA} = \mathbf{0},$$

from which, together with the previous relation $\overrightarrow{AA} = \mathbf{0}$, we find $\overrightarrow{AB} + \overrightarrow{BA} = \mathbf{0}$, that is, $\overrightarrow{AB} = -\overrightarrow{BA}$.

THEOREM 17.2. *For any points $A_1, A_2, \ldots, A_k \in X$ the following relation is valid:*

$$\overrightarrow{A_1A_2} + \overrightarrow{A_2A_3} + \ldots + \overrightarrow{A_{k-1}A_k} = \overrightarrow{A_1A_k}.$$

The *proof* follows by induction from Axiom V_2 and the relation $\overrightarrow{AB} = -\overrightarrow{BA}$.

THEOREM 17.3. *If the real numbers $\alpha^1, \ldots, \alpha^s$ have the property that*

$$\alpha^1 + \ldots + \alpha^s = 0,$$

then for any points $Q, Q', A_1, \ldots, A_s$ the following relation is valid:

$$\alpha^1\overrightarrow{QA_1} + \ldots + \alpha^s\overrightarrow{QA_s} = \alpha^1\overrightarrow{Q'A_1} + \ldots + \alpha^s\overrightarrow{Q'A_s}.$$

The *proof* follows directly from the relations

$$\overrightarrow{QA_i} = \overrightarrow{QQ'} + \overrightarrow{Q'A_i}, \qquad i = 1, \ldots, s.$$

On the basis of Theorem 17.3, we now introduce some notation which will make things easier in the following. Let $\lambda^1, \ldots, \lambda^s$ be real numbers satisfying the condition

$$\lambda^1 + \ldots + \lambda^s = 1, \tag{17.1}$$

and let $Q, A, A_1, \ldots, A_s \in X$ be points such that

$$\overrightarrow{QA} - \lambda^1\overrightarrow{QA_1} - \ldots - \lambda^s\overrightarrow{QA_s} = 0. \tag{17.2}$$

Since the sum of the coefficients on the left-hand side of (17.2) is zero, we know from Theorem 17.3 that the following relation is valid for *any* point $Q' \in X$:

$$\overrightarrow{Q'A} - \lambda^1\overrightarrow{Q'A_1} - \ldots - \lambda^s\overrightarrow{Q'A_s} = 0.$$

In other words, if equation 17.2 holds true and relation (17.1) is satisfied, then (17.2) is not actually contingent upon the choice of point Q, but rather reflects a certain relationship *only between points* $A, A_1, \ldots, A_s$. Let us write in this case

$$A = \lambda^1 A_1 + \ldots + \lambda^s A_s. \tag{17.3}$$

Consequently, relation (17.3) applies *only* when equation (17.1) is satisfied, so that (17.3) will be interpreted as meaning that for some (and thus for any) point $Q \in X$ the following equation is valid (see (17.2)):

$$\overrightarrow{QA} = \lambda^1\overrightarrow{QA_1} + \ldots + \lambda^s\overrightarrow{QA_s}. \tag{17.4}$$

THEOREM 17.4. *Let* $\lambda^1, \ldots, \lambda^s$ *be real numbers satisfying condition* (17.1), *and let* $A_1, \ldots, A_s \in X$ *be arbitrary points. Then there exists one, and only one, point* $A \in X$ *satisfying relation* (17.3).

Proof. We select (and specify) an arbitrary point $Q \in X$. Equation (17.3) is by definition equivalent to relation (17.4) (since condition (17.1) is satisfied). Hence we must show the existence and uniqueness of a point A satisfying relation (17.4). Since points $Q, A_1, \ldots, A_s$ and numbers $\lambda^1, \ldots, \lambda^s$ are known, it follows that a *known* vector appears on the right-hand side of (17.4); let us call it $\boldsymbol{a}$. Thus the object is to demonstrate the existence and uniqueness of a point A satisfying the relation

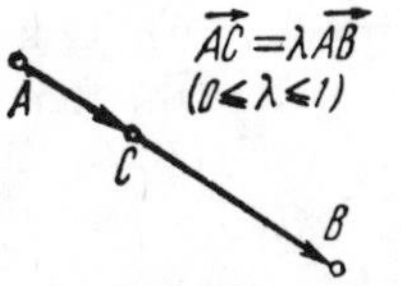

FIGURE 50

$$\overrightarrow{QA} = \boldsymbol{a},$$

where Q is a fixed point. The existence of such a point follows directly from Axiom V_1. Let us show its uniqueness.

Let A and A' be two points satisfying the condition imposed, that is $\overrightarrow{QA}=\boldsymbol{a}$, $\overrightarrow{QA'}=\boldsymbol{a}$. Then $\overrightarrow{QA}=\overrightarrow{QA'}$ and thus $\overrightarrow{AA'}=\overrightarrow{AQ}+\overrightarrow{QA'}=-\overrightarrow{QA}+\overrightarrow{QA'}=\mathbf{0}$. From Axiom V_3 it now follows that points A and A' are identical.

EXAMPLE 17.5. Let us next introduce the concept of a *line segment* in an affine space. If A and B are two different points in space X, then another *point* C *will lie on segment* $[A, B]$ if for vectors $\overrightarrow{AB}$ and $\overrightarrow{AC}$ it is true that (Figure 50)

$$\overrightarrow{AC}=\lambda\overrightarrow{AB}, \qquad \text{where } 0\leqslant\lambda\leqslant 1. \tag{17.5}$$

Obviously, points A and B both lie on segment $[A, B]$; they are called the *ends* (or *endpoints*) of segment $[A, B]$. The point C for which the number λ in (17.5) is equal to 1/2 is called the *midpoint* of segment $[A, B]$; it possesses the property that $\overrightarrow{AC}=\overrightarrow{CB}$.

Sometimes it is necessary to consider a line segment $[A, B]$ without knowing in advance whether points A and B are different. In such cases it is convenient to introduce a "segment" with coinciding ends $A = B$, a "segment" consisting by definition of a single point A. Such a line segment is called *degenerate*. This stipulation enables us to consider a "segment $[A, B]$" for *any* points A and B, coinciding or different.

Let C be an arbitrary point on segment $[A, B]$ and let us take an arbitrary point Q of space X. Then

$$\overrightarrow{AC}=\overrightarrow{QC}-\overrightarrow{QA}, \quad \overrightarrow{AB}=\overrightarrow{QB}-\overrightarrow{QA},$$

so that relation (17.5) (which holds true, since $C\in[A, B]$) becomes

$$\overrightarrow{QC}-\overrightarrow{QA}=\lambda(\overrightarrow{QB}-\overrightarrow{QA}), \qquad 0\leqslant\lambda\leqslant 1,$$

and thus

$$\overrightarrow{QC}=(1-\lambda)\overrightarrow{QB}+\lambda\overrightarrow{QA}, \qquad 0\leqslant\lambda\leqslant 1,$$

or alternatively,

$$C=(1-\lambda)A+\lambda B, \qquad 0\leqslant\lambda\leqslant 1. \tag{17.6}$$

By carrying out the calculation in reverse order, we obtain (17.5) from (17.6). Consequently, if point C satisfies relation (17.6), then $C\in[A, B]$. Therefore the following statement is true:

Point C lies on segment $[A, B]$ if, and only if, relation (17.6) *is valid.*

If A and B are different points, then all the points on segment $[A, B]$ except its ends A and B are called *interior points* of the segment. The set of all the interior points of segment $[A, B]$ is called (A, B), being known as an *interval* with ends A and B. Thus endpoints A and B do not actually belong to the interval (A, B). If *one* of its endpoints A or B is joined to interval (A, B), then we obtain the *half-intervals:* $[A, B)$ and $(A, B]$.

Let $A_0, A_1, \ldots, A_k \in X$. If any of these points, say A_0, can be expressed in terms of the rest of the points as follows (see (17.3)):

$$A_0 = \lambda^1 A_1 + \ldots + \lambda^k A_k, \quad \text{where} \quad \lambda^1 + \ldots + \lambda^k = 1,$$

then we will say that points $A_0, A_1, \ldots, A_k$ are *dependent*. On the other hand, if none of the points can be expressed in terms of the rest, then points $A_0, A_1, \ldots, A_k$ are *independent*.

THEOREM 17.6. *Points* $A_0, A_1, \ldots, A_k \in X$ *are independent if, and only if, vectors* $\overrightarrow{A_0A_1}, \ldots, \overrightarrow{A_0A_k}$ *are linearly independent.*

Proof. If points $A_0, A_1, \ldots, A_k$ are dependent, then one of them, say A_i, can be expressed in terms of the others:

$$A_i = \lambda^0 A_0 + \lambda^1 A_1 + \ldots + \lambda^k A_k,$$

where point A_i does not appear on the right-hand side (that is, $\lambda^i = 0$), and $\lambda^0 + \lambda^1 + \ldots + \lambda^k = 1$. This means (see (17.4)) that

$$\overrightarrow{QA_i} = \lambda^0 \overrightarrow{QA_0} + \lambda^1 \overrightarrow{QA_1} + \ldots + \lambda^k \overrightarrow{QA_k},$$

where Q is an arbitrary point of space X. In particular, for $Q = A_0$ we get

$$\overrightarrow{A_0A_i} = \lambda^1 \overrightarrow{A_0A_1} + \ldots + \lambda^k \overrightarrow{A_0A_k},$$

with $\lambda^i = 0$. Consequently, vectors $\overrightarrow{A_0A_1}, \ldots, \overrightarrow{A_0A_k}$ are linearly dependent.

Alternatively, let vectors $\overrightarrow{A_0A_1}, \ldots, \overrightarrow{A_0A_k}$ be linearly dependent. Then one of them, say $\overrightarrow{A_0A_j}$, can be expressed in terms of the rest:

$$\overrightarrow{A_0A_j} = \lambda^1 \overrightarrow{A_0A_1} + \ldots + \lambda^k \overrightarrow{A_0A_k},$$

where $\lambda^j = 0$. If we set $\lambda^0 = 1 - \lambda^1 - \ldots - \lambda^k$, then $\lambda^0 + \lambda^1 + \ldots + \lambda^k = 1$ and so

$$\overrightarrow{A_0A_j} = \lambda^0 \overrightarrow{A_0A_0} + \lambda^1 \overrightarrow{A_0A_1} + \ldots + \lambda^k \overrightarrow{A_0A_k},$$

that is, relation (17.4) is valid with $Q = A_0$. Consequently,

$$A_j = \lambda^0 A_0 + \lambda^1 A_1 + \ldots + \lambda^k A_k.$$

Since $\lambda^j = 0$, this means that point A_j can be expressed in terms of the others, that is, points $A_0, A_1, \ldots, A_k$ are dependent. The theorem is thereby proven.

Let $\alpha^1, \ldots, \alpha^s$ be real numbers satisfying the condition

$$\alpha^1 + \ldots + \alpha^s = 0, \tag{17.7}$$

and let $A_1, \dots, A_s$ be some points of space X. Then, in view of Theorem 17.3, the vector

$$\boldsymbol{a} = \alpha^1 \overrightarrow{QA_1} + \dots + \alpha^s \overrightarrow{QA_s} \tag{17.8}$$

does not depend on the choice of point Q. Let us agree to call this vector $\boldsymbol{a}$:

$$\boldsymbol{a} = \alpha^1 A_1 + \dots + \alpha^s A_s, \tag{17.9}$$

a notation indicating that vector $\boldsymbol{a}$ depends only on numbers $\alpha^1, \dots, \alpha^s$ and points $A_1, \dots, A_s \in X$. In other words, notation (17.9) will be used only when condition (17.7) is satisfied, and it will be understood in the sense of (17.8).

In particular, let A and B be two points of vector space X. Consider the vector

$$\boldsymbol{a} = B - A, \tag{17.10}$$

that is, vector $1B + (-1)A$ (here the sum of the coefficients is zero, meaning that condition (17.7) is satisfied). Equation (17.10) indicates by definition that $\boldsymbol{a} = \overrightarrow{QB} - \overrightarrow{QA}$. However, since $\overrightarrow{QB} - \overrightarrow{QA} = \overrightarrow{AB}$, we have

$$\overrightarrow{AB} = B - A, \tag{17.11}$$

so that, according to the notation adopted, vector $\overrightarrow{AB}$ (see Axioms V_1–V_3) can also be called $B - A$.

In accordance with the foregoing, we will also employ a notation of type $A + \boldsymbol{a}$, where $A \in X$, $\boldsymbol{a} \in R$. The meaning is as follows: the notation $B = A + \boldsymbol{a}$ indicates by definition the same thing as $\boldsymbol{a} = B - A$, that is, $\boldsymbol{a} = \overrightarrow{AB}$. Thus the notation $B = A + \boldsymbol{a}$ means that B is the point obtained if we *construct* a vector $\boldsymbol{a}$ from point A (see Axiom V_1).

THEOREM 17.7. *Let (R, X) be an affine space. Let us fix in X some point O and let us call all the points of space X vectors. The sums of these vectors and the products of these vectors times real numbers are defined as follows:*

we write $C = A + B$, if $C = A + B - O$ $(A, B, C \in X)$,

we write $D = \lambda A$, if $D = \lambda A + (1 - \lambda)O$ $(A, D \in X)$.

It turns out that the introduction of these operations transforms X into a vector space isomorphic with R. The mapping φ: $X \to R$, defined by the formula $\varphi(A) = \overrightarrow{OA}$, is an isomorphism.

Proof. First of all, we note that mapping φ: $X \to R$, defined by the formula $\varphi(A) = \overrightarrow{OA}$, is one-to-one (this follows directly from Axioms V_1 and V_3). Consequently, to prove the theorem it is sufficient to establish that

$$\varphi(A + B) = \varphi(A) + \varphi(B), \quad \varphi(\lambda A) = \lambda\varphi(A); \tag{17.12}$$

this will indicate that in X the axioms of Groups I and II are obeyed (since they are obeyed in R), that is, it will prove that X is a vector space, while at the same time φ is an isomorphism.

The notation $C = A + B$, that is, $C = A + B - O$, means that $\overrightarrow{QC} = \overrightarrow{QA} + \overrightarrow{QB} - \overrightarrow{QO}$ for any point $Q \in X$. In particular, for $Q = O$ we have $\overrightarrow{OC} = \overrightarrow{OA} + \overrightarrow{OB}$, that is, $\varphi(C) = \varphi(A) + \varphi(B)$. Hence, if $C = A + B$, then $\varphi(C) = \varphi(A) + \varphi(B)$, which establishes the first of relations (17.12).

Furthermore, the notation $D = \lambda A$, that is, $D = \lambda A + (1 - \lambda) O$, means that $\overrightarrow{QD} = \lambda\overrightarrow{QA} + (1 - \lambda)\overrightarrow{QD}$ for any point $Q \in X$. In particular, for $Q = O$ we obtain $\overrightarrow{OD} = \lambda\overrightarrow{OA}$, that is, $\varphi(D) = \lambda\varphi(A)$. This establishes the second of relations (17.12).

NOTE 17.8. Thus, if in space X some point O is fixed, then X becomes a vector space. In this case the linear combination

$$\lambda^1 A_1 + \ldots + \lambda^k A_k$$

is meaningful with *any* sum of coefficients $\lambda^1 + \ldots + \lambda^k$: we can interpret this linear combination as

$$\lambda^1 A_1 + \ldots + \lambda^k A_k + (1 - \lambda^1 - \ldots - \lambda^k) O,$$

where now the sum of the coefficients is already unity. This is precisely how we interpreted the sum $A + B$ (as $A + B - O$) and the product λA (as $\lambda A + (1 - \lambda) O$) during the formulation of Theorem 17.7. Note that the definition of the sum in Theorem 17.7, that is, the agreement to take as the sum of points A and B the point $C = A + B - O$, can be written as $C + O = A + B$, or

$$\frac{1}{2} C + \frac{1}{2} O = \frac{1}{2} A + \frac{1}{2} B,$$

and this means that the midpoint of segment $[O, C]$ coincides with the midpoint of segment $[A, B]$. Thus the operations introduced in Theorem 17.7 are in complete agreement with Example 12.1.

NOTE 17.9. The concept of an affine space described in this subsection is based on a *separate* consideration of the space of vectors R and the space of points X. However, it could also be that these spaces coincide. For instance, if R is an n-dimensional vector space, then we might agree to consider the elements of set R (that is, the vectors) as "points" as well, that is, to call each element of set R sometimes a vector, sometimes a point, depending on the circumstances. In this case if A, B are two "points" (that is, $A, B \in R$), then "vector" $\overrightarrow{AB}$ refers to the difference $A - B$ (cf. (17.11)), which is meaningful in R, since R is a vector space.

Passing to the usual notation for the elements of a vector space ($\boldsymbol{a}, \boldsymbol{b}, \boldsymbol{c}, \ldots$ instead of $A, B, C, \ldots$), let us describe the operation of vector construction as follows. To every two points $\boldsymbol{a}, \boldsymbol{b} \in R$ there corresponds a vector $\boldsymbol{b} - \boldsymbol{a} \in R$ known as the vector from point $\boldsymbol{a}$ to point $\boldsymbol{b}$. In this case Axiom V_1 signifies that for any point $\boldsymbol{a} \in R$ and any vector $\boldsymbol{c} \in R$ there exists a point $\boldsymbol{x} \in R$ such that the vector from point $\boldsymbol{a}$ to point $\boldsymbol{x}$ is equal to $\boldsymbol{c}$, that is, $\boldsymbol{x} - \boldsymbol{a} = \boldsymbol{c}$, or $\boldsymbol{x} = \boldsymbol{a} + \boldsymbol{c}$. Consequently, here the construction of vector $\boldsymbol{c}$ from point $\boldsymbol{a}$ means a transition to point $\boldsymbol{a} + \boldsymbol{c}$. We see that, for such an understanding of vectors and points, Axiom V_1 is obeyed in any vector space. It is equally simple to verify that Axioms V_2 and V_3 are obeyed.

Thus, according to this interpretation, any vector space is transformed at the same time into an affine space, that is, both points and vectors are considered in the same set R.

Another variant of the interpretation (that is, of considering both points and vectors in a given space) can be obtained on the basis of Theorem 17.7: since the stipulation of a fixed point $O \in X$ transforms X into a vector space isomorphic to R, it is enough to deal with a single set X and to consider in it both points and vectors.

The separate treatment (sets R and X) seems to be simpler and clearer. In the following it is left to the reader to choose between these two points of view. More precisely, we will refer to *points and vectors* in sets R and X, respectively (separate treatment), but the reader can, if desired, consider R and X to be coinciding sets.

18. Planes in affine space. Let (R, X) be an affine space. A set $P \subset X$ is called a *plane* (or a *linear manifold*) if for any points $A_1, \ldots, A_s \in P$ and any real numbers $\mu^1, \ldots, \mu^s$ satisfying the condition $\mu^1 + \ldots + \mu^s = 1$ a point $A = \mu^1 A_1 + \ldots + \mu^s A_s$ also belongs to set P. If in set P we can find $r + 1$ independent points, but it is not possible to find a greater number of independent points, then plane P is called *r-dimensional.* If affine space X has dimension n, then $(n - 1)$-dimensional planes in it are also called *hyperplanes.* Finally, one-dimensional planes in an affine space are called *straight lines* (or simply *lines*).

THEOREM 18.1. *Let $P \subset X$ be some r-dimensional plane. Then the set L_P of all vectors $\overrightarrow{AB}$, where A, $B \in P$, is an r-dimensional subspace of vector space R. This subspace is called the* **direction subspace** *of plane P.*

Proof. Let $\boldsymbol{a}, \boldsymbol{b} \in L_P$, that is, $\boldsymbol{a} = \overrightarrow{AB}$, $\boldsymbol{b} = \overrightarrow{CD}$, where $A, B, C, D \in P$. Since P is a plane, the point

$$E = B + D - C \tag{18.1}$$

belongs to plane P. Equation (18.1) indicates that

$$\overrightarrow{QE} = \overrightarrow{QB} + \overrightarrow{QD} - \overrightarrow{QC}$$

for any point $Q \in X$. In particular, for $Q = A$ we obtain

$$\overrightarrow{AE} = \overrightarrow{AB} + \overrightarrow{AD} - \overrightarrow{AC} = \overrightarrow{AB} + \overrightarrow{CD} = \boldsymbol{a} + \boldsymbol{b}.$$

Since A, $E \in P$, it follows that $\overrightarrow{AE} \in L_P$, that is, $\boldsymbol{a} + \boldsymbol{b} \in L_P$. Hence if $\boldsymbol{a}, \boldsymbol{b} \in L_P$, it follows that $\boldsymbol{a} + \boldsymbol{b} \in L_P$.

Now let $\boldsymbol{a} \in L_P$, that is, $\boldsymbol{a} = \overrightarrow{AB}$, where $A, B \in P$, and let λ be an arbitrary real number. Since P is a plane, the point

$$F = (1 - \lambda) A + \lambda B \tag{18.2}$$

belongs to plane P. Equation (18.2) implies that

$$\overrightarrow{QF} = (1 - \lambda) \overrightarrow{QA} + \lambda \overrightarrow{QB}$$

for any point $Q \in X$. In particular, for $Q = A$ we get

$$\overrightarrow{AF} = (1 - \lambda) \overrightarrow{AA} + \lambda \overrightarrow{AB} = \lambda \overrightarrow{AB} = \lambda \boldsymbol{a}.$$

Since $A, F \in P$, we know that $\overrightarrow{AF} \in L_P$, that is, $\lambda \boldsymbol{a} \in L_P$. Thus if $\boldsymbol{a} \in L_P$ it follows that $\lambda \boldsymbol{a} \in L_P$, so that L_P is a subspace.

Since $\dim P = r$, there must exist independent points $A_0, A_1, \ldots, A_r \in P$. According to Theorem 17.6, the vectors

$$\overrightarrow{A_0A_1}, \ldots. \overrightarrow{A_0A_r}$$

are linearly independent. However, all these vectors belong to L_P. Therefore, in L_P there are r linearly independent vectors, so that $\dim L_P \geqslant r$.

Finally, let $\boldsymbol{e}_1, \ldots, \boldsymbol{e}_s$ be some basis of subspace L_P, where $s = \dim L_P$. Then in P there exist points $B_1, \ldots, B_s, C_1, \ldots, C_s$ such that

$$\boldsymbol{e}_1 = \overrightarrow{B_1C_1}, \ldots, \boldsymbol{e}_s = \overrightarrow{B_sC_s}.$$

Consider an arbitrary point $A_0 \in P$. For each $i = 1, \ldots, s$, the point $A_i = A_0 + C_i - B_i$ belongs to plane P. This means that

$$\overrightarrow{QA_i} = \overrightarrow{QA_0} + \overrightarrow{QC_i} - \overrightarrow{QB_i}$$

for any point $Q \in X$. In particular, for $Q = A_0$ we obtain

$$\overrightarrow{A_0A_i} = \overrightarrow{A_0A_0} + \overrightarrow{A_0C_i} - \overrightarrow{A_0B_i} = \overrightarrow{A_0C_i} - \overrightarrow{A_0B_i} = \overrightarrow{B_iC_i} = \boldsymbol{e}_i.$$

Hence we have found in P points $A_0, A_1, \ldots, A_s$ such that

$$\overrightarrow{A_0A_1} = \boldsymbol{e}_1, \ldots, \overrightarrow{A_0A_s} = \boldsymbol{e}_s.$$

Since vectors $\boldsymbol{e}_1, \ldots, \boldsymbol{e}_s$ are linearly independent, it follows from Theorem 17.6 that points $A_0, A_1, \ldots, A_s$ are independent. Hence it follows that $s \leqslant r$ (since $\dim P = r$), that is, $\dim L_P \leqslant r$. These inequalities imply that $\dim L_P = r$.

THEOREM 18.2. *Let $P \subset X$ be some plane and let $L_P \subset R$ be its direction subspace. Assume also that $e_1, \ldots, e_r$ form a basis of subspace L_P and that A_0 is an arbitrary point of plane P. A point $A \in X$ will belong to plane P if, and only if, real numbers $\lambda^1, \ldots, \lambda^r$ exist such that*

$$A = A_0 + \lambda^1 e_1 + \ldots + \lambda^r e_r. \tag{18.3}$$

Proof. Equation (18.3) implies that by definition

$$\overrightarrow{A_0A} = \lambda^1 e_1 + \ldots + \lambda^r e_r \tag{18.4}$$

(see p. 117). If $A \in P$, then $\overrightarrow{A_0A} \in L_P$, that is, there exist real numbers $\lambda^1, \ldots, \lambda^r$ such that relation (18.4) is valid, and thus (18.3) as well.

Conversely, let us assume that equation (18.3) is valid, and thus (18.4) as well. Then the vector $a = \overrightarrow{A_0A}$, as a linear combination of vectors $e_1, \ldots, e_r$, belongs to subspace L_P. Thus there exist points $B, C \in P$ such that $\overrightarrow{BC} = a$. If we set

$$D = A_0 + C - B\ , \tag{18.5}$$

then point D belongs to plane P, since $A_0, B, C \in P$. Equation (18.5) implies that

$$\overrightarrow{QD} = \overrightarrow{QA_0} + \overrightarrow{QC} - \overrightarrow{QB}$$

for any point $Q \in X$. In particular, for $Q = A_0$ we get

$$\overrightarrow{A_0D} = \overrightarrow{A_0C} - \overrightarrow{A_0B} = \overrightarrow{BC} = a = \overrightarrow{A_0A}.$$

From the equation $\overrightarrow{A_0D} = \overrightarrow{A_0A}$, that is, $\overrightarrow{AD} = 0$, it follows that points A and D coincide (Axiom V_3), and thus $A \in P$. Therefore, if (18.3) is valid, then $A \in P$.

THEOREM 18.3. *Let $e_1, \ldots, e_r$ be arbitrary vectors of space R and let A_0 be an arbitrary point of space X. Then the set P of all points of form* (18.3), *where $\lambda^1, \ldots, \lambda^r$ are arbitrary real numbers, is a plane of dimension not greater than r, the subspace generated by vectors $e_1, \ldots, e_r$ being the direction subspace for this plane. If vectors $e_1, \ldots, e_r$ are linearly independent, then* $\dim P = r$.

Proof. Let $A_1, \ldots, A_s$ be an arbitrary system of points belonging to set P, that is,

$$A_i = A_0 + \lambda_i^1 e_1 + \ldots + \lambda_i^r e_r, \qquad i = 1, \ldots, s, \tag{18.6}$$

and let A be a combination of points $A_1, \ldots, A_s$, that is,

$$A = \mu^1 A_1 + \ldots + \mu^s A_s, \tag{18.7}$$

where

$$\mu^1 + \ldots + \mu^s = 1. \tag{18.8}$$

Relation (18.7) implies that

$$\overrightarrow{QA} = \mu^1 \overrightarrow{QA_1} + \ldots + \mu^s \overrightarrow{QA_s}$$

for any point $Q \in X$, while relation (18.6) implies that

$$\overrightarrow{QA_i} = \overrightarrow{QA_0} + \lambda_i^1 e_1 + \ldots + \lambda_i^r e_r, \qquad i = 1, \ldots, s.$$

Hence we obtain (in view of (18.8))

$$\overrightarrow{QA} = \sum_{i=1}^{s} \mu^i (\overrightarrow{QA_0} + \lambda_i^1 e_1 + \ldots + \lambda_i^r e_r) =$$
$$= \overrightarrow{QA_0} + \left(\sum_{i=1}^{s} \mu^i \lambda_i^1\right) e_1 + \ldots + \left(\sum_{i=1}^{s} \mu^i \lambda_i^r\right) e_r.$$

This relation is of type (18.3), so that $A \in P$. Thus, from (18.7) and (18.8), it follows that $A \in P$, that is, P is a plane.

Let L_P be the direction subspace of this plane. Relation (18.3), which can be written in form (18.4), implies that any vector of type $\lambda^1 e_1 + \ldots + \lambda^r e_r$ belongs to subspace L_P. Conversely, if $a \in L_P$, that is, $a = \overrightarrow{AB}$, where $A, B \in P$, then

$$\overrightarrow{A_0A} = \lambda^1 e_1 + \ldots + \lambda^r e_r, \quad \overrightarrow{A_0B} = \nu^1 e_1 + \ldots + \nu^r e_r,$$

so that

$$a = \overrightarrow{AB} = \overrightarrow{A_0B} - \overrightarrow{A_0A} = (\nu^1 - \lambda^1) e_1 + \ldots + (\nu^r - \lambda^r) e_r.$$

Thus, vector a belongs to subspace L_P if, and only if, a is a linear combination of vectors $e_1, \ldots, e_r$. In other words, L_P is a subspace generated by vectors $e_1, \ldots, e_r$ (see Theorem 14.6).

Consequently, we are led to the inequality $\dim P = \dim L_P \leqslant r$ in the general case and to the equation $\dim P = \dim L_P = r$ in the case of linear independence of vectors $e_1, \ldots, e_r$.

NOTE 18.4. Theorems 18.2 and 18.3 indicate that any r-dimensional plane can be described by equation (18.3), where $e_1, \ldots, e_r$ are linearly independent vectors, while $\lambda^1, \ldots, \lambda^r$ run through (independently of one another) all real values, and conversely, any such equation describes some r-dimensional plane.

Equation (18.3) is called the *vector parametric equation* of an r-dimensional plane. In particular, for $r = 1$ equation (18.3) becomes

$$A = A_0 + \lambda e; \tag{18.9}$$

this is the vector parametric equation of a straight line; here e is a nonzero vector (of space R), A_0 is an arbitrary point of space X, and λ runs through all real values.

Any line can be written in form (18.9), and, conversely, any such equation represents a straight line. Vector $\boldsymbol{e}$ is called the *direction vector* of this line.

If P is some plane, L_P is its direction subspace, and $\boldsymbol{e}_1, \ldots, \boldsymbol{e}_r$ form a basis of subspace L_P, then the system of vectors $\boldsymbol{e}_1, \ldots, \boldsymbol{e}_r$ is also called a *basis of plane P*. Therefore, in order to write the vector equation of plane P (see (18.3)), we have to know the basis $\boldsymbol{e}_1, \ldots, \boldsymbol{e}_r$ of this plane and any single point A_0 belonging to this plane.

DEFINITION 18.5. Two planes P, P' (considered in the same affine space X) are called *parallel* (notation: $P \parallel P'$) if their direction subspaces coincide.

This definition can be rephrased by saying that two planes are parallel if a basis of one of them is at the same time a basis of the other. It should be noted that, according to this definition, two *coinciding* planes are considered parallel. Note also that, according to this definition, only planes having *the same* dimension can be parallel, and that the parallelism relation (for r-dimensional planes) possesses the properties of reflexivity, symmetry, and transitivity. Sometimes, by the way, a different terminology is used, whereby a line can be parallel to a plane and, in general, a plane with one number of dimensions can be parallel to a plane with some other number of dimensions. For instance, two planes P, Q are then considered parallel if for their direction subspaces L_P, L_Q the inclusion $L_P \supset L_Q$ (or $L_P \subset L_Q$) is valid. However, we will not use this terminology.

THEOREM 18.6. *A single plane parallel to a given plane $P \subset X$ passes through any given point $B_0 \in X$. Moreover, any two planes which are parallel to one another either coincide or else have no common points. Consequently, if a plane $P \subset X$ is specified, then all of space X is filled with mutually exclusive (disjoint) planes parallel to plane P.*

Proof. Assume that the given plane P has vector parametric equation (18.3), that is, $A_0 \in P$, and also that vectors $\boldsymbol{e}_1, \ldots, \boldsymbol{e}_r$ constitute a basis of plane P. Then a plane having the equation

$$A = B_0 + \lambda^1 \boldsymbol{e}_1 + \ldots + \lambda^r \boldsymbol{e}_r$$

is parallel to plane P and passes through point B_0. This proves the first statement of the theorem.

We assume further that P_1, P_2 are two parallel planes having a common point C_0. Let us consider some basis $\boldsymbol{e}_1, \ldots, \boldsymbol{e}_r$ of plane P_1. Since planes P_1, P_2 are parallel, the vectors $\boldsymbol{e}_1, \ldots, \boldsymbol{e}_r$ are also a basis of plane P_2. Consequently, planes P_1, P_2 have a common point C_0 and a common basis $\boldsymbol{e}_1, \ldots, \boldsymbol{e}_r$. It follows that both of these planes are described by the same vector parametric equation

$$A = C_0 + \lambda^1 \boldsymbol{e}_1 + \ldots + \lambda^r \boldsymbol{e}_r$$

and thus coincide.

THEOREM 18.7. *Let P be a plane of dimension r in space X and let $A_0, A_1, \ldots, A_r$ be an independent system of points situated in this plane. A point $A \in X$ will* *belong to plane P if, and only if, there exist real numbers* $\lambda^0, \lambda^1, \ldots, \lambda^r$ *satisfying the conditions*

$$\lambda^0 + \lambda^1 + \ldots + \lambda^r = 1, \tag{18.10}$$

$$A = \lambda^0 A_0 + \lambda^1 A_1 + \ldots + \lambda^r A_r. \tag{18.11}$$

These numbers $\lambda^0, \lambda^1, \ldots, \lambda^r$ *are defined uniquely by the point* $A \in P$.

Proof. Since points $A_0, A_1, \ldots, A_r$ are independent, vectors

$$e_1 = \overrightarrow{A_0A_1}, \ldots, e_r = \overrightarrow{A_0A_r}$$

are linearly independent. All these vectors belong to direction subspace L_P of plane P, and thus they constitute a *basis* of subspace L_P and plane P. Consequently, a point $A \in X$ belongs to plane P if, and only if, it can be described in the form

$$A = A_0 + \lambda^1 e_1 + \ldots + \lambda^r e_r,$$

that is, if

$$\overrightarrow{A_0A} = \lambda^1 \overrightarrow{A_0A_1} + \ldots + \lambda^r \overrightarrow{A_0A_r}$$

(here the numbers $\lambda^1, \ldots, \lambda^r$ are defined uniquely by point $A \in P$). If we now take some arbitrary point $Q \in X$, then we can rewrite this relation as

$$\overrightarrow{QA} - \overrightarrow{QA_0} = \lambda^1 (\overrightarrow{QA_1} - \overrightarrow{QA_0}) + \ldots + \lambda^r (\overrightarrow{QA_r} - \overrightarrow{QA_0}),$$

or

$$\overrightarrow{QA} = (1 - \lambda^1 - \ldots - \lambda^r) \overrightarrow{QA_0} + \lambda^1 \overrightarrow{QA_1} + \ldots + \lambda^r \overrightarrow{QA_r},$$

this relation being equivalent to relations (18.10) and (18.11) (where $\lambda^0 = 1 - \lambda^1 - \ldots - \lambda^r$).

COROLLARY 18.8. *Let P be some line and let* A_0, A_1 *be two different points on it. A point* $A \in X$ *will belong to (lie on) line P if, and only if, there exists a real number* λ *satisfying the condition*

$$A = (1 - \lambda) A_0 + \lambda A_1.$$

This corollary also implies that one, and only one, line passes through any two different points $A_0, A_1 \in X$. Let us call this "line A_0A_1." It is obvious that for any two points $A_0, A_1 \in X$ the segment $[A_0, A_1]$ (see Example 17.5) lies wholly on line A_0A_1.

THEOREM 18.9. *The intersection of two (or any number of) planes is itself a plane.*

Proof. Let P_1 and P_2 be two planes and let $P = P_1 \cap P_2$ be their intersection. Further, let $A_1, \ldots, A_s$ be arbitrary points of set P and let $\lambda^1, \ldots, \lambda^s$ be real numbers satisfying the condition

$$\lambda^1 + \ldots + \lambda^s = 1. \tag{18.12}$$

Since $A_1, \ldots, A_s \in P \subset P_1$ and since P_1 is a plane, the point

$$A = \lambda^1 A_1 + \ldots + \lambda^s A_s \tag{18.13}$$

belongs to plane P_1. On the basis of similar considerations, we know that point A belongs to plane P_2. Consequently, $A \in P_1 \cap P_2 = P$. Thus, any point having form (18.13), where $A_1, \ldots, A_s \in P$ and numbers $\lambda^1, \ldots, \lambda^s$ satisfy condition (18.12), belongs to set P, so that P is a plane. Similar reasoning can be applied for the intersection of any number of planes.

Theorem 18.9 implies that for any set $Q \subset X$ there exists in X a *least* plane containing set Q, that is, a plane P_Q containing set Q, such that, if some plane P contains set Q, then $P \supset P_Q$ (cf. p. 89). It is called the plane generated by set Q.

THEOREM 18.10. *Let Q be a set contained in X, and let P_Q be the plane generated by this set. A point $A \in X$ belongs to plane P_Q if, and only if, there exist points $A_1, \ldots, A_s \in Q$ and real numbers $\lambda^1, \ldots, \lambda^s$ satisfying condition* (18.12), *such that equation* (18.13) *is valid.*

Proof. Consider in Q the largest number of independent points, that is, $r+1$ points $B_0, B_1, \ldots, B_r \in Q$ so chosen that they are independent, whereas each $r+2$ points of set Q are dependent. If P is a plane of dimension r containing points $B_0, B_1, \ldots, B_r$, then according to Theorem 18.7 plane P consists of all points B having the form

$$B = \lambda^0 B_0 + \lambda^1 B_1 + \ldots + \lambda^r B_r, \tag{18.14}$$

where $\lambda^0, \lambda^1, \ldots, \lambda^r$ are real numbers satisfying the condition

$$\lambda^0 + \lambda^1 + \ldots + \lambda^r = 1. \tag{18.15}$$

Since plane P_Q contains set Q and thus contains points $B_0, B_1, \ldots, B_r$, it follows that P_Q contains any point of form (18.14), where $\lambda^0, \lambda^1, \ldots, \lambda^r$ satisfy condition (18.15), that is, $P_Q \supset P$.

Note that the vectors

$$e_1 = \overrightarrow{B_0B_1}, \ldots, e_r = \overrightarrow{B_0B_r} \tag{18.16}$$

constitute a basis of plane P. Now, if B is an arbitrary point of set Q, then points $B_0, B, B_1, \ldots, B_r$ are dependent, that is, the vectors

$$\overrightarrow{B_0B}, \quad \overrightarrow{B_0B_1}, \ldots, \overrightarrow{B_0B_r}$$

are linearly dependent. Consequently, vector $\overrightarrow{B_0B}$ can be linearly expressed in terms of vectors (18.16) (since vectors (18.16) are linearly independent):

$$\overrightarrow{B_0B} = \mu^1 e_1 + \ldots + \mu^r e_r,$$

that is,

$$B = B_0 + \mu^1 e_1 + \ldots + \mu^r e_r.$$

Hence, in view of Theorem 18.2, it follows that $B \in P$. Thus $Q \subset P$, so that $P_Q \subset P$.

The specified inclusions $P_Q \supset P$, $P_Q \subset P$ indicate that $P_Q = P$. Thus it follows from (18.14) and (18.15) that each point in plane P_Q is written in the required form (see (18.12), (18.13)). Conversely, it is also true that, if point A takes the form (18.12), (18.13), where $A_1, \ldots, A_s \in Q$ and thus $A_1, \ldots, A_s \in P_Q$, then $A \in P_Q$.

THEOREM 18.11. *Let P be a plane of dimension r in an affine space X, and let O be a point in plane P. If we transform X into a vector space as indicated in Theorem 17.7, then P will be an r-dimensional subspace of this vector space.*

Proof. Let A, $B \in P$. Then a point $C = A + B - O$ belongs to P (since P is a plane). However, in the vector space being considered point C is the sum of points A and B:

$$C = A + B.$$

Hence, if A, $B \in P$, then $A + B \in P$. Similarly, it can be shown that if $A \in P$, then $\lambda A \in P$ for any real λ. Consequently, P is a subspace.

THEOREM 18.12. *Let P_1 and P_2 be two planes in an affine space X which have at least one common point, and let $P = P_1 \cap P_2$ be the intersection of these planes, while P^* is the plane generated by the set $P_1 \cup P_2$. Then*

$$\dim P^* = \dim P_1 + \dim P_2 - \dim P.$$

Proof. If we take any point $O \in P$ and transform X into a vector space as indicated in Theorem 17.7, then P_1, P_2, P, P^* will be subspaces of this vector space X. The relation to be proven now follows directly from Theorem 14.9.

DEFINITION 18.13. Let (R, X) be an affine space. Planes P, $Q \subset X$ are called *mutually complementary* if their direction subspaces L_P, L_Q are direct complements of one another, that is, if $R = L_P \oplus L_Q$.

This definition implies that, if P_1 and P_2 are mutually complementary planes and if $P_1' \| P_1$, $P_2' \| P_2$, then planes P_1' and P_2' are also mutually complementary. Moreover, in view of Theorem 14.10, for the mutually complementary planes P_1, P_2 of space X it will be true that

$$\dim P_1 + \dim P_2 = \dim X.$$

THEOREM 18.14. *Two mutually complementary planes in an affine space have only one common point.*

Proof. Let (R, X) be an affine space and let P, Q be two mutually complementary planes in X. Let the direction subspaces of planes P, Q be L_P and L_Q, respectively. If $\dim L_P = p$, $\dim L_Q = q$ and $\dim R = n$, then $p + q = n$. Consider a basis $e_1, \ldots, e_p$ in subspace L_P and a basis $e_{p+1}, \ldots, e_n$ in subspace L_Q; the note to Theorem 14.10 implies that vectors $e_1, \ldots, e_p, e_{p+1}, \ldots, e_n$ constitute a basis of vector space R.

If we now consider the arbitrary points $A \in P$, $B \in Q$, then by constructing the vector $\overrightarrow{AB}$ using the vectors of basis $e_1, \ldots, e_n$, we get

$$\overrightarrow{AB} = \alpha^1 e_1 + \ldots + \alpha^p e_p + \alpha^{p+1} e_{p+1} + \ldots + \alpha^n e_n. \tag{18.17}$$

Let $C \in X$ be the point defined by the equation

$$\overrightarrow{AC} = \alpha^1 e_1 + \ldots + \alpha^p e_p. \tag{18.18}$$

From equations (18.17) and (18.18) it follows that

$$\overrightarrow{BC} = \overrightarrow{AC} - \overrightarrow{AB} = -\alpha^{p+1} e_{p+1} - \ldots - \alpha^n e_n. \tag{18.19}$$

Taking into account inclusions $A \in P$, $B \in Q$ and relations (18.18), (18.19), we find from Theorem 18.2 that $C \in P$, $C \in Q$. Thus planes P and Q have a common point C.

If D is another point common to these planes, then $\overrightarrow{CD} \in L_P$, $\overrightarrow{CD} \in L_Q$, so that in view of the relation $R = L_P \oplus L_Q$ we have $\overrightarrow{CD} = \mathbf{0}$, or $C = D$. The theorem is thereby proven.

In conclusion, let us introduce the concept of a ray,* which will be useful in the following. If $A_0 \in X$, $e \in R$, then the *ray with an endpoint A_0 and a direction vector e* is defined as the set of all points of the form

$$A = A_0 + \lambda e, \qquad \lambda \geqslant 0. \tag{18.20}$$

This relation differs from equation (18.9) in that here λ takes on *nonnegative* values only, rather than all real values. Thus the ray with an endpoint A_0 and a direction vector e is wholly contained in the line passing through point A_0 and having the direction vector e. This same line also contains a ray with an endpoint A_0 and a direction vector $-e$, known as the *negative* of ray (18.20). No third ray with an endpoint A_0, aside from these two, is contained in line (18.9). Therefore, the two indicated rays are called the *rays defined by point A_0 on line* (18.9).

Note that, if k is an arbitrary positive number, then ray (18.20) *coincides* with the ray having the same endpoint A_0 and a direction vector ke. In other words, if l is some ray with an endpoint A_0 and if B is any point on this ray different from A_0, then the vector $\overrightarrow{A_0B}$ is a direction vector of ray l; in this case it is also said that l is the *ray originating at point A_0 and passing through point B*. Consequently, the

* A "ray" is also known as a "half-line" (Translator).

ray originating at point A_0 and passing through point B constitutes the set of all points A for which $\overrightarrow{A_0A} = \lambda\overrightarrow{A_0B}$, where $\lambda \geqslant 0$ (Figure 51).

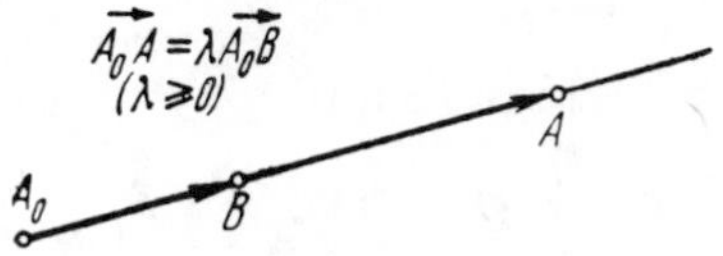

FIGURE 51

19. Affine mappings. Let (R, X) and (S, Y) be affine spaces. A mapping $\varphi\colon X \to Y$ is called an *affine* (or *linear*) mapping if validity of the relation

$$A = (1-\lambda) B + \lambda C,$$

where $A, B, C \in X$, implies that

$$\varphi(A) = (1-\lambda)\varphi(B) + \lambda\varphi(C).$$

THEOREM 19.1. *Let* $\varphi\colon X \to Y$ *be an affine mapping. Then validity of the relation*

$$A = \lambda^1 A_1 + \ldots + \lambda^s A_s, \qquad (19.1)$$

where $A, A_1, \ldots, A_s \in X$ *and*

$$\lambda^1 + \ldots + \lambda^s = 1, \qquad (19.2)$$

implies that

$$\varphi(A) = \lambda^1\varphi(A_1) + \ldots + \lambda^s\varphi(A_s). \qquad (19.3)$$

Proof. We carry out induction on s. For $s = 2$ the relation to be proven follows directly from the definition of an affine mapping.

Assume that the relation to be proven is established for the case where there are fewer than s terms on the right-hand side and also assume that relations (19.1) and (19.2) are satisfied, where $s \geqslant 3$. Obviously, at least one of the numbers $\lambda^1, \ldots, \lambda^s$ is different from unity (otherwise (19.2) would not be valid). Assume also, in order to be definite, that $\lambda^s \neq 1$, that is, $1 - \lambda^s \neq 0$. Consider the point

$$B = \frac{\lambda^1}{1-\lambda^s} A_1 + \ldots + \frac{\lambda^{s-1}}{1-\lambda^s} A_{s-1}; \qquad (19.4)$$

here the sum of the coefficients on the right is equal to unity:

$$\frac{\lambda^1}{1-\lambda^s} + \ldots + \frac{\lambda^{s-1}}{1-\lambda^s} = \frac{\lambda^1 + \ldots + \lambda^{s-1}}{1-\lambda^s} = \frac{1-\lambda^s}{1-\lambda^s} = 1.$$

By induction, the following relation is valid:

$$\varphi(B) = \frac{\lambda^1}{1-\lambda^s}\varphi(A_1) + \ldots + \frac{\lambda^{s-1}}{1-\lambda^s}\varphi(A_{s-1}). \tag{19.5}$$

From (19.1) and (19.4) it then follows that

$$A = (1-\lambda^s)B + \lambda^s A_s,$$

so that, in view of the definition of an affine mapping,

$$\varphi(A) = (1-\lambda^s)\varphi(B) + \lambda^s\varphi(A_s). \tag{19.6}$$

Relations (19.5) and (19.6) imply relation (19.3).

COROLLARY 19.2. *Let* φ: $X \to Y$ *be an affine mapping and let* $\varphi(A_i) = \overset{*}{A_i}$, $\varphi(B_i) = \overset{*}{B_i}$, $i = 1, \ldots, k$ *(where* $A_i, B_i \in X$*). Then validity of the relation*

$$\alpha^1 \overrightarrow{A_1B_1} + \ldots + \alpha^k \overrightarrow{A_kB_k} = 0 \tag{19.7}$$

implies that

$$\alpha^1 \overrightarrow{\overset{*}{A_1}\overset{*}{B_1}} + \ldots + \alpha^k \overrightarrow{\overset{*}{A_k}\overset{*}{B_k}} = 0. \tag{19.8}$$

Proof. Equation (19.7) implies that

$$\alpha^1(B_1 - A_1) + \ldots + \alpha^k(B_k - A_k) = 0,$$

that is,

$$Q = Q + \alpha^1 B_1 + \ldots + \alpha^k B_k - \alpha^1 A_1 - \ldots - \alpha^k A_k$$

(where Q is an arbitrary point of space X). Consequently, according to Theorem 19.1,

$$Q^* = Q^* + \alpha^1 B_1^* + \ldots + \alpha^b B_k^* - \alpha^1 A_1^* - \ldots - \alpha^k A_k^*$$

where $Q^* = \varphi(Q)$), that is,

$$\alpha^1(B_1^* - A_1^*) + \ldots + \alpha^k(B_k^* - A_k^*) = 0,$$

which is relation (19.8).

THEOREM 19.3. *Let* $\varphi: X \to Y$ *be an affine mapping and let* $P \subset X$ *be some plane. Then its image* $\varphi(P)$ *is a plane in space* Y, *with* $\dim \varphi(P) \leqslant \dim P$.

Proof. Let $B_1, \ldots, B_s$ be arbitrary points of set $\varphi(P)$ and let $\lambda^1, \ldots, \lambda^s$ be real numbers satisfying the condition

$$\lambda^1 + \ldots + \lambda^s = 1.$$

Let us consider a point

$$B = \lambda^1 B_1 + \ldots + \lambda^s B_s$$

of space Y and let us show that $B \in \varphi(P)$. Since $B_i \in \varphi(P)$, $i = 1, \ldots, s$, there must exist some point $A_i \in P$ such that $\varphi(A_i) = B_i$. Thus we obtain points $A_1, \ldots, A_s$ of plane P. Consequently, the point

$$A = \lambda^1 A_1 + \ldots + \lambda^s A_s$$

also belongs to plane P. In view of Theorem 19.1, we have

$$\varphi(A) = \varphi(\lambda^1 A_1 + \ldots + \lambda^s A_s) = \lambda^1 \varphi(A_1) + \ldots + \lambda^s \varphi(A_s) = \\ = \lambda^1 B_1 + \ldots + \lambda^s B_s = B.$$

Since $A \in P$, it follows that $B = \varphi(A) \in \varphi(P)$. Hence $\varphi(P)$ is a plane.

If points $B_1, \ldots, B_s$ of plane $\varphi(P)$ are independent, then points $A_1, \ldots, A_s$ satisfying conditions $\varphi(A_i) = B_i$, $i = 1, \ldots, s$, will obviously also be independent. From this the inequality $\dim \varphi(P) \leqslant \dim P$ also follows.

Similarly, we can prove

THEOREM 19.4. *The inverse image* $\varphi^{-1}(P)$ *of any plane* $P \subset Y$, *for an affine mapping* $\varphi: X \to Y$, *is a plane of space* X.

THEOREM 19.5. *Let* (R, X) *and* (S, Y) *be affine spaces and let* φ: $X \to Y$ *be an affine mapping. Then there exists (albeit only one) homomorphism* ψ: $R \to S$ *possessing the property that validity of the relation*

$$A = B + \boldsymbol{a} \qquad (A, B \in X, \ \boldsymbol{a} \in R) \tag{19.9}$$

implies that

$$\varphi(A) = \varphi(B) + \psi(\boldsymbol{a}). \tag{19.10}$$

Proof. We specify in space X some point O. Now if $\boldsymbol{a}$ is an arbitrary vector of space R, then we consider the point $A \in X$ for which $\overrightarrow{OA} = \boldsymbol{a}$ and we set

$$\psi(\boldsymbol{a}) = \varphi(A) - \varphi(O) = \overrightarrow{\varphi(O)\varphi(A)} \in S. \tag{19.11}$$

Thus some mapping ψ: $R \to S$ is defined.

Let us show that ψ is a homomorphism. If $\boldsymbol{a}, \boldsymbol{b} \in R$, then we can select points $A, B, C \in X$ such that

$$\overrightarrow{OA} = \boldsymbol{a}, \quad \overrightarrow{OB} = \boldsymbol{b}, \quad \overrightarrow{OC} = \boldsymbol{a} + \boldsymbol{b}.$$

Then $\overrightarrow{OC} - \overrightarrow{OA} - \overrightarrow{OB} = \boldsymbol{0}$, so that in view of Corollary 19.2

$$\overrightarrow{\varphi(O)\varphi(C)} - \overrightarrow{\varphi(O)\varphi(A)} - \overrightarrow{\varphi(O)\varphi(B)} = \boldsymbol{0},$$

that is, $\psi(\boldsymbol{a} + \boldsymbol{b}) - \psi(\boldsymbol{a}) - \psi(\boldsymbol{b}) = \boldsymbol{0}$. Therefore

$$\psi(\boldsymbol{a} + \boldsymbol{b}) = \psi(\boldsymbol{a}) + \psi(\boldsymbol{b}) \tag{19.12}$$

for any vectors $\boldsymbol{a}, \ \boldsymbol{b} \in R$.

Further, we assume that $\boldsymbol{a} \in R$ and that λ is a real number. If we so choose points $A, D \in X$ that

$$\overrightarrow{OA} = \boldsymbol{a}, \quad \overrightarrow{OD} = \lambda \boldsymbol{a},$$

then $\overrightarrow{OD} - \lambda\overrightarrow{OA} = \mathbf{0}$, and according to Corollary 19.2

$$\overrightarrow{\varphi(O)\varphi(D)} - \lambda\overrightarrow{\varphi(O)\varphi(A)} = \mathbf{0},$$

that is, $\psi(\lambda\boldsymbol{a}) - \lambda\psi(\boldsymbol{a}) = \mathbf{0}$. Therefore,

$$\psi(\lambda\boldsymbol{a}) = \lambda\psi(\boldsymbol{a}) \tag{19.13}$$

for any $\boldsymbol{a} \in R$ and any real λ. It follows from (19.12) and (19.13) that ψ is a homomorphism.

If relation (19.9) holds true, then in view of (19.11) we have

$$\varphi(B) - \varphi(A) = (\varphi(B) - \varphi(O)) - (\varphi(A) - \varphi(O)) = \psi(\overrightarrow{OB}) - \\ - \psi(\overrightarrow{OA}) = \psi(\overrightarrow{OB} - \overrightarrow{OA}) = \psi(\overrightarrow{AB}) = \psi(B - A) = -\psi(\boldsymbol{a}),$$

that is, the constructed homomorphism ψ is the desired one.

Uniqueness follows from the fact that relation (19.11) is a *corollary* of relations (19.9) and (19.10) (the corollary is obtained if in (19.9) and (19.10) point B is assumed to coincide with O). Thus any homomorphism satisfying the imposed requirements (see (19.9) and (19.10)) must, in particular, satisfy relation (19.11), that is, it must be identical to ψ.

The following theorem is, in a certain sense, the converse of the preceding one.

THEOREM 19.6. *Let (R, X) and (S, Y) be affine spaces and let $\psi\colon R \to S$ be some homomorphism. In addition, O is some point of space X and O' is some point of space Y. Then the mapping $\varphi\colon X \to Y$, defined by the formula*

$$\varphi(A) = O' + \psi(\overrightarrow{OA}),$$

is affine.

Proof. If $A = (1 - \lambda)B + \lambda C$ (where $A, B, C \in X$), then

$$\overrightarrow{OA} = (1 - \lambda)\overrightarrow{OB} + \lambda\overrightarrow{OC},$$

and thus

$$\varphi(A) = O' + \psi(\overrightarrow{OA}) = O' + \psi((1 - \lambda)\overrightarrow{OB} + \lambda\overrightarrow{OC}) = \\ = O' + (1 - \lambda)\psi(\overrightarrow{OB}) + \lambda\psi(\overrightarrow{OC}) = (1 - \lambda)(O' + \psi(\overrightarrow{OB})) + \\ + \lambda(O' + \psi(\overrightarrow{OC})) = (1 - \lambda)\varphi(B) + \lambda\varphi(C).$$

THEOREM 19.7. *Let φ: $X \to Y$ be an affine mapping. We select an arbitrary point $O \in X$ and let O' be the point $\varphi(O)$ of space Y. With point O fixed, we now transform X into a vector space and, with point O' fixed, we transform Y into a vector space as well. Then mapping φ will be a homomorphism of vector space X into vector space Y.*

Proof. Let A, $B \in X$; if we consider a point $C = A + B - O$, then C is the sum of points A and in vector space X. From the relation $C = A + B - O$ it follows, in view of the affineness of mapping φ, that $\varphi(C) = \varphi(A) + \varphi(B) - \varphi(O) = \varphi(A) + \varphi(B) - O'$, and this means that point $\varphi(C)$ is in vector space Y the sum of points $\varphi(A)$ and $\varphi(B)$. Thus mapping φ: $X \to Y$ preserves the operation of addition. The validity of the relation $\varphi(\lambda A) = \lambda\varphi(A)$ is established similarly.

It should be noted that Theorem 19.7 provides us with new proofs for Theorems 19.3 and 19.4. The following theorem, which is the converse of Theorem 19.7, is also readily proven.

THEOREM 19.8. *Let X, Y be affine spaces and let φ: $X \to Y$ be some mapping. We select an arbitrary point $O \in X$ and we set $O' = \varphi(O)$. As in the previous theorem, we transform X and Y into vector spaces, having fixed points O and O'. If φ is a homomorphism of these vector spaces, then φ is an affine mapping of space X into space Y.*

THEOREM 19.9. *Let X and Y be affine spaces. If we assume that $\dim X = n$ and that $A_0, A_1, \ldots, A_n$ form an independent system of points in X, then for any points $B_0, B_1, \ldots, B_n$ of space Y there must exist one, and only one, affine mapping φ: $X \to Y$ satisfying the condition*

$$\varphi(A_0) = B_0, \quad \varphi(A_1) = B_1, \ \ldots, \ \varphi(A_n) = B_n. \tag{19.14}$$

Proof. Let A be an arbitrary point of space X. Then, in view of Theorem 18.7 (as applied to the case $P = X$)

$$A = \lambda^0 A_0 + \lambda^1 A_1 + \ldots + \lambda^n A_n, \tag{19.15}$$

where $\lambda^0, \lambda^1, \ldots, \lambda^n$ are real numbers satisfying the relation

$$\lambda^0 + \lambda^1 + \ldots \lambda^n = 1.$$

Here the numbers $\lambda^0, \lambda^1, \ldots, \lambda^n$ are defined uniquely by point A. Let us set

$$\varphi(A) = \lambda^0 B_0 + \lambda^1 B_1 + \ldots + \lambda^n B_n. \tag{19.16}$$

Thus we obtain some mapping φ: $X \to Y$ obviously satisfying relations (19.14).

Now let us fix in X a point $O = A_0$ and in Y a point $O' = \varphi(A_0) = B_0$, having thereby transformed X and Y into vector spaces. Then, in view of the operations carried out in these vector spaces, relations (19.15) and (19.16) become

$$A = \lambda^1 A_1 + \ldots + \lambda^n A_n,$$
$$\varphi(A) = \lambda^1 \varphi(A_1) + \ldots + \lambda^n \varphi(A_n).$$

Hence it follows that φ is a homomorphism of the given vector spaces, and from Theorem 19.8 we know that φ is an affine mapping.

Uniqueness is implied by the fact that, for any affine mapping $\varphi: X \to Y$, relation (19.16) follows from (19.14) and (19.15).

DEFINITION 19.10. Let (R, X) be an affine space and let P, Q be mutually complementary planes in X. If we draw through an arbitrary point $A \in X$ a plane Q' parallel to plane Q (Theorem 18.6), then P and Q' will also be mutually complementary planes in X, so that P and Q' have a single common point (Theorem 18.14). Let us call this common point $\pi(A)$. Therefore, we obtain some mapping $\pi: X \to X$. It is called the *projection* of space X onto plane P in the direction of plane Q (cf. Definition 15.6).

THEOREM 19.11. *Let (R, X) be an affine space and let P, Q be mutually complementary planes in X. The projection of space X onto plane P in the direction of plane Q is then an affine mapping.*

Proof. Let O be the common point of planes P and Q (Theorem 18.14). Having fixed a point O, we transform X into a vector space. Then P and Q will be subspaces of this space (Theorem 18.11). Since intersection $P \cap Q$ contains only one point O (that is, it is a trivial subspace) and since $\dim P + \dim Q = \dim X$, it follows that $X = P \oplus Q$ (Theorem 14.10).

Consequently, we can consider the *projection* of vector space X onto subspace P in the direction of subspace Q (see Theorem 15.7). Let us call this projection π^*. According to Theorem 15.7, the mapping π^* of vector space X into itself is a homomorphism. Obviously, $\pi^*(O) = O$. Now, from Theorem 19.8 we know that the mapping π^* of affine space X into itself is affine. Thus to complete the proof it is sufficient to establish that mappings π and π^* are identical.

Let A be an arbitrary point of space X and let $B = \pi(A)$. Then B is a common point of planes P and Q', where Q' is parallel to plane Q and passes through point A. Since $A \in Q'$ and $B \in Q'$, we know that $\overrightarrow{BA} \in L_{Q'}$, where $L_{Q'}$ is the direction subspace of plane Q' (and thus of plane Q). If C is a point such that $\overrightarrow{OC} = \overrightarrow{BA}$, then we know from Theorem 18.2 that $C \in Q$. The equation $\overrightarrow{OC} = \overrightarrow{BA}$ implies that

$$A = B + \overrightarrow{OC}, \quad \text{that is,} \quad A = B + C - O,$$

so that in vector space X the relation $A = B + C$ is valid (see Theorem 17.7). From the relations $A = B + C$, $B \in P$, $C \in Q$ valid in vector space X, it follows that $B = \pi^*(A)$ (see Definition 15.6). Hence $\pi^*(A) = \pi(A)$ for any point $A \in X$, that is, mappings π and π^* are identical.

THEOREM 19.12. *Let (R, X) be an affine space and let P, Q be mutually complementary planes in X. If π is the projection of space X onto plane P in the direction of plane Q, then for any plane $N \subset P$ the inverse image $\pi^{-1}(N)$ is a plane in X having a dimension $\dim Q + \dim N$. In particular, if N is a hyperplane of affine space P, then $\pi^{-1}(N)$ is a hyperplane of plane X.*

Proof. The fact that $\pi^{-1}(N)$ is a plane follows directly from Theorems 19.11 and 19.4. If O is the common point of planes P and Q and if we transform X into a vector space, having fixed point O, then π will be the projection of vector space X onto subspace P in the direction of subspace Q (see proof of Theorem 19.11), so that from Theorem 15.7 we know that $\operatorname{Ker}\pi = Q$ and $\operatorname{Im}\pi = P$.

Now, if π_1 is the mapping π considered in subspace $\pi^{-1}(N)$, that is, if π_1 is the projection of space $\pi^{-1}(N)$ onto N in the direction of subspace Q, then $\operatorname{Im}\pi_1 = N$, $\operatorname{Ker}\pi_1 = \operatorname{Ker}\pi = Q$, and from Theorem 15.3 it follows that

$$\dim \pi^{-1}(N) = \dim(\operatorname{Im}\pi_1) + \dim(\operatorname{Ker}\pi_1) = \dim Q + \dim N.$$

If, in particular, N is a hyperplane of affine space P, that is, $\dim N = \dim P - 1$, then

$$\dim \pi^{-1}(N) = \dim N + \dim Q = (\dim P - 1) + \dim Q = \dim X - 1,$$

that is, $\pi^{-1}(N)$ is a hyperplane of space X.

THEOREM 19.13. *Let P_1 and P_2 be two parallel k-dimensional planes in X. Then there exists in X a plane P of dimension $k+1$, containing P_1 and P_2 (here if $P_1 \neq P_2$ plane P is unique).*

Proof. Let Q be a plane which is a direct complement of P_1. If π is the projection of space X onto plane Q in the direction of plane P_1, then, in view of the definition of the projection operation, $\pi(P_1) = A_1$ is *one point* of plane Q and $\pi(P_2) = A_2$ is also *one point* of plane Q. If we draw a line N through points A_1 and A_2, then $N \subset Q$, so that according to Theorem 19.12 the set $\pi^{-1}(N)$ is a plane in X of dimension

$$\dim \pi^{-1}(N) = \dim N + \dim P_1 = 1 + k.$$

It is also clear that $\pi^{-1}(N) \supset P_1$ and $\pi^{-1}(N) \supset P_2$. Thus plane $\pi^{-1}(N)$ is the desired plane.

If $P_1 \neq P_2$, then set $P_1 \cup P_2$ is not contained in a single k-dimensional plane, that is, the plane P generated by the set $P_1 \cup P_2$ is of dimension not less than $k+1$ (and not greater than $k+1$, in view of what was proven). Therefore, $\dim P = k+1$, so that any $(k+1)$-dimensional plane containing $P_1 \cup P_2$ must contain the $(k+1)$–dimensional plane P and thus must be identical to P.

EXAMPLE 19.14. Let (R, X) be an affine space, let $O \in X$ be some point, and let ν be a real number. By setting

$$f(A) = O + \nu\overrightarrow{OA}, \quad A \in X,$$

we obtain some mapping f of space X into itself. It is called a *homothety* with a center O and a coefficient ν. Since the mapping φ: $R \to R$ defined by the equation $\varphi(\boldsymbol{a}) = \nu\boldsymbol{a}$ is obviously a homomorphism, we know from Theorem 19.6 that a homothety is an *affine* mapping of space X into itself.

A homothety with a center O and a coefficient $\nu = -1$ is called a *symmetry around point* O. This symmetry transforms a point $A \in X$ into a point A' such that

$A' = O - \overrightarrow{OA}$, that is, $\overrightarrow{OA'} = -\overrightarrow{OA}$. This means that O is the midpoint of segment $[A, A']$.

If the coefficient of a homothety is different from zero, then each plane $P \subset X$ is transformed by the homothety into a plane parallel to it. Now, if $A_0 \in P$ and if $\boldsymbol{e}_1, \dots, \boldsymbol{e}_k$ constitute a basis of plane P ,an arbitrary point $A \in P$ has the form

$$A = A_0 + \lambda^1 \boldsymbol{e}_1 + \dots + \lambda^k \boldsymbol{e}_k,$$

so that

$$\overrightarrow{OA} = \overrightarrow{OA_0} + \lambda^1 \boldsymbol{e}_1 + \dots + \lambda^k \boldsymbol{e}_k.$$

Consequently,

$$f(A) = O + \nu\overrightarrow{OA} = O + \nu(\overrightarrow{OA_0} + \lambda^1 \boldsymbol{e}_1 + \dots + \lambda^k \boldsymbol{e}_k) =$$
$$= O + \nu\overrightarrow{OA_0} + \nu(\lambda^1 \boldsymbol{e}_1 + \dots + \lambda^k \boldsymbol{e}_k) = f(A_0) + \nu(\lambda^1 \boldsymbol{e}_1 + \dots + \lambda^k \boldsymbol{e}_k).$$

From this it is clear that, when A runs through plane P (that is, when $\lambda^1, \dots, \lambda^k$ run through, independently of one another, the real values), point $f(A)$ runs through a plane passing through point $f(A_0)$ and having the *same* basis $\boldsymbol{e}_1, \dots, \boldsymbol{e}_k$. Thus $f(P)$ is a plane passing through point $f(A_0)$ and lying parallel to plane P.

20. Affine functions. Let (R, X) be an affine space. The affine mapping $f\colon X \to D$, where D is the set of all real numbers (considered as an affine space; see Note 17.9), is called an *affine (or linear) function* in space X. In other words, a function f (with real values) defined in X is called affine if from the relation

$$A = (1 - \lambda) B + \lambda C,$$

valid in X, it follows that

$$f(A) = (1 - \lambda) f(B) + \lambda f(C).$$

According to Theorem 19.1, validity in X of the relations

$$A = \lambda^1 A_1 + \dots + \lambda^s A_s, \qquad \lambda^1 + \dots + \lambda^s = 1$$

implies for an affine function $f\colon X \to D$ that

$$f(A) = \lambda^1 f(A_1) + \dots + \lambda^s f(A_s).$$

Moreover, according to Theorem 19.5, for an affine function $f\colon X \to D$ it will be true that

$$f(B) = f(A) + \varphi(\overrightarrow{AB}), \qquad A, B \in X, \tag{20.1}$$

where $\varphi\colon R \to D$ is some homomorphism, that is, a linear functional (defined uniquely by affine function f). The converse of this follows from Theorem 19.6, namely that, if mapping $f\colon X \to D$ satisfies condition (20.1), where φ is a linear functional in R, then f is an affine function. Moreover, if mapping $f\colon X \to D$ satisfies the condition

$$f(B) = \lambda + \varphi(\overrightarrow{QB}), \qquad B \in X, \tag{20.2}$$

where Q is a fixed point of space X, λ is a real number, and $\varphi\colon R \to D$ is a linear functional, then it follows from Theorem 19.6 that f is an affine function.

If functional φ (see (20.1)) maps all of space R into zero, that is, if $\varphi(\boldsymbol{a}) = 0$ for any vector $\boldsymbol{a} \in R$, then $f(A) = f(B)$ for any two points $A, B \in X$, that is, the affine function f is *constant* over the entire space X. If, on the other hand, functional φ is *not* the zero functional, namely if there exists a vector $\boldsymbol{a} \in R$ such that $\varphi(\boldsymbol{a}) \neq 0$, then function f will not be constant. In this case for any number $\lambda \in D$ there exists in X a point B such that $f(B) = \lambda$. Actually, if we set $B = A + x\boldsymbol{a}$, then we get

$$f(B) = f(A) + x\varphi(\boldsymbol{a});$$

consequently, by taking an arbitrary point $A \in X$ and determining the number x from the equation

$$f(A) + x\varphi(\boldsymbol{a}) = \lambda$$

(this equation has a solution, since $\varphi(\boldsymbol{a}) \neq 0$), we can find the point B for which $f(B) = \lambda$. Thus, *if an affine function $f\colon X \to D$ is nonconstant, then it assumes in X all real values.*

It should also be noted that if $f\colon X \to D$ is an affine function, then for any number $\lambda \in D$ the function $f_1\colon X \to D$ defined by the relation $f_1(A) = f(A) + \lambda$ is also affine, and the same linear functional $\varphi\colon R \to D$ corresponds to it as to function f (this is implied directly by (20.1)).

THEOREM 20.1. *Let $f\colon X \to D$ be a nonconstant affine function. Then the set of all points $A \in X$ satisfying the condition $f(A) = 0$ (this set is called the* **kernel** *of function f, being designated as $\operatorname{Ker} f$) is a hyperplane of space X.*

Proof. In view of Theorem 19.4, the set $\operatorname{Ker} f = f^{-1}(0)$ is a *plane* in X (since O is a zero-dimensional plane in D). In order to calculate the dimension of this plane, we select in X a point O such that $f(O) = 0$ (such a point exists, since function f is assumed to be nonconstant) and, after fixing point O, we transform X into a vector space (see Theorem 17.7). Then from Theorem 19.7 we know that mapping $f\colon X \to D$ will be a homomorphism, while $\operatorname{Ker} f$ of affine mapping f will be the kernel of this homomorphism, the whole straight line R being the image $\operatorname{Im} f$ of homomorphism f (since function f is nonconstant), that is, $\dim(\operatorname{Im} f) = 1$. From Theorem 15.3 we now get

$$\dim(\operatorname{Ker} f) = \dim X - \dim(\operatorname{Im} f) = \dim X - 1,$$

so that $\operatorname{Ker} f$ is a hyperplane of affine space X.

THEOREM 20.2. *For each hyperplane* Γ *of an affine space* (R, X) *there exists a (nonconstant) affine function* $f\colon X \to D$ *such that* $\Gamma = \operatorname{Ker} f$.

Proof. Let $\dim X = n$, so that $\dim \Gamma = n-1$. If we select in Γ any n independent points $A_0, A_1, \ldots, A_{n-1}$ and if we let A_n be an arbitrary point of space X not lying in hyperplane Γ, then points $A_0, A_1, \ldots, A_{n-1}, A_n$ are independent. According to Theorem 19.9, there exists one, and only one, affine function $f\colon X \to D$ satisfying the conditions

$$f(A_0) = f(A_1) = \ldots = f(A_{n-1}) = 0, \quad f(A_n) = 1.$$

It is obvious that function f is nonconstant. $\operatorname{Ker} f$, the kernel of this affine function, contains all the points $A_0, A_1, \ldots, A_{n-1}$, and thus it contains the plane generated by these points, that is, $\operatorname{Ker} f \supset \Gamma$. If hyperplane $\operatorname{Ker} f$ contained any point $B \notin \Gamma$, then it would contain $n+1$ independent points $A_0, A_1, \ldots, A_{n-1}, B$, which is impossible, since $\dim(\operatorname{Ker} f) = n-1$. Hence $\operatorname{Ker} f$ does not contain any points not belonging to hyperplane Γ, and it follows that $\operatorname{Ker} f = \Gamma$.

THEOREM 20.3. *Let* $f_1, f_2\colon X \to D$ *be two affine functions differing from one another by a constant. Then hyperplanes* $\operatorname{Ker} f_1$, $\operatorname{Ker} f_2$ *are parallel.*

Proof. Since functions f_1, f_2 differ by a constant, it follows that the same linear functional $\varphi\colon R \to D$ will correspond to them both (see previous page):

$$f_1(B) = f_1(A) + \varphi(\overrightarrow{AB}), \quad f_2(B) = f_2(A) + \varphi(\overrightarrow{AB}).$$

Consequently, the direction subspaces for the kernels, $\operatorname{Ker} f_1$ and $\operatorname{Ker} f_2$, of these affine functions are identical (meaning that they are identical to the kernel, $\operatorname{Ker} \varphi$, of linear functional φ). However, this also indicates that hyperplanes $\operatorname{Ker} f_1$, $\operatorname{Ker} f_2$ are parallel.

THEOREM 20.4. *Let* $f_1, \ldots, f_k$ *be nonconstant affine functions defined in an* n*-dimensional affine space* (R, X), *and let* $\Gamma_1, \ldots, \Gamma_k$ *be the kernels of these affine functions, while* $\varphi_1, \ldots, \varphi_k$ *are the corresponding linear functionals defined in vector space* R:

$$f_i(B) = f_i(A) + \varphi_i(\overrightarrow{AB}), \qquad i = 1, \ldots, k.$$

If $\varphi_1, \ldots, \varphi_k$, *considered as vectors of the conjugate space* R^*, *are linearly independent, then the intersection* $\Gamma_1 \cap \ldots \cap \Gamma_k$ *of hyperplanes* $\Gamma_1, \ldots, \Gamma_k$ *is an* $(n-k)$*-dimensional plane in* X. *Any* $(n-k)$*-dimensional plane of space* X *can be represented in this form.*

Proof. Let us augment vectors $\varphi_1, \ldots, \varphi_k$ until the basis $\varphi_1, \ldots, \varphi_k, \varphi_{k+1}, \ldots, \varphi_n$ of space R^* is reached. Moreover, in space R we select a basis $e_1, \ldots, e_n$ such that (see Theorem 15.12)

$$\varphi_i(e_j) = \begin{cases} 0 & \text{for } i \neq j, \\ 1 & \text{for } i = j. \end{cases}$$

If we now take an arbitrary point $A \in X$ and assume that

$$\lambda^1 = -f_1(A), \ldots, \lambda^k = -f_k(A),$$
$$A_0 = A + \lambda^1 e_1 + \ldots + \lambda^k e_k,$$

then we have (for $i = 1, \ldots, k$)

$$f_i(A_0) = f_i(A) + \varphi_i(\lambda^1 e_1 + \ldots + \lambda^k e_k) =$$
$$= f_i(A) + \lambda^1 \varphi_i(e_1) + \ldots + \lambda^k \varphi_i(e_k) = f_i(A) + \lambda_i = 0.$$

Therefore, point A_0 belongs to hyperplane $\operatorname{Ker} f_i$, where $i = 1, \ldots, k$, that is, $A_0 \in \Gamma_1 \cap \ldots \cap \Gamma_k$.

Next we let B be an arbitrary point of space X, which can be written as

$$B = A_0 + x^1 e_1 + \ldots + x^n e_n.$$

For any $i = 1, \ldots, k$ we have

$$f_i(B) = f_i(A_0) + \varphi_i(x^1 e_1 + \ldots + x^n e_n) = x^i. \tag{20.3}$$

If point B belongs to the intersection $\Gamma_1 \cap \ldots \cap \Gamma_k$, that is, if $f_i(B) = 0$ for $i = 1, \ldots, k$, then it follows that $x^i = 0$, $i = 1, \ldots, k$, so that point B has the form

$$B = A_0 + x^{k+1} e_{k+1} + \ldots + x^n e_n. \tag{20.4}$$

Conversely, if point B has form (20.4), then it follows from (20.3) that $f_i(B) = 0$ for $i = 1, \ldots, k$, that is, $B \in \Gamma_1 \cap \ldots \cap \Gamma_k$. Hence $\Gamma_1 \cap \ldots \cap \Gamma_k$ is the set of all points of type (20.4), that is, this set is an $(n-k)$-dimensional plane passing through point A_0 and having as a direction subspace the $(n-k)$-dimensional subspace generated in P by vectors $e_{k+1}, \ldots, e_n$.

Now let us prove the latter statement of the theorem. Let P be an arbitrary $(n-k)$-dimensional plane in X. We consider a basis $e_{k+1}, \ldots, e_n$ of plane P and we augment it with vectors $e_1, \ldots, e_k$ until basis $e_1, \ldots, e_n$ of space R is reached. We also specify in P an arbitrary point A_0. Next we define the linear functionals $\varphi_1, \ldots, \varphi_k$ in R, assuming for any vector $x = x^1 e_1 + \ldots + x^n e_n \in R$ that

$$\varphi_i(x) = x^i, \qquad i = 1, \ldots, k.$$

Finally, we define affine functions $f_1, \ldots, f_k$ in X, assuming that

$$f_i(A) = \varphi_i(\overrightarrow{A_0 A}), \qquad i = 1, \ldots, k.$$

If

$$A = A_0 + x^1 e_1 + \ldots + x^n e_n$$

is an arbitrary point of space X, then

$$\overrightarrow{A_0A} = x^1e_1 + \ldots + x^ne_n,$$

and thus

$$f_i(A) = \varphi_i(\overrightarrow{A_0A}) = \varphi_i(x^1e_1 + \ldots + x^ne_n) = x^i, \qquad i = 1, \ldots, k.$$

Consequently, the system of equations

$$f_1(A) = 0, \ldots, f_k(A) = 0$$

is equivalent to saying that point A has the form

$$A = A_0 + x^{k+1}e_{k+1} + \ldots + x^ne_n,$$

that is, it belongs to plane P. In other words.

$$P = (\operatorname{Ker} f_1) \cap \ldots \cap (\operatorname{Ker} f_k).$$

DEFINITION 20.5. Let Γ be a hyperplane in affine space (R, X) and let A, B be two points of space X not lying in hyperplane Γ. If segment $[A, B]$ does not intersect (that is, does not have points in common with) hyperplane Γ, then it is said that points A, B lie *on one side of* hyperplane Γ. If, on the other hand, segment $[A, B]$ intersects hyperplane Γ, then it is said that points A, B lie *on opposite sides of* hyperplane Γ.

THEOREM 20.6. *Let Γ be a hyperplane in an affine space (R, X), and let f be an affine function in X, the kernel of which is hyperplane Γ. In addition, let A, B be two points of space X which do not lie in Γ. If the numbers $f(A)$ and $f(B)$ have the same sign, then points A, B lie on one side of Γ, whereas if these numbers have opposite signs points A, B lie on opposite sides of Γ.*

Proof. Let C be an arbitrary point on segment $[A, B]$, that is, $C = (1-\lambda)A + \lambda B$, where $0 \leqslant \lambda \leqslant 1$. Because of the affineness of mapping f, we have

$$f(C) = (1-\lambda)f(A) + \lambda f(B). \tag{20.5}$$

If the two numbers $f(A)$, $f(B)$ have the same sign, then, since $\lambda \geqslant 0$ and $1-\lambda \geqslant 0$ (where λ and $1-\lambda$ do not go to zero at the same time), the number $f(C)$ also has this sign and thus does not go to zero. Consequently, $f(C) \neq 0$ for any point $C \in [A, B]$, that is, segment $[A, B]$ does not intersect hyperplane $\Gamma = \operatorname{Ker} f$.

Now let us assume that numbers $f(A)$ and $f(B)$ have opposite signs. If we set

$$\lambda^* = \frac{f(A)}{f(A) - f(B)} = \frac{f(A)}{f(A) + (-f(B))} = \frac{|f(A)|}{|f(A)| + |f(B)|},$$

then obviously $0 < \lambda^* < 1$, so that point $C^* = (1-\lambda^*)A + \lambda^*B$ belongs to segment $[A, B]$. It follows directly from (20.5) that $f(C^*) = 0$, that is, $C^* \in \Gamma$. Thus in this case segment $[A, B]$ has a (single) point C^* in common with hyperplane Γ.

THEOREM 20.7. *Let Γ be a hyperplane in an affine space (R, X). Then all points of space X not lying in hyperplane Γ can (albeit in only one way) be distributed in two nonintersecting (disjoint) sets Π_1, Π_2 possessing the following properties: if points A, B belong to the same set, then they lie on one side of hyperplane Γ, whereas if points A, B belong to different sets, then they lie on opposite sides of Γ.*

Proof. We consider an affine function $f: X \to D$, the kernel of which is hyperplane Γ. Let Π_1 be the set of all points $A \in X$ satisfying the condition $f(A) > 0$, and let Π_2 be the set of all points $A \in X$ satisfying the condition $f(A) < 0$. It is obvious that any point A not lying in hyperplane Γ belongs to one, and only one, of the sets Π_1, Π_2. It follows directly from Theorem 20.6 that sets Π_1, Π_2 possess the properties indicated in Theorem 20.7.

It now remains to prove the uniqueness of the distribution. Let Π_1', Π_2' be two other sets possessing these same properties. Obviously, set Π_1' is contained completely in *one* of the sets Π_1, Π_2 (otherwise, in Π_1' there would be two points lying on opposite sides of Γ). To be more definite, let $\Pi_1' \subset \Pi_1$. Similarly, set Π_2' will be wholly contained in one of the sets Π_1, Π_2. If inclusion $\Pi_2' \subset \Pi_1$ were valid, then points from *different* sets Π_1', Π_2' would lie in Π_1, that is, on one side of Γ, which is impossible. Consequently, $\Pi_2' \subset \Pi_2$. Since $\Pi_1 \cup \Pi_2 = \Pi_1' \cup \Pi_2'$, inclusions $\Pi_1' \subset \Pi_1$, $\Pi_2' \subset \Pi_2$ then imply that $\Pi_1' = \Pi_1$ and $\Pi_2' = \Pi_2$.

DEFINITION 20.8. Sets Π_1, Π_2 introduced in Theorem 20.7 are called *open half-spaces* defined in X by hyperplane Γ (or open half-spaces into which hyperplane Γ *divides* space X). Thus

$$\Pi_1 \cap \Pi_2 = \varnothing, \quad \Pi_1 \cap \Gamma = \varnothing, \quad \Pi_2 \cap \Gamma = \varnothing, \quad \Pi_1 \cup \Pi_2 \cup \Gamma = X.$$

Each of sets $\overline{\Pi}_1 = \Pi_1 \cup \Gamma$, $\overline{\Pi}_2 = \Pi_2 \cup \Gamma$ is called a *closed half-space.* Thus we have

$$\overline{\Pi}_1 \cap \overline{\Pi}_2 = \Gamma, \quad \overline{\Pi}_1 \cup \overline{\Pi}_2 = X.$$

Sometimes, taking into account the connections between half-spaces and affine functions, the terms "positive half-space" and "negative half-space" are also introduced. For instance, if $f: X \to D$ is a nonconstant affine function, the kernel of which is hyperplane Γ, and if Π_1, Π_2 are the same as in the proof of Theorem 20.7 (that is, at points of set Π_1 function f is positive, while at points of set Π_2 it is negative), then Π_1 can be called an open *positive* half-space, while Π_2 is an open *negative* half-space. Similarly, $\overline{\Pi}_1$ is called a closed positive half-space, and $\overline{\Pi}_2$ is a closed negative half-space.

However, it should be kept in mind that these terms are arbitrary. Actually, affine function $f_1 = -f$ also has hyperplane Γ as its kernel, but relative to this affine function half-space Π_2 is, on the other hand, positive, while half-space Π_1 is negative. Consequently, with respect to hyperplane Γ the two half-spaces are equivalent and *either* of them could be called "positive" (after selecting properly an affine function having hyperplane Γ as its kernel).

THEOREM 20.9. *Let Γ be a hyperplane of space X and let $A, B \in X$ be points such that vector $\overrightarrow{AB}$ belongs to the direction subspace of hyperplane Γ (in this case it is also said that vector $\overrightarrow{AB}$ is* **parallel** *to hyperplane Γ). Then, either both points A, B belong to hyperplane Γ or they are situated on one side of it.*

Proof. Let C be an arbitrary point of hyperplane Γ and let $D \in X$ be a point such that $\overrightarrow{CD} = \overrightarrow{AB}$. Then $D \in \Gamma$ (since vector $\overrightarrow{AB}$ belongs to the direction subspace of hyperplane Γ). If we select an affine function f whose kernel is hyperplane Γ, then $f(C) = f(D) = 0$.

Relation $\overrightarrow{CD} = \overrightarrow{AB}$ can be written as

$$D = C + \overrightarrow{AB} = C + B - A;$$

from this relation it follows that

$$f(D) = f(C) + f(B) - f(A),$$

that is, $f(B) = f(A)$. Now all we have to do is apply Theorem 20.6.

THEOREM 20.10. *Let (R, X) be an affine space and let P, Q be mututally complementary planes in X. If π is the projection of space X onto plane P in the direction of plane Q, then for an open (closed) half-space Π of space P the inverse image $\pi^{-1}(\Pi)$ will be an open (closed) half-space of space X.*

Proof. Let Γ be a hyperplane of space P bounding half-space Π. Then, in view of Theorem 19.12, $\pi^{-1}(\Gamma)$ is a hyperplane of space X. Let A, B be points of plane P not lying in Γ, while A', B' are points of space X such that $\pi(A') = A$ and $\pi(B') = B$. Then, because of the affineness of mapping π, we have $\pi([A', B']) = [A, B]$.

If points A, B of plane P are situated on one side of hyperplane Γ, that is, if segment $[A, B]$ does not intersect Γ, then segment $[A', B']$ will not intersect $\pi^{-1}(\Gamma)$, that is, points A', B' are situated in X on one side of hyperplane $\pi^{-1}(\Gamma)$. If, on the other hand, A, B are situated in P on opposite sides of Γ, then A', B' are situated in X on opposite sides of hyperplane $\pi^{-1}(\Gamma)$. The statement to be proven follows directly from this.

21. Euclidean space. An affine space (R, X) is called a *Euclidean space* if a scalar product is introduced in R, that is, if R is a Euclidean vector space. Thus, the pair (R, X) is an *n-dimensional* Euclidean space if the vectors (elements of set R) and points (elements of set X) obey *all* five groups of axioms (I through V).

Hence it is possible to introduce in set X *metric concepts* (distances, angles, etc.). For example, if A, B are two points of space X, then the *distance* between A and B is defined as the length of vector $\overrightarrow{AB}$. The distance will be denoted by one of the symbols $\rho(A, B)$ or $|A - B|$:

$$\rho(A, B) = |A - B| = |\overrightarrow{AB}|.$$

Moreover, let A, B, C be three points of space X, where $B \neq A, B \neq C$. In this case vectors $\overrightarrow{BA}$, $\overrightarrow{BC}$ are different from the zero vector. The angle between these vectors (see Note 16.6) is designated as $\angle ABC$. For this definition of the angle, the following inequality is always satisfied:

$$0 \leqslant \angle ABC \leqslant \pi.$$

Finally, if l_1 and l_2 are two rays with a common endpoint B, then the angle $\angle(l_1, l_2)$ between these rays is defined as the angle formed by the direction vectors of rays l_1, l_2, that is, the angle $\angle ABC$, where $A \in l_1$, $C \in l_2$, with $A \neq B$, $C \neq B$ (Figure 52).

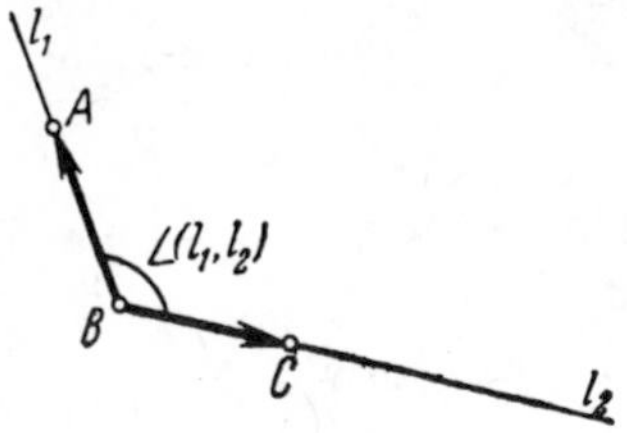

FIGURE 52

Other metric concepts, such as lengths of curves, areas of figures, and volumes of bodies, will not be required by us.

THEOREM 21.1. *Any Euclidean space is a metric space, that is, the distance introduced in a Euclidean space possesses the following three properties (for any points* A, B, C*):*

1° $\rho(A, A) = 0$; $\rho(A, B) > 0$ *for* $A \neq B$;
2° $\rho(A, B) = \rho(B, A)$;
3° $\rho(A, B) + \rho(B, C) \geqslant \rho(A, C)$.

Proof. Relation $\rho(A, A) = 0$ means that $|\overrightarrow{AA}| = 0$, that is, the zero vector has zero length (see p. 105). The inequality $\rho(A, B) \geq 0$ for $A \neq B$ means that $|\boldsymbol{a}| > 0$ for $\boldsymbol{a} = \overrightarrow{AB} \neq \boldsymbol{0}$ (see Axiom IV$_4$). Consequently, property 1° is valid.

Moreover, relation $\rho(A, B) = \rho(B, A)$ means that $|\overrightarrow{AB}| = |\overrightarrow{BA}|$, that is, $(-\boldsymbol{a})^2 = \boldsymbol{a}^2$; this relation is also valid.

Finally, inequality $\rho(A, B) + \rho(B, C) \geqslant \rho(A, C)$ means that $|\overrightarrow{AB}| + |\overrightarrow{BC}| \geqslant |\overrightarrow{AC}|$. Since $\overrightarrow{AC} = \overrightarrow{AB} + \overrightarrow{BC}$, the inequality to be proven can be written as

$$|\overrightarrow{AB}| + |\overrightarrow{BC}| \geqslant |\overrightarrow{AB} + \overrightarrow{BC}|.$$

This relation follows directly from Theorem 16.5 (see (16.3)).

THEOREM 21.2. *For any three points* A, B, C *of a Euclidean space the following relation is valid (for* $B \neq A$, $B \neq C$*):*

$$(\rho(A, C))^2 = (\rho(A, B))^2 + (\rho(B, C))^2 - \\ - 2\rho(A, B)\rho(B, C)\cos(\angle ABC). \qquad (21.1)$$

In particular, if $\angle ABC$ *is a right angle, then*

$$(\rho(A, C))^2 = (\rho(A, B))^2 + (\rho(B, C))^2. \qquad (21.2)$$

Proof. If we set $\overrightarrow{BA} = \boldsymbol{a}$ and $\overrightarrow{BC} = \boldsymbol{c}$, then we can write

$$\overrightarrow{AC} = \overrightarrow{BC} - \overrightarrow{BA} = \boldsymbol{c} - \boldsymbol{a},$$

so that

$$\begin{aligned}(\rho(A, C))^2 &= |\overrightarrow{AC}|^2 = (\overrightarrow{AC})^2 = (\boldsymbol{c} - \boldsymbol{a})^2 = \boldsymbol{a}^2 + \boldsymbol{c}^2 - 2\boldsymbol{ac} = \\ &= |\boldsymbol{a}|^2 + |\boldsymbol{c}|^2 - 2|\boldsymbol{a}|\cdot|\boldsymbol{c}|\cos(\angle(\boldsymbol{a}, \boldsymbol{c})) = \\ &= |\overrightarrow{BA}|^2 + |\overrightarrow{BC}|^2 - 2|\overrightarrow{BA}|\cdot|\overrightarrow{BC}|\cos(\angle ABC) = \\ &= (\rho(A, B))^2 + (\rho(B, C))^2 - 2\rho(A, B)\rho(B, C)\cos(\angle ABC).\end{aligned}$$

NOTE 21.3. Using Theorem 21.2 and some propositions established earlier, it is easy to show that a two-dimensional Euclidean space has the same properties as the plane familiar to us from elementary plane geometry, while a three-dimensional Euclidean space corresponds to the space of solid geometry.

For example, on p. 125 we established that one, and only one, straight line passes through two different points, and on p. 123 we showed that through each point there passes a single line parallel to the given one, and also that two lines parallel to a third line are parallel to one another (transitivity of the parallelism concept), etc. Moreover, if instead of $\rho(A, B)$ we use the ordinary "school" notation AB for the length of a segment with its ends at points A and B, then relation 3° in Theorem 21.1 becomes the familiar statement that each side of a triangle is smaller than the sum of the other two sides, while relations (21.1) and (21.2) become, respectively, the cosine formula for a triangle and the Pythagorean theorem. The other theorems of elementary geometry can be proven similarly. Therefore, the five foregoing axiom groups, I–V, provide (for $n = 2$ and $n = 3$) a complete axiomatic description of elementary geometry.*

Let P be a plane in space X. A vector $\boldsymbol{a} \in R$ is called *orthogonal* to plane P if for any points $A, B \in P$ vectors $\boldsymbol{a}$ and $\overrightarrow{AB}$ are orthogonal. If $\boldsymbol{e}_1, \ldots, \boldsymbol{e}_k$ form a basis of plane P, then for vector $\boldsymbol{a}$ to be orthogonal to plane P it is obviously necessary and sufficient that vector $\boldsymbol{a}$ be orthogonal to each of vectors $\boldsymbol{e}_1, \ldots, \boldsymbol{e}_k$.

Two planes $P_1, P_2 \subset X$ are said to be *orthogonal* if for any points $A, B \in P_1$; $C, D \in P_2$ vectors $\overrightarrow{AB}$ and $\overrightarrow{CD}$ are orthogonal.** If in this case $\dim P_1 + \dim P_2 = \dim X$, then each of planes P_1, P_2 is called an *orthogonal complement* to the other plane.

THEOREM 21.4. *Through any point $A_0 \in X$ it is possible to construct (albeit in only one way) a plane which is an orthogonal complement to a given plane $P \subset X$. Moreover, if a plane P' is an orthogonal complement of plane $P \subset X$, then planes P and P' are mutually complementary (and thus have a single common point). Finally, if vector $\overrightarrow{A_0B}$ is orthogonal to plane P, then point B belongs to plane P', which passes through A_0 and is an orthogonal complement of plane P.*

* This axiomatics for geometry was proposed by the outstanding German geometrician H. Weyl.

** Note that this definition does not jibe with the long-standing terminology according to which two planes that are "orthogonal" to one another have a line in common. According to the terminology adopted here, two (two-dimensional) planes in three-dimensional space will never be orthogonal.

Proof. Let $P \subset X$, $A_0 \in X$. For a dimension k of plane P we take an orthonormal basis $\boldsymbol{e}_1, \ldots, \boldsymbol{e}_k$ of plane P. Moreover, we augment the orthonormal system of vectors $\boldsymbol{e}_1, \ldots, \boldsymbol{e}_k$ until an orthonormal basis $\boldsymbol{e}_1, \ldots, \boldsymbol{e}_k, \boldsymbol{e}_{k+1}, \ldots, \boldsymbol{e}_n$ of space X is reached.

Next let P_1 be the set of all points of form

$$B = A_0 + \lambda^{k+1}\boldsymbol{e}_{k+1} + \ldots + \lambda^n \boldsymbol{e}_n, \tag{21.3}$$

where $\lambda^{k+1}, \ldots, \lambda^n$ are arbitrary real numbers. Thus P_1 is an $(n-k)$-dimensional plane in X containing point A_0. It is also obvious that

$$\dim P + \dim P_1 = k + (n-k) = n = \dim X.$$

Finally, for any points $A, B \in P_1$; $C, D \in P$ we have

$$\overrightarrow{AB} = \mu^{k+1}\boldsymbol{e}_{k+1} + \ldots + \mu^n \boldsymbol{e}_n, \quad \overrightarrow{CD} = \mu^1 \boldsymbol{e}_1 + \ldots + \mu^k \boldsymbol{e}_k,$$

so that vectors $\overrightarrow{AB}$ and $\overrightarrow{CD}$ are orthogonal. Consequently, P_1 is an orthogonal complement of plane P, existence of the plane thereby being established.

Now let us prove uniqueness. Let P_2 also be a plane passing through point A_0 and constituting an orthogonal complement of P. Let us show that planes P_1 and P_2 are identical. If A is an arbitrary point of plane P_2, then

$$\overrightarrow{A_0A} = \lambda^1 \boldsymbol{e}_1 + \ldots + \lambda^n \boldsymbol{e}_n. \tag{21.4}$$

Since P_2 is an orthogonal complement of P, the vector $\overrightarrow{A_0A}$ must be orthogonal to each of vectors $\boldsymbol{e}_1, \ldots, \boldsymbol{e}_k$. Hence it follows that all the coefficients $\lambda^1, \ldots, \lambda^k$ in (21.4) go to zero. Thus

$$\overrightarrow{A_0A} = \lambda^{k+1}\boldsymbol{e}_{k+1} + \ldots + \lambda^n \boldsymbol{e}_n,$$

or

$$A = A_0 + \lambda^{k+1}\boldsymbol{e}_{k+1} + \ldots + \lambda^n \boldsymbol{e}_n,$$

which implies that $A \in P_1$. Therefore $P_2 \subset P_1$. Since

$$\dim P + \dim P_1 = \dim P + \dim P_2 = \dim X,$$

we know that $\dim P_1 = \dim P_2$, so that the inclusion $P_2 \subset P_1$ implies that $P_2 = P_1$. Thus uniqueness is proven.

Moreover, since direction subspace L_P of plane P is generated by vectors $\boldsymbol{e}_1, \ldots, \boldsymbol{e}_k$, while direction subspace L_{P_1} of plane P_1 is generated by vectors $\boldsymbol{e}_{k+1}, \ldots, \boldsymbol{e}_n$, it follows that $L_P \oplus L_{P_1} = R$, that is, planes P and P_1 are mutually complementary.

Finally, if vector $\overrightarrow{A_0B}$ is orthogonal to plane P, then this vector satisfies relation (21.3), indicating that $B \in P_1$.

THEOREM 21.5. *A single perpendicular can be dropped from a given point $A_0 \in X$ to a given plane $P \subset X$. In other words, in plane P there exists one, and only one, point B possessing the property that vector $\overrightarrow{A_0B}$ is orthogonal to plane P. The distance $\rho(A_0, B)$ is called the distance from point A_0 to plane P.*

Proof. Let us construct through point A_0 a plane P_1 which is an orthogonal complement of plane P, and let us call B the point common to planes P and P_1. Since $A_0, B \in P_1$, the vector $\overrightarrow{A_0B}$ is orthogonal to plane P, that is, point B is the desired point.

If B' is another point of plane P, the vector $\overrightarrow{A_0B'}$ being orthogonal to plane P, then in view of Theorem 21.4 we have $B' \in P_1$, that is, B' is the common point of planes P and P_1, so that $B' = B$.

Point B, whose existence and uniqueness were attested to in Theorem 21.5, is known as the *orthogonal projection* of point A_0 onto plane P. Thus the *orthogonal projecting* onto plane* P is a projecting of space X onto plane P in the direction of plane P_1, where P_1 is the orthogonal complement of plane P. Consequently, the statements in Theorems 19.11, 19.12, and 20.10 are valid for the orthogonal projection.

EXAMPLE 21.6. Let P_1 and P_2 be two parallel planes of Euclidean space X. Then the distance from a point $A_1 \in P_1$ to plane P_2 does not depend on the choice of point $A_1 \in P_1$ and it is equal to the distance from an arbitrary point $B \in P_2$ to plane P_1. This distance is called the *distance between parallel planes* P_1 *and* P_2.

Let $A_1, A_2 \in P_1$, and let B_1 be the orthogonal projection of point A_1 onto plane P_2. Vector $\overrightarrow{A_1A_2}$ belongs to the direction subspace of plane P_1, and thus to the direction subspace of plane P_2 as well (since $P_1 \| P_2$). Consequently, the point

$$B_2 = B_1 + \overrightarrow{A_1A_2}$$

belongs to plane P_2. It follows from the above equation that $\overrightarrow{A_2B_2} = \overrightarrow{A_1B_1}$. Since B_1 is the orthogonal projection of point A_1 onto plane P_2, vector $\overrightarrow{A_1B_1}$ is orthogonal to plane P_2. Hence vector $\overrightarrow{A_2B_2}$ is also orthogonal to plane P_2, and since $B_2 \in P_2$ it follows that point B_2 is the orthogonal projection of point A_2 onto plane P_2. Now it is clear that the distances from points A_1 and A_2 to plane P_2 are the same:

$$\rho(A_1, B_1) = |\overrightarrow{A_1B_1}| = |\overrightarrow{A_2B_2}| = \rho(A_2, B_2).$$

Therefore, the distance from point $A \in P_1$ to plane P_2 does not depend on the choice of the point $A \in P_1$.

Finally, note that vector $\overrightarrow{A_1B_1}$ is orthogonal to plane P_1 as well as to plane P_2. that is, A_1 is the orthogonal projection of point B_1 onto plane P_1. Consequently, the distance from point A_1 to plane P_2 (that is, $\rho(A_1, B_1)$) is equal to the distance from point B_1 to plane P_1.

THEOREM 21.7. *Let (R, X) and (S, Y) be Euclidean spaces and let $f: X \to Y$ be an affine mapping. Then there exists a number $M > 0$ such that the following relation*

* Here we differentiate between the Russian terms "proektsiya" (projection) and "proektirovanie" (act of projecting). (Translator).

is valid for any points $A,\ B \in X$:

$$\rho(f(A),\ f(B)) = |f(B) - f(A)| \leqslant M|B - A| = M\rho(A,\ B).$$

Proof. In view of (20.1), we have

$$f(B) - f(A) = \varphi(\overrightarrow{AB}) = \varphi(B - A).$$

Thus the theorem to be proven follows directly from Theorem 16.15.

Now let us consider the notation used for hyperplanes, half-spaces, and planes in a Euclidean space.

THEOREM 21.8. *Let $(R,\ X)$ be a Euclidean space. For any affine function $f\colon\ X \to D$ there exists (albeit only one) vector $\boldsymbol{n}$ satisfying the relation*

$$f(B) = f(A) + \boldsymbol{n}\overrightarrow{AB}, \qquad A,\ B \in X; \tag{21.5}$$

let us call this vector $\operatorname{grad} f$. *Moreover, if Q is a fixed point of space X, $\boldsymbol{n}$ is some vector, and λ is a real number, then the formula*

$$f(B) = \lambda + \boldsymbol{n}\overrightarrow{QB}, \qquad B \in X, \tag{21.6}$$

defines an affine function in X, where $\operatorname{grad} f = \boldsymbol{n}$.

Proof. Let $f\colon\ X \to D$ be an affine function. Then for any points $A, B \in X$ the following relation is valid (see (20.1)):

$$f(B) = f(A) + \varphi(\overrightarrow{AB}),$$

where $\varphi\colon\ R \to D$ is some linear functional. In view of Theorem 16.16 we have

$$\varphi(\overrightarrow{AB}) = \boldsymbol{n}\overrightarrow{AB},$$

where $\boldsymbol{n} = \operatorname{grad}\varphi$ is the gradient of linear function φ. Thus relation (21.5) is valid.

If $\boldsymbol{n}'$ is another vector possessing the same property:

$$f(B) = f(A) + \boldsymbol{n}'\overrightarrow{AB}, \qquad A,\ B \in X,$$

then by subtracting this expression from (21.5) we obtain

$$(\boldsymbol{n} - \boldsymbol{n}')\,\overrightarrow{AB} = 0$$

for any vector $\overrightarrow{AB} \in R$. Hence it follows that $\boldsymbol{n} - \boldsymbol{n}' = 0$, that is, $\boldsymbol{n} = \boldsymbol{n}'$.

The affineness of function (21.6) is implied by the fact that $\varphi(\boldsymbol{a}) = \boldsymbol{na}$ is a linear functional in R (cf. (20.2)). The theorem is thereby proven.

Now let Γ be an arbitrary hyperplane of space X and let $f\colon\ X \to D$ be a nonconstant affine function whose kernel is hyperplane Γ. The gradient of function f is denoted by the letter $\boldsymbol{n}$. For any two points $A, B \in \Gamma$ we then have $f(A) = 0$, $f(B) = 0$, so that in view of (21.5) $\boldsymbol{n}\overrightarrow{AB} = 0$. Consequently, vector $\boldsymbol{n}$ is *orthogonal* to hyperplane Γ.

Thus, *if* Γ *is a hyperplane and if* f *is a nonconstant affine function whose kernel is hyperplane* Γ, *then a vector* $\boldsymbol{n} = \operatorname{grad} f$ *is orthogonal to hyperplane* Γ. This vector is called a *normal* (or a *normal vector*) of hyperplane Γ.

A normal vector of hyperplane Γ can also be obtained in the following manner. If we take an orthonormal basis $\boldsymbol{e}_1, \ldots, \boldsymbol{e}_{n-1}$ of hyperplane Γ and augment it with vector $\boldsymbol{e}_n$ to form an orthonormal basis $\boldsymbol{e}_1, \ldots, \boldsymbol{e}_n$ of space X, then the vector $\lambda\boldsymbol{e}_n$ for any $\lambda \neq 0$ is different from zero and orthogonal to each of vectors $\boldsymbol{e}_1, \ldots, \boldsymbol{e}_{n-1}$, that is, it is orthogonal to hyperplane Γ. In other words, vector $\lambda\boldsymbol{e}_n$ $(\lambda \neq 0)$ is a normal vector of hyperplane Γ. It is easy to see that using this notation the inverse statement is true as well: if $\boldsymbol{n}$ is a normal vector of hyperplane Γ, then $\boldsymbol{n} = \lambda\boldsymbol{e}_n$ $(\lambda \neq 0)$.

THEOREM 21.9. *Let* Γ *be some hyperplane, let* A_0 *be some point of hyperplane* Γ, *and let* $\boldsymbol{n}$ *be a normal vector to this hyperplane. Then hyperplane* Γ *is defined by the equation*

$$\boldsymbol{n}\overrightarrow{A_0A} = 0, \tag{21.7}$$

that is, point $A \in X$ *belongs to hyperplane* Γ *if, and only if, it satisfies condition* (21.7). *Moreover, the open half-spaces set off by hyperplane* Γ *are defined by the inequalities*

$$\boldsymbol{n}\overrightarrow{A_0A} > 0, \quad \boldsymbol{n}\overrightarrow{A_0A} < 0, \tag{21.8}$$

that is, point $A \in X$ *belongs to the positive (negative) half-space if, and only if, vector* $\overrightarrow{A_0A}$ *forms an acute (obtuse) angle with vector* $\boldsymbol{n}$. *Finally, the closed half-spaces set off by hyperplane* Γ *are defined by the inequalities*

$$\boldsymbol{n}\overrightarrow{A_0A} \geqslant 0, \quad \boldsymbol{n}\overrightarrow{A_0A} \leqslant 0. \tag{21.9}$$

Proof. Let us set

$$f(A) = \boldsymbol{n}\overrightarrow{A_0A}, \qquad A \in X.$$

The function f: $X \to D$ is affine and $\operatorname{grad} f = \boldsymbol{n}$. If A is an arbitrary point of hyperplane Γ, then $\boldsymbol{n}\overrightarrow{A_0A} = 0$ (since $\boldsymbol{n}$ is a normal vector of hyperplane Γ), that is, $f(A) = 0$. Therefore, $\Gamma \subset \operatorname{Ker} f$. However, since the two planes Γ and $\operatorname{Ker} f$ are of the same dimension $n - 1$, it follows that $\Gamma = \operatorname{Ker} f$. Thus, Γ is the kernel of affine function f, that is, Γ is defined by the equation $f(A) = 0$ or, what comes to the same thing, by equation (21.7).

Moreover, the open half-spaces Π_1, Π_2 set off by hyperplane Γ are defined by the inequalities $f(A) > 0$, $f(A) < 0$ (see proof of Theorem 20.7), that is, by inequalities (21.8).

Finally, the inequality $n\overrightarrow{A_0A} \geqslant 0$ is satisfied by points for which $n\overrightarrow{A_0A} > 0$ and points for which $n\overrightarrow{A_0A} = 0$. Thus the first of inequalities (21.9) describes the set $\Pi_1 \cup \Gamma = \overline{\Pi}_1$, that is, the corresponding closed half-space. The same can be said of the second of inequalities (21.9).

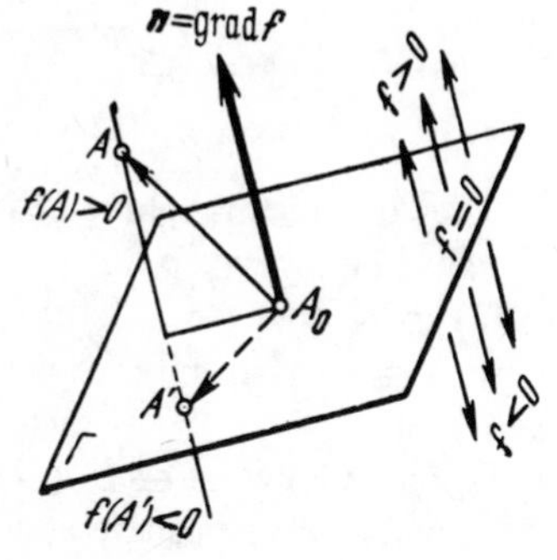

FIGURE 53

DEFINITION 21.10. Let Γ be some hyperplane, let A_0 be a point on it, and let n be its normal vector. In order to differentiate between the half-spaces set off by hyperplane Γ, let us say that *vector* n *is directed toward the half-space consisting of all points* A *satisfying the inequality* $n\overrightarrow{A_0A} > 0$ (or the inequality $n\overrightarrow{A_0A} \geqslant 0$). Consequently, if $f\colon X \to D$ is an affine function whose kernel is hyperplane Γ, then the vector $n = \operatorname{grad} f$ is directed toward the *positive* half-space (Figure 53).

THEOREM 21.11. *Let A_0 be an arbitrary point of Euclidean space X and let $n_1, \ldots, n_k \in R$ be linearly independent vectors. Then the set P of all points $A \in X$ satisfying the system of equations*

$$n_1\overrightarrow{A_0A} = 0, \ \ldots, \ n_k\overrightarrow{A_0A} = 0 \tag{21.10}$$

is an $(n-k)$-dimensional plane passing through point A_0. Any $(n-k)$-dimensional plane of space X passing through point A_0 can be represented in this form.

Proof. Let us define in X the affine functions

$$f_1(A) = n_1\overrightarrow{A_0A}, \ \ldots, \ f_k(A) = n_k\overrightarrow{A_0A}.$$

Then, by definition, the vectors

$$n_1 = \operatorname{grad} f_1, \ \ldots, \ n_k = \operatorname{grad} f_k$$

are linearly independent. If Γ_i is the kernel of affine function f_i:

$$\Gamma_i = \operatorname{Ker} f_i, \qquad i = 1, \ldots, k,$$

then from Theorem 20.4 we know that the intersection $\Gamma_1 \cap \ldots \cap \Gamma_k$ is an $(n-k)$-dimensional plane in X. However, this intersection is described by the system of relations

$$f_1(A) = 0, \ \ldots, \ f_k(A) = 0, \tag{21.11}$$

that is, by system of relations (21.10). Consequently, system (21.10) defines in X some $(n-k)$-dimensional plane. Clearly, this plane will pass through point A_0 (since A_0

satisfies system of relations (21.10)). In view of this same Theorem 20.4, any $(n-k)$-dimensional plane passing through point A_0 can be described in this form.

22. Topology of Euclidean space. Let O be an arbitrary point of Euclidean space (R, X) and let r be a positive number. The set of all points $A \in X$ satisfying the condition

$$\rho(O, A) < r$$

is called an *open sphere,* with a center O and a radius r. Let us call this sphere $U_r(O)$.

Now let H be some set in space X. Point $O \in X$ is called an *interior point* of set H if there exists a positive number r such that $U_r(O) \subset H$.

A set $G \subset X$ is called *open* (in space X) if *each* point $A \in G$ is an interior point of this set. Any empty set is also considered to be open.

EXAMPLE 22.1. *Any open sphere $U_r(O)$ is an open set in X.* Let $A \in U_r(O)$, that is, $\rho(O, A) < r$. If $r' = r - \rho(O, A)$, then the number r' is positive and we can consider a sphere $U_{r'}(A)$ with a center A and a radius r'. For any point B belonging to this sphere, we have $\rho(A, B) < r'$, so that in view of Theorem 21.1

$$\rho(O, B) \leqslant \rho(O, A) + \rho(A, B) < \rho(O, A) + r' = \\ = \rho(O, A) + (r - \rho(O, A)) = r.$$

Consequently, $\rho(O, B) < r$, that is, $B \in U_r(O)$. This proves the inclusion $U_{r'}(A) \subset U_r(O)$, indicating that sphere $U_r(O)$ is an open set.

A set $F \subset X$ is called *closed* if its *complement* $X \setminus F$ (that is, the set of all points of space X not belonging to F) is an open set.

EXAMPLE 22.2. *A closed sphere with a center O and a radius r, that is, the set of all points $A \in X$ satisfying the condition $\rho(O, A) \leqslant r$, is a closed set.* Actually, the complement of this closed sphere is the set G of all points $A \in X$ satisfying the condition $\rho(O, A) > r$. Let $A \in G$, that is, $\rho(O, A) > r$. If we set $r' = \rho(O, A) - r$, then r' is a positive number and we can consider the sphere $U_{r'}(A)$. For $B \in U_{r'}(A)$, that is, $\rho(A, B) < r'$, we have

$$\rho(O, B) = \rho(O, B) + \rho(B, A) - \rho(B, A) \geqslant \rho(O, A) - \rho(B, A) = \\ = \rho(O, A) - \rho(A, B) > \rho(O, A) - r' = \rho(O, A) - (\rho(O, A) - r) = r,$$

that is, $\rho(O, B) > r$, so that $B \in G$. This proves the inclusion $U_r\ (A) \subset G$, indicating that set G is open.

THEOREM 22.3. *A union of any (finite or infinite) number of open sets is itself an open set. An intersection of a finite number of open sets is itself an open set. An intersection of any (finite or infinite) number of closed sets is itself a closed set. A union of a finite number of closed sets is itself a closed set.*

This theorem can be proven easily on the basis of the definitions of open and closed sets.

Now let H be an arbitrary set of space X. We consider *all* the closed sets containing H and we call $\bar{H}$ the *intersection* of all these closed sets. In view of Theorem 22.3, set $\bar{H}$ is closed, it being the *smallest* closed set containing H (that is, if F is a closed set containing H, then $F \supset \bar{H}$). Set $\bar{H}$ is called the *closure* of set H.

The following readily proven theorem (the proof of which will not be given) describes the closure in somewhat different terms.

THEOREM 22.4. *Let $H \subset X$. A point $A \in X$ belongs to set $\bar{H}$ if, and only if, for any $r > 0$ sphere $U_r(A)$ contains at least one point of set H.*

Point $A \in X$ is known as a *boundary point* of set $H \subset X$ if, for any $r > 0$, sphere $U_r(A)$ contains both points belonging to set H and points not belonging to it. The set consisting of *all* the boundary points of set H is known as the *boundary* of set H, being designated as $\mathrm{bd}\, H$.

It follows from Theorem 22.4 that $\mathrm{bd}\, H \subset \bar{H}$. Similarly, $\mathrm{bd}\, H \subset \overline{X \setminus H}$, where $X \setminus H$ is the complement of set H. Thus $\mathrm{bd}\, H \subset \bar{H} \cap \overline{X \setminus H}$. It is easy to see that the inverse inclusion is valid as well, namely

$$\mathrm{bd}\, H = \bar{H} \cap \overline{X \setminus H}.$$

The following relation is also easy to prove:

$$\bar{H} = H \cup (\mathrm{bd}\, H),$$

indicating that the closure operation signifies the joining to the set of all its boundary points. In terms of the boundary, we can define open and closed sets as follows: *a set is closed if it contains its boundary:* $H \supset \mathrm{bd}\, H$; *a set is open if it does not have any points in common with its boundary:* $H \cap (\mathrm{bd}\, H) = \varnothing$.

Now let us recall the concept of a continuous mapping. Let (R, X) and (S, Y) be Euclidean spaces and let $H \subset X$. We assume that a mapping $f\colon H \to Y$ is defined, that is, each point $A \in H$ is made to correspond to some point $f(A)$ of space Y. The set H figuring into this definition is called the *domain of definition* of mapping f. The mapping $f\colon H \to Y$ is called *continuous* if for each point $A \in H$ and each number $\varepsilon > 0$ we can select a number $\delta > 0$ such that

$$f(H \cap U_\delta(A)) \subset U_\varepsilon(f(A)).$$

Hence this definition signifies that, if a point $B \in H$ is some distance less than δ away from A (that is, $|A - B| < \delta$, $B \in H$), then the image $f(B)$ of point B is a distance less than ε away from $f(A)$, that is, $|f(A) - f(B)| < \varepsilon$. Note that the number δ depends in general on both point A and number ε, that is, $\delta = \delta(A, \varepsilon)$.

If space Y coincides with a number line D, then the continuous mapping $f\colon H \to D$ is also called a *continuous function*.

THEOREM 22.5. *Any affine mapping of one Euclidean space into another is continuous.*

The *proof* follows directly from Theorem 21.7. It is sufficient to set $\delta = \varepsilon/M$ and then from $\rho(A, B) < \delta$ we obtain $\rho(f(A), f(B)) < \varepsilon$.

In the given proof, for each $\varepsilon > 0$ a number $\delta > 0$ was found which did not depend on A (that is, it was applicable for all points $A \in H$). In this case mapping $f\colon H \to Y$ is called *uniformly continuous*.

The three theorems to follow will be given without proof. They are proven in textbooks on mathematical analysis.

THEOREM 22.6. *Let* (R, X) *and* (S, Y) *be Euclidean spaces. If the domain of definition* H *of a continuous mapping* $f\colon H \to Y$ *is a closed bounded set of space* X, *then mapping* f *is uniformly continuous, while the image* $f(H)$ *is a closed bounded set of space* Y.

THEOREM 22.7. *If the domain of definition* H *of a continuous function* $f\colon H \to D$ *is a closed bounded set of space* X, *then function* f *reaches on* H *the largest and smallest values, that is, there exist points* A_0, $B_0 \in H$ *such that for any point* $A \in H$ *the following inequalities are valid:*

$$f(A_0) \leqslant f(A) \leqslant f(B_0)$$

THEOREM 22.8. *Let* $f\colon X \to Y$ *be a continuous mapping, the domain of definition of which is the whole space* X. *Then the inverse image* $f^{-1}(G)$ *of any open set* $G \subset Y$ *is an open set of space* X, *while the inverse image* $f^{-1}(F)$ *of any closed set* $F \subset Y$ *is a closed set of space* X.

EXAMPLE 22.9. Let $f\colon X \to D$ be a nonconstant affine function. Since this function is continuous (see Theorem 22.5), Theorem 22.8 can be applied to it. Moreover, since the set consisting of a single point 0 is closed in D, its inverse image $f^{-1}(0) = \operatorname{Ker} f$ is according to Theorem 22.8 a closed set of space X. In view of Theorem 20.2 it thus follows that *each hyperplane* $\Gamma \subset X$ *is a closed set.* Consequently (see Theorem 20.4), *any plane* $P \subset X$ *is a closed set.*

Further, since the set $[0, \infty)$ consisting of all nonnegative numbers, and the set $(-\infty, 0]$ consisting of all nonpositive numbers, are closed sets of number line D, we know that sets $f^{-1}([0, \infty))$ and $f^{-1}((-\infty, 0])$ are closed in X. In other words, *closed half-spaces* (see (21.9)) *are closed sets of space* X. Similarly, *open half-spaces* (see (21.88)) *are open sets of space* X.

EXAMPLE 22.10. Since each hyperplane $\Gamma \subset X$ is a closed set (Example 22.9), we know from Theorem 22.3 that the union $\Gamma_1 \cup \ldots \cup \Gamma_s$ of any finite number of hyperplanes in X is also a closed set. Let us show that this closed set is *nowhere compact* in X, that is, for any point $A \in X$ and any number $r > 0$ there exists a sphere completely contained in $U_r(A)$ which does not intersect the set $\Gamma_1 \cup \ldots \cup \Gamma_s$.

Let us consider some nonconstant affine functions $f_1, \ldots, f_s$ such that $\Gamma_i = \operatorname{Ker} f_i$, $i = 1, \ldots, s$, and let us call $\boldsymbol{n}_i$ the gradient of function f_i. Let α be the largest of the numbers $|\boldsymbol{n}_1|, \ldots, |\boldsymbol{n}_s|$. Point A_1 is defined as follows: if $f_1(A) \neq 0$, then we set $A_1 = A$; if, on the other hand, $f_1(A) = 0$, then we set $A_1 = A + \frac{r}{2\alpha}\boldsymbol{n}_1$. In the former case it is obvious that $A_1 \in U_r(A)$ and $f_1(A_1) \neq 0$. In the latter case

$$\rho(A, A_1) = |\overrightarrow{AA_1}| = \left|\frac{r}{2\alpha}\boldsymbol{n}_1\right| = \frac{r}{2\alpha}|\boldsymbol{n}_1| \leqslant \frac{r}{2\alpha}\cdot\alpha = \frac{r}{2},$$

that is, $A_1 \in U_r(A)$. Moreover, in this case

$$f_1(A_1) = f_1(A) + \boldsymbol{n}_1\overrightarrow{AA_1} = \boldsymbol{n}_1\overrightarrow{AA_1} = \boldsymbol{n}_1\left(\frac{r}{2\alpha}\boldsymbol{n}_1\right) \neq 0.$$

Consequently, in any case we find a point $A_1 \in U_r(A)$ such that $f_1(A_1) \neq 0$. In other words, $A_1 \in U_r(A)$ and $A_1 \notin \Gamma_1$, that is, $A_1 \in X \setminus \Gamma_1$, so that $A_1 \in U_r(A) \cap (X \setminus \Gamma_1)$.

Each of sets $U_r(A)$, $X \setminus \Gamma_1$ is open (see Examples 22.1 and 22.9), so that their intersection is open as well. Thus A_1 is an *interior* point of set $U_r(A) \cap (X \setminus \Gamma_1)$, that is, there exists a number $r_1 > 0$ such that

$$U_{r_1}(A_1) \subset U_r(A) \cap (X \setminus \Gamma_1).$$

In other words, $U_{r_1}(A_1) \subset U_r(A)$ and $U_{r_1}(A_1) \subset X \setminus \Gamma_1$. The second of these inclusions signifies that sphere $U_{r_1}(A_1)$ *does not intersect* hyperplane Γ_1. Hence there exists a sphere $U_{r_1}(A_1)$ contained in $U_r(A)$ which does not intersect hyperplane Γ_1 either.

Similarly, we find a sphere $U_{r_2}(A_2) \subset U_{r_1}(A_1)$ not intersecting hyperplane Γ_2 (and not intersecting Γ_1 either, in view of inclusion $U_{r_2}(A_2) \subset U_{r_1}(A_1)$). Thus we have found a sphere $U_{r_2}(A_2) \subset U_r(A)$ which does not intersect $\Gamma_1 \cup \Gamma_2$. Now let us find (in a similar manner) a sphere $U_{r_3}(A_3) \subset U_r(A)$ not intersecting $\Gamma_1 \cup \Gamma_2 \cup \Gamma_3$, etc. By continuing this process, we arrive at a sphere contained in $U_r(A)$ which does not intersect $\Gamma_1 \cup \ldots \cup \Gamma_s$.

EXAMPLE 22.11. Let $H \subset X$ be some set and let $A_0 \in X$ be a fixed point. We define in X a function f such that

$$f(A) = \rho(A_0, A), \qquad A \in H.$$

It is easy to see that this function is continuous. Let us set (for any point $A \in H$) $\delta = \varepsilon$. If $\rho(A, B) < \delta$, then

$$f(B) = \rho(A_0, B) \leqslant \rho(A_0, A) + \rho(A, B) < \rho(A_0, A) + \delta = f(A) + \varepsilon,$$
$$f(A) = \rho(A_0, A) \leqslant \rho(A_0, B) + \rho(B, A) < \rho(A_0, B) + \rho(A, B) < f(B) + \varepsilon.$$

Therefore,

$$f(B) - f(A) < \varepsilon, \quad f(A) - f(B) < \varepsilon,$$

that is, $|f(B) - f(A)| < \varepsilon$.

The following statement can now be made on the basis of the continuity of function f and Theorem 22.7: *if H is a nonempty closed bounded set in X and if $A_0 \in X$, then the function $f(A) = \rho(A_0, A)$, considered in set H, reaches its largest and smallest values, that is, in H there exists a point closest to A_0 and a point furthest from A_0.*

Note that, if the closed set $H \subset X$ is not assumed to be bounded, then the statement concerning the existence of a point *closest* to A_0 remains true. If A is an arbitrary point of set H and if Σ is a closed sphere with a center A_0 and a radius

$r = \rho(A_0, A)$ (Figure 54), then the set $H \cap \Sigma$ is closed and nonempty (for example, it contains point A). Consequently, in $H \cap \Sigma$ there exists a point B_0 closest to A_0. Obviously, B_0 will also be the point of set H closest to A_0 (since points of set H not belonging to set $H \cap \Sigma$ lie at distances greater than $\rho(A_0, A)$ from A_0).

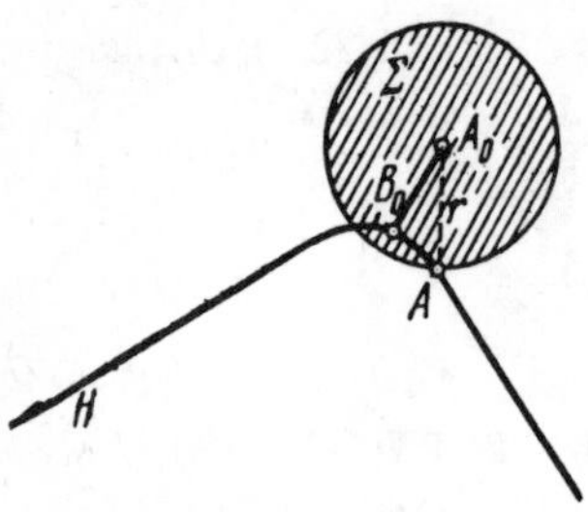

FIGURE 54

23. Coordinates. Let us now give the coordinate notation for some of the relations considered in the foregoing subsections. First of all, we examine the facts associated with an affine space (not assumed to be Euclidean).

Let (R, X) be an affine space of dimension n. We select a basis $\boldsymbol{e}_1, \ldots, \boldsymbol{e}_n$ of vector space R and in addition we consider a point $O \in X$. Let us say that $(O; \boldsymbol{e}_1, \ldots, \boldsymbol{e}_n)$ is a *system of coordinates* in the given affine space X. Point O is called the *coordinate origin,* and vectors $\boldsymbol{e}_1, \ldots, \boldsymbol{e}_n$ are called the *basis vectors.* Each point $C \in X$ can then be expressed uniquely as

$$C = O + x^1\boldsymbol{e}_1 + \ldots + x^n\boldsymbol{e}_n.$$

The numbers $x^1, \ldots, x^n$ entering into this relation are called the *coordinates* of point C in the considered system of coordinates; we write

$$C = (x^1, \ldots, x^n).$$

Moreover, any vector $\boldsymbol{a} \in R$ can be expressed uniquely (in this same basis $\boldsymbol{e}_1, \ldots, \boldsymbol{e}_n$) as

$$\boldsymbol{a} = y^1\boldsymbol{e}_1 + \ldots + y^n\boldsymbol{e}_n.$$

The numbers $y^1, \ldots, y^n$ are called the *coordinates* of vector $\boldsymbol{a}$ in the system of coordinates considered; we write

$$\boldsymbol{a} = \{y^1, \ldots, y^n\}.$$

Therefore, given the system of coordinates $(O; \boldsymbol{e}_1, \ldots, \boldsymbol{e}_n)$, we can consider the coordinates of points and vectors. In the following, the system of coordinates $(O; \boldsymbol{e}_1, \ldots, \boldsymbol{e}_n)$ will be assumed to be *fixed* (this will not be mentioned later).

THEOREM 23.1. *If*

$$C = (x^1, \ldots, x^n), \quad D = (y^1, \ldots, y^n),$$

then

$$\overrightarrow{CD} = \{y^1 - x^1, \ldots, y^n - x^n\}.$$

Proof. Since we have

$$C = O + x^1\boldsymbol{e}_1 + \ldots + x^n\boldsymbol{e}_n, \quad D = O + y^1\boldsymbol{e}_1 + \ldots + y^n\boldsymbol{e}_n,$$

that is,

$$\overrightarrow{OC} = x^1 e_1 + \ldots + x^n e_n, \quad \overrightarrow{OD} = y^1 e_1 + \ldots + y^n e_n,$$

it follows that

$$\overrightarrow{CD} = \overrightarrow{OD} - \overrightarrow{OC} = (y^1 - x^1) e_1 + \ldots + (y^n - x^n) e_n.$$

COROLLARY 23.2. *If*

$$C = (x^1, \ldots, x^n), \quad a = \{y^1, \ldots, y^n\},$$

then

$$C + a = (x^1 + y^1, \ldots, x^n + y^n).$$

THEOREM 23.3. *If*

$$a = \{x^1, \ldots, x^n\}, \quad b = \{y^1, \ldots, y^n\},$$

then

$$a + b = \{x^1 + y^1, \ldots, x^n + y^n\}, \quad \lambda a = \{\lambda x^1, \ldots, \lambda x^n\}.$$

The proof is obvious.

THEOREM 23.4. *If*

$$C_i = (x_i^1, \ldots, x_i^n), \qquad i = 1, \ldots, s,$$

and if point C is defined by the relations

$$C = \lambda^1 C_1 + \ldots + \lambda^s C_s, \text{ where } \lambda^1 + \ldots + \lambda^s = 1, \tag{23.1}$$

then

$$C = \left(\sum_{i=1}^{s} \lambda^i x_i^1, \ldots, \sum_{i=1}^{s} \lambda^i x_i^n \right). \tag{23.2}$$

Proof. Relation (23.1) signifies that

$$\overrightarrow{OC} = \lambda^1 \overrightarrow{OC_1} + \ldots + \lambda^s \overrightarrow{OC_s}.$$

Thus relation (23.2) follows from the obvious equation $O = (0, 0, \ldots, 0)$, Theorems 23.1 and 23.3, and Corollary 23.2.

From Theorem 23.4 it is easy to derive the coordinate formulations of a number of theorems (18.7, 18.8, 18.10, and others). These will not be given here.

THEOREM 23.5. *A mapping φ of one affine space into another is affine if, and only if, it is expressed in terms of coordinates using linear nonhomogeneous formulas.*

Proof. Let (R, X), (S, Y) be affine spaces and let us introduce in X the coordinate system $(O; e_1, \ldots, e_n)$ and in Y the coordinate system $(O'; e'_1, \ldots, e'_m)$. We set

$$C_0 = O, \quad C_1 = O + e_1, \quad C_2 = O + e_2, \ldots, C_n = O + e_n,$$

so that

$$\begin{aligned} C_0 &= (0, 0, 0, \ldots, 0), \quad C_1 = (1, 0, 0, \ldots, 0), \\ C_2 &= (0, 1, 0, \ldots, 0), \ldots, C_n = (0, 0, 0, \ldots, 1). \end{aligned} \tag{23.3}$$

Now let $\varphi\colon X \to Y$ be an affine mapping and let the points

$$D_0 = \varphi(C_0), \quad D_1 = \varphi(C_1), \ldots, D_n = \varphi(C_n)$$

have in the system $(O'; e'_1, \ldots, e'_m)$ the following coordinates:

$$D_i = (b_i^1, \ldots, b_i^m), \qquad i = 0, 1, \ldots, n.$$

For an arbitrary point $C = (x^1, \ldots, x^n) \in X$, let the coordinates of its image $D = \varphi(C)$ be $y^1, \ldots, y^m$:

$$D = (y^1, \ldots, y^m). \tag{23.4}$$

Our aim now is to express coordinates $y^1, \ldots, y^m$ in terms of $x^1, \ldots, x^n$. Since

$$\begin{aligned} C &= O + x^1 e_1 + \ldots + x^n e_n = C_0 + x^1(C_1 - C_0) + \ldots + x^n(C_n - C_0) = \\ &= (1 - x^1 - \ldots - x^n) C_0 + x^1 C_1 + \ldots + x^n C_n, \end{aligned}$$

we know from Theorems 19.1 and 23.4 that

$$\begin{aligned} D &= (1 - x^1 - \ldots - x^n) D_0 + x^1 D_1 + \ldots + x^n D_n = \\ &= ((1 - x^1 - \ldots - x^n) b_0^1 + x^1 b_1^1 + \ldots + x^n b_n^1, \ldots \\ &\qquad \ldots, (1 - x^1 - \ldots - x^n) b_0^m + x^1 b_1^m + \ldots + x^n b_n^m). \end{aligned}$$

A comparison of this with (23.4) gives

$$y^i = (1 - x^1 - \ldots - x^n) b_0^i + x^1 b_1^i + \ldots + x^n b_n^i, \qquad i = 1, \ldots, m. \tag{23.5}$$

In this way, mapping φ is written in coordinates using linear formulas.

Conversely, let some mapping $\varphi\colon X \to Y$ be expressed in coordinates using linear formulas:

$$y^i = a_1^i x^1 + \ldots + a_n^i x^n + c^i, \qquad i = 1, \ldots, m, \tag{23.6}$$

that is, a point $C = (x^1, \ldots, x^n) \in X$ is transformed by mapping φ into a point

$D = (y^1, ..., y^m) \in Y$, the coordinates of which are calculated using formulas (23.6). From Theorem 23.4 it follows directly that, if $C = (1-\lambda)C' + \lambda C''$, then $\varphi(C) = (1-\lambda)\varphi(C') + \lambda\varphi(C'')$, so that mapping φ is affine.

COROLLARY 23.6. *Any affine function* $f\colon X \to D$ *is written in coordinates as*

$$f(x^1, \ldots, x^n) = (1 - x^1 - \ldots - x^n)\, b_0 + x^1 b_1 + \ldots + x^n b_n, \tag{23.7}$$

where $b_i = f(C_i)$, $i = 0, 1, ..., n$ (see (23.3), (23.5)).

THEOREM 23.7. *Any hyperplane* $\Gamma \subset X$ *is expressed in coordinates using the equation*

$$A_1 x^1 + A_2 x^2 + \ldots + A_n x^n + B = 0, \tag{23.8}$$

where at least one of coefficients $A_1, ..., A_n$ *is different from zero. Conversely, any equation of this form defines in* X *some hyperplane. Moreover, the open half-spaces defined by hyperplane* (23.8) *are expressed in coordinate form using the inequalities*

$$\left.\begin{aligned} A_1 x^1 + A_2 x^2 + \ldots + A_n x^n + B > 0, \\ A_1 x^1 + A_2 x^2 + \ldots + A_n x^n + B < 0, \end{aligned}\right\} \tag{23.9}$$

while the corresponding closed half-spaces are expressed using the inequalities

$$\left.\begin{aligned} A_1 x^1 + A_2 x^2 + \ldots + A_n x^n + B \geqslant 0, \\ A_1 x^1 + A_2 x^2 + \ldots + A_n x^n + B \leqslant 0. \end{aligned}\right\} \tag{23.10}$$

Proof. Let Γ be a hyperplane and let $f\colon X \to D$ be a nonconstant affine function, the kernel of which is hyperplane Γ. Function f is expressed in coordinate form using formula (23.7), that is, using the formula

$$f(x^1, \ldots, x^n) = A_1 x^1 + A_2 x^2 + \ldots + A_n x^n + B,$$

where we have set

$$A_1 = b_1 - b_0, \ldots, A_n = b_n - b_0, \quad B = b_0.$$

If all the coefficients $A_1, \ldots, A_n$ were equal to zero, then obviously function f would be *constant*, which contradicts the assumption. Thus at least one of coefficients $A_1, \ldots, A_n$ is different from zero.

Now the equation $f(A) = 0$ describing hyperplane Γ takes form (23.8), while the inequalities $f(A) > 0$, $f(A) < 0$ describing the open half-spaces and the inequalities $f(A) \geqslant 0$, $f(A) \leqslant 0$ describing the closed half-spaces take forms (23.9) and (23.10).

Conversely, if an equation of form (23.8) is given, in which at least one of coefficients $A_1, \ldots, A_n$ is different from zero, then in view of Theorem 23.5 the left-hand side of this equation defines a nonconstant affine function in X, so that according to Theorem 20.1 equation (23.8) defines a hyperplane.

THEOREM 23.8. *Two hyperplanes*

$$A_1x^1 + \dots + A_nx^n + B = 0, \tag{23.11}$$
$$A_1'x^1 + \dots + A_n'x^n + B' = 0 \tag{23.12}$$

are identical if, and only if, there exists a number $k \neq 0$ *such that*

$$A_1 = kA_1', \quad A_2 = kA_2', \quad \dots, \quad A_n = kA_n', \quad B = kB'. \tag{23.13}$$

Proof. Let f, f' be the affine functions appearing on the left-hand sides of (23.11) and (23.12). Thus equation (23.11) defines a hyperplane $\Gamma = \operatorname{Ker} f$, and equation (23.12) defines a hyperplane $\Gamma' = \operatorname{Ker} f'$. Obviously, if relations (23.13) are valid, that is, if $f = kf'$, it follows that hyperplanes Γ and Γ' are identical.

Let us prove the converse. Assume that $\Gamma = \Gamma'$. We select n independent points $C_1, C_2, \dots, C_n$ of hyperplane Γ and we let C_0 be a point not lying in this hyperplane. Then

$$f(C_1) = \dots = f(C_n) = 0, \quad f(C_0) \neq 0,$$
$$f'(C_1) = \dots = f'(C_n) = 0, \quad f'(C_0) \neq 0.$$

If we set $k = \dfrac{f(C_0)}{f'(C_0)}$, we find that the following relations are valid:

$$f(C_i) = kf'(C_i), \qquad i = 0, 1, \dots, n.$$

Thus the affine function $f - kf'$ goes to zero at all points $C_0, C_1, \dots, C_n$. The affine function identically equal to zero is also zero at all these points. However, then in view of the uniqueness (Theorem 19.9) these affine functions must be identical (since system of points $C_0, C_1, \dots, C_n$ is independent), that is, $f - kf' = 0$. In other words,

$$(A_1 - kA_1')x^1 + \dots + (A_n - kA_n')x^n + (B - kB') = 0$$

for any point $(x^1, \dots, x^n)$. Consequently, relations (23.13) are valid.

THEOREM 23.9. *Hyperplanes* (23.11), (23.12) are parallel if, and only if, there exists a number $k \neq 0$ such that

$$A_1 = kA_1', \quad \dots, \quad A_n = kA_n'. \tag{23.14}$$

Proof. Consider the same affine functions f, f' as in the proof of the previous theorem. Obviously, if relations (23.14) are valid, the hyperplanes $\Gamma = \operatorname{Ker} f$ and $\Gamma' = \operatorname{Ker} f'$ are parallel (see Theorem 20.3).

Let us prove the converse. If $\Gamma \| \Gamma'$ and if we choose an arbitrary point $C \in \Gamma'$ and set $\lambda = f(C)$, then the affine function $f_1 = f - \lambda$ possesses the property that $f_1(C) = f(C) - \lambda = 0$, that is, hyperplane $\operatorname{Ker} f_1$ passes through point C. Moreover, hyperplane $\operatorname{Ker} f_1$ is parallel to hyperplane Γ (Theorem 20.3) and thus to hyperplane Γ' as well. Consequently, hyperplanes $\Gamma' = \operatorname{Ker} f'$ and $\operatorname{Ker} f_1$ are parallel and have a common point; therefore $\operatorname{Ker} f_1 = \operatorname{Ker} f'$ (Theorem 18.6). In other words, hyperplanes

$$A_1x^1 + \dots + A_nx^n + (B - \lambda) = 0,$$
$$A_1'x^1 + \dots + A_n'x^n + B' = 0$$

are identical, so that in view of Theorem 23.8 relations (23.14) are valid.

Now let us describe the facts associated with a *Euclidean* space in coordinate form.

THEOREM 23.10. *If a system of coordinates* $(O;\ \boldsymbol{e}_1, \dots, \boldsymbol{e}_n)$ *is orthonormal (that is, if basis* $\boldsymbol{e}_1, \dots, \boldsymbol{e}_n$ *is orthonormal), then the distance between points*

$$C = (x^1, \dots, x^n), \quad D = (y^1, \dots, y^n)$$

is determined by the formula

$$\rho(C, D) = \sqrt{(y^1 - x^1)^2 + \dots + (y^n - x^n)^2}.$$

The proof follows directly from Theorems 23.1 and 16.9.

THEOREM 23.11. *In Euclidean space* (R, X) *we define an orthonormal system of coordinates* $(O;\ \boldsymbol{e}_1, \dots, \boldsymbol{e}_n)$. *Then the gradient of the affine function*

$$f(x^1, \dots, x^n) = A_1x^1 + \dots + A_nx^n + B$$

is equal to $\operatorname{grad} f = \{A_1, \dots, A_n\}$.

Proof. We designate the vector $\{A_1, \dots, A_n\}$ as $\boldsymbol{n}$. Then for any two points

$$C = (x^1, \dots, x^n), \quad D = (y^1, \dots, y^n)$$

of space X we have

$$f(D) - f(C) = (A_1y^1 + \dots + A_ny^n + B) - (A_1x^1 + \dots + A_nx^n + B) =$$
$$= A_1(y^1 - x^1) + \dots + A_n(y^n - x^n) = \boldsymbol{n}\overrightarrow{CD}$$

(see Theorems 23.1 and 16.9), and this means that $\boldsymbol{n} = \operatorname{grad} f$ (see (21.5)).

COROLLARY 23.12. *In an orthonormal coordinate system we define the hyperplane*

$$A_1x^1 + \dots + A_nx^n + B = 0.$$

Then vector $\boldsymbol{n} = \{A_1, \dots, A_n\}$ *is orthogonal to this hyperplane.*

THEOREM 23.13. *In an* n*-dimensional Euclidean space* (R, X) *an orthonormal coordinate system is introduced and the following* k *hyperplanes are defined:*

$$A_{1i}x^1 + \dots + A_{ni}x^n + B_i = 0, \qquad i = 1, \dots, k. \tag{23.15}$$

If the vectors

$$\boldsymbol{n}_i = \{A_{1i}, \dots, A_{ni}\}, \qquad i = 1, \dots, k,$$

are linearly independent, then the intersection of all of hyperplanes (23.15) *is an* $(n-k)$ *-dimensional plane in* X. *Any* $(n-k)$*-dimensional plane of space* X *can be represented in this form.*

The *proof* follows directly from Theorems 23.11 and 20.4.

NOTE 23.14. Theorem 23.13 has an *affine* nature, meaning that it is true in any affine space and not just in Euclidean space. Actually, if in affine space (R, X) we introduce coordinate system $(O; e_1, \ldots, e_n)$, then we can *introduce* the scalar product in R using the formula

$$\{x^1, \ldots, x^n\}\{y^1, \ldots, y^n\} = x^1y^1 + \ldots + x^ny^n$$

(where the coordinates of the vectors are in the given coordinate system). Then (R, X) is transformed into a Euclidean space and Theorem 23.13 is applicable.

THEOREM 23.15. *An open sphere with its center at a point* $O = (a^1, \ldots, a^n)$ *and a radius* r *is defined by the inequality*

$$(x^1 - a^1)^2 + \ldots + (x^n - a^n)^2 < r^2,$$

and the corresponding closed sphere by the inequality

$$(x^1 - a^1)^2 + \ldots + (x^n - a^n)^2 \leqslant r^2$$

(the coordinate system is orthonormal).

This follows directly from Theorem 23.10.

In conclusion, let us consider the coordinate notation used for continuous mappings. Let (R, X), (S, Y) be Euclidean spaces with some number of dimensions and let us specify some mapping $f\colon H \to Y$, where $H \subset X$. Coordinate system $(O; e_1, \ldots, e_n)$ is introduced in space X and system $(O'; e'_1, \ldots, e'_m)$ in space Y. Finally, we take an arbitrary point $C = (x^1, \ldots, x^n) \in H$ and we call $y^1, \ldots, y^m$ the coordinates of point $f(C) \in Y$.

By specifying the point $C = (x^1, \ldots, x^n) \in H$, we define unambiguously the point $f(C) = (y^1, \ldots, y^m)$, that is, we obtain unambiguous values for the numbers $y^1, \ldots, y^m$. In other words, each of coordinates $y^1, \ldots, y^m$ of point $f(C)$ is a function defined on set H:

$$y^i = g^i(x^1, \ldots, x^n), \qquad i = 1, \ldots, m. \tag{23.16}$$

Thus, by specifying mapping $f\colon H \to Y$, we define the m functions $g^i\colon H \to D$ (see (23.16)).

The converse is also obviously true: if arbitrary functions (23.16) are defined on set H, then for each point $(x^1, \ldots, x^n) \in H$ this gives us an unambiguous determination of some point

$$f(x^1, \ldots, x^n) = (g^1(x^1, \ldots, x^n),\ g^2(x^1, \ldots, x^n),\ \ldots,\ g^m(x^1, \ldots, x^n))$$

of space Y, that is, the mapping $f\colon H \to Y$ is defined.

Thus it is evident that *specifying mapping* $f\colon H \to Y$ *is equivalent to specifying functions* (23.16) *on set* H. The set of functions (23.16) constitutes the *coordinate notation* for the mapping $f\colon H \to Y$.

Let Γ_i be the hyperplane of space Y passing through point O' and having a basis $e'_1, \ldots, e'_{i-1}, e'_{i+1}, \ldots, e'_m$, and let P_i be the line passing through point O' and having e'_i as its direction vector. On line P_i we consider the coordinate system $(O'; e'_i)$. Planes Γ_i and P_i are mutually complementary.

If the projection of space Y onto line P_i in the direction of hyperplane Γ_i is called π_i, then for any point $(y^1, \ldots, y^m) \in Y$ we have

$$\pi_i(y^1, \ldots, y^m) = \pi_i(O' + y^1 e'_1 + \ldots + y^m e'_m) = O' + y^i e'_i = (y^i),$$

that is, the projection of point $(y^1, \ldots, y^m)$ has the coordinate y^i on line P_i. Hence it follows that the number $g^i(x^1, \ldots, x^n)$, that is, the i-th coordinate of point $f(C) = f(x^1, \ldots, x^n)$, is equal to the coordinate (on line P_i) of point $\pi_i(f(C)) = (\pi_i \circ f)(C)$. Since mapping π_i is continuous (see Theorems 19.11 and 22.5), it follows that, if mapping $f\colon H \to Y$ is also continuous, functions (23.16) must be continuous as well. The converse can also be proven without difficulty.

Consequently, *a mapping* $f\colon H \to Y$ *is continuous if, and only if, functions* (23.16), *which constitute the coordinate notation for this mapping, are continuous.*

Chapter III. ELEMENTS OF THEORY OF CONVEX SETS

§6. Convex Sets

24. Definition of convex set. In the following, $E^n = (R, X)$ will denote an n-dimensional Euclidean space, and for points and for vectors we will write the inclusions $A \in E^n$, $\boldsymbol{a} \in E^n$ (instead of $A \in X$, $\boldsymbol{a} \in R$).

Let N be some set of points of space E^n. Set N is called *convex* if for any two points A, B of this set the segment $[A, B]$ belongs wholly to set N (Figure 55). For $n = 2$ (that is, in the case where E^n is a Euclidean plane) we can cite the following examples of convex sets: triangle, parallelogram, trapezium, circle, and ellipse. The figure shown in Figure 56, on the other hand, is not convex.

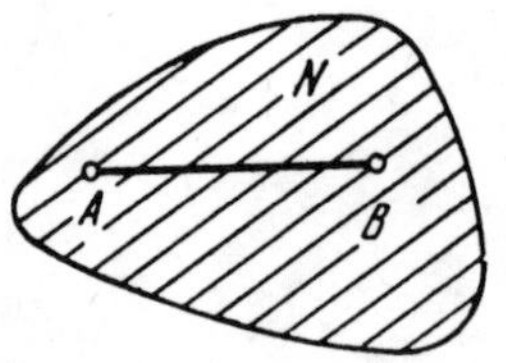

FIGURE 55

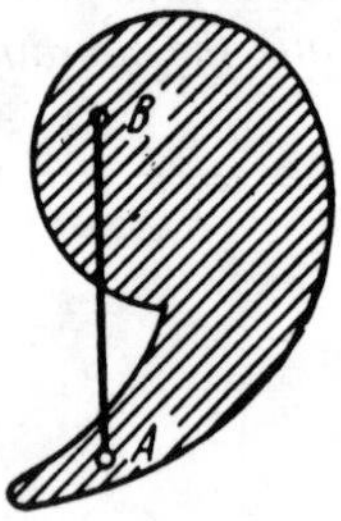

FIGURE 56

Let us list some examples of convex sets which will be important in the following.

Any plane (of any number of dimensions) is a convex set of space E^n. If A and B are two arbitrary points of plane P, then for any real number λ the point $(1 - \lambda) A + \lambda B$ also belongs to [lies on] plane P (see Subsection 18, above). This means that $[A, B] \subset P$, that is, set P is convex.

Any half-space is a convex set of space E^n. We assume that Π is some (closed) half-space, Γ is its bounding hyperplane, Q is an arbitrary point of hyperplane Γ, and $\boldsymbol{n}$ is a vector orthogonal to hyperplane Γ and directed toward half-space Π. Moreover, let A, B be two arbitrary points of half-space Π. Then $\boldsymbol{n}\overrightarrow{QA} \geqslant 0$, $\boldsymbol{n}\overrightarrow{QB} \geqslant 0$. Now if C is an arbitrary point of segment $[A, B]$, then relation (17.6) is valid, where $0 \leqslant \lambda \leqslant 1$, so that

$$\boldsymbol{n}\overrightarrow{QC} = \boldsymbol{n}((1-\lambda)\overrightarrow{QA} + \lambda\overrightarrow{QB}) = (1-\lambda)\boldsymbol{n}\overrightarrow{QA} + \lambda\boldsymbol{n}\overrightarrow{QB} \geqslant 0.$$

Consequently, $C \in \Pi$. Thus we have shown that $[A, B] \subset \Pi$, that is, that set Π is convex (it can be shown similarly that any *open* half-space is also a convex set).

A closed sphere with a center Q and a radius $r > 0$ is a convex set. Assume that A, B are two points of the sphere in question, that is, $\rho(Q, A) \leqslant r$, $\rho(Q, B) \leqslant r$. Moreover, let C be an arbitrary point of segment $[A, B]$, so that relation (17.6) is valid, with $0 \leqslant \lambda \leqslant 1$. Taking into account the relations $|\overrightarrow{QA}| = \rho(Q, A) \leqslant r$, $|\overrightarrow{QB}| = \rho(Q, B) \leqslant r$ and relation (16.2), we obtain

$$
\begin{aligned}
|\overrightarrow{OC}|^2 &= \overrightarrow{QC}^2 = ((1-\lambda)\,\overrightarrow{QA} + \lambda\overrightarrow{OB})^2 = \\
&= (1-\lambda)^2\,\overrightarrow{QA}^2 + 2\lambda(1-\lambda)\overrightarrow{QA}\,\overrightarrow{QB} + \lambda^2\overrightarrow{QB}^2 \leqslant \\
&\leqslant (1-\lambda)^2|\overrightarrow{QA}|^2 + 2\lambda(1-\lambda)\,|\overrightarrow{QA}|\,|\overrightarrow{QB}| + \lambda^2|\overrightarrow{QB}|^2 \leqslant \\
&\leqslant (1-\lambda)^2 r^2 + 2\lambda(1-\lambda)\,r^2 + \lambda^2 r^2 = r^2.
\end{aligned}
$$

Thus $|\overrightarrow{QC}| \leqslant r$, that is, point C also belongs to the sphere being considered. This proves that all of segment $[A, B]$ belongs to the sphere, that is, that the sphere is a convex set (it can be shown similarly that an *open* sphere $U_r(Q)$ with a center Q and a radius r is a convex set).

THEOREM 24.1. *The intersection of any number of convex sets is also a convex set.*

Proof. Consider the intersection of *two* convex sets (the intersection of any number of convex sets, including the intersection of an *infinite* number of convex sets, is considered in exactly the same way). Let M and N be convex sets, let A and B be two arbitrary points of the set $P = M \cap N$ and let C be some point of segment $[A, B]$. Since $A \in P$, point A belongs to *each* of sets M, N and, similarly, point B belongs to each of sets M, N. Since $A \in M$, $B \in M$, it follows from the convexity of set M that the entire segment $[A, B]$ belongs to set M, so that $C \in M$. In exactly the same way, since $A \in N$, $B \in N$, the convexity of set N implies that the entire segment $[A, B]$ belongs to set N, so that $C \in N$. Therefore, point C belongs to each of sets M, N, that is, $C \in M \cap N = P$. This shows that any point C of segment $[A, B]$ belongs to set P, that is, $[A, B] \subset P$, so that set P is convex. The theorem is proven.

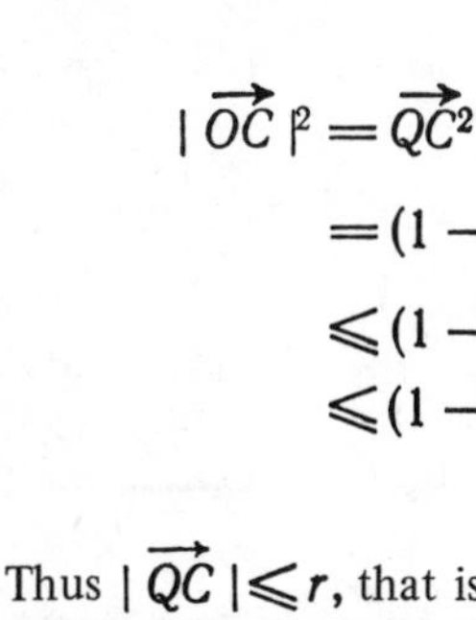

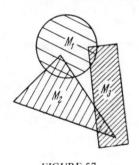

FIGURE 57

Note that the intersection of several convex sets may also turn out to be the empty set. For instance, in Figure 57 every pair of the convex sets M_1, M_2, M_3 has a nonempty intersection, while the intersection $M_1 \cap M_2 \cap M_3$ of all three of the sets is empty.

Since any half-space is a convex set of space E^n, it follows from the proven theorem that the intersection of any number of half-spaces is also a convex set. Let us now consider the intersection of a finite number of half-spaces.

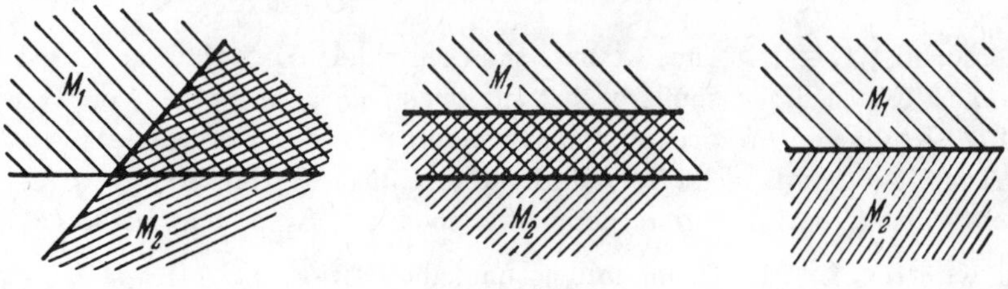

a b c

FIGURE 58

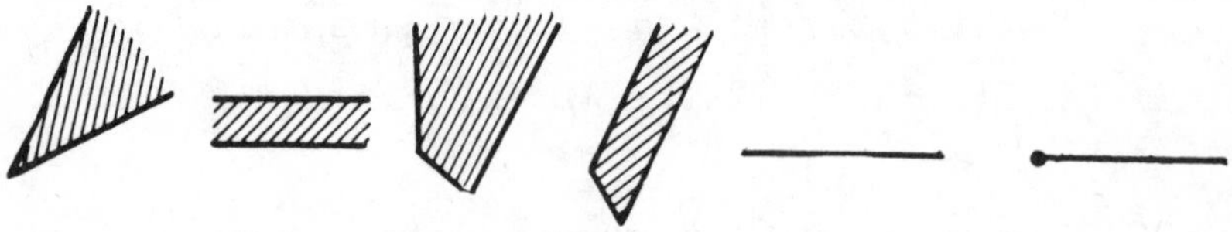

FIGURE 59

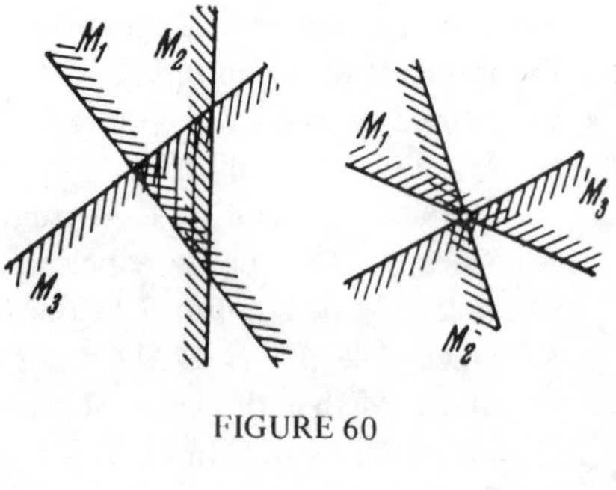

FIGURE 60

FIGURE 61

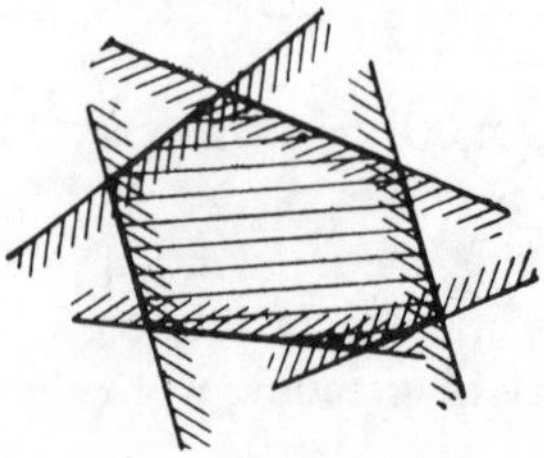

FIGURE 62

First we take the case $n = 2$, that is, the case of figures in a plane. Here we will refer to the intersection of a finite number of *half-planes*. The intersection of *two* half-planes may be an *angle* (in particular, a half-plane), a *band*, or a *straight line* (Figure 58). Thus the intersection of two half-planes (if it is nonempty) is always an *unbounded* figure. The intersection of *three* half-planes is either an unbounded figure (Figure 59) or else it may be a triangle or point (Figure 60). *Four* half-planes may intersect (aside from the above-listed cases) to form a convex quadrangle or segment (Figure 61).

In general, an intersection of several half-planes may be either an unbounded figure or else a *point, segment*, or *convex polygon*. Accordingly, *any* convex polygon can be represented as the intersection of a finite number of half-planes (Figure 62), it being sufficient to take as many half-planes as there are sides of the polygon.

Let us call a point a *zero-dimensional* polyhedron, a line segment a *one-dimensional* convex polyhedron, and a convex polygon a *two-dimensional* polyhedron. Then we can say that *the intersection of a finite number of half-planes, provided it is nonempty and forms a bounded figure, is a convex polyhedron* (zero-dimensional, one-dimensional, or two-dimensional).

The situation is similar for a three-dimensional space: the intersection of a finite number of half-spaces (provided it is nonempty and bounded) is either a zero-dimensional polyhedron (point), a one-dimensional convex polyhedron (segment), a two-dimensional convex polyhedron (that is, a convex polyhedron lying in some plane), or a three-dimensional convex polyhedron (that is, a convex polyhedron in the ordinary sense of the word).

In an n-dimensional space with $n > 3$ we do not have any illustrative geometrical representations similar to those which aid us in considering the figures in a plane or in three-dimensional space. Thus the word "polyhedron" does not convey a visual picture of a body in n-dimensional space. In view of this, the foregoing statement is taken to be the *definition* of a convex polyhedron for an n-dimensional space: the intersection of a finite number of closed half-spaces, one of which is a nonempty bounded set, is called a *convex polyhedron*.

Any convex polyhedron is a *convex set*, that is, in addition to any two points it contains all of the line segment joining them (since a half-space is convex, while an intersection of convex sets is also, according to Theorem 24.1, a convex set). The converse is naturally not true, however: not every convex set is a convex polyhedron. For instance, a sphere in n-dimensional space is a convex set, but it is not (for $n > 1$) a convex polyhedron.

One of the simplest convex polyhedrons in n-dimensional space is an *n-dimensional parallelepiped*, defined in coordinate form by the inequalities:

$$a^1 \leqslant x^1 \leqslant b^1, \quad a^2 \leqslant x^2 \leqslant b^2, \quad \ldots, \quad a^n \leqslant x^n \leqslant b^n, \tag{24.1}$$

where $a^1 < b^1, \quad a^2 < b^2, \ldots, \quad a^n < b^n$. Note that for $n = 2$ points (x^1, x^2) whose coordinates satisfy inequalities (24.1) fill a rectangle (Figure 63); for $n = 3$ inequalities (24.1) define a parallelepiped in the space of variables x^1, x^2, x^3.

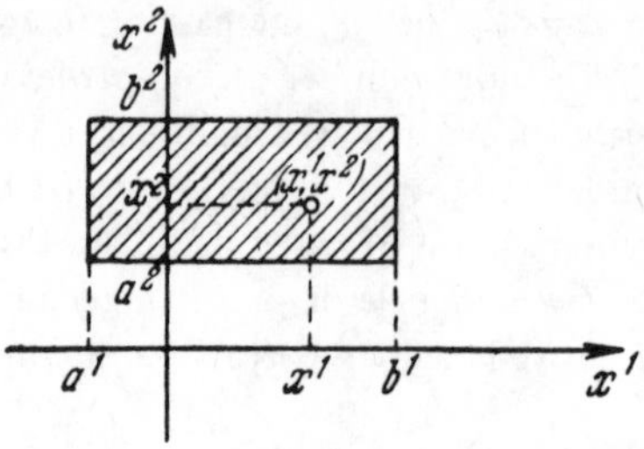

FIGURE 63

The fact that an n-dimensional parallelepiped is actually a convex polyhedron is easy to understand if we write inequalities (24.1) in the following form:

$$x^1 \geqslant a^1, \quad x^2 \geqslant a^2, \ldots, \quad x^n \geqslant a^n, \quad x^1 \leqslant b^1, \quad x^2 \leqslant b^2, \ldots, \quad x^n \leqslant b^n.$$

Each of these inequalities defines some half-space (see Theorem 23.7), and since an n-dimensional parallelepiped consists of points satisfying all these inequalities it must be an *intersection* of $2n$ half-spaces. In addition, an n-dimensional parallelepiped is obviously a bounded set. Thus it is a convex polyhedron.

25. **Convex hull.** The *convex hull* of some set F is defined as the *smallest* convex set containing F. Such a smallest convex set must exist, since if we consider *all* the convex sets containing F their intersection will be the smallest convex set containing F. In the following the convex hull of a set F will be called conv F.

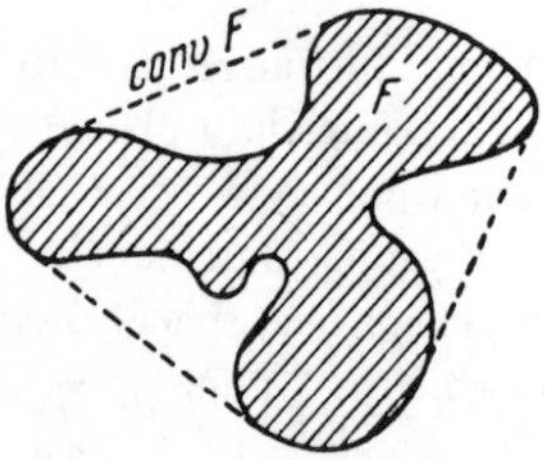

FIGURE 64

In Figure 64 set F is shaded and the line bounding the convex hull is shown dashed. The convex hull of three points not lying on the same line is a *triangle.* A taut "rubber" band pulled around the periphery of a plane figure F provides a graphic example of the convex hull of this figure (cf. Figure 64); the figure bounded by the taut band is also a convex hull. In three-dimensional space a similar representation is provided by a stretched elastic film ("bubble") enveloping a set F; the body bounded by the stretched film is also a convex hull. Naturally, a graphic representation like this can never replace an accurate description of a convex hull; moreover, it is not applicable for $n > 3$.

The following theorem provides an accurate algebraic description of the convex hull of an arbitrary set F.

THEOREM 25.1. *In order for a point C to belong to the convex hull of a set $F \subset E^n$, it is necessary and sufficient that there exist points $A_0, A_1, \ldots, A_n$ of set F (not necessarily all different) and numbers $\lambda^0, \lambda^1, \ldots, \lambda^n$, all such that the following relations are satisfied:*

$$\lambda^0 \geqslant 0, \quad \lambda^1 \geqslant 0, \ldots, \quad \lambda^n \geqslant 0, \quad \lambda^0 + \lambda^1 + \ldots + \lambda^n = 1, \tag{25.1}$$

$$C = \lambda^0 A_0 + \lambda^1 A_1 + \ldots + \lambda^n A_n. \tag{25.2}$$

Proof. Let F^* be the set of all points C for each of which there exist a natural number m, points $A_0, A_1, \ldots, A_m$ of set F, and numbers $\lambda^0, \lambda^1, \ldots, \lambda^m$, all such that the following relations are satisfied:

$$\lambda^0 \geqslant 0, \quad \lambda^1 \geqslant 0, \ldots, \lambda^m \geqslant 0, \quad \lambda^0 + \lambda^1 + \ldots + \lambda^m = 1, \tag{25.3}$$

$$C = \lambda^0 A_0 + \lambda^1 A_1 + \ldots + \lambda^m A_m. \tag{25.4}$$

Note that in relations (25.3) and (25.4) the number m is in no way related to the dimension n of the space being considered, and in addition that for each point C of set F^* it may well be necessary to choose an *appropriate* value of m.

Let us prove that set F^* is convex. If C and D are two arbitrary points of set F^*, then for point C there will exist a number m, points $A_0, A_1, \ldots, A_m$, and numbers $\lambda^0, \lambda^1, \ldots, \lambda^m$ satisfying relations (25.3) and (25.4), while for point D there will exist a natural number p, points $B_0, B_1, \ldots, B_p$, and numbers $\mu^0, \mu^1, \ldots, \mu^p$ satisfying the relations

$$\mu^0 \geqslant 0, \quad \mu^1 \geqslant 0, \ldots, \mu^p \geqslant 0, \quad \mu^0 + \mu^1 + \ldots + \mu^p = 1,$$

$$D = \mu^0 B_0 + \mu^1 B_1 + \ldots + \mu^p B_p.$$

Now if E is an arbitrary point of segment $[C, D]$, we have

$$E = (1 - \nu) C + \nu D$$

(cf. (17.6)), where the number ν satisfies the inequalities $0 \leqslant \nu \leqslant 1$, so that the numbers ν and $1 - \nu$ are both nonnegative. We then have

$$\begin{aligned} E = (1-\nu)(\lambda^0 A_0 + \lambda^1 A_1 + \ldots + \lambda^m A_m) + \nu (\mu^0 B_0 + \mu^1 B_1 + \ldots + \mu^p B_p) = \\ = \lambda^0 (1-\nu) A_0 + \lambda^1 (1-\nu) A_1 + \ldots + \lambda^m (1-\nu) A_m + \\ + \mu^0 \nu B_0 + \mu^1 \nu B_1 + \ldots + \mu^p \nu B_p \end{aligned} \tag{25.5}$$

(if some of points A_i and B_j are identical, then on the right-hand side of (25.5) there may be similar terms, which is however not important for us). It is easy to see that all the coefficients on the right-hand side of (25.5) are nonnegative, and that their sum is unity:

$$\begin{aligned} \lambda^0 (1-\nu) + \lambda^1 (1-\nu) + \ldots + \lambda^m (1-\nu) + \mu^0 \nu + \mu^1 \nu + \ldots + \mu^p \nu = \\ = (1-\nu)(\lambda^0 + \lambda^1 + \ldots + \lambda^m) + \nu (\mu^0 + \mu^1 + \ldots + \mu^p) = \\ = (1-\nu) \cdot 1 + \nu \cdot 1 = 1. \end{aligned}$$

However, then equation (25.5) indicates that point E belongs to set F^*. Thus any point E of segment $[C, D]$ belongs to set F^*, that is, $[C, D] \subset F^*$, so that set F^* is convex.

Moreover, it is obvious that set F^* contains F: for any point $C \in F$ we can set $m = 0$, $A_0 = C$, $\lambda^0 = 1$ and then write $C = \lambda^0 A_0$, so that for point C relations (25.3), (25.4) are valid.

Now let us prove that any convex set M containing F also contains F^*. Let $M \supset F$ and let C be an arbitrary point of set F^*, so that for it relations (25.3) and (25.4) are valid. It can be assumed in this case (after eliminating, if necessary, excess terms in (25.4)) that all the numbers $\lambda^0, \lambda^1, \ldots, \lambda^m$ are *positive*. Next let B_i be the point defined by the relation

$$B_i = \frac{1}{\lambda^0 + \lambda^1 + \ldots + \lambda^i}(\lambda^0 A_0 + \lambda^1 A_1 + \ldots + \lambda^i A_i), \qquad (25.6)$$

$$i = 0, 1, \ldots, m.$$

A direct calculation shows the following relations to be valid:

$$B_{i+1} = \nu_i B_i + (1 - \nu_i) A_{i+1}, \quad \text{where } \nu_i = \frac{\lambda^0 + \lambda^1 + \ldots + \lambda^i}{\lambda^0 + \lambda^1 + \ldots + \lambda^{i+1}}, \qquad (25.7)$$

$$i = 0, 1, \ldots, m-1.$$

Here it is clear that ν_i satisfies the condition $0 < \nu_i < 1$. Since point B_0 is obviously identical to A_0 (see (25.6)), for $i=0$ relation (25.7) must become $B_1 = \nu_0 A_0 + (1 - \nu_0) A_1$. Consequently, $B_1 \in [A_0, A_1]$, but since points A_0 and A_1 both belong to set F, they must also belong to set M (because $M \supset F$). Thus $[A_0, A_1] \subset M$ (since set M is convex); in particular, $B_1 \in M$. For $i=1$ relation (25.7) becomes $B_2 = \nu_1 B_1 + (1 - \nu_1) A_2$, so that $B_2 \in [B_1, A_2]$.

However, $[B_1, A_2] \subset M$ (since $B_1 \in M$, $A_2 \in F \subset M$, and set M is convex). Consequently $B_2 \in M$. By continuing in this manner, we can establish that all the points $B_0, B_1, B_2, \ldots, B_m$ belong to set M. On the other hand, it follows from (25.3), (25.4), (25.6) that $B_m = C$. This means that any point C of set F^* belongs to set M, that is, $F^* \subset M$.

We have shown that F^* is a convex set containing F, and that F^* is contained in *any* convex set M which contains F. In other words, F^* is the *smallest* convex set containing F, that is, $F^* = \operatorname{conv} F$.

This implies that the condition formulated in Theorem 25.1 is *sufficient;* if point C satisfies relations (25.1), (25.2) (that is, if it satisfies relations (25.3), (25.4) for $m = n$), then $C \in F^*$, that is, point C belongs to the convex hull of set F.

Let us next prove *necessity.* If C is an arbitrary point belonging to the convex hull ($\operatorname{conv} F$) of set F, then, in view of the foregoing, point C satisfies relations (25.3), (25.4) (where points $A_0, A_1, \ldots, A_m$ belong to set F). If $m = n$, then relations (25.3) and (25.4) are the relations sought (that is, they have the required form (25.1), (25.2)). If $m < n$, then to arrive at the required form it is sufficient to add several zero terms to the right-hand side of (25.4), that is, to set $\lambda^{m+1} = \lambda^{m+2} = \ldots = \lambda^n = 0$ and to take as $A_{m+1}, \ldots, A_n$ any points of set M (for instance, $A_{m+1} = A_{m+2} = \ldots = A_n = A_0$).

It only remains to consider the case $m > n$. Thus let point C satisfy relations (25.3), (25.4), where $m > n$. Here we can assume that all the numbers $\lambda^0, \lambda^1, \ldots, \lambda^m$

are positive. For the vectors

$$\overrightarrow{A_0A_1}, \ldots, \overrightarrow{A_0A_m},$$

we know that there are $m > n$ of these vectors and that they are linearly independent:

$$\alpha^1 \overrightarrow{A_0A_1} + \ldots + \alpha^m \overrightarrow{A_0A_m} = 0,$$

at least one of the numbers $\alpha^1, \ldots, \alpha^m$ being different from zero. Since $\overrightarrow{A_0A_i} = A_i - A_0$, this relation can be rewritten as

$$\alpha^0 A_0 + \alpha^1 A_1 + \ldots + \alpha^m A_m = 0, \tag{25.8}$$

where $\alpha^0 = -\alpha^1 - \ldots - \alpha^m$, that is,

$$\alpha^0 + \alpha^1 + \ldots + \alpha^m = 0. \tag{25.9}$$

Obviously, some of the numbers $\alpha^0, \alpha^1, \ldots, \alpha^m$ are different from zero. If we want to consider the largest of the numbers $|\alpha^0/\lambda^0|$, $|\alpha^1/\lambda^1|$, $\ldots$, $|\alpha^m/\lambda^m|$, then we can always assume (after changing, if necessary, the numbering of points $A_0, \ldots, A_m$) that the number $|\alpha^m/\lambda^m|$ is the largest:

$$\left|\frac{\alpha^0}{\lambda^0}\right| \leqslant \left|\frac{\alpha^m}{\lambda^m}\right|, \quad \left|\frac{\alpha^1}{\lambda^1}\right| \leqslant \left|\frac{\alpha^m}{\lambda^m}\right|, \quad \ldots, \quad \left|\frac{\alpha^{m-1}}{\lambda^{m-1}}\right| \leqslant \left|\frac{\alpha^m}{\lambda^m}\right|. \tag{25.10}$$

The number α^m has to be different from zero (otherwise, in view of (25.10) all the numbers $\alpha^0, \alpha^1, \ldots, \alpha^m$ would be zero). Here we can assume that $\alpha^m = \lambda^m$ (for this it is sufficient to multiply all the numbers $\alpha^0, \alpha^1, \ldots, \alpha^m$ by λ^m/α^m, which does not violate relations (25.8), (25.9)). Then (in view of (25.10))

$$\left|\frac{\alpha^0}{\lambda^0}\right| \leqslant 1, \quad \left|\frac{\alpha^1}{\lambda^1}\right| \leqslant 1, \ldots, \left|\frac{\alpha^{m-1}}{\lambda^{m-1}}\right| \leqslant 1. \tag{25.11}$$

Next we subtract (25.8) from (25.4) to obtain

$$\begin{aligned} C &= \lambda^0 A_0 + \lambda^1 A_1 + \ldots + \lambda^m A_m - (\alpha^0 A_0 + \alpha^1 A_1 + \ldots + \alpha^m A_m) = \\ &= \lambda^0\left(1 - \frac{\alpha^0}{\lambda^0}\right) A_0 + \lambda^1\left(1 - \frac{\alpha^1}{\lambda^1}\right) A_1 + \ldots + \lambda^{m-1}\left(1 - \frac{\alpha^{m-1}}{\lambda^{m-1}}\right) A_{m-1}. \end{aligned} \tag{25.12}$$

According to (25.11), all the coefficients on the right-hand side are nonnegative. Moreover, according to (25.3) and (25.9), the sum of these coefficients is unity:

$$\lambda^0\left(1 - \frac{\alpha^0}{\lambda^0}\right) + \lambda^1\left(1 - \frac{\alpha^1}{\lambda^1}\right) + \ldots + \lambda^{m-1}\left(1 - \frac{\alpha^{m-1}}{\lambda^{m-1}}\right) =$$

$$=(\lambda^0+\lambda^1+\dots+\lambda^{m-1})-(\alpha^0+\alpha^1+\dots+\alpha^{m-1})=$$
$$=(\lambda^0+\lambda^1+\dots+\lambda^m)-(\alpha^0+\alpha^1+\dots+\alpha^m)=1-0=1. \quad (25.13)$$

Relations (25.12) and (25.13) have the same form as (25.3) and (25.4), the only difference being that in (25.12) and (25.13) the value of m is less by unity. Thus, if $m-1=n$, the relations obtained (that is, (25.12), (25.13)) are the ones sought (they have form (25.1), (25.2)). If we still find that $m-1>n$, then the number $m-1$ can be reduced similarly by unity, etc. After a finite number of such steps, we are able to reduce relations (25.3), (25.4) to form (25.1), (25.2).

EXAMPLE 25.2. Let k be any of the numbers $1, \dots, n$. In space E^n we take an arbitrary $k+1$ independent points $A_0, A_1, \dots, A_k$ (points not lying in one $(k-1)$-dimensional plane). The convex hull of the set of all these points, which is designated as $[A_0, A_1, \dots, A_k]$, is called a *k-dimensional simplex.* According to the definition of a convex hull, a k-dimensional simplex is a convex set.

The convex hull of two different points, that is, a *line segment,* is a one-dimensional simplex. A two-dimensional simplex is the convex hull of *three* points not lying along the same straight line, that is, a two-dimensional simplex is a *triangle.* A three-dimensional simplex is the convex hull of four points not lying in the same (two-dimensional) plane, that is, it is a *tetrahedron.* For $k>3$ a simplex of dimension k is a multi-dimensional extension of the segment, triangle, and tetrahedron.

According to Theorem 25.1, *a point C belongs to a simplex $[A_0, A_1, \dots, A_k]$ if, and only if,*

$$C=\lambda^0A_0+\lambda^1A_1+\dots+\lambda^kA_k,$$

where the numbers $\lambda^0, \lambda^1, \dots, \lambda^k$ satisfy the relations

$$\lambda^0\geqslant 0,\quad \lambda^1\geqslant 0,\ \dots,\ \lambda^k\geqslant 0,\quad \lambda^0+\lambda^1+\dots+\lambda^k=1.$$

Let us call a set M in a space E^n a *convex field* if this set is convex and contains at least one interior point.

EXAMPLE 25.3. *Any n-dimensional simplex is a convex field in space E^n.*

Let us consider in simplex $[A_0, A_1, \dots, A_n]$ an arbitrary point $C=\lambda^0A_0+\lambda^1A_1+ +\dots+\lambda^nA_n$ for which all the numbers $\lambda^0, \lambda^1, \dots, \lambda^n$ are positive:

$$\lambda^0>0,\quad \lambda^1>0,\ \dots,\ \lambda^n>0,\quad \lambda^0+\lambda^1+\dots+\lambda^n=1. \quad (25.14)$$

Let ε be a positive number such that

$$\lambda^0\geqslant\varepsilon,\quad \lambda^1\geqslant\varepsilon,\ \dots,\ \lambda^n\geqslant\varepsilon. \quad (25.15)$$

Vectors $e_1=\overrightarrow{A_0A_1}, \dots, e_n=\overrightarrow{A_0A_n}$ are linearly independent (see Theorem 17.6), that is, they form a basis of space E^n.

If we take the arbitrary numbers $x^1, \ldots, x^n$ satisfying conditions

$$|x^1| \leqslant \frac{\varepsilon}{n}, \ldots, |x^n| \leqslant \frac{\varepsilon}{n}, \tag{25.16}$$

then the point

$$D = C + x^1 e_1 + \ldots + x^n e_n \tag{25.17}$$

also belongs to the simplex in question. Actually,

$$\begin{aligned} D &= (\lambda^0 A_1 + \lambda^1 A_1 + \ldots + \lambda^n A_n) + x^1 \overrightarrow{A_0 A_1} + \ldots + x^n \overrightarrow{A_0 A_n} = \\ &= \lambda^0 A_0 + \lambda^1 A_1 + \ldots + \lambda^n A_n + x^1 (A_1 - A_0) + \ldots + x^n (A_n - A_0) = \\ &= (\lambda^0 - x^1 - \ldots - x^n) A_0 + (\lambda^1 + x^1) A_1 + \ldots + (\lambda^n + x^n) A_n. \end{aligned}$$

It is obvious that the sum of the coefficients on the right-hand side of this relation is equal to unity (see (25.14)), while in view of (25.15) and (25.16) all these coefficients are nonnegative. Consequently, point D belongs to the simplex.

Now let $f_1, \ldots, f_n$ be some orthonormal basis of space E^n. Each vector f_i can be portrayed as a linear combination of vectors $e_1, \ldots, e_n$:

$$f_i = a_i^1 e_1 + \ldots + a_i^n e_n, \qquad i = 1, \ldots, n.$$

Let a be the largest of the numbers $|a_i^j|$, so that

$$|a_i^j| \leqslant a, \qquad i, j = 1, \ldots, n,$$

and let $r = \varepsilon/(n^2 a)$. We will now show that a sphere of radius r with its center at point C is completely contained in the simplex $[A_0, A_1, \ldots, A_n]$, from which it also follows that this simplex is a convex field.

If D is an arbitrary point of this sphere, that is, if $\rho(C, D) \leqslant r$, then we have

$$\overrightarrow{CD} = y^1 f_1 + \ldots + y^n f_n,$$

where $(y^1)^2 + \ldots + (y^n)^2 = (\rho(C, D))^2 \leqslant r^2$, so that

$$|y^1| \leqslant r, \ldots, |y^n| \leqslant r.$$

Now we can write:

$$\begin{aligned} \overrightarrow{CD} &= y^1 f_1 + \ldots + y^n f_n = y^1 (a_1^1 e_1 + \ldots + a_1^n e_n) + \\ &+ y^2 (a_2^1 e_1 + \ldots + a_2^n e_n) + \ldots + y^n (a_n^1 e_1 + \ldots + a_n^n e_n) = \\ &= (a_1^1 y^1 + \ldots + a_n^1 y^n) e_1 + \ldots + (a_1^n y^1 + \ldots + a_n^n y^n) e_n = \\ &= x^1 e_1 + \ldots + x^n e_n, \end{aligned}$$

where

$$x^i = a_1^i y^1 + \ldots + a_n^i y^n.$$

Since obviously

$$|x^i| \leqslant |a_1^i||y^1| + \ldots + |a_n^i||y^n| \leqslant nar = na \cdot \frac{\varepsilon}{n^2 a} = \frac{\varepsilon}{n},$$

it follows that point D satisfies relations (25.16), (25.17) and thus belongs to simplex $[A_0, A_1, \ldots, A_n]$. Therefore, a sphere of radius r with its center at point C belongs wholly to the simplex being considered.

THEOREM 25.4. *Any convex set of space E^n either is a convex field or is situated wholly in some plane of dimension less than n.*

Proof. Let M be an arbitrary convex set in E^n. We fix in M an arbitrary point A_0 and we consider all possible vectors of form $\overrightarrow{A_0A}$, where A runs through set M. Let us select the *maximum* number of linearly independent vectors $\overrightarrow{A_0A}$, these being called $\overrightarrow{A_0A_1}, \ldots, \overrightarrow{A_0A_k}$. There are now two possibilities: $k = n$ or $k < n$. Let us examine these separately.

If $k = n$, then in view of the linear independence of vectors $\overrightarrow{A_0A_1}, \ldots, \overrightarrow{A_0A_n}$ points $A_0, A_1, \ldots, A_n$ do not lie in a single hyperplane, that is, they are vertices of the n-dimensional simplex $[A_0, A_1, \ldots, A_n]$. Since set M contains all points $A_0, A_1, \ldots, A_n$, it must also contain the convex hull of these points, that is, it contains the simplex $[A_0, A_1, \ldots, A_n]$. However, as we know, an n-dimensional simplex is a convex field, that is, it contains interior points. Even more so, set M contains interior points, meaning that it is a convex field.

Now let $k < n$. If we take an arbitrary point $C \in M$, then the vectors $\overrightarrow{A_0A_1}, \ldots, \overrightarrow{A_0A_k}, \overrightarrow{A_0C}$ are linearly dependent (since there are more than k of them), so that from Theorem 13.3 we know that vector $\overrightarrow{A_0C}$ can be expressed linearly in terms of $\overrightarrow{A_0A_1}, \ldots, \overrightarrow{A_0A_k}$:

$$\overrightarrow{A_0C} = \alpha^1 \overrightarrow{A_0A_1} + \ldots + \alpha^k \overrightarrow{A_0A_k},$$

that is, point C is located in a k-dimensional plane P passing through point A_0 and having vectors $\overrightarrow{A_0A_1}, \ldots, \overrightarrow{A_0A_k}$ as its basis. This will be the case for any point $C \in M$, so that all of set M is located in the k-dimensional plane P.

The plane P constructed during the proof of Theorem 25.4 (for $k < n$) possesses some important properties. It is the plane of *least dimension* containing set M. Actually, since set M contains points $A_0, A_1, \ldots, A_k$, vectors $\overrightarrow{A_0A_1}, \ldots, \overrightarrow{A_0A_k}$ being linearly independent, it follows that points $A_0, A_1, \ldots, A_k$ are independent, that is, these points (and all the more so, set M) cannot be in a plane of dimension less than k.

Moreover, plane P is *uniquely defined* by convex set M as the plane of least dimension containing M. Actually, if there existed two different k-dimensional planes containing M, then set M would be contained in their intersection, that is, in a plane of lower dimension, which is impossible.

Finally, note that, if plane P is taken to be a k-dimensional Euclidean space, then set M is a convex *field relative to this space,* that is, M contains interior points relative to space P. We know that the convex hull of points $A_0, A_1, \ldots, A_k$ is a k-dimensional simplex $[A_0, A_1, \ldots, A_k]$ in k-dimensional space P and thus contains interior points in P; even more so, set M contains interior points in space P.

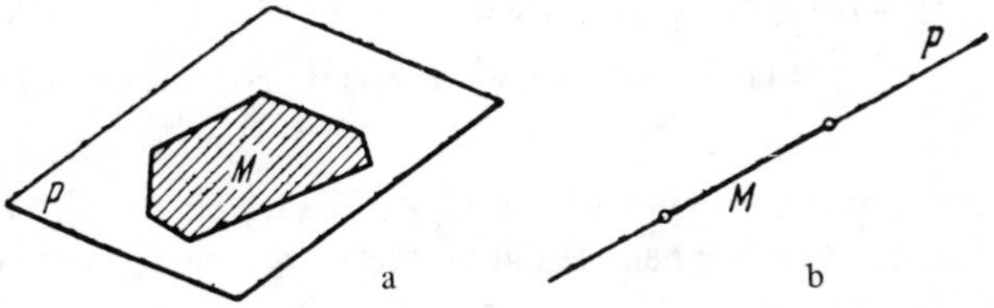

FIGURE 65

Plane P (Figure 65) is called the *carrier plane* of convex set M (provided the latter is not a convex *field* in E^n). If, on the other hand, M is a convex field of space E^n, then its *carrier plane* is taken to be space E^n itself. A comparison of all the foregoing leads us to the following statement.

COROLLARY 25.5. *For any convex set M of space E^n there exists a single plane of least dimension containing M. It is known as the carrier plane of convex set M. Set M is a convex field of space E^n if, and only if, its carrier plane coincides with space E^n. Any convex set is a convex field relative to its carrier plane.*

In conclusion, let us introduce one more definition: a convex set is called *k-dimensional* if its carrier plane is of dimension k.

26. Boundary of convex field. THEOREM 26.1. *The closure $\overline{M}$ of a convex set M is also a convex set.*

Proof. Let A and B be two arbitrary points of set $\overline{M}$. We will show that the entire segment $[A, B]$ is also contained in $\overline{M}$, that is, that each point P

$$C = (1 - \lambda) A + \lambda B, \quad \text{where } 0 \leqslant \lambda \leqslant 1,$$

belongs to set $\overline{M}$. To do this, it must be established that any sphere centered on C intersects set M.

Let us take a sphere of radius r with its center at point C. Since $A \in \overline{M}$, we know that a sphere of radius r with its center at A intersects set M, that is, there exists a point $A_1 \in M$ such that $\rho(A, A_1) \leqslant r$. Similarly, there exists a point $B_1 \in M$ such that $\rho(B, B_1) \leqslant r$. We consider a point

$$C_1 = (1 - \lambda) A_1 + \lambda B_1.$$

Then $C_1 \in [A_1, B_1]$, so that in view of the convexity of set M we know that point C_1 belongs to M. Moreover,

$$\overrightarrow{CC_1} = C_1 - C = (1-\lambda) A_1 + \lambda B_1 - (1-\lambda) A - \lambda B = \\ = (1-\lambda)(A_1 - A) + \lambda (B_1 - B) = (1-\lambda) \overrightarrow{AA_1} + \lambda \overrightarrow{BB_1}.$$

Hence we obtain (taking into account that λ and $1-\lambda$ are nonnegative)

$$\rho(C, C_1) = |\overrightarrow{CC_1}| = |(1-\lambda)\overrightarrow{AA_1} + \lambda\overrightarrow{BB_1}| \leqslant \\ \leqslant |(1-\lambda)\overrightarrow{AA_1}| + |\lambda\overrightarrow{BB_1}| = (1-\lambda)|\overrightarrow{AA_1}| + \lambda|\overrightarrow{BB_1}| = \\ = (1-\lambda)\rho(A, A_1) + \lambda\rho(B, B_1) \leqslant (1-\lambda) r + \lambda r = r.$$

Consequently, we have found a point $C_1 \in M$ such that $\rho(C, C_1) \leqslant r$, that is, we have established that a sphere of radius r centered on C intersects set M. The theorem is proven.

In the following, we will as a rule consider only *closed* convex sets, that is, we will assume (unless expressly stated to the contrary) that each convex set considered contains all its boundary points. Note, in particular, that each convex polyhedron is (by definition) a closed set.

Let M be an arbitrary convex set of space E^n. The set of all its interior points, designated as $\operatorname{int} M$, is called the *interior* of set M. The set of all boundary points of set M, designated as $\operatorname{bd} M$, is called the *boundary* of set M (Figure 66, $n=2$). In accordance with the above stipulation, we will assume (unless stated to the contrary) that $\operatorname{bd} M \subset M$, that is, $M = (\operatorname{int} M) \cup (\operatorname{bd} M)$.

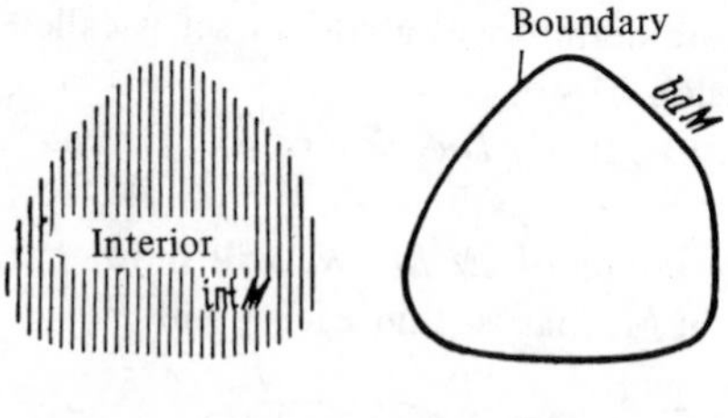

FIGURE 66

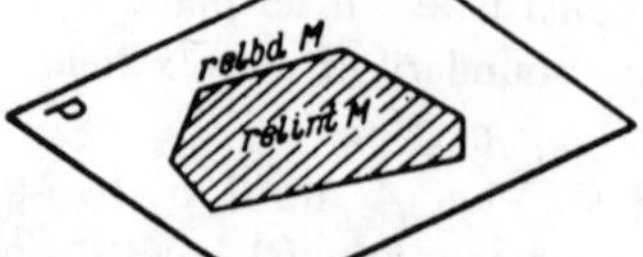

FIGURE 67

It should be noted that, if convex set M is not a convex field (meaning that it does not contain any interior points at all), then all of its points will be boundary points, that is, in this case convex set M is identical to its boundary. In other words, a study of the properties of the boundary and the boundary points is meaningful only for a convex *field*. Moreover, any convex set of space E^n constitutes a convex field relative to its carrier plane. Thus we can speak of the interior and the boundary of a convex set M *relative to its carrier plane* (Figure 67, $n=3$); these will be designated respectively as **relint** M and **relbd** M. In accordance with what has been stated, in the rest of this subsection we will restrict our discussion to the boundary of a convex *field.*

THEOREM 26.2. *If A and B are interior points of convex field M, then each point of segment $[A, B]$ is also an interior point of this field. If A is a boundary point and B is an interior point of convex field M, then each point of the half-open interval $(A, B]$ is an interior point of this field. Finally, if A and B are boundary points of convex field M, then either all the points of interval (A, B) are interior points of this field or else all the points of this interval are boundary points for field M.*

Proof. If A and B are *interior* points of convex field M, then a sphere of some radius r_1 with its center at A is completely contained in field M and a sphere of some radius r_2 with its center at B is completely contained in field M. If r is the smallest of the positive numbers r_1, r_2, then field M contains a sphere of radius r centered on A and a sphere of radius r centered on B. Let us show that, if C is an arbitrary point of segment $[A, B]$, then a sphere of radius r centered on C is completely contained in field M (Figure 68).

We have

$$C=(1-\lambda)A+\lambda B,$$

where

$$0\leqslant\lambda\leqslant 1.$$

Let C_1 be an arbitrary point belonging to the sphere of radius r centered on C, that is, $|\overrightarrow{CC_1}|\leqslant r$. If A_1 and B_1 are points such that $\overrightarrow{AA_1}=\overrightarrow{BB_1}=\overrightarrow{CC_1}$, then $|\overrightarrow{AA_1}|=|\overrightarrow{CC_1}|\leqslant r$, that is, point A_1 belongs to the sphere of radius r centered on A. However, since this sphere is wholly contained in field M, it follows that point A_1 belongs to field M. Similarly, point B_1 belongs to field M as well. Moreover:

$$\begin{aligned}C_1&=C+\overrightarrow{CC_1}=(1-\lambda)A+\lambda B+\overrightarrow{CC_1}=\\&=(1-\lambda)A+\lambda B+(1-\lambda)\overrightarrow{CC_1}+\lambda\overrightarrow{CC_1}=\\&=(1-\lambda)A+\lambda B+(1-\lambda)\overrightarrow{AA_1}+\lambda\overrightarrow{BB_1}=\\&=(1-\lambda)(A+\overrightarrow{AA_1})+\lambda(B+\overrightarrow{BB_1})=(1-\lambda)A_1+\lambda B_1.\end{aligned}$$

This relation shows that point C_1 lies on segment $[A_1, B_1]$ (since $0\leqslant\lambda\leqslant 1$). However, since both points A_1, B_1 belong to field M, the convexity of the latter implies that all of segment $[A_1, B_1]$ belongs to this field as well. In particular, $C_1\in M$. This shows that a sphere of radius r with its center at C is completely contained in field M, that is, C is an interior point of this field. The first part of the theorem is thereby established.

Now let us prove the second part. If A is a *boundary* point of convex field M and B is an *interior* point, then a sphere of some radius r centered on B will be wholly contained in M. Let C be an arbitrary point in the half-open interval $(A, B]$:

$$C=(1-\lambda)A+\lambda B, \text{ where } 0<\lambda\leqslant 1;$$

let us show that a sphere of radius λr centered on C is wholly contained in field M (Figure 69).

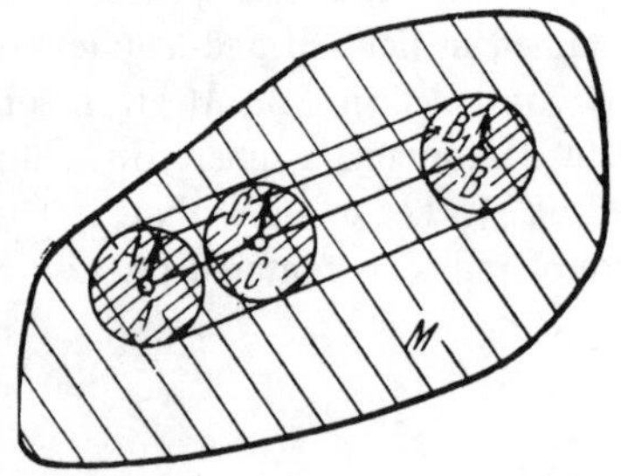

FIGURE 68

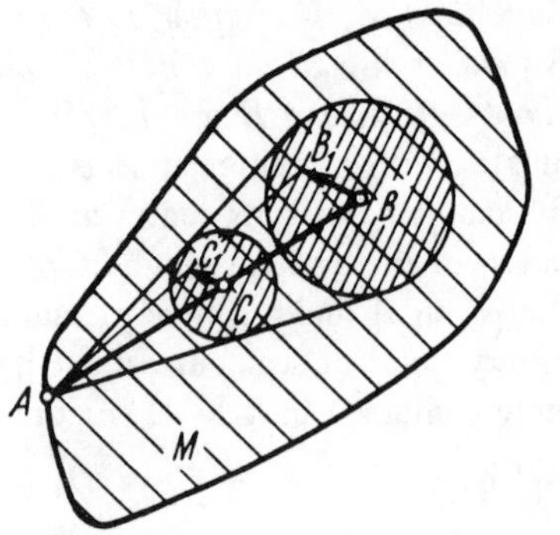

FIGURE 69

If C_1 is an arbitrary point of this sphere, so that $|\overrightarrow{CC_1}|\leqslant\lambda r$, and if B_1 is a point such that $\overrightarrow{BB_1}=(1/\lambda)\overrightarrow{CC_1}$, then $|\overrightarrow{BB_1}|=(1/\lambda)|\overrightarrow{CC_1}|\leqslant r$, that is, point B_1 belongs to a sphere of radius r with its center at B. However, since this sphere is wholly contained in field M, it follows that point B_1 is also in field M. Moreover,

$$C_1=C+\overrightarrow{CC_1}=C+\lambda\overrightarrow{BB_1}=(1-\lambda)A+\lambda B+\lambda\overrightarrow{BB_1}=$$
$$=(1-\lambda)A+\lambda(B+\overrightarrow{BB_1})=(1-\lambda)A+\lambda B_1.$$

This relation indicates that point C_1 lies on segment $[A, B_1]$. However, since points A and B_1 are both in field M (recall that we have agreed to consider only *closed* convex sets, so that each boundary point, and point A in particular, is in field M), this means that the entire segment $[A, B_1]$ is in this field as well. In particular, $C_1\in M$. This proves that a sphere of radius λr centered on C is entirely contained in field M, that is, C is an interior point of this field. The second part of the theorem is thereby proven.

Finally, let A and B be *boundary* points of field M. It may be that all the points of segment $[A, B]$ are boundary points of field M (Figure 70). On the other hand, if there exists at least one point $C\in[A, B]$ which is an interior point of field M (Figure 71), then according to the foregoing proof all the points of the half-open interval $(A, C]$ are interior and all the points of the half-open interval $(B, C]$ are interior, that is, all the points of interval (A, B) are interior points of field M.

THEOREM 26.3. *Let Q be an interior point of convex field M and let l be a ray starting from point Q. Ray l will then either lie completely in field M* (Figure 72) *or else it will cross the boundary of field* M *just at a single point* A (Figure 73), *where in the latter case ray* l *intersects field* M *along segment* $[Q, A]$.

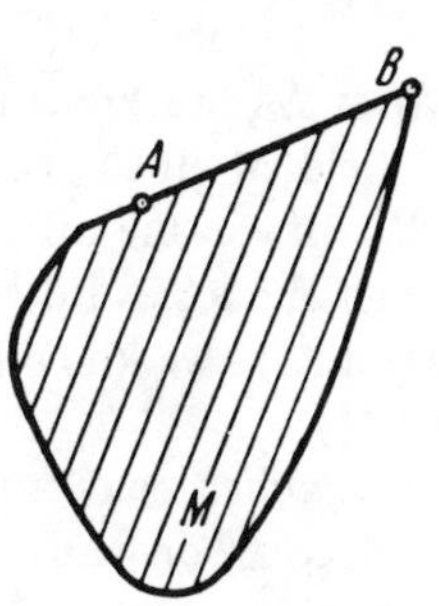

FIGURE 70

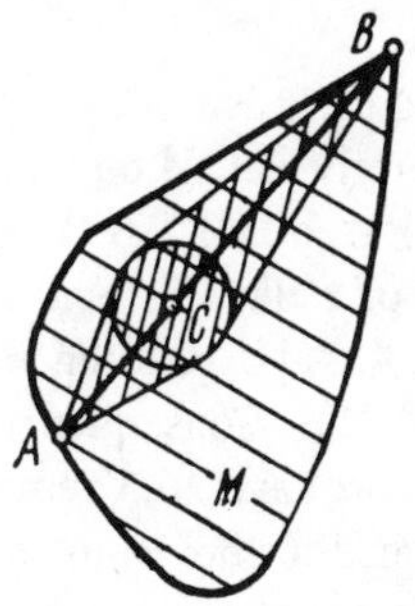

FIGURE 71

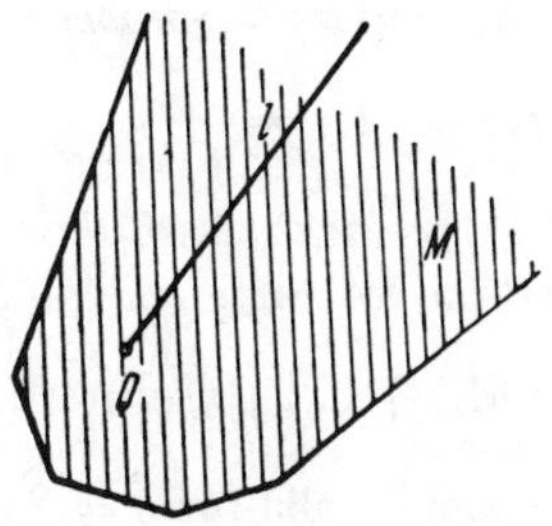

FIGURE 72

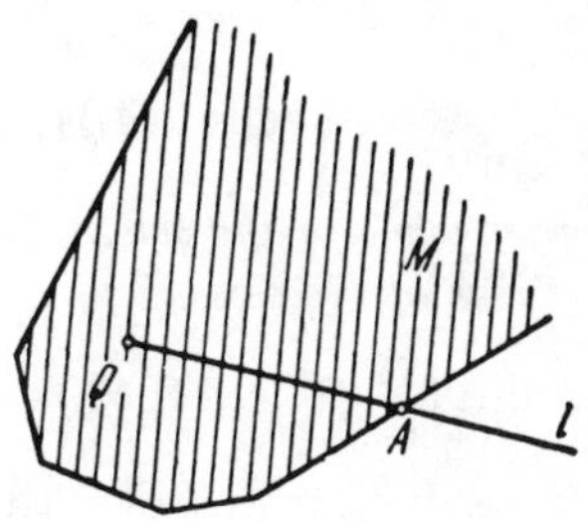

FIGURE 73

Proof. Let us assume that ray l does not lie wholly in field M. Then there will exist on ray l some point B not in field M. Obviously, if point C lies somewhere beyond point B on ray l (that is, $B \in [Q, C]$), then point C will not be in field M (otherwise the entire segment $[Q, C]$ would have to be in field M, so that point B would also be in M). Consequently, the intersection $l \cap M$ lies wholly on segment $[Q, B]$.

Let A be the point of set $l \cap M$ furthest from Q (such a point exists, since M is a closed set, so that $l \cap M$ is also a closed set). Then $l \cap M \subset [Q, A]$. However, since points A and Q both belong to field M, it follows that $[Q, A] \subset M$, and thus that $[Q, A] \subset l \cap M$. Accordingly, $l \cap M = [Q, A]$, and since Q is an interior point of field M, we know from Theorem 26.2 that all the points of half-open interval $[Q, A)$ are interior points of M.

It is obvious that A is a boundary point of field M, since arbitrarily close to A there will exist both points belonging to field M (for instance, points of the half-open interval $[Q, A)$) and points not belonging to M (for instance, points of the half-open interval $(A, B]$).

COROLLARY 26.4. *A line passing through interior point Q of convex field M crosses the field boundary at no more than two points. If field M is bounded, then such a line will cross the field boundary at just two points.*

COROLLARY 26.5. *A bounded convex field coincides with the convex hull of its boundary.*

Let M be a bounded convex field. Since $\text{bd}\, M \subset M$, we know that $\text{conv}(\text{bd}\, M) \subset M$. Conversely, let M be an arbitrary point of field M. If A is a boundary point of M, then $A \in \text{bd}\, M$, so that $A \in \text{conv}(\text{bd}\, M)$. If, on the other hand, A is an interior point of field M, then a line passing through A crosses $\text{bd}\, M$ at two points B, C, where $A \in [B, C]$; since $[B, C] \subset \text{conv}(\text{bd}\, M)$, we know that in this case $A \in \text{conv}(\text{bd}\, M)$. Consequently, $M \subset \text{conv}(\text{bd}\, M)$.

Note that Corollary 26.5 remains valid for an *unbounded* convex field M as well, provided that it does not coincide with the entire space E^n or with the half-space; in this case the proof is somewhat more complicated.

In conclusion, let us prove one more theorem, which we will need at the end of this chapter.

THEOREM 26.6. *If convex sets* $M_1, \ldots, M_s$ *possess the property that*

$$(\text{relint}\, M_1) \cap \ldots \cap (\text{relint}\, M_s) \neq \varnothing,$$

then the carrier plane of the set $M_1 \cap \ldots \cap M_s$ *coincides with the intersection of the carrier planes of the sets* $M_1, \ldots, M_s$ *and the following relation is valid:*

$$\text{relint}\,(M_1 \cap \ldots \cap M_s) = (\text{relint}\, M_1) \cap \ldots \cap (\text{relint}\, M_s).$$

Proof. It is sufficient to show that the theorem is valid for $s = 2$ (after which an obvious induction can be carried out). Consider therefore two convex sets M_1, M_2 satisfying the condition

$$(\text{relint}\, M_1) \cap (\text{relint}\, M_2) \neq \varnothing.$$

The carrier planes of sets M_1 and M_2 are designated as P_1 and P_2, respectively, and the carrier plane of set $M = M_1 \cap M_2$ is designated as P.

Since $M_1 \subset P_1$ and $M_2 \subset P_2$, it follows that $M_1 \cap M_2 \subset P_1 \cap P_2$, so that $P \subset P_1 \cap P_2$. Moreover, if $A \in (\text{relint}\, M_1) \cap (\text{relint}\, M_2)$, then all the points of plane P_1 close enough to A will belong to set M_1, while all the points of plane P_2 close enough to A will belong to set M_2. Consequently, all the points of plane $P_1 \cap P_2$ close enough to A will belong to set $M = M_1 \cap M_2$. This implies first that $P_1 \cap P_2 \subset P$, and thus that $P = P_1 \cap P_2$, that is, the carrier plane of set $M_1 \cap M_2$ coincides with the intersection of the carrier planes of sets M_1 and M_2. The second implication is that $A \in \text{relint}\, M$, that is,

$$(\text{relint}\, M_1) \cap (\text{relint}\, M_2) \subset \text{relint}\,(M_1 \cap M_2).$$

It only remains to prove the converse inclusion. Let B be an arbitrary point of the set $\text{relint}\,(M_1 \cap M_2)$, and as previously let $A \in (\text{relint}\, M_1) \cap (\text{relint}\, M_2)$.

Then $A \in M_1 \cap M_2$, so that there exists a point $C \in M_1 \cap M_2$ such that B is an *interior* point of segment $[A, C]$. Since $[A, C] \subset M_1$ and $A \in \operatorname{relint} M_1$, it follows that $B \in \operatorname{relint} M_1$ (Theorem 26.2). Similarly, $B \in \operatorname{relint} M_2$, so that

$$\operatorname{relint}(M_1 \cap M_2) \subset (\operatorname{relint} M_1) \cap (\operatorname{relint} M_2).$$

27. Convex polyhedron. A convex polyhedron was defined above (p. 165). Now let us prove some theorems revealing the "structure" of a convex polyhedron.

THEOREM 27.1. *The boundary of a convex n-dimensional polyhedron is the union of a finite number of $(n-1)$-dimensional polyhedrons, no two of which lie in the same hyperplane. These $(n-1)$-dimensional polyhedrons are called the principal faces of the initial n-dimensional polyhedron.*

Proof. Let M be some n-dimensional convex polyhedron. By definition, it can be represented as

$$M = \Pi_1 \cap \ldots \cap \Pi_k, \tag{27.1}$$

where $\Pi_1, \ldots, \Pi_k$ are half-spaces in E^n. Let us call a half-space Π_i *superfluous* in notation (27.1) if the intersection of all the other half-spaces coincides with M. In Figure 74 (for the case $n = 2$) half-spaces Π_2 and Π_6 are superfluous, since the intersection of the four half-planes Π_1, Π_3, Π_4, Π_5 already yields polygon M.

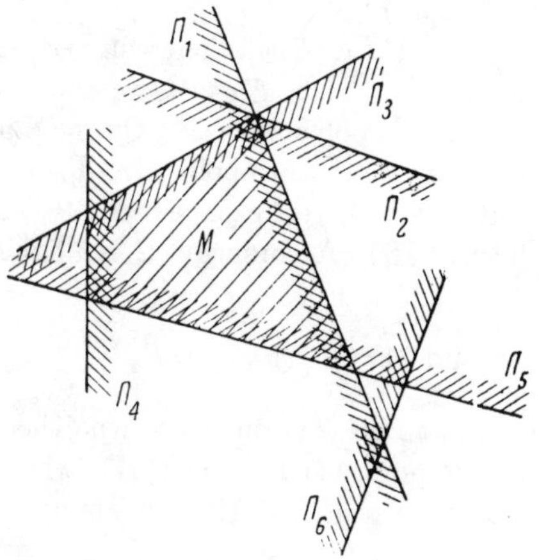

FIGURE 74

It can be assumed that in (27.1) all the superfluous half-spaces have already been eliminated, or, in other words, that none of the half-spaces $\Pi_1, \ldots, \Pi_k$ are superfluous. This means that for each $i = 1, \ldots, k$ there exists a point A_i which is not in half-space Π_i (and thus is not in polyhedron M), but which is in each of the other half-spaces. Let us select such a point for each i, that is, let us find points $A_1, \ldots, A_k$ such that $A_i \notin \Pi_i$, but $A_i \in \Pi_j$ with $i \neq j$ $(i, j = 1, \ldots, k)$.

Let Γ_i be the hyperplane bounding half-space Π_i, and also let

$$F_1 = \Gamma_1 \cap \Pi_2 \cap \Pi_3 \cap \dots \cap \Pi_{k-1} \cap \Pi_k,$$
$$F_2 = \Pi_1 \cap \Gamma_2 \cap \Pi_3 \cap \dots \cap \Pi_{k-1} \cap \Pi_k, \dots$$
$$\dots, F_k = \Pi_1 \cap \Pi_2 \cap \Pi_3 \cap \dots \cap \Pi_{k-1} \cap \Gamma_k.$$

We will show that sets $F_1, F_2, \dots, F_k$ are also the sought $(n-1)$-dimensional polyhedrons constituting the boundary of polyhedron M.

If C is an arbitrary boundary point of polyhedron M, then $C \in M$, that is, point C belongs to each of the half-spaces $\Pi_1, \dots, \Pi_k$ (see (27.1)). If, on the other hand, C were an *interior* point of each of the half-spaces $\Pi_1, \dots, \Pi_k$, then for every $i = 1, \dots, k$ we would be able to find a number $r_i > 0$ such that a sphere of radius r_i with its center at C would lie wholly in half-space Π_i. Then, however, if r were the least of the numbers $r_1, r_2, \dots, r_k$, we would find the sphere of radius r centered on C to be contained in each of the half-spaces $\Pi_1, \dots, \Pi_k$, that is, this sphere would be contained in M, so that C would be an *interior* point of polyhedron M, which contradicts our initial assumption. Thus there exists (at least one) number i such that point C is not an interior point of half-space Π_i, that is, $C \in \Gamma_i$. Since, moreover, point C belongs to all the other half-spaces (because $C \in M$), it follows that $C \in F_i$, so that $C \in F_1 \cup \dots \cup F_k$. Consequently, each boundary point of field M belongs to set $F_1 \cup \dots \cup F_k$.

Conversely, let $C \in F_1 \cup \dots \cup F_k$. Then there will exist (at least one) number i such that $C \in F_i$, so that $C \in \Gamma_i$. Since $C \in M$, we know that C is either an interior point or a boundary point of polyhedron M. On the other hand, C cannot be an interior point, since arbitrarily close to point $C \in \Gamma_i$ there will always exist points not belonging to polyhedron M (for example, points lying on the side of hyperplane Γ_i opposite to half-space Π_i). Accordingly, C is a boundary point of field M.

Thus,

$$\text{bd}\, M = F_1 \cup \dots \cup F_k.$$

Next let us show that F_i is an $(n-1)$-dimensional polyhedron whose carrier plane coincides with Γ_i. First of all, note that $\Gamma_i = \Pi_i \cap \Pi_i^*$, where Π_i^* is the second (also closed) half-space defined by hyperplane Γ_i. Thus we can write

$$F_i = (\Pi_1 \cap \dots \cap \Pi_k) \cap \Pi_i^*.$$

Therefore F_i is the intersection of a finite number of half-spaces. In addition, set F_i is bounded (since $F_i \subset M$). Consequently, F_i is a convex polyhedron; moreover, it is obvious that $F_i \subset \Gamma_i$. Thus we only have to prove that F_i contains interior points relative to hyperplane Γ_i.

Let B be an arbitrary interior point of polyhedron M. Since points A_i and B lie on opposite sides of hyperplane Γ_i (because $A_i \in \Pi_i^*$ and $B \in \Pi_i$, but neither A_i nor B lies in hyperplane Γ_i), it follows that segment $[A_i, B]$ intersects this hyperplane.

In other words, there exists a number λ such that $0<\lambda<1$ and the point

$$C=(1-\lambda)A_i+\lambda B$$

belongs to hyperplane Γ_i. If r is a positive number such that the sphere of radius r centered on B lies wholly in polyhedron M, let us show that each point D of hyperplane Γ_i lying some distance from C not exceeding λr belongs to polyhedron F_i.

Actually, let $D\in\Gamma_i$, with $\rho(C, D)\leqslant\lambda r$. If E is a point such that $\overrightarrow{BE}=$ $=(1/\lambda)\overrightarrow{CD}$, then

$$\rho(B, E)=|\overrightarrow{BE}|=\left|\frac{1}{\lambda}\overrightarrow{CD}\right|=\frac{1}{\lambda}|\overrightarrow{CD}|=\frac{1}{\lambda}\rho(C, D)\leqslant\frac{1}{\lambda}\cdot\lambda r=r,$$

that is, point E lies in the sphere of radius r with its center at B, so that $E\in M$. Furthermore,

$$D=C+\overrightarrow{CD}=(1-\lambda)A_i+\lambda B+\lambda\overrightarrow{BE}=$$
$$=(1-\lambda)A_i+\lambda(B+\overrightarrow{BE})=(1-\lambda)A_i+\lambda E,$$

from which it follows that $D\in[A_i, E]$. However, point E belongs to all the half-spaces $\Pi_1, \ldots, \Pi_k$ (since $E\in M$), while point A_i belongs to all of these half-spaces except Π_i. Consequently, all of segment $[A_i, E]$ (and point D in particular) is contained in the intersection of all the half-spaces $\Pi_1, \ldots, \Pi_k$ except Π_i. Moreover, by definition, $D\in\Gamma_i$, so that

$$D\in\Pi_1\cap\ldots\cap\Pi_{i-1}\cap\Gamma_i\cap\Pi_{i+1}\cap\ldots\cap\Pi_k,$$

that is, $D\in F_i$. Hence each point of hyperplane Γ_i lying a distance away from C not exceeding λr belongs to polyhedron F_i. This means that F_i contains interior points relative to hyperplane Γ_i, that is, it is an $(n-1)$-dimensional polyhedron with a carrier plane Γ_i.

To complete the proof, we need only establish that hyperplanes $\Gamma_1, \ldots, \Gamma_k$, which are carrier planes of polyhedrons $F_1, \ldots, F_k$, are pairwise different. Let us assume that hyperplanes Γ_i and Γ_j coincide. Then half-space Π_j should coincide with one of the two half-spaces defined by hyperplane Γ_i, that is, with one of the half-spaces Π_i, Π_i^*. But the equation $\Pi_j=\Pi_i$ signifies that one of the two half-spaces Π_i, Π_j is superfluous (contrary to the assumption), while the equation $\Pi_j=\Pi_i^*$ signifies that

$$M=\Pi_1\cap\ldots\cap\Pi_k\subset\Pi_i\cap\Pi_j=\Pi_i\cap\Pi_i^*=\Gamma_i,$$

which is also impossible, since polyhedron M is n-dimensional. Consequently, all the hyperplanes $\Gamma_1, \ldots, \Gamma_k$ are different.

The theorem proven actually pertains to polyhedrons of any dimension rather than just to *n-dimensional* polyhedrons.

Let M be a polyhedron of dimension r, where $r < n$. Polyhedron M can be portrayed as the intersection of a finite number of half-spaces (of space E^n): $M = \Pi_1 \cap \ldots \cap \Pi_q$. If P is the carrier plane of polyhedron M, then since $M \subset P$ we can write:

$$M = P \cap (\Pi_1 \cap \ldots \cap \Pi_q) = (\Pi_1 \cap P) \cap \ldots \cap (\Pi_q \cap P). \tag{27.2}$$

Let us examine each of the intersections $\Pi_i \cap P$ $(i = 1, \ldots, q)$. Such an intersection is nonempty (since $M \subset \Pi_i \cap P$). There are two alternatives: either P is wholly contained in half-space Π_i or P is not wholly contained in Π_i. In the former case $(P \subset \Pi_i)$ intersection $\Pi_i \cap P$ coincides with P and thus can be neglected (see (27.2)); in the latter case (P not contained in Π_i) intersection $\Pi_i \cap P$ is *a half-space of the Euclidean space* P.

Accordingly, polyhedron M can be represented as

$$M = \Pi'_1 \cap \ldots \cap \Pi'_s,$$

where $\Pi'_1, \ldots, \Pi'_s$ are certain half-spaces of Euclidean space P, that is, M is an r-dimensional polyhedron in the r-dimensional Euclidean space $E^r = P$. However, in this case Theorem 27.1 is directly applicable to polyhedron M, so that we can formulate the following statement: *the boundary of an r-dimensional polyhedron M (relative to its carrier plane) is the union of a finite number of $(r-1)$-dimensional polyhedrons, no two of which lie in a single $(r-1)$-dimensional plane. These $(r-1)$-dimensional polyhedrons are called the principal faces of the r-dimensional polyhedron M.*

Now we can describe in greater detail the construction of the boundary of a polyhedron. Let M be some r-dimensional polyhedron of space E^n (in particular, it may be that $r = n$). The principal faces of this polyhedron are called its $(r-1)$*-dimensional* faces. For each of these faces (which form the $(r-1)$-dimensional polyhedron) we can in turn consider principal faces. These are also taken to be *faces* of the initial r-dimensional polyhedron M (namely its $(r-2)$*-dimensional* faces). Similarly, we can define the $(r-3)$*-dimensional* faces of polyhedron M (as the principal faces of its $(r-2)$-dimensional faces), etc.

Hence for every r-dimensional polyhedron there exist faces of dimensions $r-1$, $r-2, \ldots, 1, 0$. The zero-dimensional faces of a polyhedron are called its *vertices* (these being *points*); the one-dimensional faces of a polyhedron are called its *edges* (these being *line segments*).

For instance, for $r = 1$, the case of a one-dimensional polyhedron (segment) there will exist *zero-dimensional* faces, which are the vertices (or ends) of the given segment. The *boundary* of a one-dimensional polyhedron consists of two points (the ends of the segment). For $r = 2$, the case of a *two-dimensional* convex polyhedron (convex polygon, see Figure 65a), there will exist *one-dimensional* faces, the *edges* (sides of the polygon) and *zero-dimensional* faces, the *vertices* of the polygon. In the given case the *contour* of the polygon constitutes the boundary. For $r = 3$, the case of a three-dimensional convex polyhedron (Figure 75), there will exist two-dimensional ("conventional") faces, one-dimensional faces (edges), and zero-dimensional faces (vertices). The surface of the polyhedron being considered constitutes the boundary.

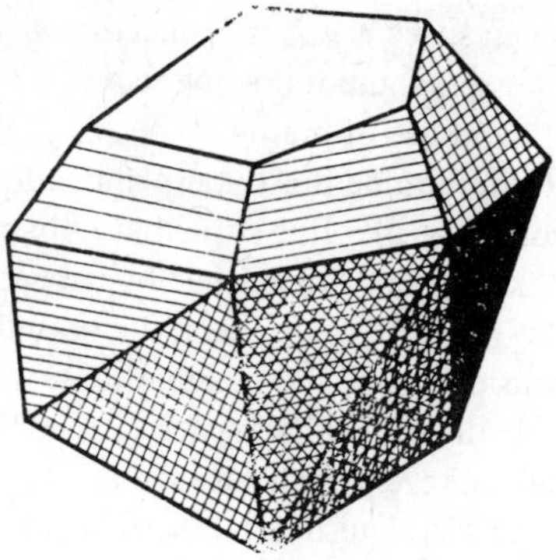

FIGURE 75

THEOREM 27.2. *Any convex polyhedron is the convex hull of its vertices. A set M is a convex polyhedron if, and only if, it is the convex hull of a finite set of points.*

Proof. The first part of the theorem is proven by induction on the dimension r of the given polyhedron M. For $r=1$, that is, if M is some segment $[A, B]$, polyhedron M will have two vertices A and B, and M will be their convex hull. Therefore, for $r=1$ the first part of the theorem is valid.

Let us assume that the first part of the theorem has already been proven for $r<m$ and let M be some polyhedron of dimension m. Let $F_1, \ldots, F_k$ be the principal faces of polyhedron M, so that $\text{bd}\, M = F_1 \cup \ldots \cup F_k$, and let M^* be the convex hull of all the vertices of polyhedron M. Obviously, $M^* \subset M$ (since convex set M contains all the vertices, and thus their convex hull as well). On the other hand, each polyhedron F_i (having dimension $m-1$) is according to the premise of the induction the convex hull of its vertices. Thus $F_i \subset M^*$ (since convex set M^* contains all the vertices of polyhedron M and, in particular, all the vertices of polyhedron F_i). Since this is true for any $i=1, \ldots, k$, it follows that $\text{bd}\, M \subset M^*$, meaning that M^* also contains the convex hull of the set $\text{bd}\, M$, that is, all of polyhedron M (see Corollary 26.5). Hence $M^* \supset M$, so that $M^* = M$, that is, M coincides with the convex hull of its vertices. The induction carried out also proves the first part of the theorem.

Since each polyhedron has a finite number of vertices, it follows from the foregoing proof that any convex polyhedron can be represented as the convex hull of a finite number of points. We have now only to prove the converse, namely that the convex hull of any finite number of points is a convex polyhedron.

Let $N = \{A_1, \ldots, A_l\}$ be a finite set of points and let $M = \text{conv}\, N$ be its convex hull. If P is the plane of least dimension containing all the points $A_1, \ldots, A_l$, then since P is a convex set containing points $A_1, \ldots, A_l$ it follows that P also contains the convex hull of these points, that is, $P \supset M$. At the same time, no plane of less dimension can wholly contain set M (nor can it even contain all the points $A_1, \ldots, A_l$). Consequently, P is the carrier plane of convex set M. If r is the dimension of plane P, let us show that M is a convex polyhedron in the Euclidean space $E^r = P$.

A half-space Π of Euclidean space P will be called *distinguished* (relative to the set of points $A_1, \ldots, A_l$) if it possesses the following two properties: 1) all points $A_1, \ldots, A_l$ belong to half-space Π; 2) those of points $A_1, \ldots, A_l$ which lie in the boundary hyperplane of half-space Π determine this hyperplane uniquely (that is, they do not all lie in one plane of dimension $r-2$). It is clear that only a *finite* number of distinguished half-spaces exist. Actually, a distinguished half-space is determined uniquely if it is indicated which of points $A_1, \ldots, A_l$ lie in the boundary hyperplane of this half-space. Consequently, by considering all the possible subsets of set $\{A_1, \ldots, A_l\}$, we will obviously not miss a single distinguished half-space (although a distinguished half-space does not correspond to each and every such subset).

If $\Pi_1, \ldots, \Pi_s$ are all the distinguished half-spaces, then let us prove that

$$M = \Pi_1 \cap \ldots \cap \Pi_s.$$

Actually, since each half-space Π_i $(i=1, \ldots, s)$ contains all the points $A_1, \ldots, A_l$, the intersection $\Pi_1 \cap \ldots \cap \Pi_s$ also contains all these points, so that it must contain their convex hull as well:

$$\Pi_1 \cap \ldots \cap \Pi_s \supset M.$$

Now let us prove the converse of this inclusion. Let Q be a point of space P not belonging to convex field M. Of points $A_1, \ldots, A_l$, we select any $r-1$ points and we consider the plane of least dimension containing point Q and the selected $r-1$ points. The dimension of this plane cannot exceed $r-1$. Let us construct all such planes, selecting differently the $r-1$ points from $A_1, \ldots, A_l$ and designating the constructed planes as $R_1, \ldots, R_p$. If we take any interior point B of field M not lying in a single one of planes $R_1, \ldots, R_p$ (see Example 22.10), then none of the points of interval (Q, B) will lie in any of planes $R_1, \ldots, R_p$ (if some point $C \in (Q, B)$ were to lie in some plane R_i, then since $Q \in R_i$ the entire line QC would have to be in plane R_i, and thus point B would also belong to plane R_i). Ray BQ crosses the boundary of field M at some point C (Theorem 26.3), where C lies on the interval (Q, B), since $B \in \operatorname{int} M$ and $Q \notin M$.

Because $C \in M$ (it is easy to see that the convex hull of a finite set of points is a closed set), we know from Theorem 25.1 that there must exist $r+1$ points of the set $A_1, \ldots, A_l$ (let these be points $A_1, \ldots, A_{r+1}$) and numbers $\lambda^1, \ldots, \lambda^{r+1}$, such that

$$C = \lambda^1 A_1 + \ldots + \lambda^{r+1} A_{r+1},$$

$$\lambda^1 \geqslant 0, \ldots, \lambda^{r+1} \geqslant 0, \quad \lambda^1 + \ldots + \lambda^{r+1} = 1.$$

If all the numbers $\lambda^1, \ldots, \lambda^{r+1}$ were positive, then point C would have to be an *interior* point of the simplex $[A_1, \ldots, A_{r+1}]$ (see Example 25.3), and thus an interior point of field M as well, which contradicts the choice of point C. Consequently, at least one of the numbers $\lambda^1, \ldots, \lambda^{r+1}$ is zero; to be more explicit, assume that

$\lambda^{r+1}=0$, so that

$$C=\lambda^1 A_1+\ \dots\ +\lambda^r A_r. \tag{27.3}$$

It is easy to see that all the coefficients $\lambda^1, \dots, \lambda^r$ are positive. Actually, if, for example, we had $\lambda^r=0$, then point C would lie in the plane containing points $A_1, \dots, A_{r-1}$, that is, in one of the planes $R_1, \dots, R_p$, which is impossible since $C \in (Q, B)$.

Accordingly, the numbers $\lambda^1, \dots, \lambda^r$ in relation (27.3) satisfy the relations

$$\lambda^1>0, \dots, \lambda^r>0, \quad \lambda^1+\ \dots\ +\lambda^r=1. \tag{27.4}$$

Clearly, points $Q, A_1, \dots, A_r$ do not lie in a single hyperplane of space P (otherwise the plane of least dimension containing these points would be one of planes $R_1, \dots, R_p$ and point C would lie in this plane). Hence it follows that points $A_1, \dots, A_r$ determine uniquely some hyperplane Γ of space P.

Let Π be the half-space of space P set off by hyperplane Γ and containing point B (that is, not containing point Q). We will show that all the points $A_1, \dots, A_t$ lie in half-space Π.

Actually, assume that some point A_j does not lie in half-space Π. Then it is clear that $j>r$ (since all the points $A_1, \dots, A_r$ lie in hyperplane $\Gamma \subset \Pi$). Since points $Q, A_1, \dots, A_r$ are independent, it follows that

$$A_j=\mu^0 Q+\mu^1 A_1+\ \dots\ +\mu^r A_r, \tag{27.5}$$

where $\mu^0+\mu^1+\ \dots\ +\mu^r=1$. If we set

$$D=\mu^0 Q+(\mu^1+\ \dots\ +\mu^r)\,C, \tag{27.6}$$

then

$$\overrightarrow{A_jD}=D-A_j=\mu^1(C-A_1)+\dots+\mu^r(C-A_r)=\mu^1\overrightarrow{A_1C}+\dots+\mu^r\overrightarrow{A_rC}.$$

This implies that vector $\overrightarrow{A_jD}$ belongs to the direction subspace of hyperplane Γ, so that point D, like A_j, does not belong to half-space Π. From (27.6) we know that

$$D=\mu^0 Q+(1-\mu^0)\,C=C+\mu^0\overrightarrow{CQ},$$

so it follows that $\mu^0>0$. From (27.5) we now get

$$Q=\frac{1}{\mu^0}A_j-\frac{\mu^1}{\mu^0}A_1-\ \dots\ -\frac{\mu^r}{\mu^0}A_r.$$

Consider the point

$$C_1=\varepsilon Q+(1-\varepsilon)\,C=$$
$$=\varepsilon\left(\frac{1}{\mu^0}A_j-\frac{\mu^1}{\mu^0}A_1-\ \dots\ -\frac{\mu^r}{\mu^0}A_r\right)+(1-\varepsilon)(\lambda^1 A_1+\dots+\lambda^r A_r)=$$

$$= \frac{\varepsilon}{\mu^0} A_l + \left(\lambda^1 (1-\varepsilon) - \frac{\varepsilon\mu^1}{\mu^0}\right) A_1 + \ldots + \left(\lambda^r (1-\varepsilon) - \frac{\varepsilon\mu^r}{\mu^0}\right) A_r. \tag{27.7}$$

Since $\mu_0 > 0$, we know from (27.4) that all the coefficients on the right-hand side of equation (27.7) will be *positive,* provided $\varepsilon > 0$ is sufficiently small. For a number ε satisfying this condition, we know from Theorem 25.1 that point C_1 belongs to set M. Moreover, $C_1 \in (Q, C)$, so that $C \in (C_1, B)$. Hence it follows (in view of Theorem 26.2) that C is an *interior* point of field M, contrary to the construction. This contradiction shows that all the points $A_1, \ldots, A_l$ lie in half-space Π.

Thus half-space Π contains all the points $A_1, \ldots, A_l$. Moreover, the boundary hyperplane Γ of this half-space contains points $A_1, \ldots, A_r$ (not lying in a single $(r-2)$-dimensional plane), that is, Γ is determined uniquely by those of points $A_1, \ldots, A_l$ which lie in it. In other words, Π is a distinguished half-space, meaning that Π is one of the half-spaces $\Pi_1, \ldots, \Pi_s$. Since $Q \notin \Pi$, it follows that $Q \notin \Pi_1 \cap \ldots \cap \Pi_s$.

Consequently, if $Q \notin M$, then $Q \notin \Pi_1 \cap \ldots \cap \Pi_s$, that is, $\Pi_1 \cap \ldots \cap \Pi_s \subset M$. Together with the inclusion established previously, this means that $\Pi_1 \cap \ldots \cap \Pi_s = M$, so that M is a convex polyhedron of Euclidean space P.

Now if $r = n$ (that is, if P coincides with E^n), it follows that M is a convex polyhedron. If, on the other hand, $r < n$, then we can so choose the half-spaces $\Pi_1^*, \ldots, \Pi_s^*$ of space E^n that $\Pi_i^* \cap P = \Pi_i$ $(i = 1, \ldots, s)$; in addition, we can so choose the half-spaces $\Pi_{s+1}^*, \ldots, \Pi_m^*$ that $\Pi_{s+1}^* \cap \ldots \cap \Pi_m^* = P$ (since in view of Theorem 20.4 P can be represented as the intersection of several hyperplanes, and each hyperplane as the intersection of two half-spaces). Then we have

$$M = \Pi_1 \cap \ldots \cap \Pi_s = (\Pi_1^* \cap P) \cap \ldots \cap (\Pi_s^* \cap P) =$$
$$= \Pi_1^* \cap \ldots \cap \Pi_s^* \cap P = \Pi_1^* \cap \ldots \cap \Pi_s^* \cap \Pi_{s+1}^* \cap \ldots \cap \Pi_m^*,$$

that is, in this case as well, M is a convex polyhedron of space E^n.

As a complement to Theorem 27.2, it should be noted that, if $A_1, \ldots, A_l$ are arbitrary points of space E^n and M is the convex hull of these points, then the *vertices* of polyhedron M can only be the selected points $A_1, \ldots, A_l$, but possibly not all of these points: some of them may turn out to be on the faces or else in the interior of the polyhedron obtained (Figure 76).

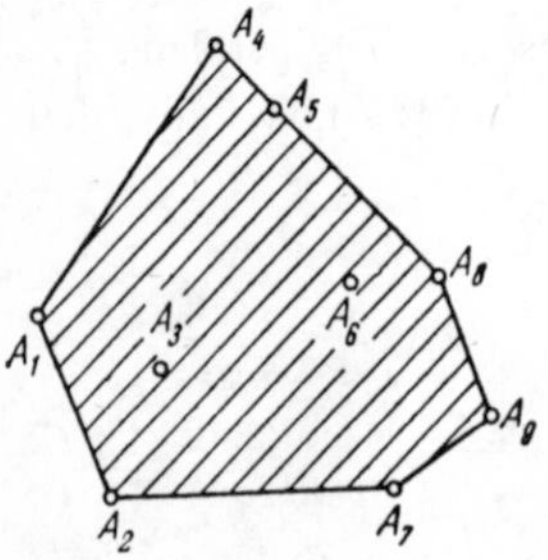

FIGURE 76

The proof of this fact is obtained with the aid of the following statement (which states a characteristic property of the vertices of a polyhedron): *a point A of polyhedron M will not be its vertex if, and only if, there exist points B, $C \in M$ such that $A \in (B, C)$.* The proof of this will be omitted.

EXAMPLE 27.3. Let $[A_0, A_1, \ldots, A_n]$ be an arbitrary n-dimensional simplex in E^n. Let Γ_i be the hyperplane containing all the points $A_0, A_1, \ldots, A_n$ except A_i $(i = 0, 1, \ldots, n)$, and let Π_i be the half-space defined by hyperplane Γ_i and containing point A_i. It is obvious from the proof of Theorem 27.2 that $\Pi_0, \Pi_1, \ldots, \Pi_n$ are distinguished half-spaces (relative to the set of points $A_0, A_1, \ldots, A_n$), so that

$$[A_0, A_1, \ldots, A_n] = \Pi_0 \cap \Pi_1 \cap \ldots \cap \Pi_n.$$

Consequently (cf. the proof of Theorem 27.1), the principal faces of simplex $[A_0, A_1, \ldots, A_n]$ will be the $(n-1)$-dimensional simplexes

$$[A_1, A_2, \ldots, A_{n-1}, A_n], [A_0, A_2, \ldots, A_{n-1}, A_n], \ldots \\ \ldots, [A_0, A_1, A_2, \ldots, A_{n-1}]. \quad (27.8)$$

Further, we can find all the $(n-2)$-dimensional faces of simplex $[A_0, A_1, \ldots, A_n]$ (as the principal faces of simplexes (27.8)), then the $(n-3)$-dimensional faces, etc. This leads us to the following statements. *Each face of the simplex $T = [A_0, A_1, \ldots, A_n]$ is also a simplex of some dimension.* If we consider only some of points $A_0, A_1, \ldots, A_n$, then the convex hull of these points (that is, the simplex whose vertices are the points selected) will be a face of simplex T. In this way *all* the faces of simplex T can be obtained, that is, each face of polyhedron T is a simplex whose vertices are some of points $A_0, A_1, \ldots, A_n$. In particular, *n-dimensional simplex T has $n+1$ faces of dimension $n-1$* (each of these being obtained by taking all the points $A_0, A_1, \ldots, A_n$ except some specified one; see (27.8)). *The zero-dimensional faces (vertices) of simplex $[A_0, A_1, \ldots, A_n]$ will be the points $A_0, A_1, \ldots, A_n$, and only these points.*

Of course, similar statements will also be valid for a simplex of any dimension, since a k-dimensional simplex $[B_0, B_1, \ldots, B_k]$ can be considered in k-dimensional Euclidean space (its carrier plane).

§7. Supporting Properties of Convex Sets

28. Supporting cone. Since an affine mapping $f\colon E^n \to E^m$ of one Euclidean space into another is continuous (see Theorem 22.5), it follows from Theorem 22.6 that the image $f(M)$ of a closed bounded set $M \subset E^n$ will also be a closed bounded set. However, an affine mapping may carry a closed *unbounded* set into an unclosed set. For example, an orthogonal projection onto the abscissa axis transforms the closed set defined in the (x, y) plane by the inequalities

$$xy \geqslant 1, \quad x > 0$$

(the "interior" of one branch of a hyperbola; Figure 77) into an open half-line $x > 0$, that is, into an unclosed set.

THEOREM 28.1. *The image $f(M)$ of a convex set $M \subset E^n$, for an affine mapping $f: E^n \to E^m$, is a convex (possibly unclosed) set of space E^m. If M is bounded, then $f(M)$ is bounded as well; if M is closed and bounded, then $f(M)$ is also closed and bounded.*

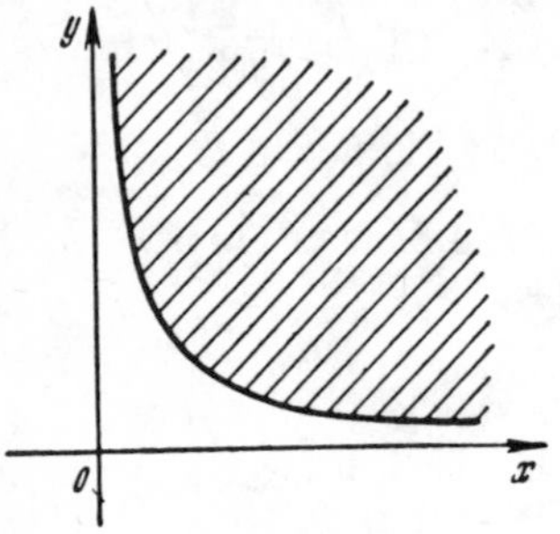

FIGURE 77

Proof. Let A_1 and B_1 be two arbitrary points of set $f(M)$. Then there must exist points $A, B \in M$ such that $A_1 = f(A)$ and $B_1 = f(B)$. Now let C_1 be an arbitrary point of segment $[A_1, B_1]$, that is,

$$C_1 = (1-\lambda) A_1 + \lambda B_1, \qquad (28.1)$$

where λ is some real number satisfying the inequalities $0 \leqslant \lambda \leqslant 1$.

If C is a point of space E^n defined by the relation

$$C = (1-\lambda) A + \lambda B, \qquad (28.2)$$

then $C \in [A, B]$, so that from the convexity of set M we know that C belongs to this set. Since $f(A) = A_1$ and $f(B) = B_1$, it follows from (28.1) and (28.2) (and from the definition of an affine mapping) that $f(C) = C_1$, so that $C_1 \in f(M)$. Thus any point C_1 of segment $[A_1, B_1]$ belongs to set $f(M)$, that is, set $f(M)$ is convex.

THEOREM 28.2. *If M is an arbitrary set of space E^n, then* $f(\text{conv}\, M) = \text{conv}\, f(M)$ (here, as previously, $f: E^n \to E^m$ is an affine mapping).

Proof. Since $f(\text{conv}\, M)$ is a convex set (Theorem 28.1) containing $f(M)$, it follows that $f(\text{conv}\, M) \supset \text{conv}\, f(M)$.

Let us prove the converse inclusion. Let D be an arbitrary point of the set $f(\text{conv}\, M)$, that is, $D = f(C)$, where $C \in \text{conv}\, M$. According to Theorem 25.1, there will exist points $A_0, A_1, \ldots, A_n$ of set M and numbers $\lambda^0, \lambda^1, \ldots, \lambda^n$ such that

$$C = \lambda^0 A_0 + \lambda^1 A_1 + \ldots + \lambda^n A_n;$$

$$\lambda^0 \geqslant 0, \quad \lambda^1 \geqslant 0, \ldots, \lambda^n \geqslant 0; \quad \lambda^0 + \lambda^1 + \ldots + \lambda^n = 1.$$

According to Theorem 19.1, we have

$$D = f(C) = \lambda^0 f(A_0) + \lambda^1 f(A_1) + \ldots + \lambda^n f(A_n).$$

Since all the points $f(A_0), f(A_1), \ldots, f(A_n)$ belong to set $f(M)$, it follows that $D \in \text{conv}\, f(M)$. Consequently, each point D of the set $f(\text{conv}\, M)$ also belongs to $\text{conv}\, f(M)$, that is $f(\text{conv}\, M) \subset \text{conv}\, f(M)$.

COROLLARY 28.3. *The image of a convex polyhedron in an affine mapping is also a convex polyhedron.*

This follows directly from Theorems 28.2 and 27.2.

As an example, Figure 78 shows the image of a three-dimensional parallelepiped for an affine mapping $f: E^3 \to E^2$. The eight vertices of the parallelepiped are carried by this mapping into eight points of a plane, while the parallelepiped itself is carried into the *convex hull* of these eight points. The figure also indicates the line segments into which the edges of the parallelepiped are carried by this mapping.

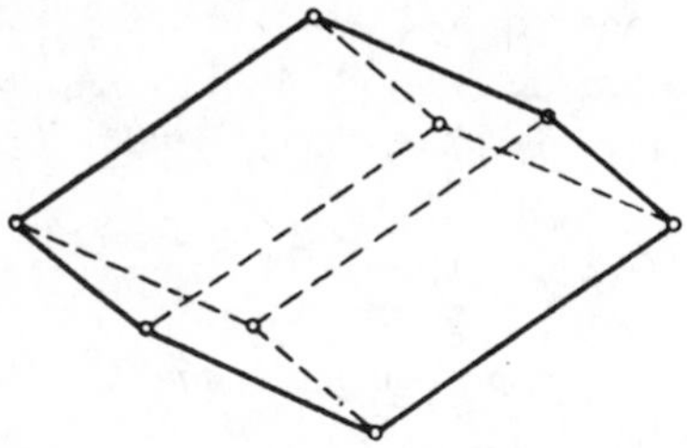

FIGURE 78

DEFINITION 28.4. A set M of space E^n is called a *cone with a vertex at point* Q if, in addition to each point A different from Q, set M also contains the entire ray originating at Q and passing through A. If set M is convex and if it is also a cone with a vertex at point Q, then it is called a *convex cone* (with a vertex at point Q). If, moreover, set M is closed, then it is known as a *closed convex cone.*

EXAMPLE 28.5. Let $P \subset E^n$ be a closed half-space and let Γ be the hyperplane bounding it, where $Q \in \Gamma$. Then P is a convex cone with a vertex Q.

Actually, the convexity of set P has already been established (p. 162). Let us show that P is a cone with a vertex Q. If $\boldsymbol{n}$ is some vector (different from the zero vector) orthogonal to hyperplane Γ and directed toward half-space P, then for any (different from Q) point $A \in P$ the relation $\boldsymbol{n}\overrightarrow{QA} \geqslant 0$ will be valid. Now, if C is an arbitrary point of the ray l originating at Q and passing through A, then $\overrightarrow{QC} = \lambda\overrightarrow{QA}$, where $\lambda \geqslant 0$, so that $\boldsymbol{n}\overrightarrow{QC} = \boldsymbol{n}(\lambda\overrightarrow{QA}) = \lambda(\boldsymbol{n}\overrightarrow{QA}) \geqslant 0$, that is, $C \in P$. Consequently, all of ray l belongs to half-space P, meaning that P is a cone with a vertex Q.

EXAMPLE 28.6. Let l be some ray in E^n originating at point Q and let K be the set of all points $A \in E^n$ such that vector $\overrightarrow{QA}$ forms with ray l an angle not exceeding φ, where $0 < \varphi < \pi/2$. Let us show that K is a convex cone in E^n with a vertex Q.

If $\boldsymbol{a}$ is the vector specifying the direction of ray l and if we define $\lambda = \cos\varphi$ (so that $\lambda > 0$), then point A will belong to set K if, and only if, vectors $\boldsymbol{a}$ and $\overrightarrow{QA}$ make an angle not exceeding φ, that is, if $\boldsymbol{a}\overrightarrow{QA} \geqslant \lambda|\boldsymbol{a}||\overrightarrow{QA}|$. Let $A, B \in K$, that is, assume that

$$\boldsymbol{a}\overrightarrow{QA} \geqslant \lambda|\boldsymbol{a}||\overrightarrow{QA}|, \quad \boldsymbol{a}\overrightarrow{QB} \geqslant \lambda|\boldsymbol{a}||\overrightarrow{QB}|,$$

in which case for any numbers $\alpha \geqslant 0$, $\beta \geqslant 0$ we will have

$$\begin{aligned}\boldsymbol{a}(\alpha\overrightarrow{QA} + \beta\overrightarrow{QB}) = \alpha(\boldsymbol{a}\overrightarrow{QA}) + \beta(\boldsymbol{a}\overrightarrow{QB}) \geqslant \alpha\lambda|\boldsymbol{a}||\overrightarrow{QA}| + \beta\lambda|\boldsymbol{a}||\overrightarrow{QB}| = \\ = \lambda|\boldsymbol{a}||\alpha\overrightarrow{QA}| + \lambda|\boldsymbol{a}||\beta\overrightarrow{QB}| = \lambda|\boldsymbol{a}|(|\alpha\overrightarrow{QA}| + |\beta\overrightarrow{QB}|) \geqslant \\ \geqslant \lambda|\boldsymbol{a}||\alpha\overrightarrow{QA} + \beta\overrightarrow{QB}|\end{aligned}$$

(see (16.3)). This means that a point C defined by the equation $\overrightarrow{QC} = \alpha\overrightarrow{QA} + \beta\overrightarrow{QB}$

belongs to set K (for any $\alpha \geqslant 0,\ \beta \geqslant 0$). Consequently, K is a convex cone with a vertex Q.

Note that in three-dimensional space the set K described in Example 28.6 is a cone of rotation with an axis l and an angle φ between the axis and the generatrix.

THEOREM 28.7. *If M is a convex cone with a vertex at point Q, $A_1, \ldots, A_k$ are points on cone M, and $\lambda^1, \ldots, \lambda^k$ are nonnegative numbers, then the point*

$$A = Q + \lambda^1\overrightarrow{QA_1} + \ldots + \lambda^k\overrightarrow{QA_k}$$

also belongs to cone M. If, moreover, A_1 is an interior point of cone M and $\lambda^1 > 0$, then A will be an interior point of cone M.

Proof. A point $B_i = Q + k\lambda^i\overrightarrow{QA_i}$ belongs to ray QA_i, and thus to cone M as well ($i = 1, \ldots, k$). Hence, because of its convexity, set M also includes the following point (see Theorem 25.1):

$$\begin{aligned}\frac{1}{k}B_1 + \ldots + \frac{1}{k}B_k &= \\ &= \frac{1}{k}(Q + k\lambda^1\overrightarrow{QA_1}) + \ldots + \frac{1}{k}(Q + k\lambda^k\overrightarrow{QA_k}) = \\ &= Q + \lambda^1\overrightarrow{QA_1} + \ldots + \lambda^k\overrightarrow{QA_k} = A.\end{aligned}$$

Now let A_1 be an interior point of convex set M and let $\lambda^1 > 0$ (where, as previously, $A_2, \ldots, A_k \in M$ and $\lambda^2 \geqslant 0, \ldots, \lambda^k \geqslant 0$). If $A_1 = Q$, then M coincides with E^n and *all* the points of set M are interior. If, on the other hand, $A_1 \neq Q$, then all the points of ray QA_1 except Q are also *interior* points of field M (Theorem 26.2). In particular, the point $B_1 = Q + k\lambda^1\overrightarrow{QA_1}$ is interior. From the relation

$$\begin{aligned}A = \frac{1}{k}B_1 + \frac{1}{k}B_2 + \ldots + \frac{1}{k}B_k &= \\ &= \frac{1}{k}B_1 + \frac{k-1}{k}\left(\frac{1}{k-1}B_2 + \ldots + \frac{1}{k-1}B_k\right)\end{aligned}$$

it follows that A is an interior point of the segment joining points B_1 and $C = \frac{1}{k-1}B_2 + \ldots + \frac{1}{k-1}B_k \in M$, so that, in view of Theorem 26.2, A is an interior point of field M. The theorem is thereby proven.

It is easy to see that on a line (that is, for $n = 1$) the only convex cones are the entire line, a ray [half-line] and a point. On a plane ($n = 2$) the only convex cones are the following sets: point, ray, line, angle not exceeding π, half-plane, entire plane (Figure 79). In spaces having more dimensions than this (even for $n = 3$) the convex cones may be considerably more complicated in structure. If, for instance, M is an arbitrary convex set situated in a hyperplane Γ of space E^n, while Q is a point not lying in this hyperplane, then all the possible rays originating at point Q and passing through the points of set M will fill a convex cone having its vertex at Q (Figure 80).

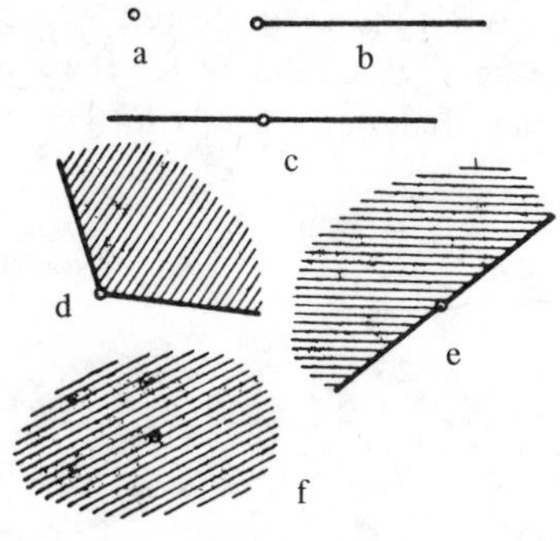

FIGURE 79

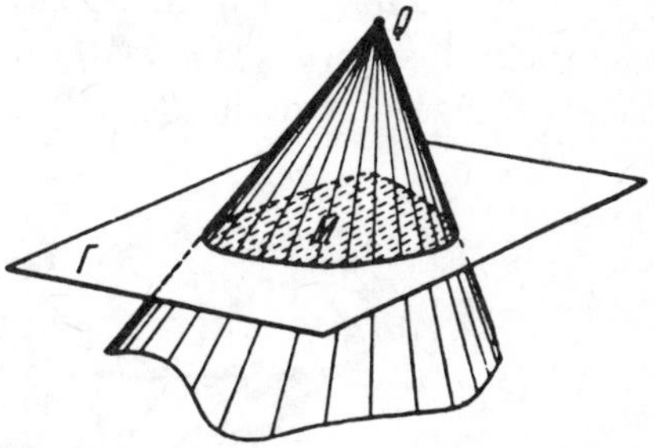

FIGURE 80

THEOREM 28.8. *If* $M \subset E^n$ *is a convex cone with a vertex* Q *and if* $f\colon E^n \to E^m$ *is an affine mapping, then* $f(M)$ *is a convex cone with a vertex* $f(Q)$.

Proof. Let M be a convex cone with a vertex Q, situated in space E^n. According to Theorem 28.1, set $f(M)$ is convex. Let us show that it is a cone with a vertex $Q_1 = f(Q)$.

Let A_1 be an arbitrary point of set $f(M)$ different from point Q_1. Since $A_1 \in f(M)$, there will exist a point $A \in M$ such that $f(A) = A_1$. Obviously, point A will differ from Q (since $A_1 \neq Q_1$, that is, $f(A) \neq f(Q)$). Now let C_1 be an arbitrary point on the ray originating at Q_1 and passing through A_1. Then, according to the definition of a ray, $C_1 = \lambda A_1 + (1-\lambda) Q_1$, where $\lambda \geqslant 0$. Let C be a point of space E^n such that $C = \lambda A + (1-\lambda) Q$; then

$$f(C) = \lambda f(A) + (1-\lambda) f(Q) = \lambda A_1 + (1-\lambda) Q_1 = C_1.$$

Consequently, $C_1 \in f(M)$, that is, the entire ray starting at Q_1 and passing through A_1 is contained in set $f(M)$.

EXAMPLE 28.9. In three-dimensional space E^3 with coordinates x, y, z consider the set M described by the inequalities

$$x^2 + y^2 - z^2 \leqslant 0, \quad z \geqslant 0. \tag{28.3}$$

This set is a closed convex cone in E^3 having its vertex at the coordinate origin. It is formed by all the possible rays issuing from the origin and making with the positive z semiaxis an angle not exceeding $\pi/4$, that is, it constitutes (cf. Example 28.6) a right circular cone whose axis makes an angle of $\pi/4$ with its generatrices (Figure 81).

The ray starting from the coordinate origin and passing through point $A(0, 1, 1)$ is one of the generatrices of cone M. Consider the projection f of space E^3 onto the plane E^2 of variables x and y parallel to this generatrix. It will be the affine mapping determined by the correspondence

$$(x,\ y,\ z) \to (x,\ y - z)$$

(that is, transforming a point $(x,\ y,\ z)$ of space E^3 into a point $(x',\ y')$ of plane E^2 having coordinates $x' = x$ and $y' = y - z$). It is easy to see from (28.3) that for any point $(x,\ y,\ z)$ of cone M the inequality $y \leqslant z$ is valid, so that for the affine mapping being considered the image $f(M)$ of cone M lies completely in the half-plane defined in E^2 by the inequality $y \leqslant 0$.

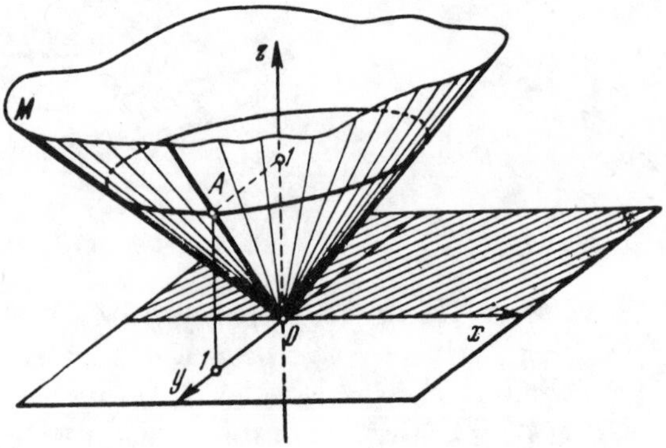

FIGURE 81

Obviously, all the points of the *open* half-plane $y < 0$ belong to set $f(M)$ (for example, a point $\left(x_0,\ \frac{x_0^2}{2h},\ \frac{x_0^2}{2h} + h\right)$ belonging, as can easily be shown, to cone M turns into a point $(x_0,\ y_0)$ of plane E^2 satisfying the condition $y_0 = -h$, where $h > 0$). At the same time, on the line $y = 0$ forming the boundary of this open half-plane, there will not be a single point of set $f(M)$ except the coordinate origin (actually, at point $(x,\ 0)$ only points of form $(x,\ -h,\ h)$ will be transformed by mapping f, and such a point belongs to cone M only for $x = 0$). Thus, set $f(M)$ contains all the points of the xy plane having a negative coordinate y, and it also contains the coordinate origin (see Figure 81). Set $f(M)$ is (in full agreement with Theorem 28.8) a convex cone, but this cone is not closed.

This example indicates that for an affine mapping the image of a closed convex cone may turn out to be a *nonclosed* convex cone.

Inspection of Figure 79 indicates that on a plane, for any convex cone M with a vertex Q, there are two possible alternatives: either cone M will coincide with the entire plane or there will exist a line passing through Q such that all of cone M lies wholly in one of the two (closed) half-planes into which the plane is divided by this line. As shown by the following theorem, the situation will be similar for spaces of higher dimension.

THEOREM 28.10. *Let $M \subset E^n$ be a convex cone with a vertex Q. If M does not coincide with all of space E^n, then there will exist in E^n a hyperplane Γ passing* *through point Q such that cone M lies wholly in one of the two (closed) half-spaces defined by hyperplane* Γ (in this theorem convex cone M may also be nonclosed).

Proof. Let us carry out induction on the dimension n of the space E^n in which cone M lies. We assume the theorem to be valid for Euclidean spaces of dimension less than n, and we consider a convex cone M with a vertex Q located in space E^n and not coinciding with all of this space. Then there will exist a point $A \in E^n$ not belonging to cone M. If cone M has dimension less than n, then it will lie completely in some hyperplane, and thus in any of the closed half-spaces defined by this hyperplane, so that in this case the theorem is valid. Hence we will assume that cone M is a convex *field* in E^n, that is, it contains interior points.

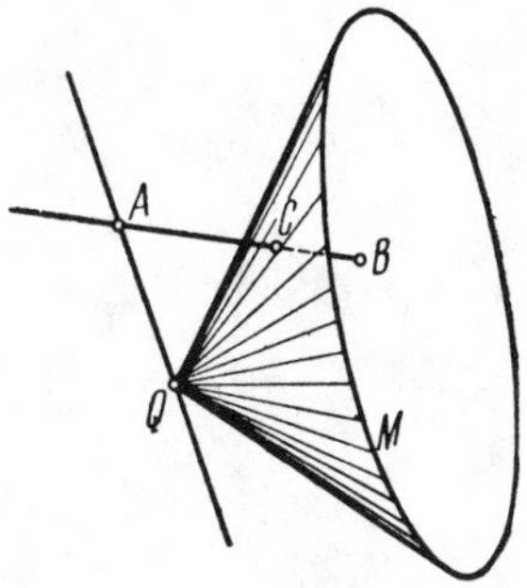

FIGURE 82

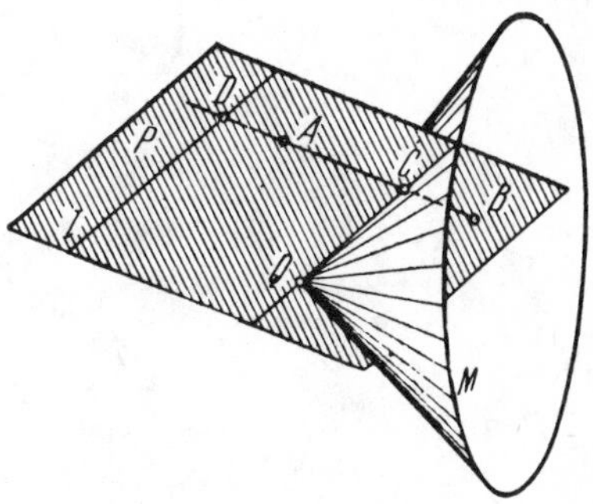

FIGURE 83

Let B be an interior point of cone M which does not lie on line AQ (Figure 82) (note that line AQ may cross the interior of cone M). Since ray BA does not lie wholly in cone M (because $A \notin M$), it will have to contain a boundary point C of cone M (which may also coincide with A, if cone M is nonclosed). Here point C cannot coincide with Q, since if it did point B would lie on line AQ.

On line AB let us take some point D lying beyond A, that is, a point $A \in (D, B)$, and let us draw through D a line l parallel to QC. We will prove that not a single point of cone M lies on line l. First let us assume that a point $E \in M$ lies on line l, in which case all the points Q, B, C, A, D, E would lie in one two-dimensional plane P (Theorem 19.13), where B and D lie in this plane on opposite sides of line QC, since segment BD crosses this line (Figure 83). Moreover, since $DE \parallel QC$, points D and E lie on the same side of line QC (Theorem 20.9). Consequently, points B and E lie on opposite sides of line QC, that is, segment BE crosses line QC at some point F. Since $E \in M$ and B is an interior point of field M, F must also be an interior point of this field (Theorem 26.2). Hence all points of ray QF (except Q) are interior points of field M as well (Theorem 26.3). However, since point C (lying on line QF) is not an interior point, it cannot be on ray QF, that is, it lies on the opposite ray originating at point Q. In other words, $Q \in (C, F)$. However, this makes Q an interior point of field M (Theorem 26.2), which is impossible. This contradiction indicates that line l does not intersect cone M.

Now let Γ be an arbitrary hyperplane crossing line l (for instance, a hyperplane orthogonal to line l). If π is the projection of space E^n onto hyperplane Γ parallel to line l, then this projection will be an *affine* mapping of space E^n onto the $(n-1)$-dimensional Euclidean space Γ. Consequently, the image $\pi(M)$ of cone M for this projection will also be a cone with a vertex $Q_1 = \pi(Q)$. Obviously, any line parallel to l will be carried into a point of space Γ by mapping π. In particular, mapping π will carry line l into a single point $D_1 \in \Gamma$. Moreover, a point K of hyperplane Γ will belong to cone $\pi(M)$ if, and only if, a line parallel to l and passing through K intersects cone M. But since, in view of what was proven above, line l itself does not intersect cone M, it follows that point D_1 does not belong to cone $\pi(M)$. Consequently, convex cone $\pi(M)$, located in space Γ, does not coincide with this entire space.

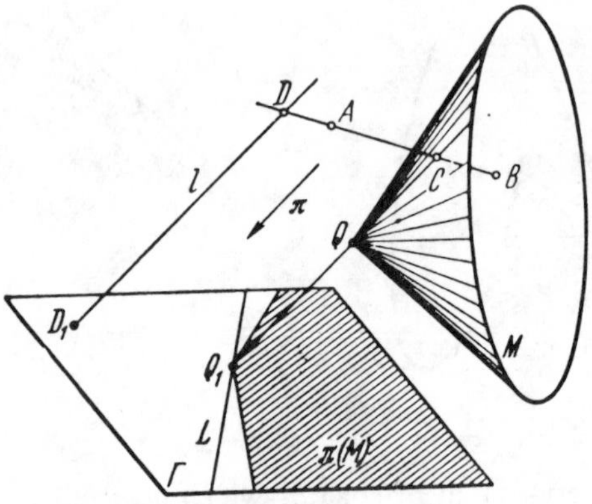

FIGURE 84

According to induction there will exist in the $(n-1)$-dimensional Euclidean space Γ a hyperplane L, passing through point Q_1, such that cone $\pi(M)$ lies wholly in one closed half-space Π^* defined in Γ by hyperplane L (note that L is an $(n-2)$-dimensional plane of the initial space E^n; Figure 84). All lines of space E^n which are parallel to l and pass through points of plane L fill a hyperplane $\Gamma^* = \pi^{-1}(L)$ of space E^n. Similarly, all the lines parallel to l and passing through points of the half-space $\Pi^* \subset \Gamma$ fill a half-space $P^* = \pi^{-1}(\Pi^*)$ of E^n, for which Γ^* constitutes a boundary hyperplane (Figure 85). Since $f(M) \subset \Pi^*$, it follows that $M \subset \pi^{-1}(\Pi^*) = P^*$. Accordingly, hyperplane Γ^* (obviously passing through point Q) possesses the property that cone M is wholly contained in one of the half-spaces defined by this hyperplane, that is, hyperplane Γ^* is the desired one.

Thus, if Theorem 28.10 is valid for cones situated in an $(n-1)$-dimensional Euclidean space, then it will be valid for cones in an n-dimensional space as well. On the other hand, for $n=2$ the theorem, as mentioned above, is valid; all the convex cones of a two-dimensional space are shown in Figure 79 (by the way, it is even simpler as a beginning of induction to take the case $n=1$, where the theorem is completely trivial).

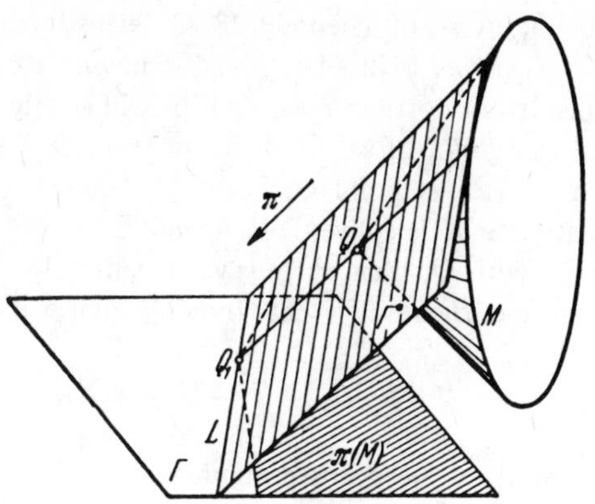

FIGURE 85

DEFINITION 28.11. Let M be some convex set of space E^n and let Q be a boundary point on it. A hyperplane passing through point Q is called a *supporting hyperplane* of set M if all of set M is situated in one of the two (closed) half-spaces into which space E^n is divided by this hyperplane.

Theorem 28.10 can now be formulated as follows: *if a convex cone* $M \subset E^n$ *does not coincide with the entire space* E^n*, then a supporting hyperplane of this cone can be drawn through its vertex.*

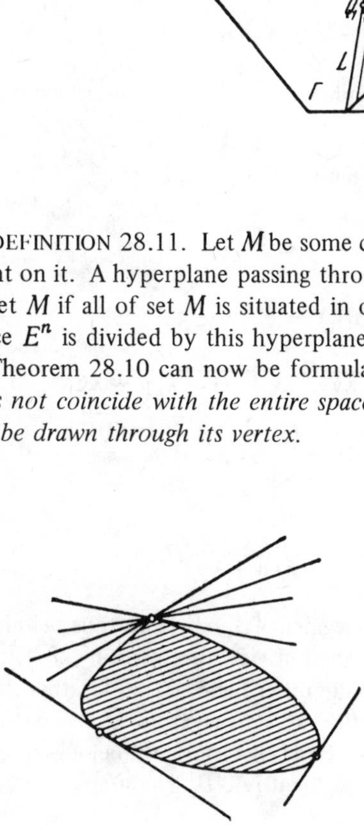

FIGURE 86

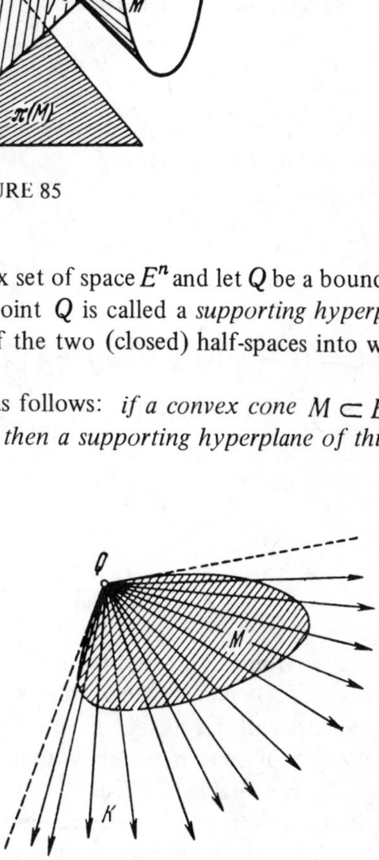

FIGURE 87

It is intuitively clear (at any rate for convex sets situated in a plane or in three-dimensional space) that at least one supporting hyperplane of a convex set can be drawn through *each* boundary point of the set (Figure 86). Theorem 28.12, to be proven below, shows that with regard to this our intuition does not deceive us. Let us stress that we can assert the existence of *at least one* supporting hyperplane passing through the specified boundary point: it may happen that through some boundary points an infinite number of supporting hyperplanes will pass.

Before we go on to the proof of Theorem 28.12, let us introduce the concept of the *supporting cone* of a convex field. Let M be some convex set and let Q be a point of this set. Consider all possible rays, each of which originates at point Q and passes through at least one point of set M different from Q (Figure 87). These rays fill some cone K with a vertex at point Q.

It is easy to see that cone K is convex. If A and B are two points of cone K and if one of them coincides with Q, then obviously segment $[A, B]$ will be in K. If, on the other hand, both A and B are different from Q, then

$$A = Q + \mu\overrightarrow{QA_1}, \quad B = Q + \nu\overrightarrow{QB_1},$$

where $A_1, B_1 \in M$ and μ, ν are positive numbers. Now, if C is an arbitrary point of segment $[A, B]$, that is,

$$C = (1-\lambda)A + \lambda B, \quad \text{where} \quad 0 \leqslant \lambda \leqslant 1,$$

then

$$\begin{aligned} C &= (1-\lambda)(Q + \mu\overrightarrow{QA_1}) + \lambda(Q + \nu\overrightarrow{QB_1}) = \\ &= Q + ((1-\lambda)\mu + \lambda\nu)\left(\frac{(1-\lambda)\mu}{(1-\lambda)\mu + \lambda\nu}\overrightarrow{QA_1} + \frac{\lambda\nu}{(1-\lambda)\mu + \lambda\nu}\overrightarrow{QB_1}\right) = \\ &= Q + ((1-\lambda)\mu + \lambda\nu)\overrightarrow{QC_1}, \end{aligned} \tag{28.4}$$

where

$$C_1 = \frac{(1-\lambda)\mu}{(1-\lambda)\mu + \lambda\nu}A_1 + \frac{\lambda\nu}{(1-\lambda)\mu + \lambda\nu}B_1. \tag{28.5}$$

It is clear from (28.5) that point C_1 lies on segment $[A_1, B_1]$ and thus belongs to convex set M. If point C_1 coincides with Q, then in view of equation (28.4) point C also coincides with Q, so that $C \in K$. If, on the other hand, $C_1 \neq Q$, then from the definition of cone K we know that the entire ray starting from Q and passing through C_1 is contained in cone K, so that in view of (28.4) point C belongs to cone K. Thus, in any case $C \in K$, meaning that all of segment $[A, B]$ is contained in cone K. This establishes the convexity of cone K.

In general, the constructed cone K may turn out to be nonclosed. The closure of cone K is called the *supporting cone of convex set M at point Q*.

Clearly, the supporting cone contains set M and is contained in the carrier plane of convex set M. It is also obvious that, if Q is an *interior* point of set M relative to its carrier plane L, then the supporting cone of set M at point Q coincides with the entire carrier plane L. If, on the other hand, Q is a *boundary* point of set M relative to its carrier plane L, then the supporting cone of set M at point Q *will not coincide with all of plane L*.

Let A be an interior point of set M relative to its carrier plane and let l be a ray originating at a point Q situated on line AQ, where point A is not on this ray (Figure 88). If some point B of ray l, different from Q, belonged to set M, then according to Theorem 26.2 point Q would have to be an interior point (relative to L) of set M, which contradicts the condition imposed. Consequently, *no point* on ray l except Q belongs to set M, so that ray l is not included in cone K. Therefore, cone K does not coincide with the entire plane L, and from Theorem 28.10 it follows that cone K is wholly contained in some half-space of Euclidean space L. Then, however, the closure of cone K (that is, the supporting cone of set M at point Q) is also contained in this half-space, that is, the supporting cone does not coincide with all of plane L.

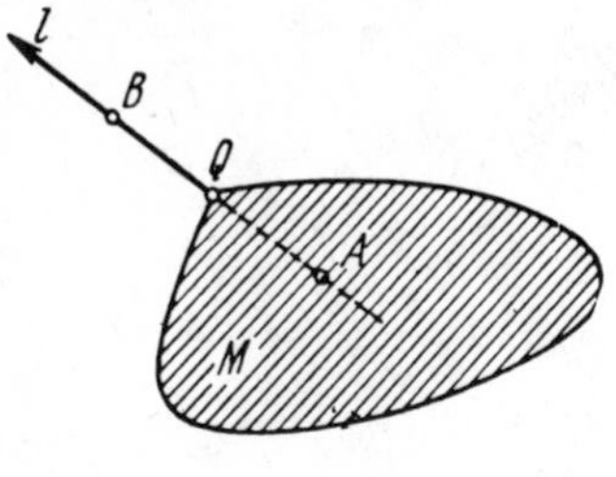

FIGURE 88

Thus, *the supporting cone of a convex set* M *at a point* Q *coincides with the entire carrier plane of set* M *if, and only if,* Q *is an interior point of set* M *relative to its carrier plane.*

THEOREM 28.12. *Through each boundary point of a convex set there passes at least one supporting hyperplane of this set.*

Proof. If convex set $M \subset E^n$ is not a convex field, then it will lie wholly in some hyperplane Γ of space E^n. Obviously, this hyperplane has to be a supporting hyperplane of set M, since M lies completely in (any) closed half-space defined by hyperplane Γ. Here supporting hyperplane Γ passes through *any* point Q of set M. Therefore, if M is not a convex field, the theorem is valid in a trivial way.

Next let M be a convex *field* of space E^n and Q a boundary point of field M. If K^* is the supporting cone of set M at point Q, this means that Q is the vertex of this cone. Since Q is a boundary point of field M, it follows from the foregoing that cone K^* does not coincide with the entire carrier plane of field M, that is, it does not coincide with all of space E^n. According to Theorem 28.10, there exists in E^n a hyperplane Γ passing through Q, such that cone K^* lies wholly in one of the two (closed) half-spaces defined by hyperplane Γ. Since $K^* \supset M$, field M must lie wholly in this same half-space. Consequently, Γ is a supporting hyperplane of field M. According to the assumption, this supporting hyperplane passes through point Q.

29. Affine functions on a convex set. Let M be an arbitrary convex set of space E^n, let Q be a boundary point of it, and let Γ be the supporting hyperplane of set M passing through point Q. The half-space (closed defined by hyperplane Γ and containing set M is called P. Finally, let C be a point not lying in half-space P, such that the vector $\boldsymbol{n} = \overrightarrow{QC}$ is orthogonal to hyperplane Γ (Figure 89). Then hyperplane Γ consists of those, and only those, points $A \in E^n$ for which the equation $\boldsymbol{n}\overrightarrow{QA} = 0$ is valid, while half-space P consists of those, and only those, points $A \in E^n$ for which the inequality $\boldsymbol{n}\overrightarrow{QA} \leqslant 0$ is valid. Since set M lies wholly in half-space P, it follows that the inequality $\boldsymbol{n}\overrightarrow{QA} \leqslant 0$ holds true for any point $A \in M$.

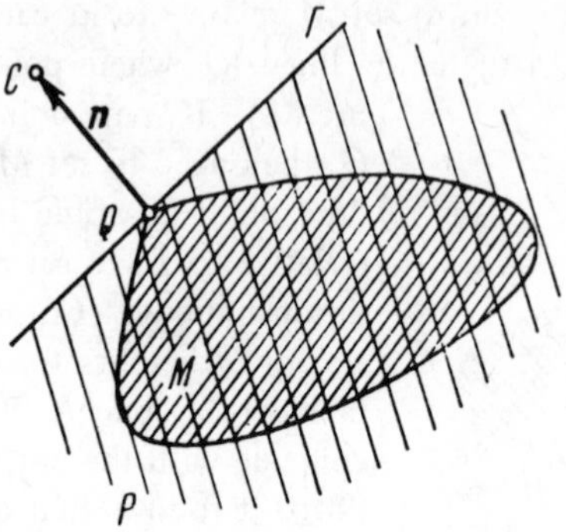

FIGURE 89

In addition, let us define an affine function f in space E^n by the equation $f(A) = \boldsymbol{n}\overrightarrow{QA}$. Then, in view of what was stated above, function f is *nonpositive* on set M, while at point Q it goes to zero. Thus we have proven the following theorem.

THEOREM 29.1. *Let M be an arbitrary convex set of space E^n and let Q be a boundary point of it. Then there exists a vector $\boldsymbol{n} \neq \boldsymbol{0}$ such that for any point A of set M the following relation is valid:*

$$\boldsymbol{n}\overrightarrow{QA} \leqslant 0.$$

Moreover, there exists a nonconstant affine function in E^n which is nonpositive on M and goes to zero at point Q.

Note that the function f, whose existence is attested to here, goes through a *maximum* (equal to zero) at point Q, provided it is considered just on set M. The following theorem is in a certain sense the converse of the above.

THEOREM 29.2. *Let $M \subset E^n$ be a convex set and let f be a nonconstant affine function defined in E^n. Assume that function f, considered only on set M, reaches a maximum at some point $Q \in M$. If f_1 is an affine function defined by the equation $f_1(A) = f(A) - f(Q)$, then the kernel of affine function f_1 will be the supporting hyperplane of set M passing through point Q. A similar statement can be made for the minimum of function f, considered just on M.*

Proof. The inequality $f_1(A) \leqslant 0$ defines some half-space P of space E^n, the boundary hyperplane of this half-space serving as the kernel of affine function f_1. Moreover, since function f, considered on M, reaches a maximum at Q, it follows that $f(Q) \geqslant f(A)$ for any point $A \in M$. In other words, $f_1(A) \leqslant 0$ for any point $A \in M$, that is, set M is wholly contained in half-space P. However, this also means that the kernel Γ of affine function f_1 is the supporting hyperplane of set M (since obviously $Q \in \Gamma$).

COROLLARY 29.3. *An affine nonconstant function, considered on a convex field $M \subset E^n$, cannot reach a maximum (or minimum) at an interior point of field M.*

Actually, if this function reaches a maximum at point $Q \in M$, then according to Theorem 29.2 the supporting hyperplane of field M passes through point Q, and thus Q cannot be an interior point.

COROLLARY 29.4. *If $M \subset E^n$ is a convex set and if Γ is some hyperplane, then there will exist no more than two supporting hyperplanes of set M parallel to Γ. If M is a bounded convex field, then the number of such supporting hyperplanes will be exactly two.*

Let f be an affine function in E^n, the kernel of which coincides with Γ (see Theorem 20.2). If at a point $C \in M$ function f, considered just on M, reaches neither a maximum nor a minimum, then the hyperplane Γ_C passing through C parallel to Γ is not a supporting hyperplane for field M, since in M there will exist a point A for which $f(A) < f(C)$, as well as a point B for which $f(B) > f(C)$, implying that points A and B of field M lie on opposite sides of hyperplane Γ_C.

Now let Q be a point at which function f, considered on M, reaches a maximum (provided such a point exists), and let Γ_Q be the hyperplane parallel to Γ passing through this point. According to Theorem 29.2, Γ_Q is a supporting hyperplane of set M (note that *all* points at which f reaches a maximum lie in this hyperplane: these, and only these, will be points of the set $\Gamma_Q \cap M$). A similar statement concerning the minimum may give us a second supporting hyperplane.

For a *bounded* convex field M (closed), function f actually reaches both a maximum and a minimum on M, although they are different; consequently, in this case there will exist exactly two supporting hyperplanes parallel to Γ.

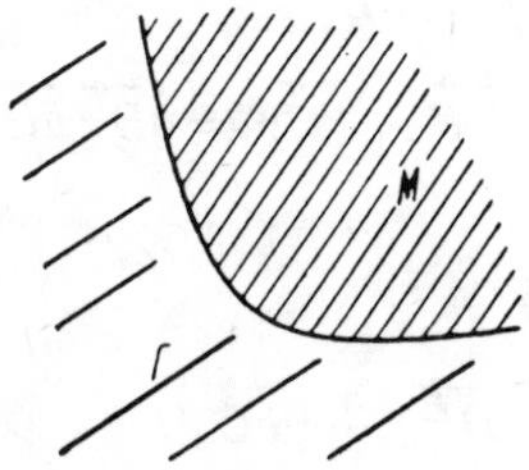

FIGURE 90

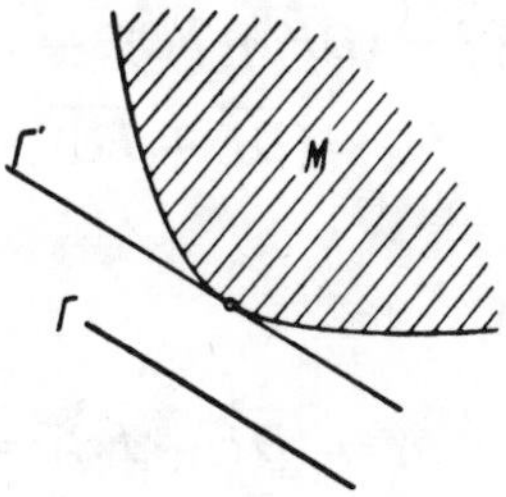

FIGURE 91

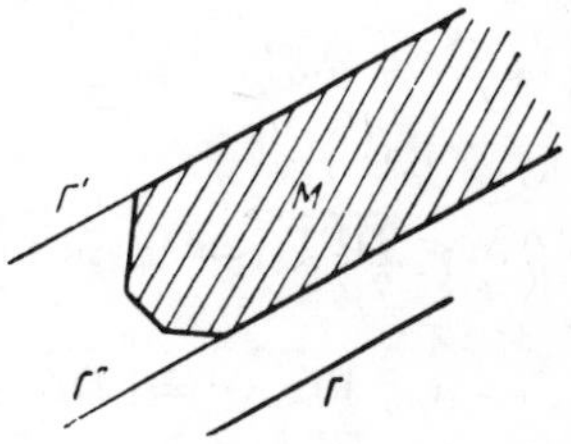

FIGURE 92

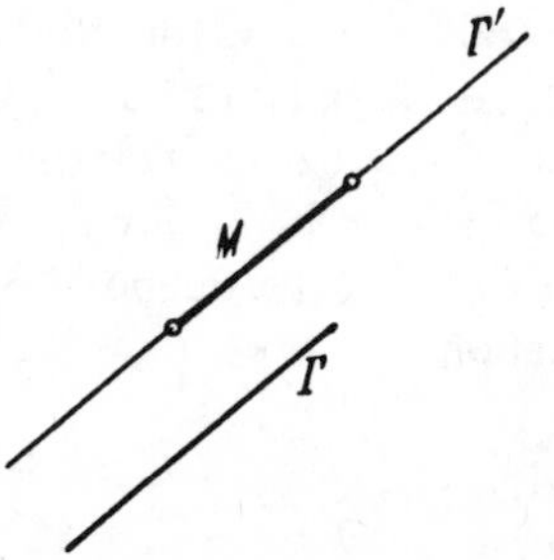

FIGURE 93

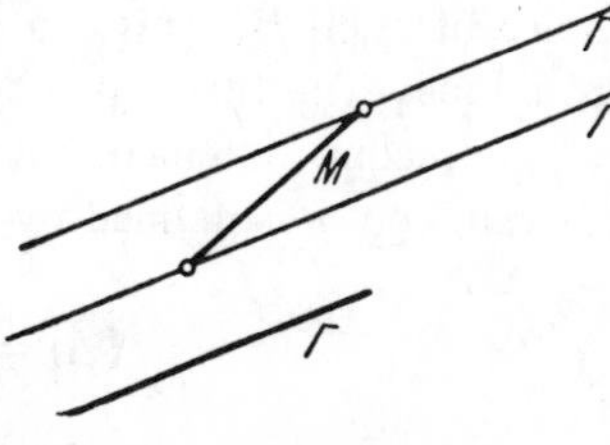

FIGURE 94

Figures 90 through 92 indicate that for unbounded convex fields the number of supporting hyperplanes parallel to Γ may be 0, 1, or 2. If, on the other hand, set M is closed and bounded but it is not a convex *field,* then the minimum and maximum of function f (considered in the proof) must be reached on M, but they may well coincide. Thus for such a set there may exist either a single supporting plane parallel to Γ (if the carrier plane of set M is contained in a hyperplane parallel to Γ; Figure 93) or two supporting planes (Figure 94).

THEOREM 29.5. *Let $M \subset E^n$ be a closed convex set and let D be a point not belonging to it. The following statements will then be valid:*

1) *in set M there exists only one point closest to D;*

2) *there exists an affine function f in E^n which is nonpositive on M and positive at point D;*

3) *there exists a supporting hyperplane of set M relative to which point D lies in one open half-space, while the entire set M lies in the other (closed) half-space.*

Proof. According to what was said in Example 22.11, there exists a point B of set M which is *closest* to D. Let us show that B is the *only* closest point. If we assume, conversely, that there exists a point $C \in M$ different from B for which $\rho(C, D) = \rho(B, D)$, and if F is a point defined by the equation $\overrightarrow{DF} = {}^1/_2\overrightarrow{DC} + {}^1/_2\overrightarrow{DB}$, then we have

$$0 = (\rho(C, D))^2 - (\rho(B, D))^2 = |\overrightarrow{DC}|^2 - |\overrightarrow{DB}|^2 = |\overrightarrow{DF} + \overrightarrow{FC}|^2 - |\overrightarrow{DF} + \overrightarrow{FB}|^2 = (\overrightarrow{DF} + \overrightarrow{FC})^2 - (\overrightarrow{DF} + \overrightarrow{FB})^2 = (\overrightarrow{DF} + \overrightarrow{FC})^2 - (\overrightarrow{DF} - \overrightarrow{FC})^2 = 4\overrightarrow{DF}\,\overrightarrow{FC}.$$

Therefore, $\overrightarrow{DF}\,\overrightarrow{FC} = 0$, so that

$$|\overrightarrow{DC}|^2 = |\overrightarrow{DF} + \overrightarrow{FC}|^2 = (\overrightarrow{DF} + \overrightarrow{FC})^2 = |\overrightarrow{DF}|^2 + |\overrightarrow{FC}|^2.$$

However, since $B \neq C$, it follows that $\overrightarrow{BC} \neq \mathbf{0}$ and thus that $\overrightarrow{FC} = {}^1/_2\overrightarrow{BC} \neq \mathbf{0}$. Consequently, $|\overrightarrow{FC}|^2 > 0$, that is, $|\overrightarrow{DC}|^2 > |\overrightarrow{DF}|^2$, or, alternatively, $\rho(C, D) > \rho(F, D)$. We see that point F lies *closer* to D than does C (or B). At the same time, point F belongs to set M (since $B \in M$, $C \in M$, and $F \in [B, C]$), which contradicts the choice of point B. This contradiction proves that the point in set M closest to D is unique. The first statement of the theorem is thereby proven.

Let us prove the second statement. As previously, let B be the point of set M closest to D. If function f is defined by the equation

$$f(A) = \overrightarrow{BD}\,\overrightarrow{BA},$$

then f is an affine function with a gradient $\boldsymbol{n} = \overrightarrow{BD}$ (Theorem 21.8), where $f(B) = 0$ and

$f(D)=\overrightarrow{BD}\,\overrightarrow{BD}=|\overrightarrow{BD}|^2>0$. It remains to prove that function f is nonpositive on M. If $A\in M$ and $A_1\in[A, B]$, that is, if $\overrightarrow{BA_1}=\lambda\overrightarrow{BA}$, where $0\leqslant\lambda\leqslant 1$, then point A_1 also belongs to set M (in view of its convexity), so that $\rho(D, B)\leqslant\leqslant\rho(D, A_1)$. In other words,

$$|\overrightarrow{BD}|^2\leqslant|\overrightarrow{DA_1}|^2=|\overrightarrow{DB}+\overrightarrow{BA_1}|^2=(\overrightarrow{DB}+\overrightarrow{BA_1})^2=(\overrightarrow{DB}+\lambda\overrightarrow{BA})^2=$$
$$=(-\overrightarrow{BD}+\lambda\overrightarrow{BA})^2=|\overrightarrow{BD}|^2-2\lambda\overrightarrow{BD}\,\overrightarrow{BA}+\lambda^2|\overrightarrow{BA}|^2,$$

and thus $2\lambda\overrightarrow{BD}\,\overrightarrow{BA}\leqslant\lambda^2|\overrightarrow{BA}|^2$ for $0\leqslant\lambda\leqslant 1$. Hence it follows that $\overrightarrow{BD}\,\overrightarrow{BA}\leqslant\leqslant\frac{\lambda}{2}|\overrightarrow{BA}|^2$, for $0<\lambda\leqslant 1$, and consequently $\overrightarrow{BD}\,\overrightarrow{BA}\leqslant 0$, that is, $f(A)\leqslant 0$. The second statement of the theorem is thereby proven.

The third statement follows directly from the second: the kernel of the constructed affine function f is also the desired hyperplane.

COROLLARY 29.6. *Any closed convex set can be represented in the form of the intersection of a finite or infinite number of half-spaces.*

Actually, consider *all* the closed half-spaces containing closed convex set M and let M^* be the intersection of all these half-spaces. Since *each* half-space considered contains M, it follows that $M^*\supset M$. Let us prove the converse inclusion. If $D\notin M$, then in view of the third statement of Theorem 29.5 there exists a closed half-space containing M but not containing point D. Consequently, $D\notin M^*$. Thus, if $D\notin M$ then it follows that $D\notin M^*$, that is, $M^*\subset M$. Therefore, $M=M^*$, meaning that M is the intersection of all the considered half-spaces.

THEOREM 29.7. *Let $M\subset E^n$ be a convex polyhedron and let Γ be its supporting hyperplane. Then the intersection $\Gamma\cap M$ either coincides with the entire polyhedron M or else is a face of it.*

Proof. Let us carry out induction on the dimension n of space E^n. For $n=1$ the theorem is trivial: the hyperplanes in the one-dimensional Euclidean space E^1 (that is, on a line) are points, while the convex polyhedrons are in this case points and line segments.

We assume the theorem to be proven for Euclidean spaces of dimension less than n, and we assume that M is a convex polyhedron in E^n, while Γ is a supporting hyperplane of this polyhedron. If m is the dimension of polyhedron M, then there are two possibilities: $m<n$ and $m=n$, which will now be considered separately.

Let $m<n$ and let L be the carrier plane of polyhedron M. If hyperplane Γ contains plane L, then $\Gamma\supset M$, so that $\Gamma\cap M=M$, that is, in this case the theorem is valid. If, on the other hand, Γ does not wholly contain plane L, then the intersection $\Gamma\cap L$ will be some hyperplane Γ^* of Euclidean space $E^m=L$. Then we have

$$\Gamma\cap M=\Gamma\cap(L\cap M)=(\Gamma\cap L)\cap M=\Gamma^*\cap M,$$

that is, in the m-dimensional Euclidean space L it is necessary to consider the intersection $\Gamma^*\cap M$ of polyhedron M and its supporting hyperplane Γ^*. Since $m<n$,

it follows from the assumption of induction that this intersection is some face of polyhedron M. Thus in this case the theorem is valid.

Finally, let us look at the case $m = n$, that is, the case where polyhedron M is a convex *field* of space E^n. Obviously, the set $N = \Gamma \cap M$ will be a convex polyhedron, since M is the intersection of a finite number of half-spaces, while Γ is the intersection of two half-spaces (defined by this hyperplane). Let A be an interior point of polyhedron N relative to its carrier plane L. Since $A \in \Gamma \cap M$, it follows that A is a boundary point of polyhedron M. According to Theorem 27.1, there exists an $(n - 1)$-dimensional face M_1 of polyhedron M containing point A. Let Γ_1 be the hyperplane which is the carrier plane of $(n - 1)$-dimensional polyhedron M. We will show that polyhedron N is contained in hyperplane Γ_1.

If $B \in N$, then since A is an interior point of polyhedron N relative to its carrier plane L, there must exist a point $C \in N$ such that $A \in (B, C)$. If point B did not belong to hyperplane Γ_1, then point C would not belong to it either, while points B and C would lie on *opposite sides* of hyperplane Γ_1. Consequently, only *one* of them can belong to polyhedron M (which lies completely in one of the two half-spaces defined by hyperplane Γ_1). But this is impossible, since both of points B and C belong to the polyhedron $N \subset M$. This contradiction indicates that $N \subset \Gamma_1$.

Now if hyperplanes Γ and Γ_1 coincide, then it follows that $N = \Gamma \cap M = \Gamma_1 \cap M = M_1$, that is, N is a face of polyhedron M. It remains to consider the case in which hyperplanes Γ and Γ_1 do not coincide. Then $\Gamma' = \Gamma \cap \Gamma_1$ is a hyperplane of an $(n - 1)$-dimensional Euclidean space $E^{n-1} = \Gamma_1$, the latter being a supporting hyperplane for polyhedron $M_1 \subset \Gamma_1$. Moreover, since $N \subset \Gamma_1$, we have

$$N = \Gamma_1 \cap N = \Gamma_1 \cap (\Gamma \cap M) = (\Gamma_1 \cap \Gamma) \cap (M \cap \Gamma_1) = \Gamma' \cap M_1.$$

Consequently, N is the intersection between the $(n - 1)$-dimensional polyhedron M_1 and its supporting hyperplane Γ' (in the $(n - 1)$-dimensional Euclidean space $E^{n-1} = \Gamma_1$). It then follows from the assumption of induction that N is some face of polyhedron M_1. However, N is by definition also a face of polyhedron M. The induction carried out thus proves the theorem.

It is not difficult to establish (continuing along the same line) that for *any* face of polyhedron M there exists a supporting hyperplane of polyhedron M such that its intersection with M gives precisely this face. If M is an n-*dimensional* polyhedron in E^n, the face in question being $(n - 1)$-dimensional, then there will exist *only one* supporting hyperplane (namely the carrier plane of this face), giving this face at the intersection with the polyhedron. In other cases there will exist an *infinite number* of supporting hyperplanes, giving the specified face at the intersection with the polyhedron.

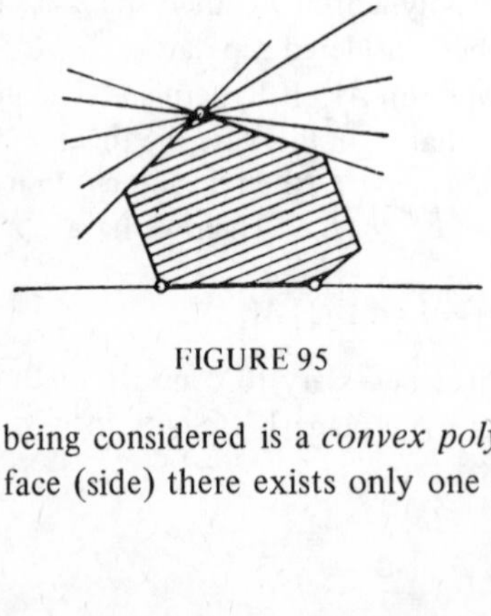

FIGURE 95

For example, if $n = 2$ and the polyhedron being considered is a *convex polyhedron*, then (Figure 95) for each *one-dimensional* face (side) there exists only one supporting hyperplane containing it (that is, in this case it is a supporting line), while for each zero-dimensional face (vertex) there will be an infinite number of supporting lines containing it. Similarly, for a three-dimensional polyhedron in three-dimensional space ($n=3$), only a single supporting plane passes through each two-dimensional face, while an infinite number pass through each edge and each vertex.

COROLLARY 29.8. *Let $M \subset E^n$ be a convex polyhedron and let f be an affine function defined in E^n. Then, either function f is constant on M or else the set of all the points at which this function (considered only on polyhedron M) has the highest value is a face of polyhedron M. In particular, for any affine function f there exists a* **vertex** *of polyhedron M at which function f, considered on M, reaches the highest value. The same will be true with regard to the least value of function f on M.*

If $Q \in M$ is the point at which function f reaches a maximum on M, then the kernel Γ of the affine function f_1 defined by the equation $f_1(A)=f(A)-f(Q)$ will be the supporting hyperplane of polyhedron M passing through point Q. At all points of hyperplane Γ (and only at these points) function f has the value $f(Q)$. Consequently, the set of all points of polyhedron M at which function f reaches its highest value on M coincides with $M \cap \Gamma$. Hence, in view of Theorem 29.7, we obtain the required statement.

THEOREM 29.9. *Let $M \subset E^n$ be a convex polyhedron and let f be an affine function defined in E^n. In addition, let M' be the face of polyhedron M on which function f (considered on M) reaches the highest value. In order for a vertex A of polyhedron M to lie on face M', it is necessary and sufficient that the relation $\mathbf{n}\overrightarrow{AB} \leqslant 0$ be satisfied for each edge $[A, B]$ of polyhedron M emerging from A, where $\mathbf{n}$ is the gradient of affine function f. In particular, in order for affine function f to reach a maximum at only a single vertex A on polyhedron M, it is necessary and sufficient that for each edge $[A, B]$ of polyhedron M emerging from A the relation $\mathbf{n}\overrightarrow{AB} < 0$ be satisfied.*

Proof. If vertex A belongs to face M' and if $[A, B]$ is some edge emerging from vertex A, then $f(A) \geqslant f(B)$, that is, $\mathbf{n}\overrightarrow{AB} \leqslant 0$ (see (21.5)). If, however, vertex B *does not lie on* face M', then $f(A) > f(B)$, that is, $\mathbf{n}\overrightarrow{AB} < 0$. The necessity of the conditions formulated is thereby established.

Let us now prove sufficiency. The dimension of polyhedron M is assumed to be not less than two, since otherwise the theorem would be obvious. Let A be some vertex of polyhedron M not belonging to face M'. We consider an arbitrary point C of face M' and we let D be an arbitrary point of polyhedron M, not lying on line AC (such a point D exists, since the dimension of polyhedron M was assumed to be not less than two). For a two-dimensional plane Π passing through points A, C, and D, we set $M^* = \Pi \cap M$. Then M^* is a two-dimensional polyhedron (that is, a convex polygon) for which A is a vertex. Note that function f is not constant on M^*, since point A does not lie on face M', so that $f(A) < f(C)$.

Let E and F be the vertices of polygon M^* adjacent to A. If the inequalities $f(E) \leqslant f(A)$ and $f(F) \leqslant f(A)$ were satisfied, then at all points on rays AE and AF

function f would have values not exceeding $f(A)$, so that at all points of the angle formed by these rays the values of function f would not exceed $f(A)$. However, then at all points of polygon M^* as well (lying inside this angle, see Figure 96) the values of function f would not exceed $f(A)$, which is impossible, since $f(C) > f(A)$. Consequently, at least one of the inequalities $f(E)>f(A)$ and $f(F)>f(A)$ should be satisfied.

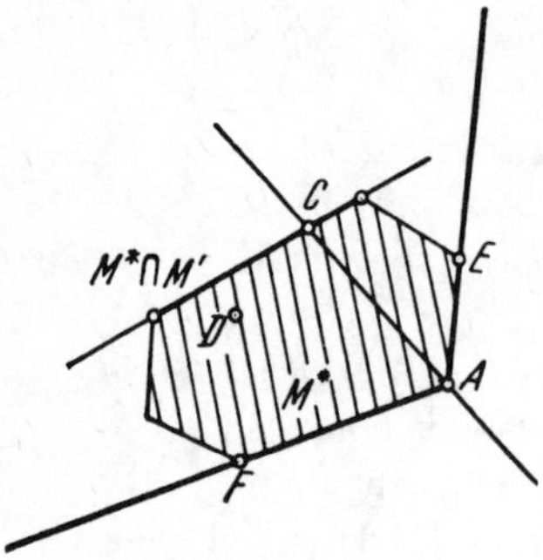

FIGURE 96

To be more definite, let $f(E)>f(A)$. If we take an arbitrary point K on side $[A, E]$ of polygon M^* then K belongs to polygon M^* and thus to polyhedron M as well. It will be a *boundary* point of polyhedron M (if it were an interior point for M, then it would have to be interior for M^* too). Consequently, in view of Theorem 27.1, there exists an $(m-1)$-dimensional *face* M_1 of polyhedron M containing point K, where m is the dimension of polyhedron M. Since $K \in (A, E)$, it follows that all of segment $[A, E]$ lies on face M_1. Here function f (considered on M_1) will not reach its highest value at point A, since $f(A)<f(E)$. Thus, if at vertex $A \in M$ function f does not reach its highest value, the dimension of polyhedron M will be reduced by unity (where, as before, function f does not reach its highest value at vertex A).

By reducing the dimension of the polyhedrons in this manner, we obtain a *one-dimensional* face $[A, B]$ of polyhedron M, where as before function f, considered on edge $[A, B]$, will not reach its highest value at point A, that is, $f(A)<f(B)$. However, then $\boldsymbol{n}\overrightarrow{AB} > 0$. Consequently, if vertex A does not lie on face M', there will exist an edge $[A, B]$ emerging from this vertex such that $\boldsymbol{n}\overrightarrow{AB} > 0$. This proves the sufficiency of the conditions formulated in the theorem.

§8. Theorems on Separability of Convex Cones

30. Separation of convex sets. Let M_1 and M_2 be two convex sets of space E^n. Sets M_1 and M_2 are called *separable* in E^n if there exists a hyperplane $\Gamma \subset E^n$ such that set M_1 lies in one closed half-space defined by hyperplane Γ and set M_2 lies in the other. A hyperplane Γ possessing this property is called a *separating hyperplane* for sets M_1 and M_2 (or a hyperplane *separating* sets M_1 and M_2).

We stress that sets M_1 and M_2 do not have to lie in *open* half-spaces defined by hyperplane Γ (as in Figure 97); they may have points in common with the hyperplane Γ separating them (Figure 98). In particular, if sets M_1 and M_2 lie in one hyperplane Γ (Figure 99), then this hyperplane is a separating hyperplane for sets M_1 and M_2. It should also be noted that the separability of sets depends quite considerably on the particular Euclidean space in which these sets are considered. For instance, the sets M_1 and M_2 portrayed in Figure 99 are *separable* when considered in a three-dimensional Euclidean space but they will be *nonseparable* if they are considered in a *two-dimensional* Euclidean space Γ.

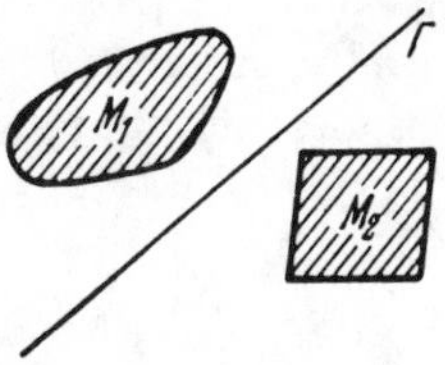

FIGURE 97

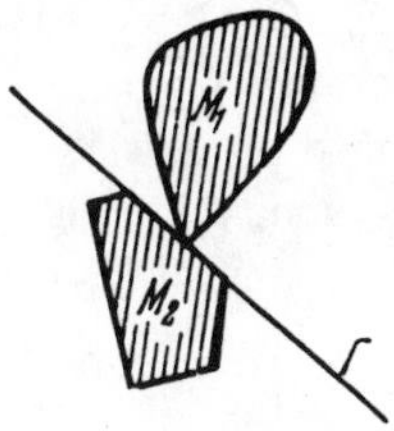

FIGURE 98

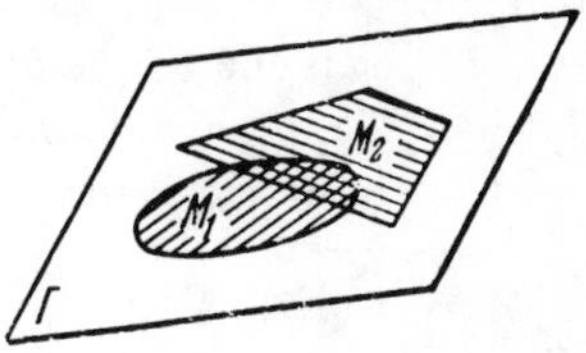

FIGURE 99

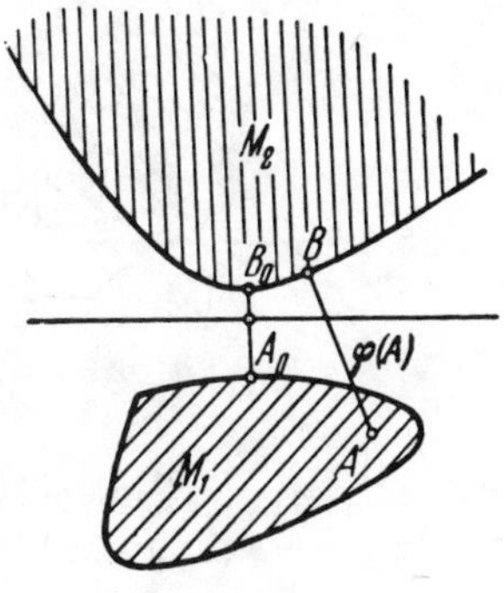

FIGURE 100

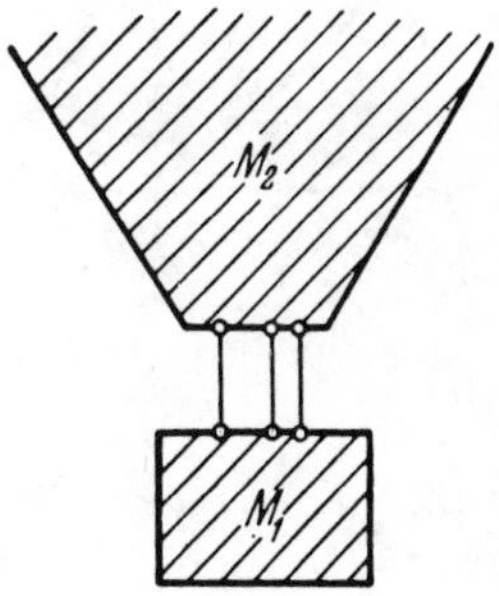

FIGURE 101

In this subsection we will find the different conditions for the separability of convex sets. The convex sets considered are always assumed to be closed. First let us examine the separability when the convex sets do not have common points and at least one of the sets is bounded. In this case the convex sets are always separable. The simplest means of constructing the separating hyperplane for such sets is as follows.

Let M_1 and M_2 be convex nonintersecting (disjoint) sets, set M_1 being bounded. For each point $A \in M_1$ we find the point B of set M_2 *closest* to it (Theorem 29.5) and we let $\varphi(A)$ be the length of segment $[A, B]$. In other words, $\varphi(A)$ is the *shortest* distance from A to the points of set M_2 (Figure 100). Function $\varphi(A)$ is defined (and obviously continuous) on set M_1. Thus there will exist a point A_0 at which this function (considered in M_1) reaches its *lowest* value. Hence we find the points $A_0 \in M_1$, $B_0 \in M_2$ ensuring the minimum distance between points of sets M_1 and M_2 (note that the pair of points A_0, B_0 for which the distance between points of M_1 and M_2 is a minimum may not be unique; Figure 101). Now it is sufficient to construct a hyperplane Γ orthogonal to segment $[A_0, B_0]$ and passing through some interior point of this segment; such a hyperplane will also be separating (see Figure 100).

Actually, if there existed in set M_1 a point C lying on the side of Γ opposite to point A_0 (Figure 102), then all of segment $[A_0, C]$ would belong to set M_1. However, then the points of segment $[A_0, C]$ close to A_0 would lie *nearer* to B_0 than point A_0 would (since the angle between segments $[A_0, B_0]$ and $[A_0, C]$ is acute), which contradicts the choice of points A_0 and B_0. Consequently, all of set M_1 lies on the same side of Γ as point A_0, indicating that set M_2 is located on the same side of Γ as point B_0. Thus Γ is a separating hyperplane.

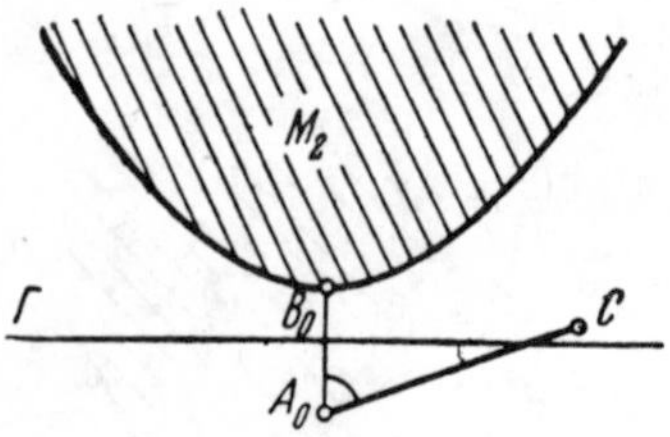

FIGURE 102

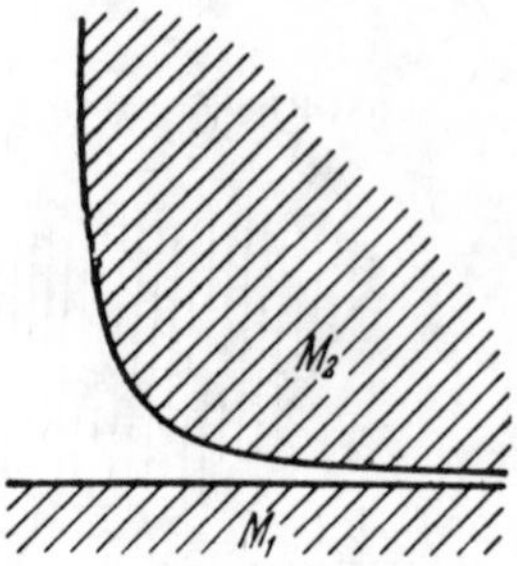

FIGURE 103

Note, though, that if nonintersecting convex sets M_1 and M_2 are both unbounded then this statement is inapplicable, since there may not exist a *least* distance between points of sets M_1 and M_2 (Figure 103). Later we will extend the result obtained to unbounded sets as well (see Corollary 30.8), but the method of the proof will be completely different: the following auxiliary proposition, which complements Theorem 25.1, will be used.

THEOREM 30.1. *Let F be an arbitrary closed set located in space E^n. In order for a point $C \in E^n$ to be an interior point of the set* conv F, *it is necessary and sufficient that there exist a natural number m such that the points $A_0, A_1, \ldots, A_m$ of set F do not lie in a single hyperplane, as well as numbers $\lambda^0, \lambda^1, \ldots, \lambda^m$ such that the following relations are valid:*

$$\lambda^0 > 0, \quad \lambda^1 > 0, \ldots, \lambda^m > 0, \quad \lambda^0 + \lambda^1 + \ldots + \lambda^m = 1, \tag{30.1}$$

$$C = \lambda^0 A_0 + \lambda^1 A_1 + \ldots + \lambda^m A_m. \tag{30.2}$$

Proof. Let F^* be the set of all points C for each of which there exists a natural number m such that the points $A_0, A_1, \ldots, A_m$ of set F do not lie in a single hyperplane, as well as numbers $\lambda^0, \lambda^1, \ldots, \lambda^m$ such that relations (30.1) and (30.2) are valid. Set F^* is convex (cf. proof of Theorem 25.1). Let us show that it is an *open* set.

Assume that $C \in F^*$, that is, that (30.1) and (30.2) are valid. Since points A_0, $A_1, \ldots, A_m$ do not lie in a single hyperplane, it follows that $m \geqslant n$ and thus that, of vectors $\overrightarrow{A_0A_1}, \ldots, \overrightarrow{A_0A_m}$, n will be linearly independent. Let us assume for simplicity that the linearly independent vectors are $\overrightarrow{A_0A_1}, \ldots, \overrightarrow{A_0A_n}$, so that A_0, $A_1, \ldots, A_n$ are the vertices of an n-dimensional simplex. If

$$D = \lambda^0 A_0 + \lambda^1 A_1 + \ldots + \lambda^{n-1} A_{n-1} + (\lambda^n + \lambda^{n+1} + \ldots + \lambda^m) A_n, \tag{30.3}$$

then in view of (30.1) point D is an *interior* point of simplex $[A_0, A_1, \ldots, A_n]$ (see Example 25.3), and hence it is an interior point of the polyhedron M serving as the convex hull of points $A_0, A_1, \ldots, A_m$. It is also clear that all the interior points of simplex $[A_0, A_1, \ldots, A_n]$ belong to set F^*.

If $D = C$, then C is an interior point of set F^*. Now let us assume that $C \neq D$, so that $m > n$ (cf. relations (30.2) and (30.3)). If we consider the point

$$E = C + \varepsilon \overrightarrow{DC},$$

where $\varepsilon > 0$ (Figure 104), then clearly point C belongs to interval (D, E). Moreover,

$$\begin{aligned} E &= C + \varepsilon (C - D) = (1+\varepsilon) C - \varepsilon D = (1+\varepsilon)(\lambda^0 A_0 + \lambda^1 A_1 + \ldots + \lambda^m A_m) - \\ &- \varepsilon(\lambda^0 A_0 + \lambda^1 A_1 + \ldots + \lambda^{n-1} A_{n-1} + (\lambda^n + \ldots + \lambda^m) A_n) = \\ &= \lambda^0 A_0 + \lambda^1 A_1 + \ldots + \lambda^{n-1} A_{n-1} + (\lambda^n - \varepsilon(\lambda^{n+1} + \ldots + \lambda^m)) A_n + \\ &\qquad + (1+\varepsilon)\lambda^{n+1} A_{n+1} + \ldots + (1+\varepsilon)\lambda^m A_m. \end{aligned}$$

For sufficiently small $\varepsilon > 0$, all the coefficients on the right-hand side will be *positive*, their sum being unity, that is, $E \in F^*$. Thus points D, E both belong to set F^*, while D is an interior point of this set. Since $C \in (D, E)$, it follows (in view of Theorem 26.2) that C is an *interior* point of set F^*. Hence we see that any point C of set F^* is an interior point of it, that is, set F^* is open.

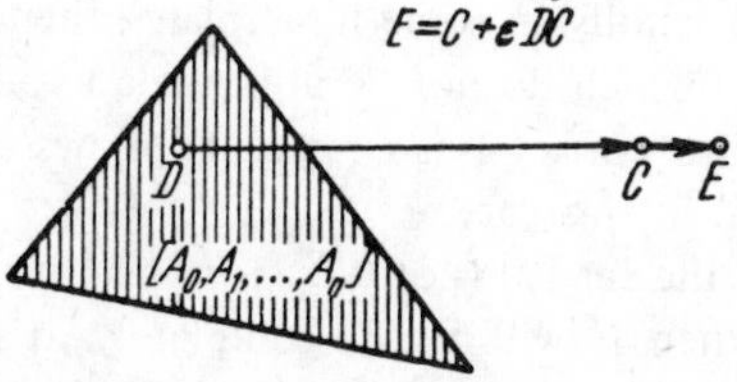

FIGURE 104

Finally, let us show that, if set F^* is nonempty, its closure will coincide with the convex hull of set F. Let C_1 be an arbitrary point of the set conv F, so that

$$C_1 = \mu^0 B_0 + \mu^1 B_1 + \ldots + \mu^k B_k,$$

where $B_0, B_1, \ldots, B_k$ are certain points of set F, while numbers $\mu^0, \mu^1, \ldots, \mu^k$ are nonnegative and satisfy the relation $\mu^0 + \mu^1 + \ldots + \mu^k = 1$. It can be assumed (after rejecting, if necessary, the excess points B_j) that all of these numbers are *positive*:

$$\mu^0 > 0, \quad \mu^1 > 0, \ldots, \mu^k > 0, \quad \mu^0 + \mu^1 + \ldots + \mu^k = 1.$$

We take an arbitrary point C of set F^* (recalling that this set was assumed to be nonempty), that is, a point satisfying relations (30.1) and (30.2), where $A_0, A_1, \ldots, A_m$ do not lie in a single hyperplane. Moreover, if K is an arbitrary point of interval (C, C_1), that is, if

$$K = (1 - \nu) C + \nu C_1, \qquad 0 < \nu < 1,$$

then we have

$$\begin{aligned} K = (1 - \nu)(\lambda^0 A_0 + \ldots + \lambda^m A_m) + \nu(\mu^0 B_0 + \mu^1 B_1 + \ldots + \mu^k B_k) = \\ = (1 - \nu)\lambda^0 A_0 + \ldots + (1 - \nu)\lambda^m A_m + \\ + \nu\mu^0 B_0 + \nu\mu^1 B_1 + \ldots + \nu\mu^k B_k, \end{aligned}$$

where all the coefficients on the right-hand side are positive, their sum being unity. In addition, since points $A_0, A_1, \ldots, A_m$ and $B_0, B_1, \ldots, B_k$ do not lie in one hyperplane (points $A_0, A_1, \ldots, A_m$ already do not lie in a single hyperplane), it follows that $K \in F^*$. Thus any point K of interval (C, C_1) belongs to set F^*, so that arbitrarily close to C_1 there will be points of set F^*, implying that C_1 belongs to the closure of set F^*. In other words, the entire set conv F is contained in the closure of set F^*, while, since the set conv F is closed and contains F^*, it follows that conv F *coincides* with the closure of F^*.

From the foregoing it is easy to see that

$$\text{int}(\text{conv } F) = F^*. \tag{30.4}$$

Actually, if the set conv F does not contain interior points, that is, if the entire set F (and conv F with it) lies wholly in some hyperplane, then in F there will not exist points $A_0, A_1, \ldots, A_m$ which do not lie in a single hyperplane, so that set F^* is empty. Thus, if the set int (conv F) is empty, it follows that F^* is also empty, that is, equation (30.4) is valid in this case.

If, on the other hand, the set int (conv F) is nonempty, meaning that conv F contains interior points, then F will not lie completely in a single hyperplane, so that there will exist in F points $A_0, A_1, \ldots, A_n$ not lying in a single hyperplane, that is, set F^* is also nonempty. In this case F^* is an open convex set, the closure of which coincides with $\operatorname{conv} F$, and thus equation (30.4) is also valid.

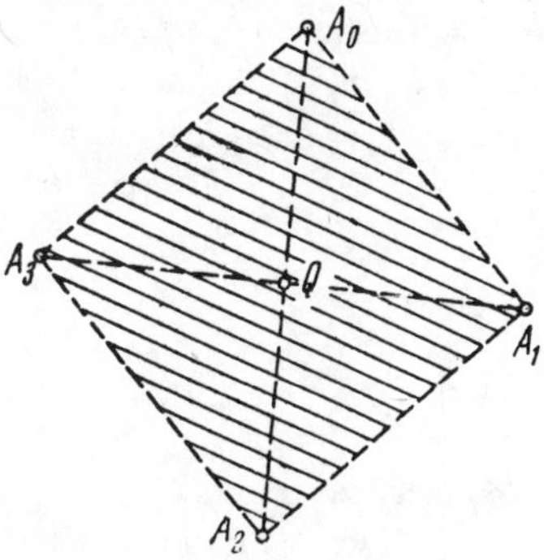

FIGURE 105

Accordingly, (30.1) and (30.2) (where points $A_0, A_1, \ldots, A_m$ do not lie in a single hyperplane) constitute the necessary and sufficient condition for point C to belong to the set $\operatorname{int}(\operatorname{conv} F)$.

NOTE 30.2. In contrast to Theorem 25.1, in Theorem 30.1 we cannot state that the number m in relations (30.1) and (30.2) is to be assumed equal to n. For instance, if $n=2$ and if F is a set consisting of the four corner points of a square (Figure 105), then the center Q of the square is an interior point of the set $\operatorname{conv} F$. In accordance with Theorem 30.1, we then have

$$Q = \lambda^0 A_0 + \lambda^1 A_1 + \lambda^2 A_2 + \lambda^3 A_3,$$

where $\lambda^0 = \lambda^1 = \lambda^2 = \lambda^3 = 1/4$. However, it is impossible to select *three* points not lying on the same line for which relations (30.1) and (30.2) are valid. Consequently, in this example $n=2$ and $m=3$.

Now let us turn to the separation of convex sets. First we consider the separation of two convex cones with a common vertex. This problem is solved using the following theorem.

THEOREM 30.3. *Let M_1 and M_2 be two convex cones in space E^n having a common vertex Q. In order for cones M_1 and M_2 not to be separable, it is necessary and sufficient that there exist points $B_1, \ldots, B_k$ of cone M_1 and points $C_1, \ldots, C_l$ of cone M_2, such that these points $B_1, \ldots, B_k$ and $C_1, \ldots, C_l$ do not lie in one hyperplane and*

$$\overrightarrow{QB_1} + \ldots + \overrightarrow{QB_k} = \overrightarrow{QC_1} + \ldots + \overrightarrow{QC_l}. \tag{30.5}$$

Proof. Let N_1 be a cone symmetric to cone M_1 relative to point Q (see Example 19.14), and let $F = N_1 \cup M_2$. If cones M_1 and M_2 are separated by a hyperplane Γ, then cones N_1 and M_2 both lie in a *single* (closed) half-space defined by hyperplane Γ. This means that all of set F lies in one half-space, so that its convex hull $\operatorname{conv} F$ lies in this same half-space. In other words, Γ is a supporting hyperplane of the convex set $\operatorname{conv} F$ and thus point Q is a *boundary* point of the set $\operatorname{conv} F$ (since the supporting hyperplane Γ of $\operatorname{conv} F$ passes through it). Conversely, if Q is a boundary point of the set $\operatorname{conv} F$, then the supporting hyperplane Γ of $\operatorname{conv} F$ may pass through point Q so that all of the set $\operatorname{conv} F$ (and thus both cones N_1 and M_2) will lie in one half-space defined by hyperplane Γ, cones M_1 and M_2 thereby being separated by this hyperplane.

Thus, cones M_1 and M_2 are separable if, and only if, Q is a boundary point of the set $\operatorname{conv} F$. This means that the cones are *nonseparable* if, and only if, Q is an *interior* point of $\operatorname{conv} F$.

Let the cones be nonseparable, that is, assume that Q is an interior point of the set conv F. According to Theorem 30.1, there exist points $A_0, A_1, \ldots, A_m$ of set F not lying in one hyperplane and numbers $\lambda^0, \lambda^1, \ldots, \lambda^m$ satisfying conditions (30.1) and (30.2), where $C=Q$. Since $F=N_1 \cup M_2$, it follows that each of points A_0, $A_1, \ldots, A_m$ belongs to at least one of cones N_1 and M_2. Let us assume, to be definite, that points $A_0, A_1, \ldots, A_{k-1}$ belong to cone N_1, where $k \leqslant m$, while the rest of the points $A_k, \ldots, A_m$ belong to cone M_2. Now, if we set $l=m-k+2$ and define points $B_1, \ldots, B_k, C_1, \ldots, C_l$ by the relations

$$\overrightarrow{QB_i} = -\lambda^{i-1}\overrightarrow{QA_{i-1}} \qquad (i=1, \ldots, k);$$
$$\overrightarrow{QC_j} = \lambda^{j+k-1}\overrightarrow{QA_{j+k-1}} \qquad (j=1, \ldots, l-1);\ C_l = Q,$$

then it is clear that points $B_1, \ldots, B_k$ belong to cone M_1, whereas points $C_1, \ldots, C_l$ belong to cone M_2. Moreover, points $B_1, \ldots, B_k$ and $C_1, \ldots, C_l$ do not lie in one hyperplane (otherwise all the points $A_0, A_1, \ldots, A_m$ would have to lie in this same hyperplane). Finally, from relation (30.2), in which $C=Q$, we obtain

$$\overrightarrow{QB_1} + \ldots + \overrightarrow{QB_k} = \overrightarrow{QC_1} + \ldots + \overrightarrow{QC_{l-1}} + \overrightarrow{QC_l}.$$

Consequently, the condition specified in Theorem 30.3 is satisfied, that is, this condition is necessary.

Next let us prove sufficiency, that is, we will assume the condition in Theorem 30.3 to be satisfied and we will show that cones M_1 and M_2 are nonseparable. First, however, we assume the cones to be separable, that is, there exists a hyperplane Γ passing through point Q such that $M_1 \subset P_1$ and $M_2 \subset P_2$, where P_1 and P_2 are closed half-spaces set off by hyperplane Γ. If point D is defined by the relation

$$\overrightarrow{QD} = \overrightarrow{QB_1} + \ldots + \overrightarrow{QB_k} = \overrightarrow{QC_1} + \ldots + \overrightarrow{QC_l},$$

then, since all the points $Q, B_1, \ldots, B_k$ belong to half-space P_1, while Q lies in the boundary hyperplane, it follows that point D also lies in this half-space. Similarly, point D also belongs to half-space P_2, and thus $D \in \Gamma$. However, then all the points $B_1, \ldots, B_k$ lie in hyperplane Γ (otherwise point D would lie in an open half-space, and not in hyperplane Γ), implying that all the points $C_1, \ldots, C_l$ lie in hyperplane Γ. On the other hand, this contradicts the fact that points $B_1, \ldots, B_k$ and $C_1, \ldots, C_l$ do not lie in a single hyperplane. This contradiction shows that cones M_1 and M_2 are nonseparable.

The following theorem states the necessary and sufficient condition for nonseparability of cones with a common vertex in a more geometrical form.

THEOREM 30.4. *If M_1 and M_2 are convex cones in E^n with a common vertex Q, then for these cones to be nonseparable it is necessary and sufficient that the following two conditions be satisfied:*

1) *there is no hyperplane containing both cones M_1 and M_2;*

2) *there exists a point A which is an interior point of each of cones M_1 and M_2 relative to its carrier plane.*

Proof. Let us show the sufficiency of the conditions formulated. We assume the opposite, namely that (with conditions 1 and 2 above satisfied) cones M_1 and M_2 are separable, with Γ as the separating hyperplane. For some arbitrary point $B \in M_1$, since A is an interior point of cone M_1 relative to its carrier plane, we can find a point $C \in M_1$ such that $A \in (B, C)$. Since point A belongs to M_1, it will lie in one (closed) half-space defined by hyperplane Γ, whereas since A is also in cone M_2 it must lie in the other half-space as well. Consequently, $A \in \Gamma$. If point B did not belong to hyperplane Γ, then point C would not lie in this hyperplane either, and B and C would be on *different* sides of Γ. However, this is impossible, since all of cone M_1 is in one of the half-spaces set off by hyperplane Γ. Consequently, $B \in \Gamma$, that is, all of cone M_1 is contained in hyperplane Γ. It can be shown similarly that $M_2 \subset \Gamma$, which however contradicts condition 1 above. This contradiction proves that the cones are nonseparable if conditions 1 and 2 are satisfied.

Now let us prove necessity. In other words, we will show that, if even one of conditions 1 or 2 is not satisfied, cones M_1 and M_2 are separable. This is obvious for condition 1 since, if cones M_1 and M_2 lie in one hyperplane, they are separable. It remains to show that cones M_1 and M_2 are separable if condition 2 is not satisfied.

Assume that no point A exists which is interior (relative to the corresponding carrier plane) for each of cones M_1 and M_2. If L_1 and L_2 are the carrier planes of cones M_1 and M_2, then point Q is not interior (relative to the carrier plane) for at least one of cones M_1 and M_2, that is, at least one of these cones does not coincide with its entire carrier plane. To be definite, let $M_1 \neq L_1$. Consider the set $\operatorname{relint} M_1$ of all points which are interior (relative to L_1) for cone M_1. By adding point Q to this set, we obtain a convex cone, which we will call $\mathring{M}_1$. Similarly, by adding point Q to the set $\operatorname{relint} M_2$, we obtain the convex cone $\mathring{M}_2$. According to assumption, cones $\mathring{M}_1$ and $\mathring{M}_2$ do not have any points in common other than Q. Here $\mathring{M}_1 \neq L_1$ (since $\mathring{M}_1 \subset M_1$). We will show that cones $\mathring{M}_1$ and $\mathring{M}_2$ are separable (which implies directly that the initial cones M_1 and M_2 are separable as well).

If, on the other hand, cones $\mathring{M}_1$ and $\mathring{M}_2$ are assumed to be nonseparable, then in view of Theorem 30.3 there will exist points $B_1, \ldots, B_k$ of cone $\mathring{M}_1$ and points $C_1, \ldots, C_l$ of cone $\mathring{M}_2$ such that these points $B_1, \ldots, B_k$ and $C_1, \ldots, C_l$ do not lie in one hyperplane, and relation (30.5) is satisfied. We define point D by the relation

$$\overrightarrow{QB_1} + \ldots + \overrightarrow{QB_k} = \overrightarrow{QC_1} + \ldots + \overrightarrow{QC_l} = \overrightarrow{QD}.$$

Since Q is the vertex of cone $\mathring{M}_1$, while points $B_1, \ldots, B_k$ belong to this cone, it follows that point D also belongs to cone $\mathring{M}_1$. Similarly, point D belongs to cone $\mathring{M}_2$. However, since these cones have a single common point Q, it follows that $D=Q$ and

$$\overrightarrow{QB_1} + \ldots + \overrightarrow{QB_k} = \overrightarrow{QC_1} + \ldots + \overrightarrow{QC_l} = \mathbf{0}. \tag{30.6}$$

Now let N_1 be the polyhedron which forms the convex hull of points $B_1, \ldots, B_k$ and let N_2 be the polyhedron which forms the convex hull of points $C_1, \ldots, C_l$. The carrier planes of these polyhedrons will be called L_1^* and L_2^*, respectively. The relation

$$Q = \frac{1}{k} B_1 + \ldots + \frac{1}{k} B_k,$$

implied directly by (30.6), shows that point Q belongs to plane L_1^* and thus (in view of Theorem 30.1) is in Euclidean space L_1^* an *interior* point of the convex hull of points $B_1, \ldots, B_k$, that is, it is an interior (relative to L_1^*) point of polyhedron N_1. However, since $N_1 \subset \mathring{M}_1$ and since $\mathring{M}_1$ is a cone with a vertex Q, it follows that all of plane L_1^* is contained in $\mathring{M}_1$. Similarly, it can be shown that $L_2^* \subset \mathring{M}_2$. Thus planes L_1^* and L_2^* have only one common point Q. Then both planes L_1^* and L_2^* cannot be contained simultaneously in any hyperplane (since this hyperplane would have to contain all points $B_1, \ldots, B_k$ and $C_1, \ldots, C_l$).

Consequently, planes L_1^* and L_2^* are mutually complementary, meaning that they have a single common point Q and that the sum of their dimensions is n. From this it is easy to see that $\mathring{M}_1 = L_1^*$. If A is an arbitrary point of cone $\mathring{M}_1$, then there will exist points $B \in L_1^*$ and $C \in L_2^*$ such that $\overrightarrow{QA} = \overrightarrow{QB} + \overrightarrow{QC}$. If B_1 is a point symmetrical to B relative to point Q, then clearly $B_1 \in L_1^*$, while $\overrightarrow{QB_1} = -\overrightarrow{QB}$. Consequently, $\overrightarrow{QC} = \overrightarrow{QA} - \overrightarrow{QB} = \overrightarrow{QA} + \overrightarrow{QB_1}$. Since $A \in \mathring{M}_1$ and $B_1 \in \mathring{M}_1$, we know that $C \in \mathring{M}_1$. Therefore, point C belongs to both cones $\mathring{M}_1$ and $\mathring{M}_2$, so that $C = Q$. The relation $\overrightarrow{QA} = \overrightarrow{QB} + \overrightarrow{QC}$ now implies that $\overrightarrow{QA} = \overrightarrow{QB}$, that is, $A = B$, and thus $A \in L_1^*$. Accordingly, cone $\mathring{M}_1$ coincides with plane L_1^*. However, then L_1^* is also a carrier plane of cone $\mathring{M}_1$, that is, $\mathring{M}_1 = L_1^* = L_1$, which, however, contradicts the assumption that $\mathring{M}_1 \neq L_1$. This contradiction indicates that cones $\mathring{M}_1$ and $\mathring{M}_2$ are separable.

Thus there exists a hyperplane Γ such that $\mathring{M}_1 \subset P_1$ and $\mathring{M}_2 \subset P_2$, where P_1 and P_2 are closed half-spaces defined by hyperplane Γ. Because of the closure of half-space P_1 it contains, along with cone $\mathring{M}_1$, all the boundary points (relative to L_1) of this cone, that is, all of cone M_1 is contained in P_1. Similarly, $M_2 \subset P_2$. Therefore, Γ is a separating hyperplane for cones M_1 and M_2.

COROLLARY 30.5. *If M_1 and M_2 are convex cones in E^n, and if they have a common vertex Q and no other common points, then either cones M_1 and M_2 are separable or else M_1 and M_2 are two mutually complementary planes (that is, the sum of their dimensions is n).*

COROLLARY 30.6. *If M_1 and M_2 are convex cones in E^n with a common vertex Q, and if cone M_1 is solid (that is, of dimension n) and cone M_2 does not intersect the interior of cone M_1, then cones M_1 and M_2 are separable. In particular, two solid cones with a common vertex and without any common interior points are separable.*

Finally, let us examine the general case. Let M_1 and M_2 be arbitrary convex sets in E^n. Consider an $(n+1)$-dimensional Euclidean space E^{n+1} containing E^n as a hyperplane. If Q is an arbitrary point of space E^{n+1} not lying in E^n, then all the possible rays starting from Q and passing through the points of set M_1 will form a convex cone in E^{n+1}, which we will call K_1. Similarly, the rays starting from Q and passing through the points of set M_2 form a convex cone $K_2 \subset E^{n+1}$ (Figure 106). Assume that $\Gamma \subset E^n$ is a hyperplane separating sets M_1 and M_2. Then hyperplane Γ^* of space E^{n+1}, which contains Γ and passes through point Q, separates cones K_1 and K_2 in E^{n+1}. Conversely, if cones K_1 and K_2 are separable in E^{n+1} and Γ^* is the separating hyperplane, then obviously hyperplane Γ^* is not parallel to hyperplane E^n in space E^{n+1} and thus must intersect it. Intersection $\Gamma^* \cap E^n$ is a hyperplane of space E^n, and this hyperplane separates sets M_1 and M_2.

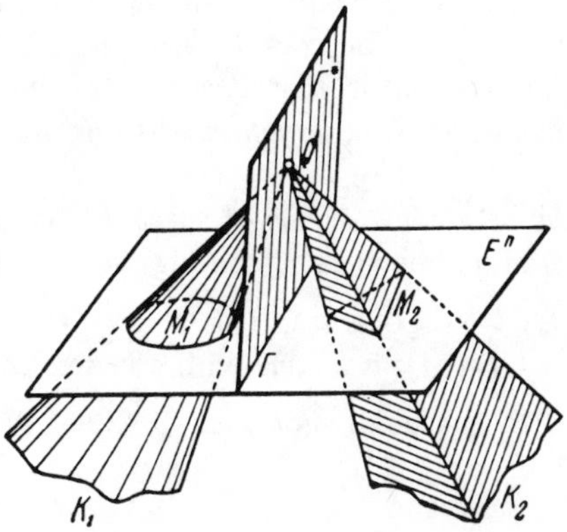

FIGURE 106

Using this procedure, the problem of the separation of arbitrary convex sets in E^n reduces to the problem of the separation of convex cones with a common vertex in E^{n+1}. Therefore, the foregoing results embodied in Theorem 30.4 and Corollaries 30.5 and 30.6, concerning the separation of convex cones, lead us directly to the following facts (where the convex sets M_1 and M_2 being considered may be bounded or unbounded, closed or nonclosed).

COROLLARY 30.7. *In order for convex sets M_1 and M_2 of space E^n to be nonseparable, it is necessary and sufficient that the following two conditions be satisfied:*

1) *no hyperplane $\Gamma \subset E^n$ exists which contains both sets M_1 and M_2;*

2) *a point A exists which is an interior point of each of sets M_1 and M_2 relative to its carrier plane.*

COROLLARY 30.8. *Any two nonintersecting (nonempty) convex sets in E^n are separable.*

COROLLARY 30.9. *Let M_1 be a convex field of space E^n and let $M_2 \subset E^n$ be a convex set not intersecting the interior of field M_1; then M_1 and M_2 are separable. In particular, two convex fields not having any common interior points are separable.*

In conclusion, let us note a very simple relationship which exists between the concept of the separability of convex sets, on the one hand, and affine functions, on the other. Let Γ be a separating hyperplane of convex sets M_1 and M_2. Consider a nonconstant affine function f, the kernel of which is hyperplane Γ. Function f is nonnegative in one of the half-spaces set off by hyperplane Γ and nonpositive in the other; thus it is nonnegative on one of sets M_1, M_2 and nonpositive on the other. Accordingly, *if sets M_1 and M_2 are separable, then there will exist a nonconstant affine function which is nonnegative on one of sets M_1 and M_2 and nonpositive on the other.* The converse is also obviously true: *if such a function exists, then sets M_1 and M_2 are separable* (the kernel of this function is the separating hyperplane).

Consequently, the problem of the separation of convex sets is *equivalent* to the problem of constructing a nonconstant affine function which is nonnegative on one set and nonpositive on the other. This makes it possible to reformulate results 30.3 through 30.9 in terms of affine functions. For example, Corollary 30.8 can be formulated as follows: *if M_1 and M_2 are nonintersecting (disjoint) nonempty convex sets in E^n, then there exists in E^n a nonconstant affine function which is nonnegative on M_1 and nonpositive on M_2.*

31. Dual cone. Let M be a convex cone of space E^n having point Q as its vertex. Let $D(M)$ be the set of all points $B \in E^n$ having the property that for *any* point $A \in M$ the inequality $\overrightarrow{QA}\overrightarrow{QB} \leqslant 0$ is satisfied.

It is easy to see that set $D(M)$ is a *cone* with a vertex Q. If $B \in D(M)$, λ is a nonnegative number, and B_1 is a point such that $\overrightarrow{QB_1} = \lambda\overrightarrow{QB}$, then for any point $A \in M$ we have

$$\overrightarrow{QA}\overrightarrow{QB_1} = \overrightarrow{QA}(\lambda\overrightarrow{QB}) = \lambda(\overrightarrow{QA}\overrightarrow{QB}) \leqslant 0, \quad \text{that is,} \quad B_1 \in D(M).$$

It is clear, moreover, that cone $D(M)$ is *convex*. If $B_1 \in D(M)$, $B_2 \in D(M)$, and C is a point on segment $[B_1, B_2]$, that is, if $\overrightarrow{QC} = (1-\lambda)\overrightarrow{QB_1} + \lambda\overrightarrow{QB_2}$, where $0 \leqslant \lambda \leqslant 1$, then for any point $A \in M$ we have

$$\overrightarrow{QA}\overrightarrow{QC} = \overrightarrow{QA}((1-\lambda)\overrightarrow{QB_1} + \lambda\overrightarrow{QB_2}) =$$
$$= (1-\lambda)\overrightarrow{QA}\overrightarrow{QB_1} + \lambda\overrightarrow{QA}\overrightarrow{QB_2} \leqslant 0, \quad \text{that is,} \quad C \in D(M).$$

Finally, note that convex cone $D(M)$ is *closed* (even if the initial cone M was nonclosed). If a point E *does not belong* to cone $D(M)$, then there exists a point $A \in M$ such that $\overrightarrow{QA}\overrightarrow{QE} > 0$. However, then for any point E' sufficiently close to E the inequality $\overrightarrow{QA}\overrightarrow{QE'} > 0$ will also be satisfied, that is, $E' \notin D(M)$. Consequently, the complement (in space E^n) to set $D(M)$ is an open set, so that $D(M)$ is a closed set. Therefore, $D(M)$ is a closed convex cone. It is called a *dual cone* for M (in space E^n).

Let us take as an example a *two-dimensional* space E^2. If cone M coincides with the entire space E^2, considered as a cone with a vertex Q, then cone $D(M)$ will obviously consist just of the single point Q. If M is a half-plane then $D(M)$ will be a ray perpendicular to the boundary line of this half-plane (Figure 107). On the other hand, if M is an angle less than π, it follows that $D(M)$ will be an angle whose sides are perpendicular to the sides of angle M (if angle M has a value of α, then angle $D(M)$ will be equal to $\pi - \alpha$; Figure 108). Finally, if M is a line, then $D(M)$ will be a line perpendicular to it (Figure 109), while if M is a ray, it follows that $D(M)$ is a half-plane (Figure 110).

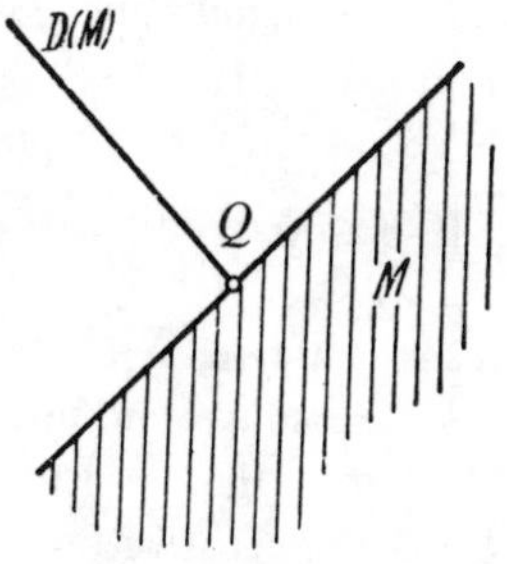

FIGURE 107

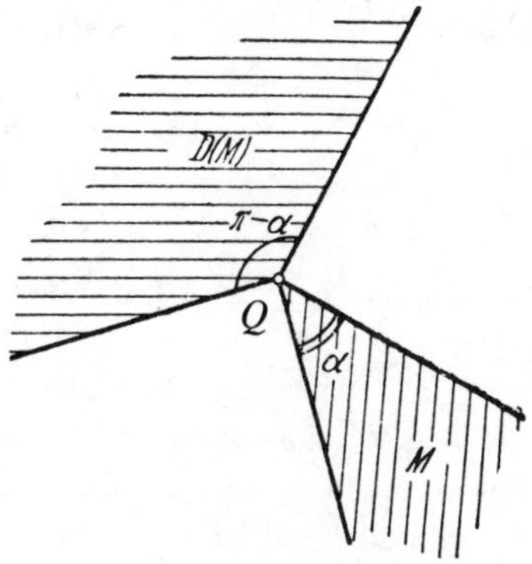

FIGURE 108

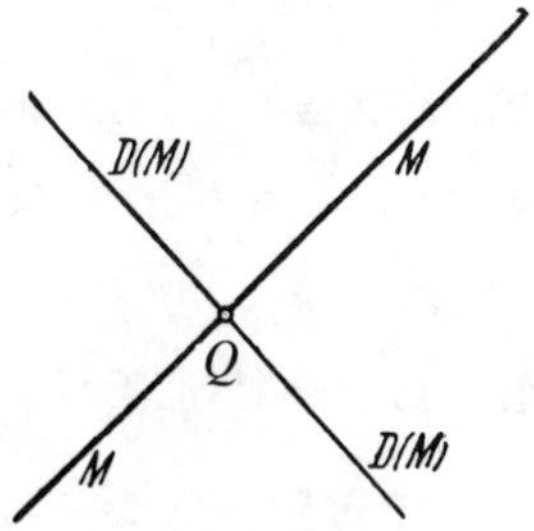

FIGURE 109

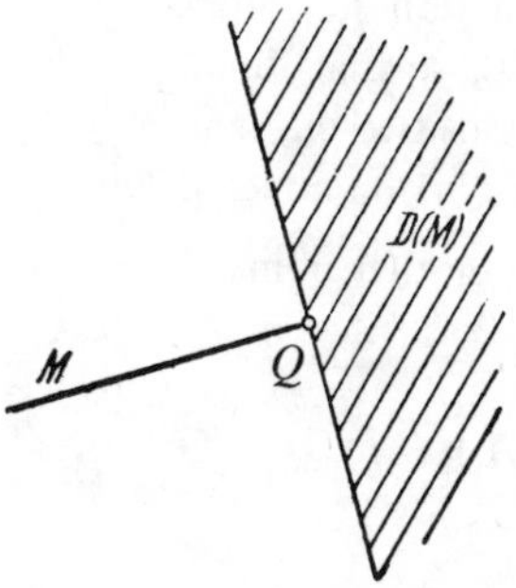

FIGURE 110

THEOREM 31.1. *Let M be a convex cone in E^n with a vertex Q. Point B will belong to dual cone $D(M)$ if, and only if, the affine function f_B, satisfying the condition*

$$f_B(Q) = 0, \quad \operatorname{grad} f_B = \overrightarrow{QB}, \tag{31.1}$$

is nonpositive on M.

Proof. We have (for any point $A \in E^n$):

$$f_B(A) = f_B(A) - f(Q) = \overrightarrow{QA} \operatorname{grad} f_B = \overrightarrow{QA}\overrightarrow{QB}$$

(see (21.5)). Hence it is obvious, if $B \in D(M)$ (that is $\overrightarrow{QA}\overrightarrow{QB} \leqslant 0$ for any point $A \in M$), that $f_B(A) \leqslant 0$ for any point $A \in M$, that is, function f_B is nonpositive on M. Conversely, if function f_B, defined by relations (31.1), is nonpositive on M, that is, if $f_B(A) \leqslant 0$ for any point $A \in M$, then it follows that $\overrightarrow{QA}\overrightarrow{QB} \leqslant 0$ for any point $A \in M$ and thus that $B \in D(M)$.

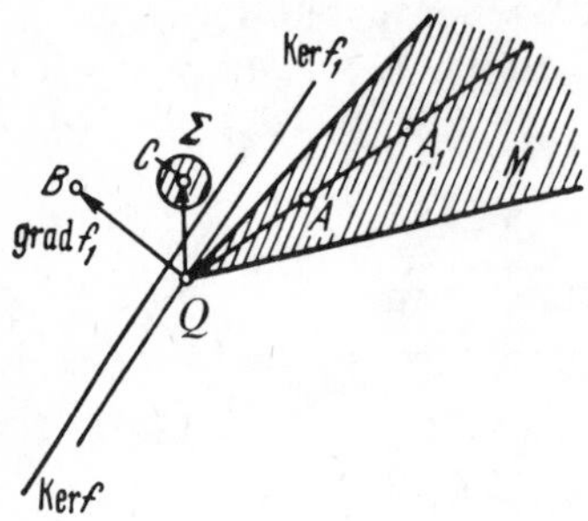

FIGURE 111

THEOREM 31.2. *If M is a closed convex cone with a vertex Q, then $D(D(M)) = M$, that is, the cone which is dual to the dual cone $D(M)$ coincides with the initial cone M.*

Proof. If $A \in M$, then for *any* point $B \in D(M)$ the inequality $\overrightarrow{QA}\overrightarrow{QB} \leqslant 0$ is valid, so that $M \subset \subset D(D(M))$.

Now let $C \notin M$. Then a sphere Σ of some radius r centered on C will not intersect M (recall that M is a closed set). Since Σ and M are convex nonintersecting sets in E^n, it follows that they are separable (Figure 111), that is, there exists an affine nonconstant function f, which is nonpositive on M and nonnegative on sphere Σ. Let us define an affine function f_1, by setting $f_1(D) = f(D) - f(Q)$ for any point $D \in E^n$. Function f_1 is nonconstant. Moreover, since $f(Q) \leqslant 0$, it follows that $f_1(D) \geqslant f(D)$, so that function f_1, like f, is nonnegative on sphere Σ. We will show that it is nonpositive on cone M.

If A is a point on cone M which is different from Q and if λ is a positive number, there will be another point A_1 satisfying the equation $\overrightarrow{QA_1} = \lambda\overrightarrow{QA}$. Then $A_1 \in M$ and $f(A_1) \leqslant 0$, so that in view of (21.5) we have

$$f(A_1) - f(Q) = \overrightarrow{QA_1}\,\mathrm{grad}\, f = \lambda\overrightarrow{QA}\,\mathrm{grad}\, f = \lambda\,(f(A) - f(Q)).$$

Since $f(A_1) \leqslant 0$, it follows that

$$\lambda\,(f(A) - f(Q)) \leqslant -f(Q).$$

This relation will be valid for any $\lambda > 0$, so that $f(A) - f(Q) \leqslant 0$. In other words, $f_1(A) \leqslant 0$, that is, function f_1 is nonpositive on M.

Thus function f_1 is nonconstant, since it is nonnegative on Σ, is nonpositive on M, and satisfies the relation $f_1(Q) = 0$. Because function f_1 is nonnegative on sphere Σ and nonconstant, the center C of sphere Σ cannot belong to the kernel of function f_1 (in view of Theorem 29.3), that is, $f_1(C) \neq 0$; consequently, $f_1(C) > 0$.

If B is a point such that $\overrightarrow{QB} = \mathrm{grad}\, f_1$, then for any point $A \in M$ we have

$$\overrightarrow{QA}\overrightarrow{QB} = \overrightarrow{QA}\,\mathrm{grad}\, f_1 = f_1(A) - f_1(Q) = f_1(A) \leqslant 0,$$

and thus $B \in D(M)$. Moreover,

$$\overrightarrow{QB}\overrightarrow{QC} = \overrightarrow{QC}\,\mathrm{grad}\, f_1 = f_1(C) - f_1(Q) = f_1(C) > 0.$$

Consequently, we have found a point $B \in D(M)$ such that $\overrightarrow{QB}\overrightarrow{QC} > 0$. From the definition of a dual cone, it follows that $C \notin D(D(M))$.

Accordingly, if $C \notin M$, then $C \notin D(D(M))$, that is, $D(D(M)) \subset M$. Together with the previously proven inclusion $M \subset D(D(M))$, this gives the relation $D(D(M)) = M$ as well.

THEOREM 31.3. *If $M_1, \ldots, M_k$ are convex cones with a common vertex Q in E^n and if M is the convex hull of set $M_1 \cup \ldots \cup M_k$, then M is also a convex cone with a vertex Q, the following relation being valid:*

$$D(M) = D(M_1) \cap \ldots \cap D(M_k). \tag{31.2}$$

Proof. The statement that M is a convex cone with a vertex Q is obvious. Let us prove relation (31.2). If $B \in D(M_1) \cap \ldots \cap D(M_k)$, this implies that $B \in D(M_i)$ for any $i = 1, \ldots, k$, so that for any point $A_i \in M_i$ the relation $\overrightarrow{QB}\overrightarrow{QA_i} \leqslant 0$ is valid. Consequently, for any point $A \in M_1 \cup \ldots \cup M_k$ we have $\overrightarrow{QB}\overrightarrow{QA} \leqslant 0$. However, then for any point $A \in M$ we know that $\overrightarrow{QB}\overrightarrow{QA} \leqslant 0$ (see Theorem 25.1) and thus that $B \in D(M)$. Therefore,

$$D(M_1) \cap \ldots \cap D(M_k) \subset D(M). \tag{31.3}$$

Now let us prove the converse inclusion. If $B \in D(M)$, then for any point $A \in M$ the relation $\overrightarrow{QB}\overrightarrow{QA} \leqslant 0$ is valid. In particular, for any point $A_i \in M_i$ we have the inequality $\overrightarrow{QB}\overrightarrow{QA_i} \leqslant 0$ $(i = 1, \ldots, k)$. But this implies that $B \in D(M_i)$. Since this inclusion is valid for any $i = 1, \ldots, k$, it follows that $B \in D(M_1) \cap \ldots \cap D(M_k)$. Therefore,

$$D(M) \subset D(M_1) \cap \ldots \cap D(M_k). \tag{31.4}$$

Inclusions (31.3) and (31.4) also prove equation (31.2).

COROLLARY 31.4. *If $M_1, \ldots, M_k$ are closed convex cones with a common vertex Q in E^n, then $D(M_1 \cap \ldots \cap M_k)$ is the convex hull of cones $D(M_1), \ldots, D(M_k)$.*

Now let $N_1 = D(M_1), \ldots, N_k = D(M_k)$ and let N be the convex hull of cones $N_1, \ldots, N_k$. Taking into account Theorems 31.3 and 31.2, we get

$$\begin{aligned} D(N) &= D(N_1) \cap \ldots \cap D(N_k) = \\ &= D(D(M_1)) \cap \ldots \cap D(D(M_k)) = M_1 \cap \ldots \cap M_k. \end{aligned}$$

Consequently,

$$N = D(D(N)) = D(M_1 \cap \ldots \cap M_k).$$

Theorem 31.3 and Corollary 31.4 indicate that, if we restrict ourselves just to *closed* convex cones, then during the transition to dual cones the operation of intersection and the operation of taking the convex hull of the union change places, that is, these operations are *dual* relative to one another.

32. Separability of system of convex cones. We will say that a system of closed convex cones $K_1, \ldots, K_s$ with a common vertex Q in E^n possesses *separability* if it can be so divided into two subsystems (each of which contains at least one cone) that the intersection of the cones of the first subsystem is separable from the intersection of the cones of the second subsystem.

Before formulating the conditions which the system of cones has to satisfy in order to be separable, let us prove the following auxiliary proposition.

LEMMA 32.1. *If $K_1, \ldots, K_l$ are convex cones in space E^n with a common vertex Q, then $A \in E^n$ will belong to the convex hull K of cones $K_1, \ldots, K_l$ if, and only if, there exist points $A_1 \in K_1, \ldots, A_l \in K_l$ such that*

$$\overrightarrow{QA} = \overrightarrow{QA_1} + \ldots + \overrightarrow{QA_l}. \tag{32.1}$$

Proof. Let point A be defined by relation (32.1), where $A_1, \ldots, A_l$ are points belonging, respectively, to cones $K_1, \ldots, K_l$. If $B_1, \ldots, B_l$ are points such that

$$\overrightarrow{QA_1} = \frac{1}{l}\overrightarrow{QB_1}, \ \ldots, \ \overrightarrow{QA_l} = \frac{1}{l}\overrightarrow{QB_l}, \tag{32.2}$$

then points $B_1, \ldots, B_l$ belong, respectively, to convex cones $K_1, \ldots, K_l$ (because of the definition of a cone). Taking (32.2) into account, we can rewrite (32.1) as

$$\overrightarrow{QA} = \frac{1}{l}\overrightarrow{QB_1} + \ldots + \frac{1}{l}\overrightarrow{QB_l},$$

and thus according to Theorem 25.1 point A belongs to the convex hull of the set $K_1 \cup \ldots \cup K_l$, that is, $A \in K$.

Conversely, let $A \in K$. Then, according to Theorem 25.1,

$$\overrightarrow{QA} = \lambda^1\overrightarrow{QC_1} + \ldots + \lambda^s\overrightarrow{QC_s}, \tag{32.3}$$

where $\lambda^1 \geqslant 0, \ldots, \lambda^s \geqslant 0$, $\lambda^1 + \ldots + \lambda^s = 1$, and all the points $C_1, \ldots, C_s$ belong to set $K_1 \cup \ldots \cup K_l$. Assume that the numbering of points $C_1, \ldots, C_s$ is such that the first q_1 of these points belong to set K_1, the next q_2 points belong to set $K_2, \ldots,$ and, finally, the last q_l points belong to set K_l (so that $q_1 + q_2 + \ldots + q_l = s$):

$$C_i \in K_j \quad \text{for} \quad q_1 + \ldots + q_j < i \leqslant q_1 + \ldots + q_j + q_{j+1}.$$

In addition, assume each of the whole numbers $q_1, \ldots, q_l$ to be *positive;* for instance, if $q_1 = 0$, then on the right-hand side of (32.3) we can add the term $\lambda\overrightarrow{QC}$, where $C \in K_1$ and $\lambda = 0$.

If $A_1, \ldots, A_l$ are the points defined by the relations

$$\overrightarrow{QA_j} = \sum_{i=q_1+\ldots+q_{j-1}+1}^{q_1+\ldots+q_j} \lambda^i \overrightarrow{QC_i}, \qquad j=1, \ldots, l, \tag{32.4}$$

then relation (32.3) is written in form (32.1), while it follows directly from (32.4), in view of Theorem 28.7, that $A_1 \in K_1, \ldots, A_l \in K_l$.

THEOREM 32.2. *In order for a system of closed convex cones $K_1, \ldots, K_s$ with a common vertex Q in E^n to possess separability, it is necessary and sufficient that there exist points $A_1 \in D(K_1), \ldots, A_s \in D(K_s)$, at least one of which is different from Q, such that*

$$\overrightarrow{QA_1} + \ldots + \overrightarrow{QA_s} = \mathbf{0}. \tag{32.5}$$

Proof. Assume that the system of closed convex cones $K_1, \ldots, K_s$ with a common vertex Q possesses separability in E^n. Without any loss of generality, we can assume (after changing, if necessary, the numbering of the cones) that cones $K_1 \cap \ldots \cap K_l$ and $K_{l+1} \cap \ldots \cap K_s$ are separable, where l is some natural number less than s. In other words,

$$K_1 \cap \ldots \cap K_l \subset P_1, \quad K_{l+1} \cap \ldots \cap K_s \subset P_2,$$

where P_1 and P_2 are two closed half-spaces defined in E^n by some hyperplane Γ (passing through the common vertex Q of cones $K_1, \ldots, K_s$). Let $\boldsymbol{n} \neq \mathbf{0}$ be a normal vector to hyperplane Γ directed toward half-space P_2, so that $\boldsymbol{n}\overrightarrow{QA} \leqslant 0$ for any point $A \in P_1$ and $\boldsymbol{n}\overrightarrow{QA} \geqslant 0$ for any point $A \in P_2$. If, furthermore, B is a point such that $\boldsymbol{n} = \overrightarrow{QB}$, then $\boldsymbol{n}\overrightarrow{QA} \leqslant 0$ for any point $A \in K_1 \cap \ldots \cap K_l$ (since $K_1 \cap \ldots \cap K_l \subset P_1$), so that $B \in D(K_1 \cap \ldots \cap K_l)$.

From Corollary 31.4 it now follows that point B belongs to the convex hull of cones $D(K_1), \ldots, D(K_l)$ and thus that (in view of Lemma 32.1)

$$\boldsymbol{n} = \overrightarrow{QB} = \overrightarrow{QA_1} + \ldots + \overrightarrow{QA_l}, \tag{32.6}$$

where $A_i \in D(K_i)$, $i=1, \ldots, l$. Similarly, point B', defined by the relation $\overrightarrow{QB'} = -\boldsymbol{n}$, belongs to the cone $D(K_{l+1} \cap \ldots \cap K_s)$, so that

$$-\boldsymbol{n} = \overrightarrow{QB'} = \overrightarrow{QA_{l+1}} + \ldots + \overrightarrow{QA_s}, \tag{32.7}$$

where $A_i \in D(K_i)$, $i = l+1, \ldots, s$. Addition of relations (32.6) and (32.7) gives equation (32.5). Note that some of vectors $\overrightarrow{QA_1}, \ldots, \overrightarrow{QA_s}$ must be different from zero, since $\boldsymbol{n} \neq \mathbf{0}$, so that at least one of points $A_1, \ldots, A_s$ is different from Q.

Conversely, let us assume that there exist points $A_1 \in D(K_1), \ldots, A_s \in D(K_s)$, satisfying relation (32.5), at least one of points $A_1, \ldots, A_s$ being different from Q.

To be definite, let $A_1 \neq Q$. In addition, let Γ be a hyperplane which passes through point Q and is orthogonal to the vector $\boldsymbol{n}=\overrightarrow{QA_1} \neq \mathbf{0}$. Since $A_1 \in D(K_1)$, it follows that cone K_1 lies in the half-space P_1 consisting of all points $A \in E^n$ for which $\boldsymbol{n}\overrightarrow{QA} \leqslant 0$. Moreover, let A' be the point defined by the relation $\overrightarrow{QA'}=-\boldsymbol{n}=$ $=-\overrightarrow{QA_1}$, so that point A' also lies in half-space P_1. Since

$$\overrightarrow{QA'}=-\overrightarrow{QA_1}=\overrightarrow{QA_2}+\ldots+\overrightarrow{QA_s},$$

it follows from Lemma 32.1 that point A' belongs to the convex hull of cones $D(K_2), \ldots, D(K_s)$, that is (in view of Corollary 31.4), point A' belongs to the cone $D(K_2 \cap \ldots \cap K_s)$. In other words, for any point $A \in K_2 \cap \ldots \cap K_s$ the relation $\overrightarrow{QA'}\,\overrightarrow{QA} \leqslant 0$ holds true, that is, $\boldsymbol{n}\overrightarrow{QA} \geqslant 0$. Consequently, cone $K_2 \cap \ldots \cap K_s$ is contained in the half-space P_2 consisting of all points $A \in E^n$ for which $\boldsymbol{n}\overrightarrow{QA} \geqslant 0$.

Thus, $K_1 \subset P_1$, $K_2 \cap \ldots \cap K_s \subset P_2$, that is, hyperplane Γ separates cones K_1 and $K_2 \cap \ldots \cap K_s$, so that the system of cones $K_1, \ldots, K_s$ possesses separability.

COROLLARY 32.3. *If none of cones $K_1, \ldots, K_s$ is separable from the intersection of the others, then the system of cones $K_1, \ldots, K_s$ does not possess separability.*

If the system of cones $K_1, \ldots, K_s$ (which, as previously, are assumed to be closed and convex, and to have a common vertex Q) possesses separability, then according to Theorem 32.2 there exist points $A_1 \in D(K_1), \ldots, A_s \in D(K_s)$ which satisfy relation (32.5) and among which there are points different from Q. However, then the second part of the proof of Theorem 32.2 shows that one of cones $K_1, \ldots, K_s$ must be separable from the intersection of the others.

COROLLARY 32.4. *In order for a system of closed convex cones $K_1, \ldots, K_s$ with a common vertex Q in E^n to possess separability, it is necessary and sufficient that there exist affine functions $f_1, \ldots, f_s$ in E^n, going to zero at point Q but not all identically zero, such that function f_i is nonpositive on cone K_i $(i=1, \ldots, s)$ and*

$$f_1+\ldots+f_s \equiv 0.$$

If this system of cones possesses separability, then in view of Theorem 32.2 there will exist points $A_1 \in D(K_1), \ldots, A_s \in D(K_s)$, at least one of which is different from Q, such that relation (32.5) is satisfied. Let $f_1, \ldots, f_s$ be affine functions in E^n which go to zero at point Q and which have vectors $\overrightarrow{QA_1}, \ldots, \overrightarrow{QA_s}$ as their gradients. Relation (32.5) then implies that $f_1+\ldots+f_s \equiv 0$. Moreover, since $A_i \in D(K_i)$, it follows that $f_i(A)=\overrightarrow{QA}\,\overrightarrow{QA_i} \leqslant 0$ for $A \in K_i$.

Conversely, if there exist functions $f_1, \ldots, f_s$ satisfying the conditions specified in Corollary 32.4, then, after defining points $A_1, \ldots, A_s$ by means of relations $\overrightarrow{QA_i}=\operatorname{grad} f_i$, $i=1, \ldots, s$, we find after carrying out the operations in inverse order that points $A_1, \ldots, A_s$ satisfy the conditions of Theorem 32.2, that is, system of cones $K_1, \ldots, K_s$ possesses separability.

THEOREM 32.5. *In order for a system of closed convex cones* $K_1, \ldots, K_s$ *with a common vertex* Q *not to possess separability, it is necessary and sufficient that the following two conditions be satisfied:*

1) $(\operatorname{relint} K_1) \cap \ldots \cap (\operatorname{relint} K_s) \neq \varnothing$;

2) *there exist subspaces* $L_1, \ldots, L_s$ *into the direct sum of which space* E^n *decomposes (where it may be that* $\dim L_i = 0$ *for some* i*), such that for any* $i \neq j$ $(i, j = 1, \ldots, s)$ L_i *is contained in a subspace parallel to the carrier plane of cone* K_j.

Proof. Assume that conditions 1 and 2 are satisfied, but at the same time, contrary to the proposition, that system of cones $K_1, \ldots, K_s$ possesses separability. Then, in view of Corollary 32.3, some one of these cones (K_1, say) will be separable from the intersection of the others, that is, there will exist a hyperplane Γ passing through Q such that $K_1 \subset P_1$, $K_2 \cap \ldots \cap K_s \subset P_2$, where P_1 and P_2 are closed half-spaces defined by hyperplane Γ.

If A is some point of the set $\bigcap_{i=1}^{s} (\operatorname{relint} K_i)$ (such a point must exist, in view of condition 1), then $A \in K_1 \subset P_1$, $A \in K_2 \cap \ldots \cap K_s \subset P_2$, so that $A \in P_1 \cap P_2 = \Gamma$. For $i \neq 1$ a plane M_i, parallel to L_i and passing through A, is contained in the carrier plane of cone   and, since A is an *interior* point of cone K_1 relative to its carrier plane, it follows that all points of plane M_i which are sufficiently close to A will belong to cone K_1, and hence to half-space P_1. This implies that $M_i \subset \Gamma$ $(i = 2, \ldots, s)$. Moreover, plane M_1, which is parallel to L_1 and passes through A, is contained in the carrier plane of each of cones $K_2, \ldots, K_s$, and all the points of plane M_1 sufficiently close to A belong to each of the cones $K_2, \ldots, K_s$, and thus to their intersection as well. Consequently, $M_1 \subset \Gamma$.

Hence we see that *all* the planes $M_1, \ldots, M_s$ are contained in Γ. Accordingly, all the subspaces $L_1, \ldots, L_s$ (and thus the space $E^n = L_1 \oplus \ldots \oplus L_s$ as well) are contained in a subspace L parallel to hyperplane Γ, which is impossible, since L is a hyperplane and thus cannot embrace all of space E^n. This contradiction shows that system of cones $K_1, \ldots, K_s$ does not possess separability.

Conversely, assume that system of cones $K_1, \ldots, K_s$ does not possess separability. Let us show that conditions 1 and 2 are satisfied. If we first assume, on the other hand, that condition 1 is not satisfied, and if we choose a natural number $l < s$ such that

$$(\operatorname{relint} K_1) \cap \ldots \cap (\operatorname{relint} K_l) \neq \varnothing, \tag{32.8}$$

$$(\operatorname{relint} K_1) \cap \ldots \cap (\operatorname{relint} K_l) \cap (\operatorname{relint} K_{l+1}) = \varnothing, \tag{32.9}$$

then relation (32.8) implies (in view of Theorem 26.6) that

$$(\operatorname{relint} K_1) \cap \ldots \cap (\operatorname{relint} K_l) = \operatorname{relint} (K_1 \cap \ldots \cap K_l),$$

and thus relation (32.9) becomes

$$(\operatorname{relint} (K_1 \cap \ldots \cap K_l)) \cap (\operatorname{relint} K_{l+1}) = \varnothing.$$

From this it follows, in view of Theorem 30.4, that cones $K_1 \cap \ldots \cap K_l$ and K_{l+1} are separable, and thus all the more that cones $K_1 \cap \ldots \cap K_l$ and $K_{l+1} \cap \ldots \cap K_s$ are separable. But this contradicts the assumption. Consequently, condition 1 is satisfied.

Finally, let us show that condition 2 is satisfied. To do this, we carry out induction on the number s of cones being considered. First let $s=2$. Since cones K_1 and K_2 were assumed to be nonseparable, we know from Theorem 30.4 that these cones do not lie in a single hyperplane, that is, subspaces A and B, which are parallel to the carrier planes of cones K_1 and K_2, possess the property that $A+B=E^n$. Theorem 14.13 states that there exists a subspace $D \subset B$ such that $A \oplus D = E^n$. Therefore, subspaces $L_1 = D$ and $L_2 = A$ are the ones sought, and the case $s=2$ has been exhaustively studied.

Now assume the validity of condition 2 to have been demonstrated for less than s cones, and consider s cones $K_1, \ldots, K_s$ (not possessing separability). Without any loss of generality, we can assume that point Q is the coordinate origin of space E^n, so that any plane passing through Q is a subspace of the corresponding vector space E^n. Let Π_s be the carrier plane of cone K_s and let Π^* be the carrier plane of cone $K_1 \cap \ldots \cap K_{s-1}$. Since cones $K_1 \cap \ldots \cap K_{s-1}$ and K_s are nonseparable, it follows from Theorem 30.4 that Π_s and Π^* do not lie in a single hyperplane, that is, $\Pi_s + \Pi^* = E^n$. Consequently, according to Theorem 14.13, there exists a subspace $L_s \subset \Pi^*$ such that $E^n = \Pi_s \oplus L_s$.

Now consider the cones

$$K_1 \cap K_s, \ \ldots, \ K_{s-1} \cap K_s \tag{32.10}$$

lying in subspace Π_s. We will show that this system of closed convex cones (with a common vertex Q) does not possess separability in space Π_s. Let us assume the opposite (see Corollary 32.3), namely that one of cones (32.10) (the first, say) is separable in Π_s from the intersection of the others, that is, that cone $K_1 \cap K_s$ is separable in Π_s from the cone $K_2 \cap \ldots \cap K_{s-1} \cap K_s$. In view of the already established condition 1, we have

$$(\text{relint}\, K_1) \cap (\text{relint}\, K_2) \cap \ldots \cap (\text{relint}\, K_s) \neq \varnothing,$$

and thus (from Theorem 26.6)

$$\text{relint}\,(K_1 \cap K_s) = (\text{relint}\, K_1) \cap (\text{relint}\, K_s),$$
$$\text{relint}\,(K_2 \cap \ldots \cap K_s) = (\text{relint}\, K_2) \cap \ldots \cap (\text{relint}\, K_s),$$

so that

$$(\text{relint}\,(K_1 \cap K_s)) \cap (\text{relint}\,(K_2 \cap \ldots \cap K_s)) \neq \varnothing.$$

Consequently, separability of cones $K_1 \cap K_s$ and $K_2 \cap \ldots \cap K_s$ signifies (according to Theorem 30.4) that the carrier planes of cones $K_1 \cap K_s$ and $K_2 \cap \ldots \cap K_s$

lie in one plane $\Gamma \subset \Pi_s$, having a dimension $\dim \Pi_s - 1$. From Theorem 26.6 we know that the carrier plane of cone $K_1 \cap K_s$ coincides with $\Pi_1 \cap \Pi_s$, where Π_1 is the carrier plane of cone K_1. Thus $\Pi_1 \cap \Pi_s \subset \Gamma$, that is, $\Pi_1 \cap \Pi_s = \Gamma \cap \Pi_1$. If Π° is the plane of least dimension containing Γ and Π_1, that is, $\Pi^\circ = \Gamma + \Pi_1$, then

$$\dim(\Pi_1 \cap \Gamma) = \dim(\Pi_1 \cap \Pi_s) \geqslant \dim \Pi_1 + \dim \Pi_s - n,$$
$$\dim \Pi^\circ = \dim \Pi_1 + \dim \Gamma - \dim(\Pi_1 \cap \Gamma) \leqslant$$
$$\leqslant \dim \Pi_1 + \dim \Gamma - (\dim \Pi_1 + \dim \Pi_s - n) = n - 1,$$

that is, Γ and Π_1 are contained in a single plane of dimension not greater than $n-1$. Since $K_1 \subset \Pi_1$, $K_2 \cap \ldots \cap K_s \subset \Gamma$, it follows that cones K_1 and $K_2 \cap \ldots \cap K_s$ are separable, which, however, contradicts the assumption.

Thus system of cones (32.10) does not possess separability in Π_s. Since the number of these cones is less than s, it follows from the assumption of induction that there exist subspaces $L_1, \ldots, L_{s-1}$, into the direct sum of which subspace Π_s decomposes, such that for any $i \neq j$ $(i, j = 1, \ldots, s-1)$ L_i is contained in the carrier plane of cone $K_j \cap K_s$. It can easily be shown that these subspaces $L_1, \ldots, L_{s-1}$, together with the previously constructed subspace L_s, satisfy condition 2, that is, they are the subspaces sought (recall that space E^n decomposes into the direct sum of subspaces L_s and Π_s, so that according to Theorem 14.12 space E^n decomposes into the direct sum of subspaces $L_1, \ldots, L_{s-1}, L_s$).

Now, by comparing the proven theorem with Corollary 32.4, we obtain the following statement.

THEOREM 32.6. *Let closed convex cones $K_1, \ldots, K_s$ with a common vertex Q be defined and let these cones satisfy condition 2 of Theorem 32.5. In order for the intersection*

$$(\operatorname{relint} K_1) \cap \ldots \cap (\operatorname{relint} K_s)$$

to be empty, it is necessary and sufficient that there exist affine functions $f_1, \ldots, f_s$ in E^n which go to zero at point Q and which are not all identically zero, such that function f_i is nonpositive on cone K_i $(i = 1, \ldots, s)$ and also such that

$$f_1 + \ldots + f_s \equiv 0.$$

In this theorem, as in Corollary 32.4, the condition of nonpositivity of function f_i on cone K_i, $i = 1, \ldots, s$, can be replaced by the condition of nonnegativity: it is enough just to change the signs of all the functions $f_1, \ldots, f_s$.

Theorems 32.5 and 32.6 become particularly simplified in the case where all the cones $K_1, \ldots, K_s$, except possibly one, are solid. Then condition 2 of Theorem 32.5 is obviously satisfied automatically. Assume that cones $K_2, \ldots, K_s$ are solid, while K_1 is an arbitrary cone. In order to verify that condition 2 is indeed satisfied, it is sufficient to take a subspace parallel to the carrier plane of cone K_1 as L_2, to take the direct complement of this subspace as L_1, and to take the trivial (zero-dimensional) subspaces as $L_3, \ldots, L_s$. Note that, cones $K_2, \ldots, K_s$ being solid, condition 1 of Theorem 32.5 is equivalent to the condition

$$K_1 \cap (\operatorname{int} K_2) \cap \ldots \cap (\operatorname{int} K_s) \neq \varnothing. \tag{32.11}$$

This leads us to the following two propositions.

COROLLARY 32.7. *Let $K_1, K_2, \ldots, K_s$ be closed convex cones with a common vertex Q, the cones $K_2, \ldots, K_s$ being solid. In order for system of cones $K_1, \ldots, K_s$ not to possess separability, it is necessary and sufficient for condition* (32.11) *to be satisfied.*

COROLLARY 32.8. *(Dubovitskii-Milyutin theorem*). Let $K_2, \ldots, K_s$ be closed convex solid cones in E^n with a common vertex Q , and let $K_1 \subset E^n$ be a plane of some given dimension passing through point Q. In order for the intersection*

$$K_1 \cap (\operatorname{int} K_2) \cap \ldots \cap (\operatorname{int} K_s) \tag{32.12}$$

to be empty, it is necessary and sufficient that there exist in E^n affine functions $f_1, \ldots, f_s$, which go to zero at point Q but which are not all identically zero, such that function f_1 is identically zero on plane K_1, function f_i is nonpositive on cone K_i, $i = 2, \ldots, s$, and the following relation is valid:

$$f_1 + \ldots + f_s \equiv 0$$

(note that nonpositivity of function f_1 on K_1 indicates that this function is identically zero on K_1.)

NOTE 32.9. The Dubovitskii-Milyutin theorem was obtained here as a particular case of more general results. In Chapter IV these general results will serve as a basis for deriving the theorems of mathematical programming. However, most of these theorems can be derived only on the basis of the Dubovitskii-Milyutin theorem, which explains the great popularity of the latter. Consequently, although this theorem will not be used by us in the following as a study tool, we will nevertheless present here a direct proof of the Dubovitskii-Milyutin theorem, one which is not based on the more complex Theorem 32.5 and the other results of this subsection.

First of all we establish necessity, that is, we demonstrate that, if intersection (31.12) is empty, there must exist the required functions $f_1, \ldots, f_s$. We carry out the proof by induction on s. For $s = 2$ intersection (32.12) has the form $K_1 \cap (\operatorname{int} K_2)$ and, since it is assumed to be empty, plane K_1 does not cross the interior of cone K_2. Consequently, according to Corollary 30.6, plane K_1 and cone K_2 are separable, that is, there exists a nonconstant affine function f in E^n, going to zero at point Q, which is nonnegative on K_2 and nonpositive on K_1. The fact that f is nonpositive on K_1 implies that it is identically zero on K_1. Consequently, by setting $f_1 = f$ and $f_2 = -f$, we get the required functions f_1 and f_2. This proves the necessity of the formulated condition for $s = 2$.

Now assume that, for fewer than s cones, necessity has been established, and consider s solid cones $K_2, \ldots, K_s$ for which intersection (32.12) is empty. If the intersection

$$K_1 \cap (\operatorname{int} K_2) \cap \ldots \cap (\operatorname{int} K_{s-1}) \tag{32.13}$$

* Dubovitskii, A.Ya., and A.A. Milyutin. *Zadachi na ekstremum pri nalichii ogranichenii (Extremum Problems in the Presence of Constraints).* – Zhurn. Vychis. Mat. Mat. Fiz., 5, No. 3. 1965.

is empty, then the assumption of induction implies that there exist functions $f_1, \ldots, f_{s-1}$ possessing the required properties, so that by setting $f_s \equiv 0$ we obtain the sought functions $f_1, \ldots, f_s$.

It only remains to consider the case when intersection (32.13) is nonempty. Then, according to Theorem 26.6, the set $(\operatorname{int} K_2) \cap \ldots \cap (\operatorname{int} K_{s-1})$ represents the *interior* of cone $K_2 \cap \ldots \cap K_{s-1}$, while set (32.13) represents the interior of cone $K_1 \cap K_2 \cap \ldots \cap K_{s-1}$ (relative to its carrier plane). Consequently, cones $K_1 \cap K_2 \cap \ldots \cap K_{s-1}$ and K_s do not have any common interior points (since intersection (32.12) is empty), so that in view of Theorem 30.4 these cones are separable. In other words, there exists a nonconstant affine function f, going to zero at point Q, which is nonnegative on K_s and nonpositive on the cone $K_1 \cap K_2 \cap \ldots \cap K_{s-1}$.

If point B is defined by the relation $\overrightarrow{QB} = \operatorname{grad} f$, then according to Theorem 31.1 point B belongs to the cone $D(K_1 \cap K_2 \cap \ldots \cap K_{s-1})$, that is, in view of Corollary 31.4, point B belongs to the convex hull of the cones $D(K_1), D(K_2), \ldots, D(K_{s-1})$. Hence it follows (from Lemma 32.1) that there must exist points

$$C_1 \in D(K_1), \quad C_2 \in D(K_2), \ldots, C_{s-1} \in D(K_{s-1}),$$

satisfying the relation

$$\overrightarrow{QB} = \overrightarrow{QC_1} + \ldots + \overrightarrow{QC_{s-1}}.$$

Next we define affine functions $f_1, \ldots, f_{s-1}$ in E^n, by setting

$$f_i(A) = \overrightarrow{QA}\,\overrightarrow{QC_i}, \qquad i = 1, \ldots, s-1. \tag{32.14}$$

Obviously, all these functions go to zero at point Q and satisfy the relation

$$f_1(A) + \ldots + f_{s-1}(A) = \overrightarrow{QA}\,(\overrightarrow{QC_1} + \ldots + \overrightarrow{QC_{s-1}}) =$$
$$= \overrightarrow{QA}\,\overrightarrow{QB} = \overrightarrow{QA} \operatorname{grad} f = f(A) - f(Q) = f(A),$$

that is,

$$f_1 + \ldots + f_{s-1} \equiv f. \tag{32.15}$$

Moreover, since $C_i \in D(K_i)$ for $i = 2, \ldots, s-1$, it follows that for any point A of cone K_i the relation $\overrightarrow{QA}\,\overrightarrow{QC_i} \leqslant 0$ is satisfied, so that in view of (32.14) function f_i is nonpositive on cone K_i. Similarly, function f_1 is nonpositive on plane K_1 so that it is identically zero on K_1. Therefore, by setting $f_s \equiv -f$, we get (in view of (32.15)) the required functions $f_1, \ldots, f_{s-1}, f_s$ (note that not all of the functions are identically zero, since, for example, function $f_s \equiv -f$ is nonconstant). The induction carried out proves the necessity of the condition formulated in the theorem.

Now let us prove sufficiency, that is, let us show that, when functions $f_1, \ldots, f_s$ in the theorem exist, intersection (32.12) is empty. First assume the opposite, namely that intersection (32.12) is nonempty and that A is some point belonging to this intersection. Since $A \in K_1$, it follows that $f_1(A) = 0$. Moreover, since $A \in \operatorname{int} K_i \subset K_i$, we know that $f_i(A) \leqslant 0$ for any $i = 2, \ldots, s$. Thus the relation $f_1 + \ldots + f_s \equiv 0$ implies that

$$f_2(A) = \ldots = f_s(A) = 0.$$

Since $f_i \leqslant 0$ on set K_i, this function f_i, considered on K_i, reaches a maximum at an *interior* point $A \in \operatorname{int} K_i$ of field K_i. Thus according to Corollary 29.3 function f_i is constant, that is $f_i \equiv 0$, $i = 2, \ldots, s$. But then from the equation $f_1 + \ldots + f_s \equiv 0$ it follows that $f_1 \equiv 0$ as well, indicating that all the functions $f_1, \ldots, f_s$ are identically zero. However, this contradicts the conditions imposed upon these functions in the theorem. This contradiction shows that intersection (32.12) is empty.

Chapter IV. EXTREMA OF FUNCTIONS

§9. Existence Theorems

33. Tangent mapping. Let M be some set of Euclidean space E^n and let Q be a fixed point of M. Moreover, let $\boldsymbol{g}$: $M \to E^m$ be a vector function defined on set M, that is, a mapping which makes some vector $\boldsymbol{g}(A)$ of Euclidean space E^m correspond to each point $A \in M$. We will write $\boldsymbol{g}(A) = o_Q(A)$ if

$$\boldsymbol{g}(Q) = \boldsymbol{0} \quad \text{and} \quad \lim_{\substack{A \in M \\ A \to Q}} \frac{|\boldsymbol{g}(A)|}{|\overrightarrow{QA}|} = 0. \tag{33.1}$$

In other words, the notation $\boldsymbol{g}(A) = o_Q(A)$ means that for any $\varepsilon > 0$ there exists a number $\delta > 0$ such that for $A \in M$ and $|\overrightarrow{QA}| < \delta$ the relation $|\boldsymbol{g}(A)| \leqslant \varepsilon |\overrightarrow{QA}|$ is valid. If E^n is an arithmetical space and $Q = (0, 0, \ldots, 0)$, then instead of $o_Q(A)$ it is customary to use the notation $o(A)$.

We will use the symbol $o_Q(A)$ as a *typical notation: different* vector functions will be denoted by one and the same symbol $o_Q(A)$, if, of course, they possess the above-indicated property. With this stipulation, we can write

$$o_Q(A) + o_Q(A) = o_Q(A),$$

that is, if $\boldsymbol{f}$: $M \to E^m$ and $\boldsymbol{g}$: $M \to E^m$ are two vector functions each of which has the form $o_Q(A)$, then it follows that vector function $\boldsymbol{f} + \boldsymbol{g}$ (defined by the equation $(\boldsymbol{f} + \boldsymbol{g})(A) = \boldsymbol{f}(A) + \boldsymbol{g}(A)$) is also $o_Q(A)$. Similarly, if $\lambda(A)$ is a real continuous function defined on set M (or a function *bounded* in some neighborhood of point Q), then $\lambda(A) \cdot o_Q(A) = o_Q(A)$.

Note that, although the symbol $o_Q(A)$ was defined in a *Euclidean* space (since in formula (33.1) the *lengths* of the vectors are used), this very definition is *affine* in nature.* More precisely, if in a given affine space the scalar product is introduced using two different methods, turning it into two (different) Euclidean spaces E_1^n and E_2^n, then a vector function $\boldsymbol{g}(A)$ having the form $o_Q(A)$ in one of these spaces also has the form $o_Q(A)$ in the other space.

The identity mapping $E_1^n \to E_2^n$ (that is, the transformation carrying a point A of a given affine space into the very same point) is obviously affine. Consequently, in

* Concerning affine spaces and affine mappings, see pp. 129 and 130.

view of Theorem 21.7, $|\overrightarrow{QA}|_1 \leqslant M|\overrightarrow{QA}|_2$, where $|\overrightarrow{QA}|_1$ denotes the length of vector $\overrightarrow{QA}$ in the metrics of space E_1^n, while $|\overrightarrow{QA}|_2$ denotes the vector length in E_2^n. However, then

$$\frac{|g(A)|}{|\overrightarrow{QA}|_2} \leqslant M \frac{|g(A)|}{|\overrightarrow{QA}|_1},$$

from which it also follows that validity of relation (33.1) in the metrics of space E_1^n implies validity of this relation in the metrics of space E_2^n.

Similarly, it can be established that the validity of (33.1) does not depend on the choice of the metrics in E^m. In other words, if E^n and E^m are affine spaces, then for any vector function $\boldsymbol{g}$: $M \to E^m$ (where $M \subset E^n$) it is reasonable to state whether or not $\boldsymbol{g}(A)$ has a value $o_Q(A)$, where Q is a specified point of set M (that is, whether relation (33.1) is valid for a *certain*, and thus for any, introduction of Euclidean metrics in spaces E^n and E^m).

In accordance with the foregoing, all the cases considered in this subsection will be *affine* in nature.

DEFINITION 33.1. Let M be some set of Euclidean space E^n and let Q be a fixed point of set M. Moreover, let φ: $M \to E^m$ be some mapping of set M into Euclidean space E^m, and, finally, let f be an affine mapping of space E^n into E^m. We will say that mapping φ has f as its *tangent affine mapping at point* Q if (for $A \in M$) the following relation is valid:

$$\varphi(A) = f(A) + o_Q(A). \tag{33.2}$$

Let us consider a particular case of this definition, one which will be important in the ensuing discussion. Let M be some convex cone in space E^n with a vertex Q. Moreover, let φ be a mapping (into space E^m) defined over the entire cone M or only close to the vertex of this cone (that is, at points A satisfying the conditions $A \in M$ and $|\overrightarrow{QA}| < h$, where h is some specified positive number). In this case we can refer to a tangent affine mapping f for mapping φ at point Q (provided relation (33.2) is satisfied).

In the general case mapping φ: $M \to E^m$ may not have a tangent affine mapping at point $Q \in M$. Alternatively, mapping φ may not have just one tangent affine mapping. However, as the next theorem indicates, if set M is "sufficiently solid" close to point Q, then the tangent affine mapping, if it exists, will be unique.

THEOREM 33.2. *If an n-dimensional simplex with one of its vertices at point Q is contained in a set $M \subset E^n$, then for mapping φ: $M \to E^m$ a tangent affine mapping at point Q (if it exists) will be unique.*

Proof. Let us assume that φ has two tangent affine mappings f_1 and f_2 at point Q. In other words,

$$\varphi(A) = f_1(A) + o_Q(A), \quad \varphi(A) = f_2(A) + o_Q(A), \qquad A \in M,$$

where f_1 and f_2 are *affine* mappings of space E^n into E^m. Then, for $A \in M$,

$$f_1(A) = f_2(A) + o_Q(A). \tag{33.3}$$

If $[Q, A_1, \ldots, A_n]$ is a simplex of dimension n contained in M, then relation (33.3) will be valid, in particular, in this simplex.

Now let $B_i = Q + \lambda \overrightarrow{QA_i}$, $i = 1, \ldots, n$. For sufficiently small $\lambda > 0$, point B_i will lie arbitrarily close to Q and will belong to the simplex being considered. Consequently, by taking an arbitrary $\varepsilon > 0$, we can so choose $\lambda > 0$ that (in view of (33.3)) the following relation is valid:

$$|f_1(B_i) - f_2(B_i)| < \varepsilon |\overrightarrow{QB_i}|. \tag{33.4}$$

On the other hand, the affineness of mappings f_1 and f_2 implies that

$$f_1(B_i) = f_1((1-\lambda)Q + \lambda A_i) = (1-\lambda) f_1(Q) + \lambda f_1(A_i),$$
$$f_2(B_i) = f_2((1-\lambda)Q + \lambda A_i) = (1-\lambda) f_2(Q) + \lambda f_2(A_i),$$

so that

$$f_1(B_i) - f_2(B_i) = \lambda (f_1(A_i) - f_2(A_i)).$$

Moreover, $|\overrightarrow{QB_i}| = |\lambda \overrightarrow{QA_i}| = \lambda |\overrightarrow{QA_i}|$. From relation (33.4) we now get

$$|f_1(A_i) - f_2(A_i)| < \varepsilon |\overrightarrow{QA_i}|.$$

Since this inequality holds for any $\varepsilon > 0$, it follows that $|f_1(A_i) - f_2(A_i)| = 0$, that is, $f_1(A_i) = f_2(A_i)$ $(i = 1, \ldots, n)$.

In addition, $f_1(Q) = f_2(Q)$ (since for $A = Q$ the quantity $o_Q(A)$ in (33.3) goes to zero, according to (33.1)). Consequently, affine mappings f_1 and f_2 coincide at all the vertices of simplex $[Q, A_1, \ldots, A_n]$, so that in view of Theorem 19.9 mappings f_1 and f_2 must be identical.

COROLLARY 33.3. *If Q is an interior point of set $M \subset E^n$, then for mapping $\varphi: M \to E^n$ a tangent affine mapping at point Q (if it exists) will be unique.*

COROLLARY 33.4. *If M is a solid convex cone with a vertex Q in E^n and if φ is a mapping (into space E^m) defined over the whole cone M or just close to its vertex, then for mapping φ a tangent affine mapping at point Q (if it exists) will be unique.*

THEOREM 33.5. *If $M \subset E^n$ and $N \subset E^m$, while $\varphi: M \to E^m$ and $\psi: N \to E^p$ are mappings such that $\varphi(M) \subset N$, and if in addition mapping φ has at point $Q \in M$ a tangent affine mapping $f: E^n \to E^m$, while mapping $\psi: N \to E^p$ has at point $Q' = \varphi(Q)$ a tangent affine mapping $g: E^m \to E^p$, then it follows that $h = g \circ f$ is a tangent affine mapping at point Q for a mapping $\xi = \psi \circ \varphi: M \to E^p$.*

Proof. Since

$$\varphi(A) = f(A) + o_Q(A), \qquad A \in M,$$
$$\psi(B) = g(B) + o_{Q'}(B), \qquad B \in N,$$

it follows that

$$\xi(A) = \psi(\varphi(A)) = g(\varphi(A)) + o_{Q'}(\varphi(A)) =$$
$$= g(f(A) + o_Q(A)) + o_{Q'}(f(A) + o_Q(A)).$$

Moreover, since $|f(A) - Q'| = |f(A) - f(Q)| \leqslant M|\overrightarrow{QA}|$ (see Theorem 21.7), we know that $o_{Q'}(f(A) + o_Q(A)) = o_Q(A)$. Also, in view of the affineness of mapping g we have $g(f(A) + o_Q(A)) = g(f(A)) + o_Q(A)$. Consequently,

$$\xi(A) = g(f(A)) + o_Q(A) = h(A) + o_Q(A).$$

Let us consider the tangent mapping in the case where space E^m (the range of values) is a number axis (that is, $m = 1$). In this case mapping φ: $M \to E^1$ is a *real function* on set M. Here we assume that set $M \subset E^n$ is *open* (in particular, it may coincide with the entire space E^n).

According to general definition 33.1, an affine function f defined in E^n is called *tangent* to function φ at a point $Q \in M$ if relation (33.2) is valid. In view of Corollary 33.3, a tangent affine function f at point Q (if it exists) is unique. The gradient ($\operatorname{grad} f$) of this tangent function is also known as the *gradient of function* φ *at point* Q, being designated as $\operatorname{grad} \varphi(Q)$. According to (33.2) and (33.1), we have $\varphi(Q) = f(Q)$. Moreover, taking into account formula (21.5), we have

$$f(A) = f(Q) + \overrightarrow{QA}\operatorname{grad} f = \varphi(Q) + \overrightarrow{QA}\operatorname{grad}\varphi(Q).$$

Consequently, relation (33.2) becomes

$$\varphi(A) = \varphi(Q) + \overrightarrow{QA}\operatorname{grad}\varphi(Q) + o_Q(A). \tag{33.5}$$

We see that, if function φ has a tangent affine function at point Q, there will exist a vector $\boldsymbol{n}$ (namely $\boldsymbol{n} = \operatorname{grad}\varphi(Q)$) such that

$$\varphi(A) = \varphi(Q) + \boldsymbol{n}\overrightarrow{QA} + o_Q(A). \tag{33.6}$$

Conversely, if function φ has form (33.6), then the affine function

$$f(A) = \varphi(Q) + \boldsymbol{n}\overrightarrow{QA}$$

is obviously tangent to it at point Q and then $\boldsymbol{n} = \operatorname{grad} f = \operatorname{grad}\varphi(Q)$. This leads us to

THEOREM 33.6. *Let* φ *be a real function defined on an open set* $M \subset E^n$. *Function* φ *has a tangent affine function at a point* $Q \in M$ *if, and only if,* φ *has form* (33.6). *Vector* $\boldsymbol{n}$, *which is defined uniquely by function* φ *and point* Q, *is called* $\operatorname{grad}\varphi(Q)$.

Let us also give the coordinate notation for this theorem. We introduce in E^n an orthonormal system of coordinates $x^1, \ldots, x^n$, making $\varphi(A) = \varphi(x^1, \ldots, x^n)$ a function of n variables. In addition, we set

$$Q = (x_0^1, \ldots, x_0^n), \quad A = (x^1, \ldots, x^n)$$

and we designate point Q as x_0 and point A as x. The coordinates of the vector $\boldsymbol{n} = \operatorname{grad} \varphi(Q) = \operatorname{grad} \varphi(x_0)$ will then be $\frac{\partial \varphi(x_0)}{\partial x^1}, \ldots, \frac{\partial \varphi(x_0)}{\partial x^n}$:

$$\operatorname{grad} \varphi(x_0) = \left\{ \frac{\partial \varphi(x_0)}{\partial x^1}, \ldots, \frac{\partial \varphi(x_0)}{\partial x^n} \right\}. \tag{33.7}$$

Using this notation, formula (33.6) becomes

$$\begin{aligned} \varphi(x) &= \varphi(x_0) + (x - x_0) \operatorname{grad} \varphi(x_0) + o_{x_0}(x) = \\ &= \varphi(x_0) + \sum_{i=1}^{n} \frac{\partial \varphi(x_0)}{\partial x^i}(x^i - x_0^i) + o(x - x_0), \end{aligned} \tag{33.8}$$

that is, it becomes the familiar formula of mathematical analysis.

NOTE 33.7. As in the previous cases, we assume that relation (33.5) is *affine* in nature. Here it should be remembered that the vector $\operatorname{grad} \varphi(Q)$ (that is, the gradient of the affine function f tangent to φ at point Q) belongs not to space E^n itself but rather to the *conjugate* space (or, as is sometimes said, it is a *covariant* vector). With regard to formula (33.5), this means that the scalar product is to be understood in the sense of Subsection 15. If, on the other hand, space E^n is Euclidean but not affine, then the conjugate vector space is identified with E^n (see end of Subsection 16), and the scalar product in formula (33.5) turns into the ordinary scalar product in Euclidean space E^n.

Formula (33.8) is also meaningful for an *affine* space E^n (in particular, for a Euclidean space it is meaningful for any, not necessarily orthonormal, coordinate system); then the coordinate notation (33.7) of the vector $\operatorname{grad} \varphi(x)$ refers not to the system of coordinates $(x^1, \ldots, x^n)$ in E^n, but rather to the coordinate system of the conjugate space, which is related to coordinate system $(x^1, \ldots, x^n)$ as indicated in Theorem 15.12.

DEFINITION 33.8. Let M be an open set of space E^n. A real function φ defined on M is called *smooth* if at each point $A \in M$ there exists a vector $\operatorname{grad} \varphi(A)$, this vector depending continuously on A, that is, in coordinate notation, the partial derivatives $\frac{\partial \varphi}{\partial x^i}$, $i = 1, \ldots, n$, exist and are continuous on set M.

DEFINITION 33.9. Let φ be a smooth (real) function defined on an open set M of space E^n. The set S of all points $A \in M$ satisfying the relation $\varphi(A) = 0$ will be called a *smooth hypersurface* in E^n if the vector $\operatorname{grad} \varphi(A)$ is different from zero at each point $A \in S$. The relation $\varphi(A) = 0$ is called the *equation* of this hypersurface. In coordinate form the equation of the hypersurface is written as

$$\varphi(x^1, \ldots, x^n) = 0; \tag{33.9}$$

here $\varphi(x^1, \ldots, x^n)$ is the coordinate notation for a smooth function defined on some open set of space E^n.

For $n = 2$ (33.9) becomes

$$\varphi(x^1, x^2) = 0,$$

and the concept of a smooth hypersurface reduces simply to a *smooth line* (in the space of variables x^1 and x^2). For $n = 3$ equation (33.9) becomes

$$\varphi(x^1, x^2, x^3) = 0,$$

and the concept of a smooth hypersurface reduces to a *smooth surface* (in the space of variables x^1, x^2, x^3).

If Q is an arbitrary point of a smooth hypersurface S defined by the equation $\varphi(A) = 0$, then the vector $\operatorname{grad} \varphi(Q)$ (for any vector $\lambda \operatorname{grad} \varphi(Q)$, where $\lambda \neq 0$) is called the *normal vector* (or simply the *normal*) of hypersurface S at point Q. The hypersurface passing through point Q and having a vector $\operatorname{grad} \varphi(Q)$ as its normal vector is called the *tangent hyperplane* of hypersurface S at point Q.

DEFINITION 33.10. Let $S_1, S_2, \ldots, S_k$ be smooth hypersurfaces defined, respectively, in space E^n by the equations

$$\varphi_1(A) = 0, \quad \varphi_2(A) = 0, \quad \ldots, \quad \varphi_k(A) = 0, \tag{33.10}$$

where each of functions $\varphi_1, \ldots \varphi_k$ is defined on some open set of space E^n. The intersection P of all these hypersurfaces (that is, the set of all points $A \in E^n$ satisfying all of equations (33.10) simultaneously) is known as an $(n - k)$-dimensional *smooth manifold* in E^n, provided the following condition is met: at each point $A \in P$ the vectors

$$\operatorname{grad} \varphi_1(A), \quad \operatorname{grad} \varphi_2(A), \quad \ldots, \quad \operatorname{grad} \varphi_k(A)$$

are linearly independent. Therefore, by definition, an r-dimensional smooth manifold in E^n is specified by a system of $n - r$ equations.* In particular, an $(n - 1)$-dimensional

* Strictly speaking, the given definition of a smooth manifold is too narrow, in comparison with the generally accepted one. The general definition states that: *close to each of its points* (that is, locally), *a smooth manifold satisfies a system of equations of form* (33.10). However, the description of the entire manifold, *as a whole*, by equations (33.10) may turn out to be impossible. However, let us put aside these fine points (a detailed consideration of which is possible only within the framework of *topology*), and let us limit our discussion here just to manifolds which can be defined by equations (33.10). Note, by the way, that all the ensuing discussion dealing with smooth manifolds will essentially be *local* in nature, that is, it is actually applicable to any smooth manifolds.

manifold is defined by a single equation. Thus the $(n-1)$-dimensional manifolds of a space E^n coincide with hypersurfaces. One-dimensional manifolds are also called *lines.*

Note that the condition of linear independence of the vectors $\operatorname{grad}\varphi_1(A), \ldots, \ldots, \operatorname{grad}\varphi_k(A)$ is equivalent (in coordinate notation) to the requirement that the rank of the functional matrix

$$\frac{\partial \varphi(A)}{\partial x} = \begin{pmatrix} \frac{\partial \varphi_1(A)}{\partial x^1} & \frac{\partial \varphi_1(A)}{\partial x^2} & \cdots & \frac{\partial \varphi_1(A)}{\partial x^n} \\ \frac{\partial \varphi_2(A)}{\partial x^1} & \frac{\partial \varphi_2(A)}{\partial x^2} & \cdots & \frac{\partial \varphi_2(A)}{\partial x^n} \\ \cdot & \cdot & \cdot & \cdot \\ \frac{\partial \varphi_k(A)}{\partial x^1} & \frac{\partial \varphi_k(A)}{\partial x^2} & \cdots & \frac{\partial \varphi_k(A)}{\partial x^n} \end{pmatrix}$$

be a maximum (that is, equal to k).

This implies that, provided functions $\varphi_1(A), \ldots, \varphi_k(A)$ go to zero at point Q and are *independent* at this point, that is, provided the vectors $\operatorname{grad}\varphi_1(Q), \ldots, \operatorname{grad}\varphi_k(Q)$ are linearly independent, system of equations (33.10) defines an $(n-k)$-dimensional manifold in some neighborhood of point Q. Actually, since the specified vectors are linearly independent, it follows that the matrix $\frac{\partial \varphi(Q)}{\partial x}$ is of rank k, that is, some determinant of order k consisting of the columns of this matrix is different from zero. This determinant depends *continuously* on A, and thus it is different from zero in a certain neighborhood of point Q. Consequently, in some neighborhood of Q the matrix $\frac{\partial \varphi(A)}{\partial x}$ has a maximum rank k, that is, the vectors $\operatorname{grad}\varphi_1(A), \ldots, \operatorname{grad}\varphi_k(A)$ are linearly independent.

DEFINITION 33.11. Let P be a smooth $(n-k)$-dimensional manifold defined in space E^n by equations (33.10) and let Q be some point in it. If L_i is the tangent hyperplane of hypersurface $\varphi_i(A)=0$ at point Q $(i=1, \ldots, k)$, then the intersection of hyperplanes $L_1, \ldots, L_k$ will be an $(n-k)$-dimensional plane in E^n (since the normal vectors $\operatorname{grad}\varphi_i(Q)$, $i=1, \ldots, k$, of these hyperplanes are linearly independent). This $(n-k)$-dimensional plane is called the *tangent plane* of manifold P at point Q. Each vector $\boldsymbol{a}$ parallel to this tangent plane is called a *tangent vector* of manifold P at point Q.

THEOREM 33.12. *Let P_1 and P_2 be smooth manifolds in E^n having a common point Q and let L_1 and L_2 be the tangent planes of manifolds P_1 and P_2 at point Q. If L_1 and L_2 do not lie in a single hyperplane of space E^n, then the intersection $P_1 \cap P_2$ close to point Q will be a smooth manifold having a tangent plane $L_1 \cap L_2$ at point Q.*

Proof. Let P_1 be defined (close to point Q) by the equations $\varphi_1(A)=0, \ldots, \varphi_k(A)=0$, the vectors $\operatorname{grad}\varphi_i(Q)$, $i=1, \ldots, k$, being linearly independent, and let manifold P_2 be defined by the equations $\psi_1(A)=0, \ldots, \psi_l(A)=0$, the vectors $\text{grad}\,\psi_j(Q), j=1, \ldots, l$, also being linearly independent. Then

$$L_1 = L_1^{(1)} \cap \ldots \cap L_1^{(k)}, \quad L_2 = L_2^{(1)} \cap \ldots \cap L_2^{(l)},$$

where $L_1^{(i)}$ is a hyperplane passing through point Q and having $\text{grad}\,\varphi_i(Q)$ as its normal vector, and $L_2^{(j)}$ is a hyperplane passing through point Q and having $\text{grad}\,\psi_j(Q)$ as its normal vector.

If we assume the vectors

$$\text{grad}\,\varphi_1(Q), \ldots, \text{grad}\,\varphi_k(Q), \text{grad}\,\psi_1(Q), \ldots, \text{grad}\,\psi_l(Q) \quad (33.11)$$

to be linearly dependent, then the following relation has to be valid:

$$\alpha_1 \,\text{grad}\,\varphi_1(Q) + \ldots + \alpha_k \,\text{grad}\,\varphi_k(Q) = \\ = \beta_1 \,\text{grad}\,\psi_1(Q) + \ldots + \beta_l \,\text{grad}\,\psi_l(Q),$$

where at least one of coefficients $\alpha_1, \ldots, \alpha_k$ is different from zero, so that the vector $\boldsymbol{a} = \alpha_1 \,\text{grad}\,\varphi_1(Q) + \ldots + \alpha_k \,\text{grad}\,\varphi_k(Q)$ is different from zero as well. This vector is orthogonal to plane L_1 and, since $\boldsymbol{a} = \beta_1 \,\text{grad}\,\psi_1(Q) + \ldots + \beta_l \,\text{grad}\,\psi_l(Q)$, it is also orthogonal to plane L_2. However, then the hyperplane passing through point Q and orthogonal to vector $\boldsymbol{a}$ will contain both planes L_1 and L_2, which contradicts the assumption of the theorem. Hence vectors (33.11) are linearly independent.

It follows from this that the equations $\varphi_1(A) = \ldots = \varphi_k(A) = \psi_1(A) = \ldots = \psi_l(A) = 0$ will define a smooth manifold in the neighborhood of point Q. Obviously, this manifold coincides (close to Q) with the intersection $P_1 \cap P_2$, while its tangent plane at point Q coincides with

$$L_1^{(1)} \cap \ldots \cap L_1^{(k)} \cap L_2^{(1)} \cap \ldots \cap L_2^{(l)} = L_1 \cap L_2.$$

THEOREM 33.13. *Let P be a smooth $(n-k)$-dimensional manifold defined in space E^n by equations* (33.10) *and let Q be some point of it. Moreover, let L be the tangent plane of manifold P at point Q and let M be a plane (k-dimensional) complementary to L. If π is the projection of space E^n onto plane L parallel to plane M, then there will exist a sphere Σ centered at Q such that the part of manifold P included in Σ will be mapped one-to-one by projection π onto some neighborhood of point Q in plane L* (Figure 112). *The following relation is then valid:*

$$\pi(A) = A + o_Q(A), \qquad A \in P. \quad (33.12)$$

Proof. If vectors $\boldsymbol{e}_1, \ldots, \boldsymbol{e}_k$ form a basis of plane M and vectors $\boldsymbol{e}_{k+1}, \ldots, \boldsymbol{e}_n$ form a basis of plane L, then $\boldsymbol{e}_1, \ldots, \boldsymbol{e}_n$ form a basis of the whole space E^n. Let us consider in E^n the system of coordinates $(Q; \boldsymbol{e}_1, \ldots, \boldsymbol{e}_n)$. Moreover, we introduce in Euclidean space E^n a *new* scalar product, after setting

$$\boldsymbol{ab} = x^1y^1 + \ldots + x^ny^n$$

for any vectors

$$a = x^1 e_1 + \dots + x^n e_n,$$
$$b = y^1 e_1 + \dots + y^n e_n$$

(cf. Example 16.3). For such an introduction of the scalar product (instead of the initial one), space E^n becomes a new Euclidean space, which we will call E_1^n.

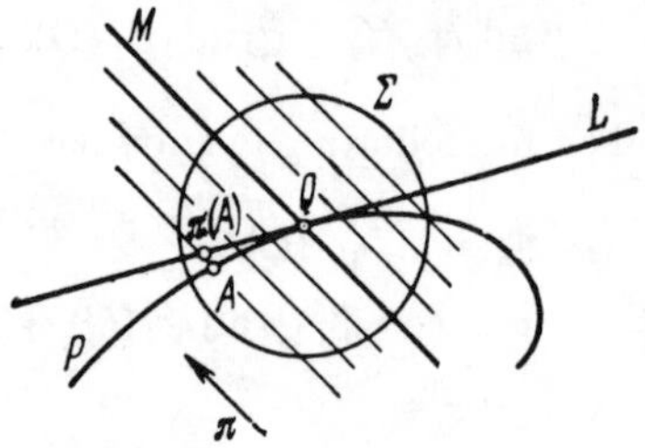

FIGURE 112

It will be sufficient to prove the theorem for space E_1^n, since all the factors indicated when formulating the theorem are affine in nature and thus do not depend on the choice of any particular scalar product. In space E_1^n the coordinate system $(Q; e_1, \dots, e_n)$ will be *orthonormal*; this coordinate system will also be used in the following discussion.

Let L_i be the tangent hyperplane of hypersurface $\varphi_i(A) = 0$ at point Q $(i = 1, \dots, k)$, so that $L = L_1 \cap \dots \cap L_k$. We can now write

$$\operatorname{grad} \varphi_i(Q) = a_i + b_i, \qquad i = 1, \dots, k,$$

where vector a_i is parallel to plane L and vector b_i is parallel to plane M. It is easy to see that vectors $b_1, \dots, b_k$ are linearly independent. Actually, if

$$\beta_1 b_1 + \dots + \beta_k b_k = 0,$$

then the vector

$$\beta_1 \operatorname{grad} \varphi_1(Q) + \dots + \beta_k \operatorname{grad} \varphi_k(Q) = \beta_1 a_1 + \dots + \beta_k a_k$$

will be *parallel* to plane L (as a linear combination of vectors $a_1, \dots, a_k$) and *orthogonal* to plane L (as a linear combination of vectors $\operatorname{grad} \varphi_i(Q)$, $i = 1, \dots, k$). Accordingly,

$$\beta_1 \operatorname{grad} \varphi_1(Q) + \dots + \beta_k \operatorname{grad} \varphi_k(Q) = 0,$$

which implies (in view of the linear independence of the vectors $\operatorname{grad} \varphi_i(Q)$, $i = 1, \dots, k$) that $\beta_1 = \dots = \beta_k = 0$.

Next we note that the equation

$$\operatorname{grad}\varphi_i(Q)=\frac{\partial\varphi_i(Q)}{\partial x^1}e_1+\ldots+\frac{\partial\varphi_i(Q)}{\partial x^n}e_n$$

leads to the relation

$$b_i=\frac{\partial\varphi_i(Q)}{\partial x^1}e_1+\ldots+\frac{\partial\varphi_i(Q)}{\partial x^k}e_k;\qquad i=1,\ \ldots,\ k. \tag{33.13}$$

The linear independence of vectors (33.13) means that the determinant

$$\det\left|\frac{\partial\varphi_i(Q)}{\partial x^j}\right|_{i,\ j=1,\ \ldots,\ k} \tag{33.14}$$

is different from zero, that is, the determinant consisting of the first k columns in matrix $(\partial\varphi_i(Q)/\partial x^j)$ is different from zero.

From the implicit-function theorem it follows that system (33.10), or, in coordinate form, the system

$$\varphi_1(x^1,\ \ldots,\ x^n)=0,\ \ldots,\ \varphi_k(x^1,\ \ldots,\ x^n)=0 \tag{33.15}$$

can in some neighborhood of point $Q=(x_0^1,\ \ldots,\ x_0^n)$ be *solved* for the unknowns $x^1,\ \ldots,\ x^k$ (note that each of functions $\varphi_1,\ \ldots,\ \varphi_k$ has an *open* set as its domain of definition, so that these functions are defined in some neighborhood of point Q). This means that we can write the relations

$$x^i=\psi^i(x^{k+1},\ \ldots,\ x^n),\qquad i=1,\ \ldots,\ k, \tag{33.16}$$

which close to point Q are equivalent to system of equations (33.15). More precisely, functions ψ^i are smooth and there exists a sphere Σ centered on Q such that any point of sphere Σ satisfying system (33.15) also satisfies system of equations (33.16), and vice versa.

Let us show that each plane parallel to M crosses the part of manifold P lying in sphere Σ at no more than one point. Actually, let $A=(x^1,\ \ldots,\ x^n)$ and $A_*=(x_*^1,\ \ldots,\ x_*^n)$ be points such that $A,\ A_*\in P\cap\Sigma$, with both points M and $A,\ A_*$ lying in one plane parallel to M, that is, vector $\overrightarrow{AA_*}$ is parallel to plane M. Inclusions $A,\ A_*\in P\cap\Sigma$ signify that point A satisfies relations (33.16), while point A_* satisfies the analogous relations

$$x_*^i=\psi^i(x_*^{k+1},\ \ldots,\ x_*^n),\qquad i=1,\ \ldots,\ k. \tag{33.17}$$

Moreover, since the vector

$$\overrightarrow{AA_*}=\{x_*^1-x^1,\ \ldots,\ x_*^n-x^n\}$$

is parallel to plane M, it follows that its coordinates numbered $k+1, \ldots, n$ go to zero, that is,

$$\underset{*}{x}^{k+1} = x^{k+1}, \ \ldots, \ \underset{*}{x}^{n} = x^{n}.$$

But then, in view of (33.16) and (33.17), we have

$$\underset{*}{x}^{1} = x^{1}, \ \ldots, \ \underset{*}{x}^{k} = x^{k},$$

that is, points A and A_* are identical. Thus set $P \cap \Sigma$ is projected *one-to-one* onto its image $\pi(P \cap \Sigma)$.

Next we note that point $Q = (0, 0, \ldots, 0)$ belongs to set $P \cap \Sigma$ and thus that, in view of (33.16),

$$\psi^i(0, 0, \ldots, 0) = 0, \qquad i = 1, \ldots, k. \tag{33.18}$$

Consequently, because of the continuity of functions ψ^i, if the numbers $x^{k+1}, \ldots, x^n$ are close enough to zero, this means that quantities (33.16) will also be close to zero. Therefore, if point $B = (0, \ldots, 0, x^{k+1}, \ldots, x^n)$ of plane L is close enough to point Q, then point $A = (x^1, \ldots, x^k, x^{k+1}, \ldots, x^n)$ defined by relations (33.16) will also be close to Q and thus belongs to set $P \cap \Sigma$. It is also obvious that this point A satisfies the relation $B = \pi(A)$. Consequently, all points of plane L close enough to Q are contained in the set $\pi(P \cap \Sigma)$, that is, this set constitutes a neighborhood of point Q in plane L.

It remains now only to show the validity of formula (33.12). Let A be an arbitrary point of manifold P. Then

$$\varphi_i(A) = \varphi_i(Q) + \overrightarrow{QA}\,\mathrm{grad}\,\varphi_i(Q) + o_Q(A),$$

and, since $\varphi_i(A) = 0$ and $\varphi_i(Q) = 0$ (because $A \in P$ and $Q \in P$), it follows that

$$\overrightarrow{QA}\,\mathrm{grad}\,\varphi_i(Q) = o_Q(A), \qquad i = 1, \ldots, k, \ A \in P.$$

Moreover, since point $B = \pi(A)$ lies in plane L, we know that vector $\overrightarrow{QB}$ is parallel to L, and thus that $\overrightarrow{QB}\,\mathrm{grad}\,\varphi_i(Q) = 0$. Accordingly,

$$\overrightarrow{AB}\,\mathrm{grad}\,\varphi_i(Q) = (\overrightarrow{QB} - \overrightarrow{QA})\,\mathrm{grad}\,\varphi_i(Q) = o_Q(A). \tag{33.19}$$

Now let us note that plane M is the orthogonal complement of plane L (since the discussion pertains to space E_1^n). Vectors $\mathrm{grad}\,\varphi_1(Q), \ldots, \mathrm{grad}\,\varphi_k(Q)$ are orthogonal to plane L, that is, they are parallel to M. Consequently, since these vectors are linearly independent, they form a basis of plane M. Therefore

$$e_i = a_{i1}\,\mathrm{grad}\,\varphi_1(Q) + \ldots + a_{ik}\,\mathrm{grad}\,\varphi_k(Q), \qquad i = 1, \ldots, k,$$

where a_{ij} are certain numbers. A comparison of the last equation with (33.19) gives

$$e_i\overrightarrow{AB}=o_Q(A), \qquad i=1, \ldots, k. \tag{33.20}$$

Finally, let

$$\overrightarrow{AB}=x^1(A)\,e_1+\ldots+x^k(A)\,e_k,$$

where vector $\overrightarrow{AB}$ is represented as a linear combination of the vectors of the basis (note that vector $\overrightarrow{AB}$ is orthogonal to plane L). Let $\xi(A)$ be the largest of the numbers $|x^1(A)|, \ldots, |x^k(A)|$. Since $x^i(A)=e_i\overrightarrow{AB}$ (basis $e_1, \ldots, e_n$ being orthonormal), we know from (33.20) that $x^i(A)=o_Q(A)$, $i=1, \ldots, k$, and thus that $\xi(A)=o_Q(A)$. On the other hand,

$$|\overrightarrow{AB}|=\sqrt{(x^1(A))^2+\ldots+(x^n(A))^2}\leqslant\sqrt{n(\xi(A))^2}=\sqrt{n}\cdot\xi(A),$$

so that $|\overrightarrow{AB}|=o_Q(A)$, or, equivalently, $\overrightarrow{AB}=o_Q(A)$. Recalling that $B=\pi(A)$, we obtain formula (33.12).

THEOREM 33.14. *If P is a smooth $(n-k)$-dimensional manifold in space E^n and if Q is some point in it, then in the vicinity of point Q we can introduce in manifold P the local coordinates $\xi^1, \ldots, \xi^{n-k}$. This means that a point $A(\xi^1, \ldots, \xi^{n-k})$ corresponds to any system of numbers $\xi^1, \ldots, \xi^{n-k}$ satisfying the condition $|\xi^i|<\delta$, $i=1, \ldots, n-k$, in manifold P, while any point $A\in P$ sufficiently close to Q can be expressed uniquely in the form $A=A(\xi^1, \ldots, \xi^{n-k})$. Moreover, we also have*

$$A(\xi^1, \ldots, \xi^{n-k})=Q+\xi^1a_1+\ldots+\xi^{n-k}a_{n-k}+o(\xi), \tag{33.21}$$

where $a_1, \ldots, a_{n-k}$ are linearly independent vectors in E^n, while $\xi=\{\xi^1, \ldots, \xi^{n-k}\}$ is a vector of the $(n-k)$-dimensional arithmetical space. Finally, all the vectors $a_1, \ldots, a_{n-k}$ are tangent vectors of manifold P at point Q, and any tangent vector of manifold P at point Q can be linearly expressed in terms of vectors $a_1, \ldots, a_{n-k}$.

Proof. We continue the arguments which were being presented during the proof of Theorem 33.13 (using the same notation). We set $A(\xi^1, \ldots, \xi^{n-k})=\psi^1(\xi^1, \ldots, \xi^{n-k})e_1+\ldots+\psi^k(\xi^1, \ldots, \xi^{n-k})e_k+\xi^1e_{k+1}+\ldots+\xi^{n-k}e_n$, that is, we call $A(\xi^1, \ldots, \xi^{n-k})$ the point which in system $(Q;\ e_1, \ldots, e_n)$ has the coordinates

$$\begin{aligned} x^1&=\psi^1(\xi^1, \ldots, \xi^{n-k}),\\ &\ldots\ldots\ldots\ldots\\ x^k&=\psi^k(\xi^1, \ldots, \xi^{n-k}),\\ x^{k+1}&=\xi^1,\\ &\ldots\ldots\ldots\ldots\\ x^n&=\xi^{n-k}, \end{aligned} \tag{33.22}$$

where $\psi^1, \ldots, \psi^k$ are the functions appearing on the right-hand sides of relations (33.16). Point $A(\xi^1, \ldots, \xi^{n-k})$ is defined for all $\xi^1, \ldots, \xi^{n-k}$ having low enough absolute values, that is, there exists a value $\delta > 0$ such that any system of numbers $\xi^1, \ldots, \xi^{n-k}$ satisfying condition $|\xi^i| < \delta$, $i = 1, \ldots, n-k$, corresponds to some point $A(\xi^1, \ldots, \xi^{n-k})$. It follows directly from (33.22) that the coordinates $x^1, \ldots, x^n$ of this point satisfy relations (33.16), so that $A(\xi^1, \ldots, \xi^{n-k}) \in P$. Then the coordinates $x^1, \ldots, x^n$ of any point $A \in P$ close enough to Q will satisfy relations (33.16), so that there must exist an (obviously uniquely determined) system of numbers $\xi^1, \ldots, \xi^{n-k}$ satisfying relations (33.22). In other words, any point $A \in P$ sufficiently close to Q can be expressed uniquely as $A = A(\xi^1, \ldots, \xi^{n-k})$.

Moreover, since coordinates $x^{k+1}, \ldots, x^n$ of point $A(\xi^1, \ldots, \xi^{n-k})$ are equal, respectively, to $\xi^1, \ldots, \xi^{n-k}$ (see (33.22)), it follows that

$$\pi A(\xi^1, \ldots, \xi^{n-k}) = Q + \xi^1 e_{k+1} + \ldots + \xi^{n-k} e_n,$$

so that, in view of (33.12), we can write

$$A(\xi^1, \ldots, \xi^{n-k}) = Q + \xi^1 e_{k+1} + \ldots + \\ + \xi^{n-k} e_n + o_Q(A(\xi^1, \ldots, \xi^{n-k})). \tag{33.23}$$

This relation implies that

$$\xi^1 e_{k+1} + \ldots + \xi^{n-k} e_n = \overrightarrow{QA(\xi^1, \ldots, \xi^{n-k})} + o_Q(A(\xi^1, \ldots, \xi^{n-k}))$$

and hence that

$$o_Q(A(\xi^1, \ldots, \xi^{n-k})) = o(\xi^1 e_{k+1} + \ldots + \xi^{n-k} e_n).$$

Since, finally, the length of vector $\xi^1 e_{k+1} + \ldots + \xi^{n-k} e_n$ is $\sqrt{(\xi^1)^2 + \ldots + (\xi^{n-k})^2}$, that is, it equals the length of the vector $\boldsymbol{\xi} = \{\xi^1, \ldots, \xi^{n-k}\}$, we know that $o(\xi^1 e_{k+1} + \ldots + \xi^{n-k} e_n) = o(\boldsymbol{\xi})$, and thus formula (33.23) becomes

$$A(\xi^1, \ldots, {}^{n-k}\xi) = Q + \xi^1 e_{k+1} + \ldots + \xi^{n-k} e_n + o(\boldsymbol{\xi}).$$

Consequently, relation (33.21) is valid, with

$$a_1 = e_{k+1}, \ \ldots, \ a_{n-k} = e_n. \tag{33.24}$$

It is also obvious from (33.24) that all the vectors $a_1, \ldots, a_{n-k}$ are tangent vectors of manifold P at point Q and that any tangent vector of manifold P at point Q can be expressed linearly in terms of vectors $a_1, \ldots, a_{n-k}$ (since vectors $e_{k+1}, \ldots, e_n$ form a basis of the tangent plane L constructed to manifold P at point Q). The theorem is thereby proven.

Now let us consider a case which is, in a certain sense, the opposite of that referred to in Theorem 33.6, namely the case where $n = 1$. Let h be some positive number (it may be that $h = \infty$), and let M be the half-open interval $[0, h)$ of number line E^1. The continuous mapping $\varphi: M \rightarrow E^m$ is called the *parametrized curve* (or simply the *curve*) *originating at the point* $Q = \varphi(0)$. Let us assume that φ has a tangent affine mapping $f: E^1 \rightarrow E^m$ at point 0. In this case the vector

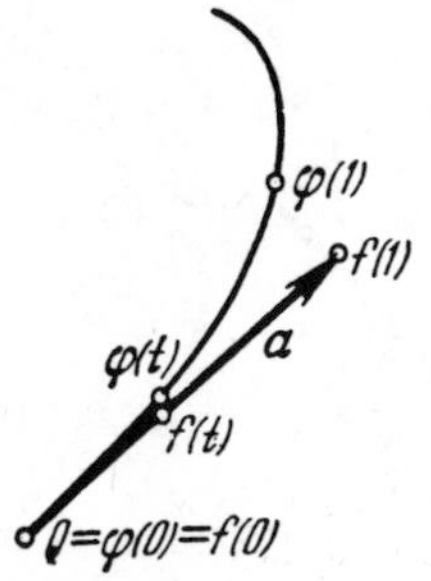

FIGURE 113

$$a = f(1) - f(0) = f(1) - Q$$

is called a *tangent vector* of curve φ at its initial point Q (Figure 113). From the definition of a tangent mapping we know that (for points t lying in the half-open interval $M = [0, h)$):

$$\varphi(t) = f(t) + o(t) = \varphi(0) + (f(t) - f(0)) + o(t) = \\ = \varphi(0) + t(f(1) - f(0)) + o(t) = \varphi(0) + ta + o(t).$$

Thus, *if the curve* φ: $[0, h) \rightarrow E^m$ *has a tangent vector* a *at point* $Q = \varphi(0)$, *then*

$$\varphi(t) = \varphi(0) + ta + o(t). \tag{33.25}$$

It is easy to see that the converse is also true: if relation (33.25) is valid for a curve φ, then it follows that a is a tangent vector of curve φ at point $Q = \varphi(0)$.

Now, if we introduce coordinates $y^1, \ldots, y^m$ in space E^m, then specifying a curve $\varphi(t)$ is equivalent to specifying m numerical functions (coordinates of point $\varphi(t)$) in the half-open interval $M = [0, h)$:

$$\varphi(t) = (y^1(t), y^2(t), \ldots, y^m(t)).$$

Tangent vector a of curve $\varphi(t)$ is in this case written (in the same coordinate system) as

$$a = \frac{d\varphi(0)}{dt} = \left\{ \frac{dy^1(0)}{dt}, \frac{dy^2(0)}{dt}, \ldots, \frac{dy^m(0)}{dt} \right\}.$$

THEOREM 33.15. *If* $\varphi(t)$ *is a curve originating at point* $Q(=\varphi(0))$ *and having at point* Q *a tangent vector* $a(=d\varphi(0)/dt)$, *and if* $\psi(A)$ *is a smooth function defined in some neighborhood of* Q, *then*

$$\psi(\varphi(t)) = \psi(Q) + t(a \operatorname{grad} \psi(Q)) + o(t), \tag{33.26}$$

or, in other words,

$$\frac{d\psi(\varphi(t))}{dt} = a \operatorname{grad} \psi(Q). \tag{33.27}$$

This implies that, if scalar product (33.27) *is positive, then for all sufficiently small* $t > 0$ *the inequality* $\psi(\varphi(t)) > \psi(\varphi(0))$ *is valid. In other words, for any point* A

different from Q, *lying on the curve in question and sufficiently close to* Q, *the inequality* $\psi(A) > \psi(Q)$ *is valid* (Figure 114). *However, if scalar product* (33.27) *is negative, then* $\psi(A) < \psi(Q)$ *for any point* $A \neq Q$ *sufficiently close to* Q *and lying on the curve in question* (Figure 115).

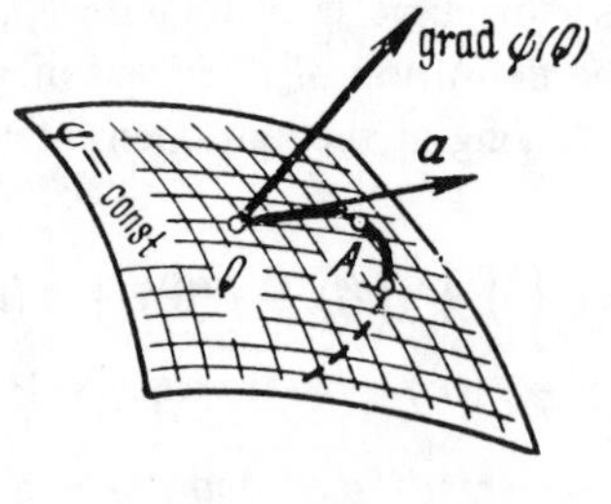

FIGURE 114

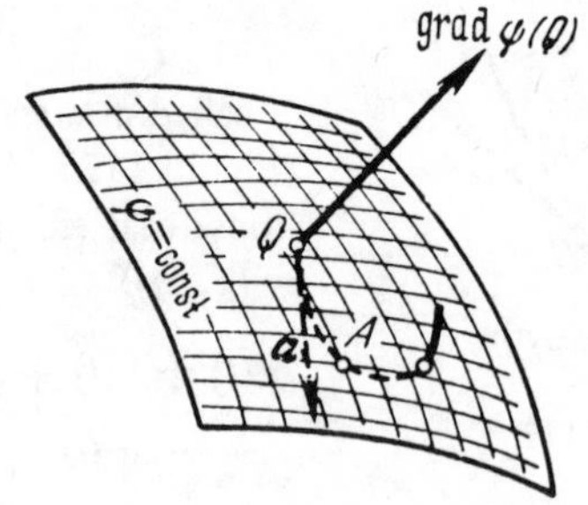

FIGURE 115

Proof. In view of (33.25), the affine mapping

$$f(t) = \varphi(0) + t\mathbf{a} = Q + t\mathbf{a}$$

(defined on the number axis) is the tangent mapping for φ at point 0. Similarly, in view of (33.5), the affine mapping

$$g(A) = \psi(Q) + \overrightarrow{QA}\ \mathrm{grad}\ \psi(Q)$$

is tangent for ψ at point Q. Consequently, according to Theorem 33.5, the affine mapping $h = g \circ f$ is tangent for $\psi \circ \varphi$, that is, $\psi(\varphi(t)) = h(t) + o(t)$. It now just remains to note that

$$h(t) = g(f(t)) = \psi(Q) + \overrightarrow{Qf(t)}\ \mathrm{grad}\ \psi(Q) = \psi(Q) + t\mathbf{a}\ \mathrm{grad}\ \psi(Q).$$

34. Covering of set. DEFINITION 34.1. Let Ω be some set in space E^n and let $Q \in \Omega$. Moreover, let M be some convex cone with a vertex Q, also in space E^n. We will call cone M a *covering* of set Ω at point Q if it is possible to find a continuous mapping ψ, defined for all points of cone M sufficiently close to Q and assuming values in space E^n, such that the following conditions are satisfied:

1) the identity mapping of space E^n is tangent for mapping ψ at point Q, that is,

$$\psi(A) = A + o_Q(A);$$

2) at all points $A \in M$ for which mapping ψ is defined, the relation $\psi(A) \in \Omega$ is valid.

Roughly speaking, this definition implies that, "accurate to $o_Q(A)$," cone M (that is, the covering) is contained in set Ω (Figure 116). Note that, if M is a covering of set Ω at point Q, then any *smaller* (that is, contained in M) convex cone with vertex Q will also be a covering of set Ω at point Q. Thus it is of interest to find the *maximum* covering of set Ω at point Q (if such a covering exists).

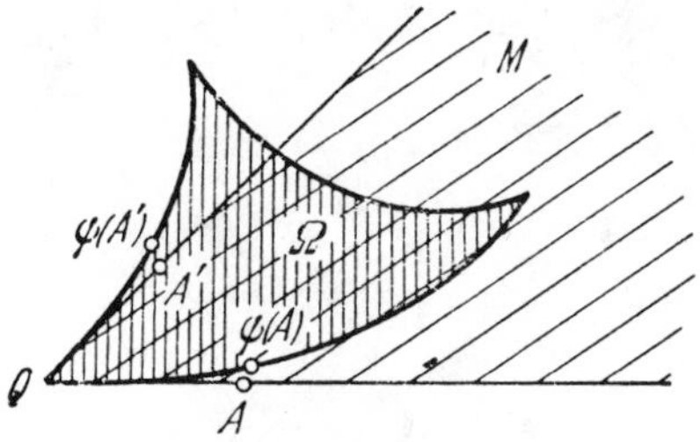

FIGURE 116

A maximum covering does not always exist. If, for instance, Ω is a part of the plane making an angle greater than π (with vertex Q), then any angle with vertex Q not exceeding π and belonging to Ω will be a covering of set Ω at point Q, but no maximum covering exists.

The next two theorems describe the covering of a set in the two cases which will be of greatest interest to us later.*

THEOREM 34.2. *The supporting cone M of a closed convex set Ω at an arbitrary point $Q \in \Omega$ is a covering of set Ω at point Q.*

Proof. Let E^n be the carrier plane of convex set Ω. Cone M also lies in this plane. According to Theorem 29.5, for any point $A \in E^n$ there exists in set Ω a unique point that is *closest* to A. This closest point will be called $\psi(A)$.

Point $\psi(A)$ coincides with A if $A \in \Omega$. If, on the other hand, point A does not belong to set Ω, then a sphere centered on A, on the boundary of which lies point $\psi(A)$, will not have any points in common with Ω other than $\psi(A)$ (Figure 117). Mapping ψ is defined and continuous throughout the entire space E^n, but we will consider it only at points of cone M.

To prove the theorem, it now remains to establish that mapping ψ satisfies conditions 1 and 2 on the previous page. The fact that for any point $A \in M$ we have $\psi(A) \in \Omega$ implies that condition 2 is satisfied.

Let us next show that condition 1 is also satisfied. Choose an arbitrary positive number ε (not exceeding unity) which will not change during the course of the discussion. Consider also a sphere centered on Q, the intersection of this sphere with cone M being called Σ (Figure 118). Let $B \in \Sigma$. If l_B is the ray originating at Q and passing through point B, and if we consider all the rays starting at Q and forming angles not exceeding ε with ray l_B, then these rays will fill a closed convex cone in space E^n (see Example 28.6), which we call $K_\varepsilon(B)$.

* Actually, in both cases the covering constructed is a maximum, but we do not show this, since this fact will not be needed subsequently.

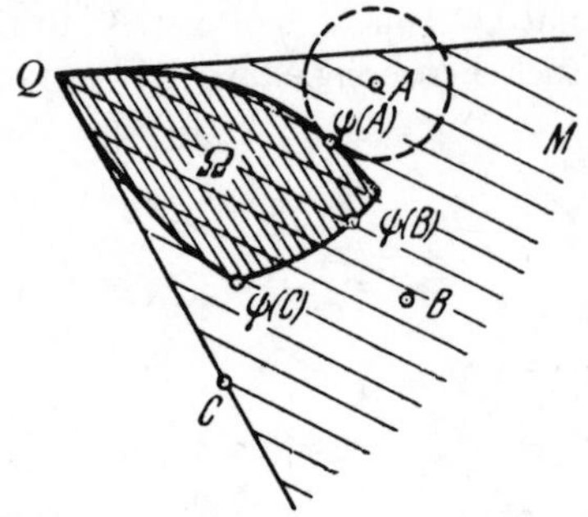

FIGURE 117

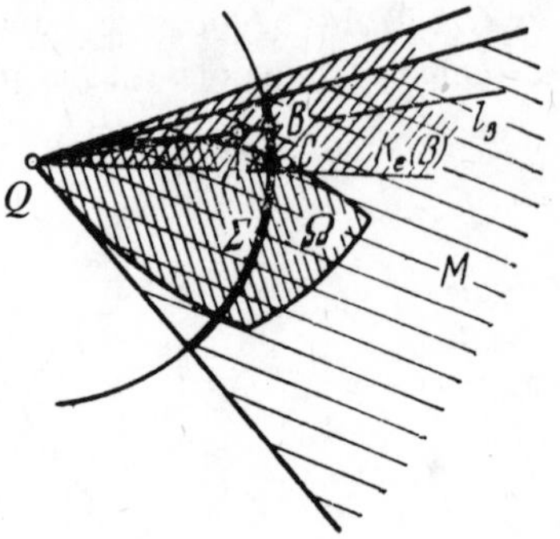

FIGURE 118

Since $B \in \Sigma$, that is, since ray l_B is contained in cone M, it follows that in cone $K_\varepsilon(B)$ there will be rays passing through points of set Ω that are different from Q. In other words, intersection $K_\varepsilon(B) \cap \Omega$ contains points other than Q. This intersection is a closed convex set. If set $K_\varepsilon(B) \cap \Omega$ has points lying outside the sphere of radius 1 centered on Q, then we set $d(B) = 1$. If, on the other hand, set $K_\varepsilon(B) \cap \Omega$ lies wholly inside this sphere, then $d(B)$ will denote the greatest distance from Q to the points of set $K_\varepsilon(B) \cap \Omega$. It is obvious that $d(B) > 0$, since set $K_\varepsilon(B) \cap \Omega$ contains points other than Q.

Therefore, $d(B)$ is a *positive* function defined on the closed bounded set Σ. It is easy to show that this function is continuous. Consequently, the minimum value of function $d(B)$ is positive, that is, there exists some $h > 0$ such that $d(B) \geqslant h$ for any point $B \in \Sigma$.

Now let A be an arbitrary point of cone M other than Q, situated at some distance less than h from Q. Let B be the point at which the ray starting at Q and passing through A intersects set Σ. Since $d(B) \geqslant h$, it follows that in set $K_\varepsilon(B) \cap \Omega$ there exists a point C separated from Q by a distance not less than h. In other words, $C \in \Omega$, and the angle between the rays originating at Q and passing through points A and C, respectively, will not exceed ε (see Figure 118). Point A', which is the orthogonal projection of point A onto the ray passing through C, belongs to set Ω (since it lies on the segment joining points Q and C). The distance between points A and A' does not exceed $|\overrightarrow{QA}| \cdot \sin \varepsilon$, that is, it is less than $\varepsilon |\overrightarrow{QA}|$. However, then the point $\psi(A)$ of set Ω *closest* to A will all the more be a distance from A less than $\varepsilon |\overrightarrow{QA}|$, that is,

$$|\psi(A) - A| < \varepsilon |\overrightarrow{QA}| \quad \text{for} \quad |\overrightarrow{QA}| < h, \qquad A \in M.$$

Since in our discussion the number $\varepsilon > 0$ is arbitrary, it follows that

$$\lim_{\substack{A \to Q \\ A \in M}} \frac{|\psi(A) - A|}{|\overrightarrow{QA}|} = 0,$$

and this (together with the obvious relation $\psi(Q)=Q$) implies that $\psi(A)=A+$ $+o_Q(A)$, that is, condition 1 is satisfied.

Before we formulate the next theorems, it will be useful to give some definitions. First let us introduce in space E^n the coordinate system $z^1, \ldots, z^n$ and let us write the coordinates of each point $A \in E^n$: $A=(z^1, \ldots, z^n)=z$. Let $g^1(z), \ldots,$ $g^k(z)$ be some system of smooth functions defined in space E^n. Moreover, let Ω be the set of all points $A=(z^1, \ldots, z^n) \in E^n$ satisfying the conditions

$$g^1(z) \leqslant 0, \ldots, g^k(z) \leqslant 0. \tag{34.1}$$

Each of the inequalities $g^i(z) \leqslant 0$ entering into the definition of set Ω is called a *constraint.*

If the relation $g^i(z_0)=0$ is valid at some point $z_0 \in \Omega$, then the constraint $g^i(z) \leqslant 0$ will be called *active* at point z_0; if, on the other hand, $g^i(z_0)<0$, then $g^i(z) \leqslant 0$ is called an *inactive* constraint at point z_0. Therefore, for any point $z_0 \in \Omega$ all the constraints (34.1) can be divided into active and inactive constraints at this point. The set of all superscripts i for which constraints $g^i(z) \leqslant 0$ are active at a point $z_0 \in \Omega$ is known as the *zone of activity* (it is denoted as $I(z_0)$, and sometimes as $J(z_0)$).

EXAMPLE 34.3. Consider the relations

$$(z^1)^2 - z^2 \leqslant 0, \quad z^1 + z^2 - 2 \leqslant 0, \; z^2 - z^1 - 2 \leqslant 0 \tag{34.2}$$

in the z^1, z^2 plane, that is, consider the case of *three* constraints (34.1), where $g^1=(z^1)^2-z^2$, $g^2=z^1+z^2-2$, $g^3=z^2-z^1-2$. Constraints (34.2) define in the z^1, z^2 plane the set Ω portrayed in Figure 119. At point a the first and second constraints are active; in other words, $I(a)=\{1, 2\}$. At point b only the second constraint is active, and at point c only the third: $I(b)=\{2\}$, $I(c)=\{3\}$. At point d all of constraints (34.2) are inactive, that is, the zone of activity $I(d)$ is the *empty* set.

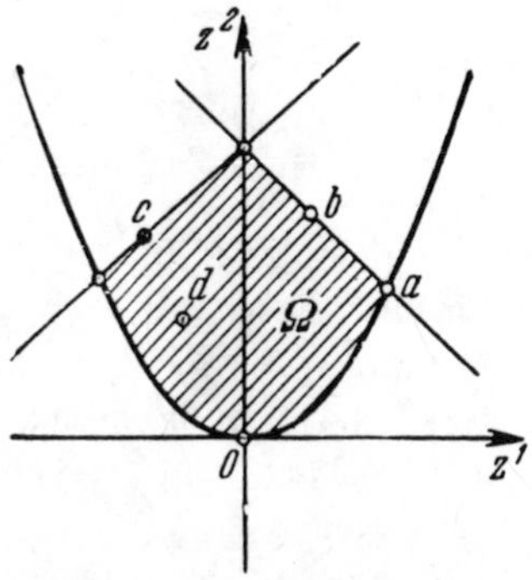

FIGURE 119

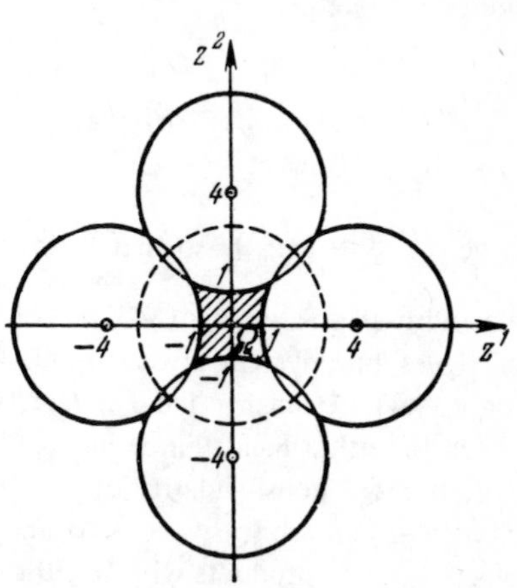

FIGURE 120

Note that some of constraints (34.1) may not be active at any point of set Ω, but nevertheless they cannot be disregarded during the description of set Ω. As an example, we cite the set Ω defined on the plane of variables z^1, z^2 by the relations:

$$\left.\begin{array}{c} 9-(z^1-4)^2-(z^2)^2\leqslant 0, \quad 9-(z^1+4)^2-(z^2)^2\leqslant 0, \\ 9-(z^1)^2-(z^2-4)^2\leqslant 0, \quad 9-(z^1)^2-(z^2+4)^2\leqslant 0, \\ (z^1)^2+(z^2)^2-9\leqslant 0 \end{array}\right\} \tag{34.3}$$

(Figure 120). Obviously, the last of constraints (34.3) is not active at any point of set Ω. However, this constraint cannot be eliminated, since the set described by the first four of constraints (34.3) will contain, in addition to Ω, the exterior of the figure represented in Figure 120 by a heavy line.

Since the functions g^i used to define constraints (34.1) are assumed to be smooth, at each point $z_0 \in \Omega$ we can consider the vectors $\operatorname{grad} g^i(z_0)$, $i=1, \ldots, k$ and, in particular, the vectors

$$\operatorname{grad} g^i(z_0), \qquad i \in I(z_0) \tag{34.4}$$

corresponding to the active constraints at point z_0. Let us say that at point $z_0 \in \Omega$ system of constraints (34.1) is *nonsingular* if there exists in E^n a vector $\boldsymbol{a}$ that forms *strictly negative* scalar products with each of vectors (34.4):

$$\boldsymbol{a} \operatorname{grad} g^i(z_0) < 0 \quad \text{for} \quad i \in I(z_0). \tag{34.5}$$

If set $I(z_0)$ is empty (as, for instance, at point d in Figure 119), then system of vectors (34.4) will not contain a single vector, and system of constraints (34.1) can be assumed, by definition, to be nonsingular.

Note that, if vectors (34.4) are *linearly independent,* then system of constraints (34.1) is nonsingular, since it is possible to find vectors $\boldsymbol{b}_i$, $i \in I(z_0)$ possessing the property that

$$\boldsymbol{b}_i \operatorname{grad} g^j(z_0) = \begin{cases} 1 & \text{for} \quad i=j; \\ 0 & \text{for} \quad i \neq j \end{cases}$$

(where $i, j \in I(z_0)$), so that the vector $\boldsymbol{a} = -\sum_{i \in I(z_0)} \boldsymbol{b}_i$ forms negative scalar products with all the vectors (34.4). For instance, at each point z_0 of the set Ω portrayed in Figure 119, system of constraints (34.2) is nonsingular, since, as is easily shown, vectors (34.4) are linearly independent.

On the other hand, linear independence of vectors (34.4) is by no means a *necessary* condition for nonsingularity of system of constraints (34.1) at point z_0. Let us consider in E^n an arbitrary vector $\boldsymbol{a}$ and let us select any vectors $\boldsymbol{e}_1, \boldsymbol{e}_2, \ldots, \boldsymbol{e}_k$ forming negative scalar products with $\boldsymbol{a}$, the number k of these vectors being taken to be *greater* than n. Moreover, let $f_1, \ldots, f_k$ be affine functions which go to zero at point z_0 and satisfy the conditions $\operatorname{grad} f_i = \boldsymbol{e}_i$, $i=1, \ldots, k$. Then system of constraints $f_1 \leqslant 0, \ldots, f_k \leqslant 0$ (these are all active at z_0) is nonsingular at point z_0, although the vectors $\operatorname{grad} f_i$, $i=1, \ldots, k$, are linearly dependent, their number being greater than the dimension n of space E^n.

Geometrically, the condition of nonsingularity signifies the following. Consider for $i \in I(z_0)$ the half-space P_i consisting of all points $z \in E^n$ for which $(z-z_0) \operatorname{grad} g^i(z_0) \leqslant 0$. Then we obtain half-spaces P_i, $i \in I(z_0)$, corresponding to the constraints (34.1) active at point z_0, where the boundary hyperplane of half-space P_i is orthogonal to the vector $\operatorname{grad} g^i(z_0)$ and passes through point z_0, that is, it is a tangent hyperplane of hypersurface $g^i(z)=0, i \in I(z_0)$. The condition of nonsingularity means that the intersection of *all* the half-spaces P_i, $i \in I(z_0)$, is a *solid* cone in E^n, that is, the corresponding *open* half-spaces have a nonempty intersection.

THEOREM 34.4. *Let Ω be a set defined in space E^n by constraint $g(z) \leqslant 0$, where $g(z)$ is a smooth function defined on an open set of space E^n. In addition, let $z_0 \in \Omega$ be a point such that $g(z_0)=0$ and $\operatorname{grad} g(z_0) \neq \mathbf{0}$. If M is the half-space consisting of all points $z \in E^n$ for which $(z-z_0) \operatorname{grad} g(z_0) \leqslant 0$, then M is a covering of set Ω at point z_0* (Figure 121).

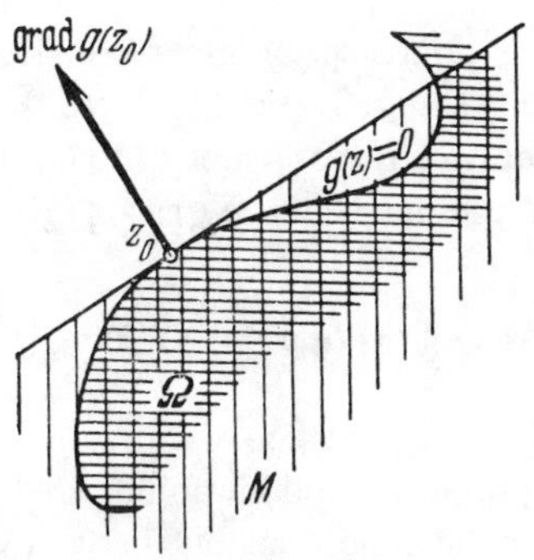

FIGURE 121

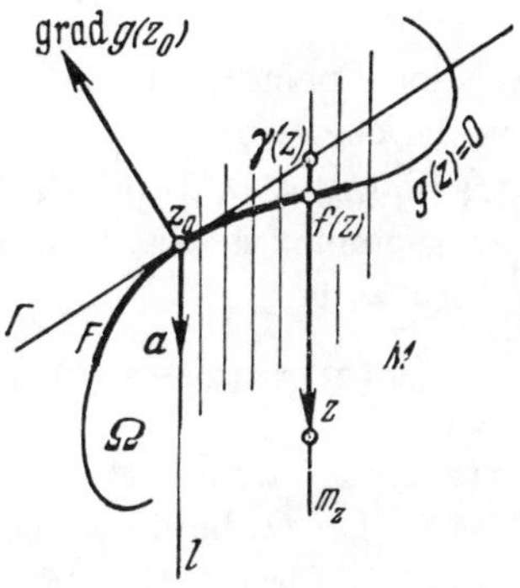

FIGURE 122

Proof. Select a ray l originating at point z_0, such that the unit vector $\boldsymbol{a}$ with the direction of this ray forms a *negative* scalar product with the vector $\operatorname{grad} g(z_0)$. Consider the hypersurface defined by the equation $g(z)=0$ (recalling that $g(z)$ is a smooth function defined on an open set; see Definition 33.9). Let F be a piece of this hypersurface close to point z_0, the piece being so small (Figure 122) that any line parallel to ray l and lying close to it will cross F at one point only (see Theorem 33.13). Moreover, let Γ be the boundary hyperplane of half-space M, that is, the hyperplane consisting of all points z for which $(z-z_0) \operatorname{grad} g(z_0)=0$.

Let z be an arbitrary point of space E^n, and let m_z be the line parallel to l and passing through z. Line m_z crosses hyperplane Γ at some point $\gamma(z)$ and, if point z is close enough to z_0, it crosses hypersurface F at some point $f(z)$. Since Γ is the tangent hyperplane of hypersurface F at point z_0, it follows that

$$f(z)-\gamma(z)=o_{z_0}(z),$$

because $\gamma(z) = \pi(f(z))$, where π is the projection of space E^n onto hyperplane Γ parallel to l (see (33.12)).

Now let us set

$$\psi(z) = z + (f(z) - \gamma(z)).$$

Mapping ψ is defined close to point z_0 and assumes values in space E^n. Let us show that mapping ψ, considered close to vertex z_0 of cone M, possesses properties 1 and 2 specified in Definition 34.1, which will also establish that M is a covering of set Ω at point z_0,

Since $\psi(z) - z = f(z) - \gamma(z) = o_{z_0}(z)$, it follows that condition 1 is satisfied. Thus it remains only to verify that condition 2 is satisfied, namely that $\psi(z) \in \Omega$ for points $z \in M$ sufficiently close to z_0. The fact that $f(z) \in F$ implies that $g(f(z)) = 0$ (since all points of hypersurface F satisfy the equation $g(z) = 0$). Consequently,

$$g(\psi(z)) = g(\psi(z)) - g(f(z)) = (\psi(z) - f(z)) \operatorname{grad} g(\xi) = \\ = (z - \gamma(z)) \operatorname{grad} g(\xi),$$

where ξ is some point on the segment joining points $\psi(z)$ and $f(z)$. However, for $z \in M$ we have $z - \gamma(z) = \lambda(z)\,a$, where $\lambda(z) \geqslant 0$ (see Figure 122). Moreover, since $a \operatorname{grad} g(z_0) < 0$, it follows that if point ξ is close enough to z_0 (which will be the case if point z is close enough to z_0) the inequality $a \operatorname{grad} g(\xi) < 0$ will be valid. Consequently,

$$g(\psi(z)) = (z - \gamma(z)) \operatorname{grad} g(\xi) = \lambda(z)(a \operatorname{grad} g(\xi)) \leqslant 0,$$

so that $\psi(z) \in \Omega$. Accordingly, condition 2 is also satisfied.

NOTE 34.5. Let Ω^* be the set containing point z_0 and all points $z \in E^n$ for which $g(z) < 0$. Clearly, $\Omega^* \subset \Omega$. Then for the conditions of Theorem 34.4 it is possible to formulate a stronger statement: *M is a covering of set Ω^* at point z_0.*

Now, if we set

$$\psi^*(z) = \psi(z) + (\gamma(z) - z_0)^2 a,$$

then it follows that $\psi^*(z) = \psi(z) + o_{z_0}(z)$, so that mapping ψ^*, like ψ, satisfies condition 1 of Definition 34.1. Moreover, if point $z \in M$ does not lie on ray l, then $(\gamma(z) - z_0)^2 > 0$, and thus

$$g(\psi^*(z)) - g(\psi(z)) = (\psi^*(z) - \psi(z)) \operatorname{grad} g(\xi) = \\ = (\gamma(z) - z_0)^2 (a \operatorname{grad}(g(\xi)) < 0$$

(here ξ is some point on the segment joining points $\psi(z)$ and $\psi^*(z)$). Consequently, $g(\psi^*(z)) < g(\psi(z)) \leqslant 0$, and thus $\psi^*(z) \in \Omega^*$. On the other hand, if point $z \in M$ lies on ray l, that is, if $z = z_0 + \lambda a$, where $\lambda \geqslant 0$, then $\psi^*(z) = \psi(z) = z$, so that again $\psi^*(z) \in \Omega^*$. Therefore, in any case $\psi^*(z) \in \Omega^*$ (for points $z \in M$ sufficiently close to z_0), that is, condition 2 is satisfied as well.

THEOREM 34.6. *Let Ω be a set defined in space E^n by constraints* (34.1), *and let system of constraints* (34.1) *be nonsingular at a point $z_0 \in \Omega$. If M is the set of all points $z \in E^n$ for which the vector $z - z_0$ forms a nonpositive scalar product with each of vectors* (34.4), *then M is a covering of set Ω at point z_0.*

Proof. First note that M is the intersection of the half-spaces P_i, $i \in I(z_0)$, where P_i is the half-space consisting of all the points $z \in E^n$ for which $(z - z_0)\,\mathrm{grad}\, g^i(z_0) \leqslant 0$. Therefore, M is a convex cone with a vertex z_0. Let s be the number of indexes entering into the zone of activity $I(z_0)$. For $s = 0$ point z_0 is an *interior* point of set Ω and the proof of the theorem is trivial. Thus in the following we assume that $s \geqslant 1$.

To be definite, let us assume that the first s constraints (34.1) are active at point z_0, the others being inactive, that is, $I(z_0) = \{1, \ldots, s\}$. Then set Ω is defined close to z_0 by the inequalities

$$g^1(z) \leqslant 0, \ \ldots, \ g^s(z) \leqslant 0, \tag{34.6}$$

and system (34.4) consists of the vectors

$$\mathrm{grad}\, g^i(z_0), \qquad i = 1, \ldots, s. \tag{34.7}$$

We now select a ray l such that a unit vector $\boldsymbol{a}$ with the direction of this ray forms a *negative* scalar product with each vector (34.7):

$$\boldsymbol{a}\,\mathrm{grad}\, g^i(z_0) < 0, \qquad i = 1, \ldots, s.$$

For each $i = 1, \ldots, s$ we construct a mapping ψ_i (defined close to point z_0), just as was done during the proof of Theorem 34.4. Therefore,

$$\psi_i(z) = z + \alpha_i(z)\,\boldsymbol{a}, \quad g^i(\psi_i(z)) \leqslant 0 \quad \text{for} \quad z \in P_i.$$

The symbol $\alpha(z)$ will denote the *largest* of the numbers $\alpha_1(z), \ldots, \alpha_s(z)$. Since functions $\alpha_1(z), \ldots, \alpha_s(z)$ are continuous, this means that function $\alpha(z)$ (defined close to point z_0) is also continuous. Now we set

$$\psi(z) = z + \alpha(z)\,\boldsymbol{a}.$$

Function $\psi(z)$ is defined close to z_0 and satisfies condition 1 of Definition 34.1 (since functions $\psi_1(z), \ldots, \psi_s(z)$ satisfy this condition).

It remains to show that mapping ψ satisfies condition 2, that is, that $\psi(z) \in \Omega$ for $z \in M$ (if point z is sufficiently close to z_0). We have $g^i(\psi(z)) - g^i(\psi_i(z)) = (\psi(z) - \psi_i(z))\,\mathrm{grad}\, g^i(\xi) = (\alpha(z) - \alpha_i(z))\,\boldsymbol{a}\,\mathrm{grad}\, g^i(\xi)$, where ξ is some point on the segment joining points $\psi(z)$ and $\psi_i(z)$. But since $\alpha(z) \geqslant \alpha_i(z)$ and $\boldsymbol{a}\,\mathrm{grad}\, g^i(\xi) \leqslant 0$ (provided that point z is close enough to z_0), it follows that $g^i(\psi(z)) - g^i(\psi_i(z)) \leqslant 0$. Since $M \subset P_i$, for any $i = 1, \ldots, s$ we will have (for $z \in M$)

$$g^i(\psi(z)) \leqslant g^i(\psi_i(z)) \leqslant 0.$$

Thus, if point $z \in M$ is sufficiently close to z_0, then for all $i \in I(z_0)$ the inequalities $g_i(\psi(z)) \leqslant 0$ are valid, implying that $\psi(z) \in \Omega$. Hence condition 2 is satisfied as well.

THEOREM 34.7. *Let Ω be some set in E^p and let M be a covering of set Ω at a point $Q \in \Omega$. Moreover, let $\varphi\colon \Omega \to E^q$ be a continuous mapping having a tangent affine mapping $f\colon E^p \to E^q$ at point Q. Finally, let C be an interior point of cone $f(M)$, relative to its carrier plane P,* *different from its vertex $Q_1 = \varphi(Q) = f(Q)$. Then there will exist in E^q a closed convex cone $N \subset f(M)$ with vertex Q_1, having P as its carrier plane, such that C is an interior point of cone N and N is a covering of set $\varphi(\Omega)$ at point Q_1.*

Proof. Since vector $e_1 = \overrightarrow{CQ_1}$ is different from zero, it can be augmented by vectors $e_2, \ldots, e_n$ until some basis $e_1, e_2, \ldots, e_n$ of plane P is reached, where $n = \dim P = \dim f(M)$. Here we can assume (after reducing, if necessary, the lengths of vectors $e_2, \ldots, e_n$ without altering their directions) that points $A_1, A_2, \ldots, A_n$, defined by the relations

$$\overrightarrow{CA_1} = -e_1 - e_2 - \ldots - e_n,$$

$$\overrightarrow{CA_i} = e_i, \qquad i = 2, \ldots, n,$$

belong to cone $f(M)$ (since $C \in \operatorname{relint} M$; Figure 123).

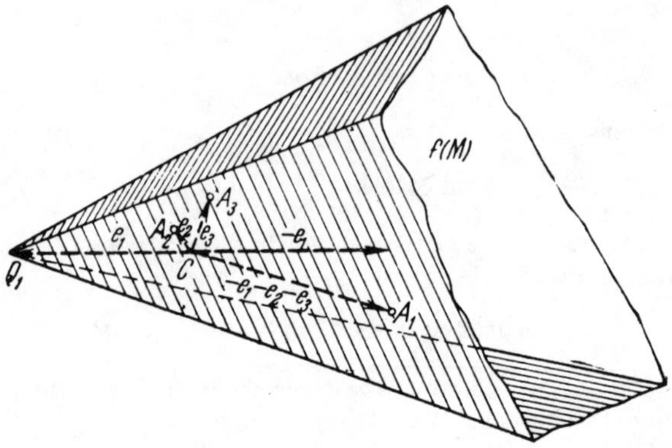

FIGURE 123

Since obviously

$$\frac{1}{n+1}\overrightarrow{CQ_1} + \frac{1}{n+1}\overrightarrow{CA_1} + \frac{1}{n+1}\overrightarrow{CA_2} + \ldots + \frac{1}{n+1}\overrightarrow{CA_n} = 0 = \overrightarrow{CC},$$

it follows (cf. (17.3) and (17.4)) that

$$C = \frac{1}{n+1}Q_1 + \frac{1}{n+1}A_1 + \frac{1}{n+1}A_2 + \ldots + \frac{1}{n+1}A_n, \tag{34.8}$$

and hence point C belongs to the convex hull of points $Q_1, A_1, \ldots, A_n$. If points $Q_1, A_1, \ldots, A_n$ were in one $(n-1)$-dimensional plane, then point C would also have to be in this plane, which is impossible, since the vectors

$$e_1 = \overrightarrow{CQ_1}, \quad e_2 = \overrightarrow{CA_2}, \ldots, e_n = \overrightarrow{CA_n}$$

are linearly independent, and thus points $C, Q_1, A_2, \ldots, A_n$ do not lie in one $(n-1)$-dimensional plane. Consequently, points $Q_1, A_1, \ldots, A_n$ do not lie in one $(n-1)$-dimensional plane, that is, they are vertices of some n-dimensional simplex T. In view of (34.8), point C is seen to be an *interior* point of this simplex (see Example 25.3).

Let N be the convex cone formed by all possible rays originating at point Q_1 and passing through the points of the $(n-1)$-dimensional simplex $[A_1, \ldots, A_n]$ (Figure 124). Let us show that this cone is the desired one. Since all the points $A_1, \ldots, A_n$ belong to the cone $f(M)$ having Q_1 as its vertex, we know that $N \subset f(M)$. In addition, it is obvious that $T \subset N$. Hence it follows that cone N has P as its carrier plane and that C is an interior point of cone N. It remains to prove that N is a covering of set $\varphi(\Omega)$ at point Q_1.

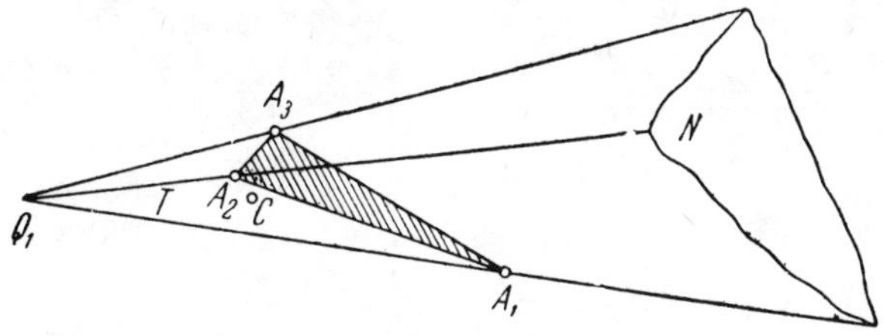

FIGURE 124

Since points $A_1, \ldots, A_n$ belong to set $f(M)$, there must exist points $B_1, \ldots, B_n$ of set M such that $f(B_i) = A_i$, $i = 1, \ldots, n$. Let g be an affine mapping of plane P into space E^p which carries points $Q_1, A_1, \ldots, A_n$ into points $Q, B_1, \ldots, B_n$, respectively (see Theorem 19.9). Note that points $Q, B_1, \ldots, B_n$ are vertices of an n-dimensional simplex in E^p (if these points were in one $(n-1)$-dimensional plane, then points $Q_1, A_1, \ldots, A_n$, the images of points Q, $B_1, \ldots, B_n$ for affine mapping f, would also, according to Theorem 19.3, lie in one $(n-1)$-dimensional plane, which is not the case). Affine mapping g carries cone N into a cone forming all possible rays originating at point Q and passing through the points of simplex $[B_1, \ldots, B_n]$. Since this simplex is contained in cone M, it follows that $g(N) \subset M$.

Since M is a covering of set Ω at point Q, there must exist a mapping $\psi\colon M \to E^p$ such that $\psi(A) \in \Omega$ for all points $A \in M$ sufficiently close to Q and such that ψ has the identity mapping of space E^p as its tangent mapping at point Q. Therefore, according to Theorem 33.5, mapping $\psi_1 = \varphi \circ \psi \circ g$ (defined for all points of cone N sufficiently close to Q_1) has $f \circ g$ as its tangent mapping at point Q_1. Since

$$(f \circ g)(Q_1) = f(g(Q_1)) = f(Q) = Q_1,$$
$$(f \circ g)(A_i) = f(g(A_i)) = f(B_i) = A_i, \qquad i = 1, \ldots, n,$$

we know from Theorem 19.9 that mapping $f \circ g$ coincides with the identity mapping of space E^q.

Consequently, mapping ψ_1, defined close to the vertex of cone N, has the identity mapping of space E^q as its tangent mapping at point Q_1. Moreover, for all points A of cone N close enough to Q_1, we have

$$\psi_1(A) = \varphi(\psi(g(A))) = \varphi(\psi(A')) \in \varphi(\Omega)$$

(where $A' = g(A) \in M$). This means that conditions 1 and 2 in Definition 34.1 are satisfied, and thus that N is a covering of set $\varphi(\Omega)$ at point Q_1.

35. Theorem on intersection. In this subsection we prove a theorem that will serve as a basis for all the subsequent discussion. The meaning of this theorem can be illustrated graphically as follows.

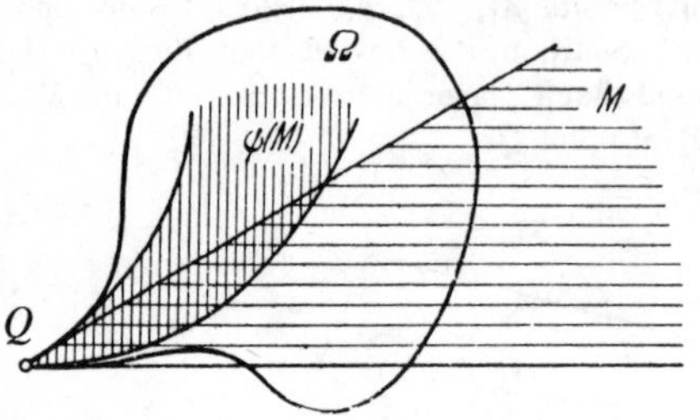

FIGURE 125

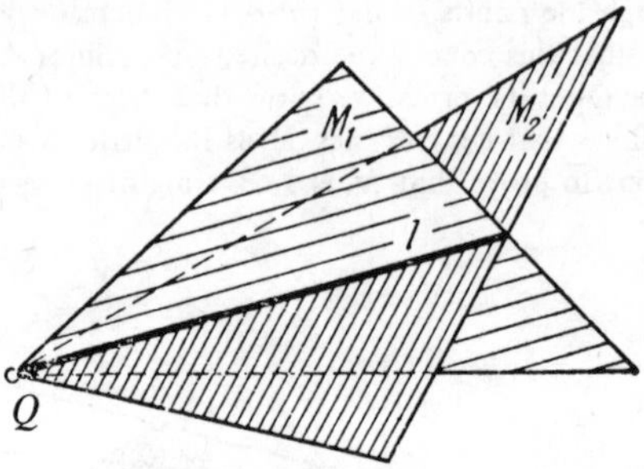

FIGURE 126

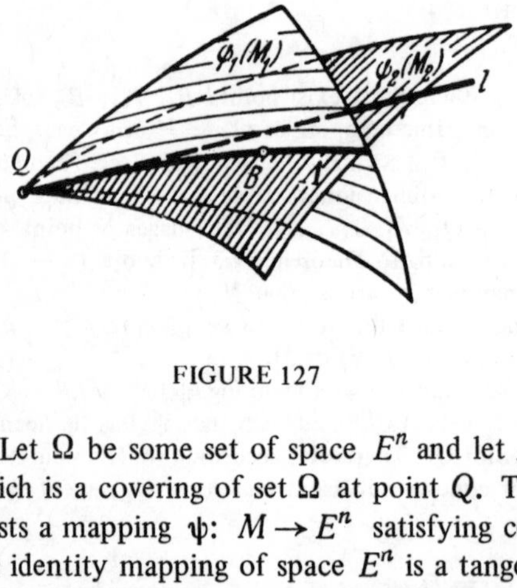

FIGURE 127

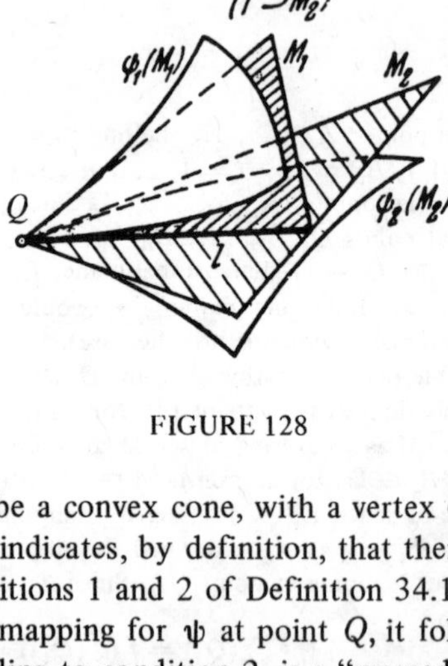

FIGURE 128

Let Ω be some set of space E^n and let M be a convex cone, with a vertex $Q \in \Omega$, which is a covering of set Ω at point Q. This indicates, by definition, that there exists a mapping ψ: $M \to E^n$ satisfying conditions 1 and 2 of Definition 34.1. Since the identity mapping of space E^n is a tangent mapping for ψ at point Q, it follows that set $\psi(M)$, which is contained in Ω according to condition 2, is a "warped cone" with a vertex Q "touching" cone M at point Q (Figure 125).

Now let Ω_1 and Ω_2 be two sets having a common point Q, and let M_1 and M_2 be coverings of these sets at point Q. Assume that cones M_1 and M_2 do not lie in one hyperplane and that they have a common ray l, which is interior for each of cones M_1 and M_2 relative to its carrier plane (Figure 126); such an arrangement of the cones will exist if they are nonseparable and if at least one of them is not a plane. Since the "warped cones" $\psi_1(M_1)$ and $\psi_2(M_2)$ (where ψ_1 and ψ_2 are the corresponding mappings satisfying conditions 1 and 2 of Definition 34.1) "touch" cones M_1 and M_2, it is likely that $\psi_1(M_1)$ and $\psi_2(M_2)$ intersect along the line Λ tangent to ray l at point Q (Figure 127). Consequently, there will exist a point $B \in \Lambda$, different from Q and belonging to both "warped cones" $\psi_1(M_1)$ and $\psi_2(M_2)$. However, since $\psi_1(M_1) \subset \Omega_1$, $\psi_2(M_2) \subset \Omega_2$, it follows that $B \in \Omega_1 \cap \Omega_2$.

Hence, *if sets* Ω_1, Ω_2 *have coverings* M_1, M_2 *at their common point* Q, *while cones* M_1, M_2 *are nonseparable and at least one of them is not a plane, there will exist a point* $B \in \Omega_1 \cap \Omega_2$ *different from* Q. This is also the gist of Theorem 35.1, to be proven below, in the case of two sets (naturally, the graphical considerations just presented do not constitute a strict proof of this theorem).

Note that nonseparability of cones M_1 and M_2 is an important condition for validity of the above statement. Actually, if there exists a hyperplane Γ dividing cones M_1 and M_2, then (even if these cones have a common ray l) the "warped cones" $\psi_1(M_1)$ and $\psi_2(M_2)$ may turn out to be on opposite sides of hyperplane Γ and they may not have any common points other than Q (Figure 128).

These considerations indicate quite clearly the meaning of Theorem 35.1. However, its proof is somewhat complicated. Thus it can be omitted during a first reading. A simpler proof (for more restrictive assumptions) will be presented in Note 35.2.

THEOREM 35.1. *Let* $\Omega_1, \ldots, \Omega_k$ *be some sets in* E^n *having a common point* Q, *and let* $M_1, \ldots, M_k$ *be coverings of sets* $\Omega_1, \ldots, \Omega_k$ *at point* Q. *If the system of convex cones* $M_1, \ldots, M_k$ *does not possess separability and if at least one of these cones is not a plane, then there will exist a point* $B \in \Omega_1 \cap \ldots \cap \Omega_k$ *different from* Q.

Proof. Since system of cones $M_1, \ldots, M_k$ does not possess separability, we know from Theorem 32.5 that there exists a point

$$O \in (\text{relint } M_1) \cap \ldots \cap (\text{relint } M_k)$$

and, in addition, there exist subspaces $L_1, \ldots, L_k$ into the direct sum of which space E^n decomposes, such that for any $i \neq j$ $(i, j = 1, \ldots, k)$ subspace L_i is contained in the carrier plane of cone M_j (here O is the zero point of space E^n, so that any plane passing through O will be a subspace). Note that $O \neq Q$, since at least one of cones $M_1, \ldots, M_k$ was assumed not to be a plane, and for this cone Q is a *boundary* point relative to its carrier plane.

Now let us select a basis $e_1, \ldots, e_n$ of space E^n and whole numbers $0 = q_0 \leqslant q_1 \leqslant q_2 \leqslant \ldots \leqslant q_k = n$, such that for any $j = 1, \ldots, k$ the vectors

$$e_{q_{j-1}+1},\ e_{q_{j-1}+2}, \ldots, e_{q_j}$$

form a basis of subspace L_j. Having selected this basis, we will retain it right up to the end of the proof. Moreover, let us assume the basis $e_1, \ldots, e_n$ to be orthonormal (this specifies the Euclidean metrics in space E^n).

Next let us introduce the concept of an r-dimensional coordinate parallelepiped in E^n. To do this, we let S be the set consisting of numbers $1, \ldots, n$ and we select an arbitrary subset $H \subset S$. Moreover, for any $i \in H$ we select some number a^i, while for any $i \in S \setminus H$ we select two numbers a^i and b^i satisfying the condition $a^i < b^i$. The *coordinate parallelepiped* is then defined as the set of all points of space E^n, the coordinates $x^1, \ldots, x^n$ of which in the system $(O'; e_1, \ldots, e_n)$, where $O' \in E^n$, satisfy the conditions (Figure 129):

$$x^i = a^i \quad \text{for} \quad i \in H;$$
$$a^i < x^i < b^i \quad \text{for} \quad i \in S \setminus H.$$

If all the numbers a^i, b^i are rational, then we can say that the coordinate parallelepiped being considered is *rational* (in system $(O'; e_1, \ldots, e_n)$). A coordinate parallelepiped constructed in this way constitutes the interior of an r-dimensional polyhedron (a closed r-dimensional parallelepiped) in E^n, where r is the number of elements of the set $S \setminus H$.

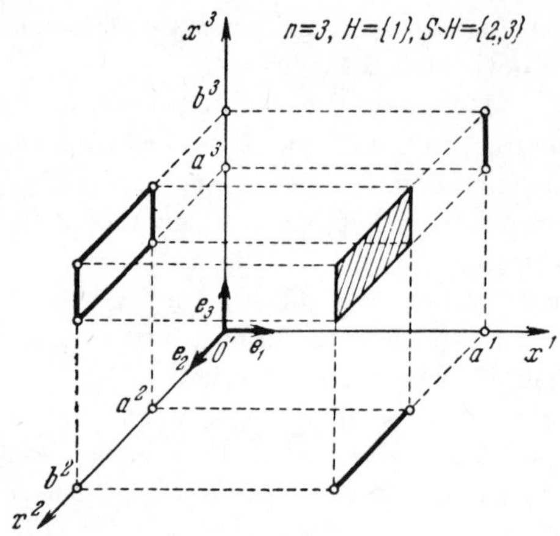

FIGURE 129

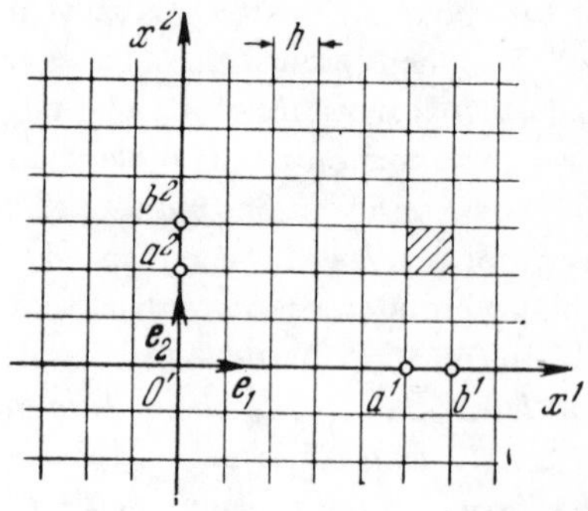

FIGURE 130

Let h be a positive number. The coordinate parallelepiped will be called a *basis h cube* in the coordinate system $(O'; e_1, \ldots, e_n)$ if for each $i = 1, \ldots, n$ we have $a^i = hq^i$, where q^i is a whole number, and also if for each $i \in S \setminus H$ the number b^i satisfies the condition $b^i = a^i + h$ (Figure 130). It is easy to verify that all the possible basis h cubes (of dimensions $0, 1, \ldots, n$) in the system $(O'; e_1, \ldots, e_n)$ are mutually disjoint, while the union of all the basis h cubes coincides with E^n. The aggregate of all the basis h cubes in system $(O'; e_1, \ldots, e_n)$ will be designated as $K_h(O')$ and the corresponding number h and point O' will be called the "*cubilage*" corresponding to the number h and the point O', (we stress especially the dependence of the cubilage on point O', since during the course of the discussion basis $e_1, \ldots, e_n$ will, as already noted, be retained unchanged, but we will need cubilages derived for a different choice of the origin O').

Next let us prove the following statement. *If $O_1, \ldots, O_k$ are points in E^n such that for any $i \neq j$ $(i, j = 1, \ldots, k)$ all the coordinates of the vector $\frac{1}{h}(O_j - O_i)$ are nonintegral, and if for each $j = 1, \ldots, k$ some basis h cube R_j is selected in cubilage $K_h(O_j)$, then the intersection $R_1 \cap \ldots \cap R_k$, if it is nonempty, is a coordinate parallelepiped of dimension $r_1 + \ldots + r_k - (k-1)n$, where $r_j = \dim R_j$, $j = 1, \ldots, k$* (Figure 131).

Let H_j be the subset of set S corresponding to h cube R_j, so that h cube R_j is defined in coordinate system $(O_j; e_1, \ldots, e_n)$ by relations of the form

$$\begin{aligned} x^i &= hq_j^i && \text{for} \quad i \in H_j; \\ hq_j^i < x^i &< h(q_j^i + 1) && \text{for} \quad i \in S \setminus H_j. \end{aligned} \tag{35.1}$$

It is easy to see that, if any two of sets $H_1, \ldots, H_k$ have a nonempty intersection, the intersection $R_1 \cap \ldots \cap R_k$ will be empty. Actually, if, for example, $i \in H_1 \cap H_2$, then the equation $x^i = hq_1^i$ will enter into the definition of h cube R_1, while the equation $x^i = hq_2^i$ will enter into the definition of h cube R_2. In other words, h cube R_1 lies in the hyperplane defined in system of coordinates $(O_1; e_1, \ldots, e_n)$ by the equation $x^i = hq_1^i$, while h cube R_2 lies in the hyperplane defined in system $(O_2;\ e_1, \ldots, e_n)$ by the equation $x^i = hq_2^i$ (Figure 132). Since numbers q_1^i and q_2^i are integral, while the ith coordinate of vector $\frac{1}{h}(O_2 - O_1)$ is nonintegral, it follows that these two hyperplanes are parallel but not coinciding, so that $R_1 \cap R_2 = \varnothing$ and thus $R_1 \cap \ldots \cap R_k = \varnothing$.

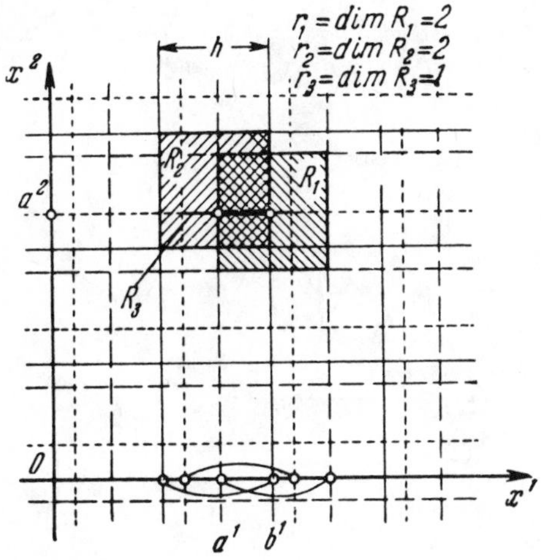

FIGURE 131

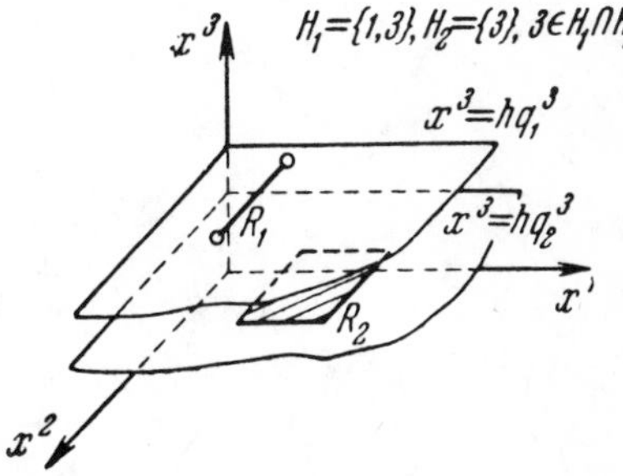

FIGURE 132

Hence if intersection $R_1 \cap \ldots \cap R_k$ is nonempty, then all the sets $H_1, \ldots, H_k$ are mutually disjoint, so that intersection $R_1 \cap \ldots \cap R_k$ is defined by as many equations as there are elements in the set

$$H = H_1 \cup \ldots \cup H_k$$

(in Figure 131, $H_1 = \varnothing$, $H_2 = \varnothing$, $H_3 = \{2\}$). However, for the rest of the coordinates (the numbers of which do not enter into set H), intersection $R_1 \cap \ldots \cap R_k$ is defined by inequalities of the type $a^i < x^i < b^i$ (thus, in Figure 131 the interval (a^1, b^1) is the *intersection* of three intervals). From this it is clear that $R_1 \cap \ldots \cap R_k$ is a coordinate parallelepiped, its dimension being $n - p$, where p is the number of elements of set H. Since $p = p_1 + \ldots + p_k$, where p_j is the number of elements of set H_j, $j = 1, \ldots, k$, it follows that

$$\begin{aligned}\dim(R_1 \cap \ldots \cap R_k) &= n - p = n - (p_1 + \ldots + p_k) = \\ &= (n - p_1) + \ldots + (n - p_k) - (k-1)\,n = \dim R_1 + \ldots + \dim R_k - (k-1)\,n,\end{aligned}$$

as was confirmed.

Now let us introduce the concept of a *chain*. For example, an *r-dimensional chain* (where r is one of the numbers $0, 1, \ldots, n$) is defined as a finite number of r-dimensional coordinate parallelepipeds written one after the other and joined by + signs. For example, we write

$$\xi^r = P_1 + P_2 + \ldots + P_s$$

and we say that ξ^r is an r-dimensional chain *consisting* of the r-dimensional coordinate parallelepipeds $P_1, P_2, \ldots, P_s$. Graphically, a chain can be represented as a "broken" r-dimensional surface (possibly consisting of several individual pieces; see Figure 133). Let us cancel out any *doubly* appearing parallelepiped in any chain, for instance,

$$P_1 + P_1 + P_2 + P_3 = P_2 + P_3,$$
$$P_1 + P_2 + P_2 + P_2 = P_1 + P_2,$$
$$P_1 + P_1 + P_2 + P_2 = 0$$

(in other words, we will, as is still said, consider *chains modulo* 2). Chains of the same dimension can be added together like polynomials (canceling out doubly occurring parallelepipeds, if they appear).

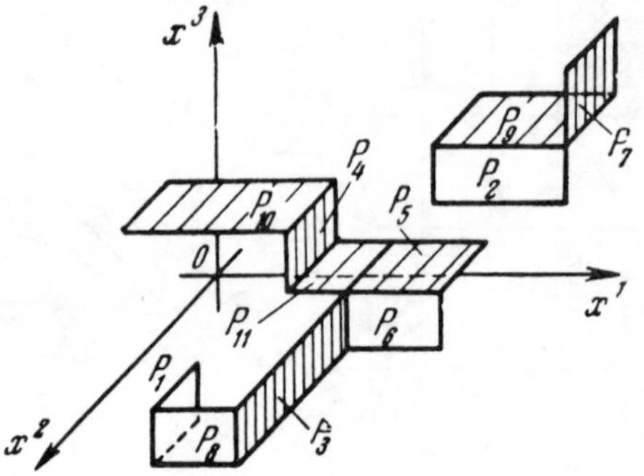

FIGURE 133

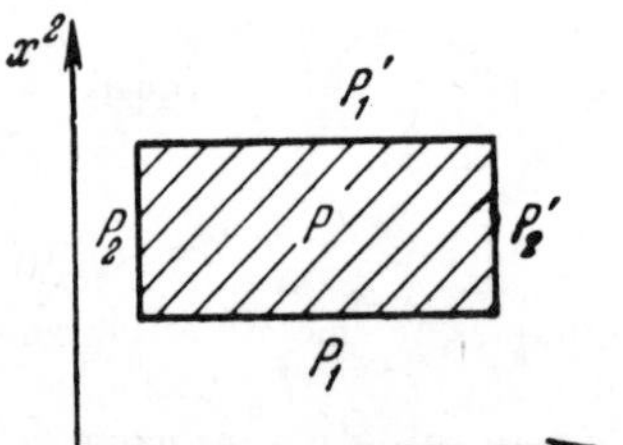

FIGURE 134

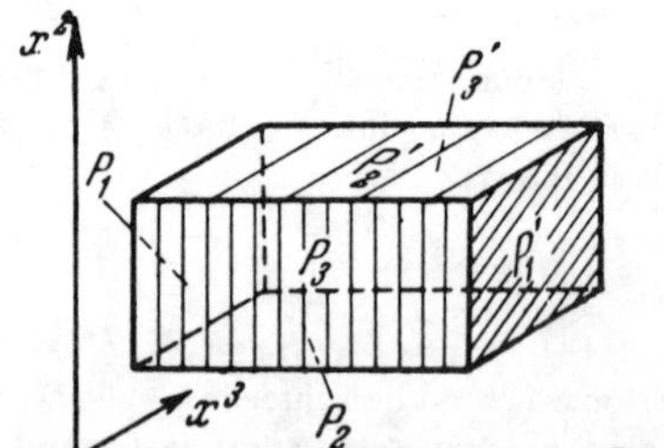

FIGURE 135

The boundary of an r-dimensional parallelepiped consists of $2r$ principal (that is, $(r-1)$-dimensional) faces, so that the boundary of any coordinate parallelepiped P can be portrayed as a chain, by writing out all of its principal faces. Let us designate this chain (the boundary of coordinate parallelepiped P) as dP. For example, in Figure 134

$$dP = P_1 + P_2 + P_1' + P_2',$$

and in Figure 135 the boundary dP of the three-dimensional parallelepiped P is expressed by the formula

$$dP = P_1 + P_2 + P_3 + P_1' + P_2' + P_3'.$$

Therefore, the boundary dP of any r-dimensional coordinate parallelepiped is an $(r-1)$-dimensional chain.

Now let us define the boundary of any chain; specifically, the *boundary* of an r-dimensional chain

$$\xi^r = P_1 + P_2 + \dots + P_s$$

is taken to be the $(r-1)$-dimensional chain

$$d\xi^r = dP_1 + dP_2 + \dots + dP_s.$$

For example, Figure 136 portrays the boundary of the chain $\xi^2 = P_1 + \dots + P_{10}$ shown in Figure 133. Note, however, that this definition has meaning only for $r > 0$. For any zero-dimensional chain the boundary is taken to be zero.

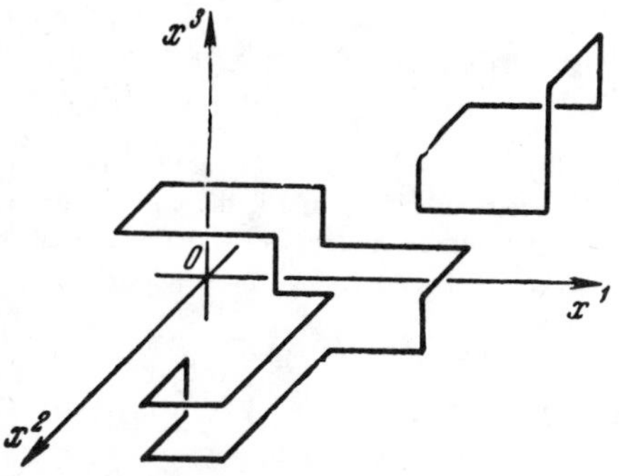

FIGURE 136

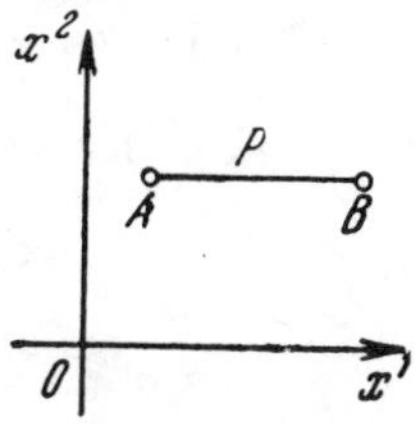

FIGURE 137

Obviously, given this definition of the boundary, the following relation will be valid for any two r-dimensional chains ξ^r, η^r:

$$d(\xi^r + \eta^r) = d\xi^r + d\eta^r.$$

Note especially the case $r = 1$, that is, the case of *one-dimensional* chains. Each one-dimensional coordinate cube constitutes an interval parallel to one of the coordinate axes. Its boundary consists of *two* points; for example, in Figure 137 we have: $dP = A + B$. Hence it follows that the *boundary of any one-dimensional chain consists of an even number of zero-dimensional parallelepipeds (points).*

Now let $O_1, \dots, O_k$ be points in E^n such that for any $i \neq j$ $(i, j = 1, \dots, k)$ all the coordinates of the vector $\frac{1}{h}(O_j - O_i)$ are nonintegral. In addition, let $r_1, \dots, r_k$ be whole numbers for which $0 \leqslant r_j \leqslant n$ and let ξ_j be some chain of dimension r_j consisting of the basis h cubes of the cubilage $K_h(O_j)$, $j = 1, \dots, k$. Here we assume that $r_1 + \dots + r_k - (k-1)n > 0$, and we take for each $j = 1, \dots, k$ some h cube R_j (of dimension r_j) entering into chain ξ_j. Then, according to what was proven earlier, intersection $R_1 \cap \dots \cap R_k$, if it is nonempty, will be a coordinate parallelepiped of dimension $r_1 + \dots + r_k - (k-1)n$. By taking the sum of *all* the coordinate parallelepipeds so obtained (where R_j runs through all the basis h cubes entering into chain ξ_j, $j = 1, \dots, k$), we obtain some chain of dimension $r_1 + \dots + r_k - (k-1)n$, which is known as the *intersection* of chains $\xi_1, \dots, \xi_k$ and is designated as $\xi_1 \times \dots \times \xi_k$.

Now we need the following formula,* which gives an expression for the boundary of the chain $\xi_1 \times \xi_2 \times \ldots \times \xi_k$:

$$d(\xi_1 \times \xi_2 \times \ldots \times \xi_k) = \\ = (d\xi_1) \times \xi_2 \times \ldots \times \xi_k + \xi_1 \times (d\xi_2) \times \xi_3 \times \ldots \times \xi_k + \ldots \\ \ldots + \xi_1 \times \xi_2 \times \ldots \times (d\xi_k). \quad (35.2)$$

It will be sufficient to prove formula (35.2) for the case in which each chain ξ_j consists only of a *single* h cube $\xi_j = R_j$, where $\dim R_j = r_j$ (the general case of formula (35.2) is obtained from this particular case by an obvious summation).

For instance, we have to prove the relation

$$d(R_1 \times R_2 \times \ldots \times R_k) = \\ = (dR_1) \times R_2 \times \ldots \times R_k + R_1 \times (dR_2) \times \ldots \times R_k + \ldots \\ \ldots + R_1 \times R_2 \times \ldots \times (dR_k), \quad (35.3)$$

where R_j is the basis h cube of the cubilage $K_h(O_j)$, $j = 1, \ldots, k$. If h cube R_j is defined in system of coordinates $(O_j;\ e_1, \ldots, e_n)$ by relations (35.1), and if the coordinates of point O_j in system $(O;\ e_1, \ldots, e_n)$ are called $(o_j^1, \ldots, o_j^n)$, then h cube R_j is defined in system $(O;\ e_1, \ldots, e_n)$ by the relations

$$\begin{aligned} x^i &= a_j^i && \text{for} \quad i \in H_j; \\ a_j^i < x^i &< a_j^i + h && \text{for} \quad i \in S \setminus H_j, \end{aligned} \quad (35.4)$$

where $a_j^i = hq_j^i + o_j^i$ (Figure 138; cf. Figure 130). Consequently, the coordinate parallelepiped $R_1 \cap \ldots \cap R_k$ is defined in system $(O;\ e_1, \ldots, e_n)$ by the relations

$$\begin{aligned} x^i &= a_j^i && \text{for} \quad i \in H_j, \quad j = 1, \ldots, k; \\ b^i < x^i &< c^i && \text{for} \quad i \in S \setminus H, \end{aligned} \quad (35.5)$$

where $H = H_1 \cup \ldots \cup H_k$, and where (b^i, c^i) denotes (for a given fixed $i \in S \setminus H$) the intersection of *all* intervals $(a_j^i, a_j^i + h)$, $j = 1, \ldots, k$. Note that b^i is (for any $i \in S \setminus H$) the end of *only one* of the intervals $(a_j^i, a_j^i + h)$, $j = 1, \ldots, k$ (the same is true for c^i). This follows from the relation

$$a_j^i - a_l^i = h(q_j^i - q_l^i) + (o_j^i - o_l^i) = h\left((q_j^i - q_l^i) + \frac{o_j^i - o_l^i}{h}\right) \neq 0,$$

which is valid because the number $q_j^i - q_l^i$ is whole, while the number $\dfrac{o_j^i - o_l^i}{h}$ is not whole, in view of the choice of points $O_1, \ldots, O_k$.

Let us consider the jth term $R_1 \times \ldots \times R_{j-1} \times (dR_j) \times R_{j+1} \times \ldots \times R_k$ on the right-hand side of the relation (35.3) which is to be proven. Chain dR_j, that is, the boundary of h cube R_j, can be portrayed as follows (Figure 139). We choose some index $i_0 \in S \setminus H_j$ and we

* It is a particular case of *Lefschetz's formula*; see: Glezerman, M.E., and L.S. Pontryagin. *Peresecheniya v mnogoobraziyakh (Intersections in Manifolds)*. – Usp. Mat. Nauk, 2, No. 1. 1947.

consider the coordinate h cube defined in system $(O;\ e_1, \ldots, e_n)$ by the relations

$$\begin{aligned} x^i &= a_j^i && \text{for} \quad i \in H_j; \\ x^i &= c && \text{for} \quad i = i_0; \\ a_j^i < x^i &< a_j^i + h && \text{for} \quad i \in S \setminus H_j, \quad i \neq i_0, \end{aligned} \tag{35.6}$$

where c is one of the numbers $a_j^{i_0}$ or $a_j^{i_0} + h$. Each such cube is a principal face of cube R_j, and all the principal faces of cube R_j can be described in this manner (where i_0 runs through the set $S \setminus H_j$ and each time the number c takes on values $a_j^{i_0}$, $a_j^{i_0} + h$).

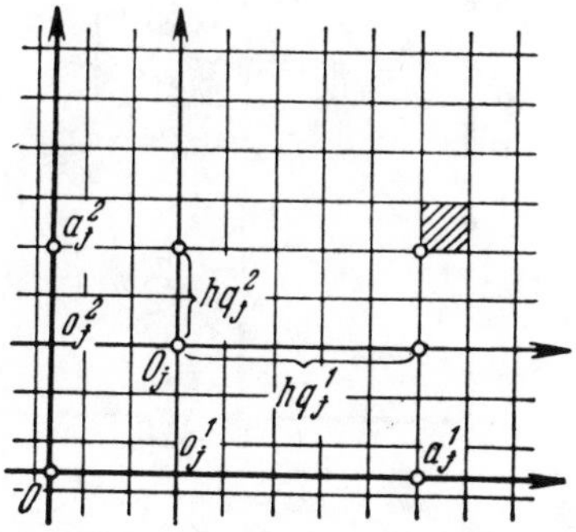

FIGURE 138

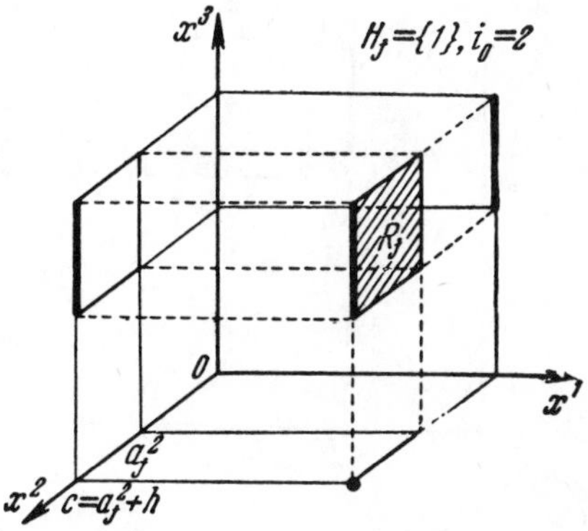

FIGURE 139

Let R'_j be one of the principal faces of cube R_j (described by relations (35.6)), and let us find the intersection $R_1 \times \ldots \times R_{j-1} \times R'_j \times R_{j+1} \times \ldots \times R_k$. To do this, let H'_j be the set containing all the elements of set H_j and the number i_0. Cube R'_j is defined by relations of the form

$$\begin{aligned} x^i &= \text{const} && \text{for} \quad i \in H'_j, \\ a_j^i < x^i &< a_j^i + h && \text{for} \quad i \in S \setminus H'_j. \end{aligned}$$

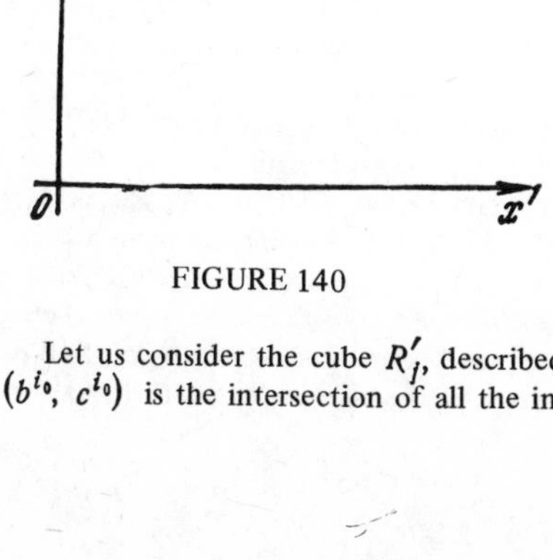

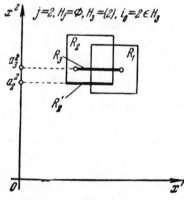

FIGURE 140

Consequently, if sets $H_1, \ldots, H_{j-1}, H'_j, H_{j+1}, \ldots, H_k$ are not mutually disjoint, then intersection $R_1 \cap \ldots \cap R'_j \cap \ldots \cap R_k$ is empty. In other words, if i_0 belongs to one of sets $H_1, \ldots, H_{j-1}, H_{j+1}, \ldots, H_k$, then intersection $R_1 \times \ldots \times R'_j \times \ldots \times R_k$ is zero (Figure 140; cf. Figure 131). Therefore, we only have to consider the case where $i_0 \in S \setminus H$.

Thus the right-hand side of relation (35.3) is the sum of terms having the form $R_1 \times \ldots \times R'_j \times \ldots \times R_k$, where R'_j is cube (35.6), with i_0 running through the set $S \setminus H$; further, $j = 1, \ldots, k$, while the number c in relations (35.6) assumes the values $a_j^{i_0}$, $a_j^{i_0} + h$.

Let us consider the cube R'_j, described by relations (35.6), where $c = a_j^{i_0}$. Since interval (b^{i_0}, c^{i_0}) is the intersection of all the intervals $(a_j^{i_0}, a_j^{i_0} + h)$, $j = 1, \ldots, k$, it follows that

$a_j^{i_0} \leqslant b^{i_0}$. It is easy to see that if (for the index j being considered) this inequality is strict: $a_j^{i_0} < b^{i_0}$, then intersection $R_1 \cap \ldots \cap R'_j \cap \ldots \cap R_k$ is empty. Actually, b^{i_0} is one of the numbers $a_1^{i_0}, \ldots, a_k^{i_0}$. Since $a_j^{i_0} < b^{i_0}$, it follows that $b^{i_0} = a_l^{i_0}$, where $l \neq j$. But the equation $x^{i_0} = a_j^{i_0}$ enters into the definition of cube (35.6), while the inequality $a_l^{i_0} < x^{i_0} < a_l^{i_0} + h$, that is, $x^{i_0} > b^{i_0}$, enters into the definition of cube R_l. Since the system of relations $x^{i_0} = a_j^{i_0}$, $x^{i_0} > b^{i_0}$ is obviously disjoint for $a_j^{i_0} < b^{i_0}$ (Figure 141), intersection $R_1 \cap \ldots \cap R'_j \cap \ldots \cap R_k$ must be empty in this case.

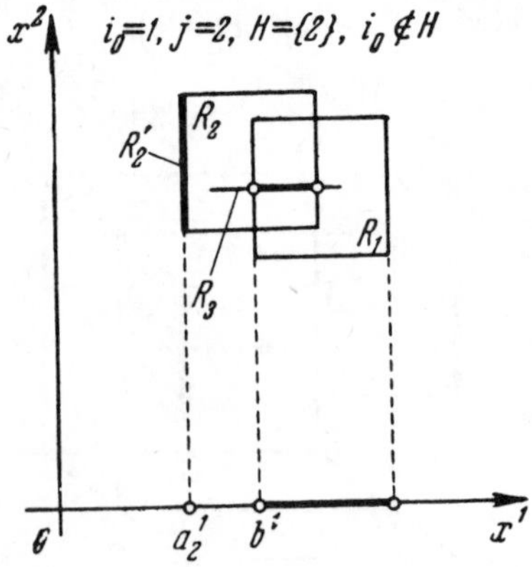

FIGURE 141

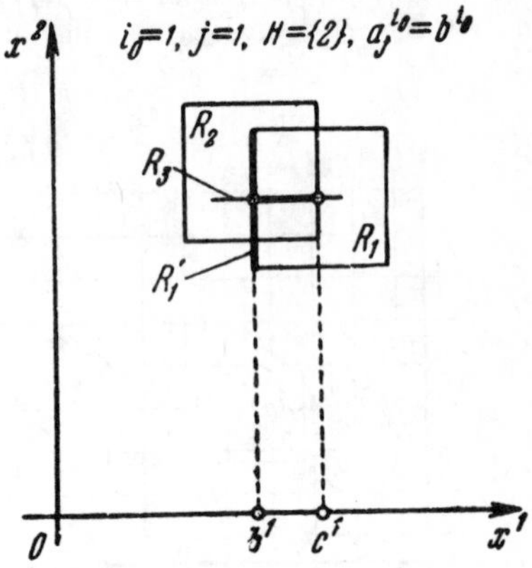

FIGURE 142

Hence if in (35.6) $c = a_j^{i_0}$, then this means that intersection $R_1 \times \ldots \times R'_j \times \ldots \times R_k$ will be different from zero only on condition that $b^{i_0} = a_j^{i_0}$. It is easy to see what this intersection is equal to when the given condition is satisfied. Cube R'_j is obtained if in the system of relations defining cube R_j (see 35.4)) the inequality $a_j^{i_0} < x^{i_0} < a_j^{i_0} + h$ is replaced by the equality $x^{i_0} = b^{i_0} = a_j^{i_0}$. Thus intersection $R_1 \times \ldots \times R'_j \times \ldots \times R_k$ is obtained if in the system of relations (35.5) defining parallelepiped $R_1 \cap \ldots \cap R_k$ the inequality $b^{i_0} < x^{i_0} < c^{i_0}$ is replaced by the equality $x^{i_0} = b^{i_0}$. Therefore, if the condition $b^{i_0} = a_j^{i_0}$ is satisfied, then intersection $R_1 \times \ldots \times R'_j \times \ldots \times R_k$ is a principal face $S^{i_0}(b^{i_0})$ of the parallelepiped $R_1 \cap \ldots \cap R_k$ defined by the condition $x^{i_0} = b^{i_0}$ (Figure 142).

Similarly, if $c = a_j^{i_0} + h$ in relations (35.6), then intersection $R_1 \times \ldots \times R'_j \times \ldots \times R_k$ will be different from zero only on condition that $c^{i_0} = a_j^{i_0} + h$, and in this case it will be a principal face $S^{i_0}(c^{i_0})$ of the parallelepiped $R_1 \cap \ldots \cap R_k$ defined by the condition $x^{i_0} = c^{i_0}$.

Now let us fix some $i_0 \in S \setminus H$. Then, as we know, there will exist *only one* value j (= 1, ..., k) for which $a_j^{i_0} = b^{i_0}$ (cf. Figure 131). Therefore, of all the terms of form $R_1 \times \ldots \times R'_j \times \ldots \times R_k$ on the right-hand side of (35.3) for which $i = i_0$ and $c = a_j^{i_0}$ ($j = 1, \ldots, k$) *only one* is different from zero and, in view of the foregoing, this nonzero term is equal to $S^{i_0}(b^{i_0})$. Similarly, of all the terms on the right-hand side of (35.3) for which $i = i_0$ and $c = a_j^{i_0} + h$ ($j = 1, \ldots, k$), *only one* is different from zero, and this nonzero term is equal to $S^{i_0}(c^{i_0})$. Thus the right-hand side of (35.3) is equal to $\Sigma(S^{i_0}(b^{i_0}) + S^{i_0}(c^{i_0}))$, where the sum is over all the $i_0 \in S \setminus H$. However, this is also obviously the boundary of parallelepiped $R_1 \times \ldots \times R_k$. Consequently, relation (35.3) (and together with it (35.2) as well) is completely proven.

Before passing directly to the proof of Theorem 35.1, let us formulate another statement regarding chains. For this formulation we introduce the concept of the *field of a chain.* Let $\xi^r = P_1 + \ldots + P_s$ be some r-dimensional chain. Its field is defined as the union of all parallelepipeds $P_1, \ldots, P_s$ and all their faces (that is, the union of the *closures* of parallelepipeds

$P_1, \ldots, P_s$). The field of chain ξ^r will be called $|\xi^r|$; it is a closed bounded set of space E^n (cf. Figure 133). Note that for any chain ξ^r the relation $|d\xi^r| \subset |\xi^r|$ is valid (cf. Figure 136). The required statement can now be formulated as follows.

Let $K_h(O')$ be some cubilage of space E^n and let ξ^r be an arbitrary r-dimensional chain composed of cubes of this cubilage. If $\gamma\ (= h\sqrt{n})$ denotes the diameter of the n-dimensional cubes of the given cubilage, and if $f\colon |\xi^r| \to E^n$ is a continuous mapping and d is a number such that $\rho(x, f(x)) \leqslant d$ for any point $x \in |\xi^r|$, then there exists in cubilage $K_h(O')$ chains ζ^{r+1}, ζ^r, η^r such that the chain ζ^{r+1} lies in the $(d+\gamma)$ neighborhood of set $|\xi^r|$, chain ζ^r lies in the γ neighborhood of set $f(|\xi^r|)$, and chain η^r lies in the $(d+\gamma)$ neighborhood of set $|d\xi^r|$, the following relations being valid:

$$d\zeta^{r+1} = \xi^r + \zeta^r + \eta^r, \quad d\zeta^r = d\xi^r + d\eta^r.$$

This statement (which in the following will be called the *approximation theorem*) is proven with the aid of algebraic topology. Let us outline it briefly. For any point $x \in |\xi^r|$ let $\Phi_t(x)$ be the point $tx + (1-t)f(x)$ (Figure 143). Then Φ_t, $0 \leqslant t \leqslant 1$, constitutes a *continuous deformation* of set $|\xi^r|$ taking place in the d neighborhood of set $|\xi^r|$, while the deformation of set $|d\xi^r| \subset |\xi^r|$ takes place in the d neighborhood of set $|d\xi^r|$. This deformation causes an $(r+1)$-dimensional *continuous* chain $D(\xi^r)$ (also known as a "deformation" chain) to originate from chain ξ^r, and a continuous r-dimensional chain $D(d\xi^r)$ to originate from chain $d\xi^r$, with

$$d(D(\xi^r)) = \xi^r + f(\xi^r) + D(d\xi^r),$$

where $f(\xi^r)$ is a continuous r-dimensional chain, the image of chain ξ^r for mapping f (Figure 144). By applying to chains $D(\xi^r)$, $f(\xi^r)$, $D(d\xi^r)$ the theorem of cell approximation (for the case where cubilage $K_h(O')$ is taken to be a cell division (a complex) of space E^n), we also obtain from them the chains ζ^{r+1}, ζ^r, η^r satisfying the required conditions (Figure 145). For the theorem of cell approximation and the other details of this proof, we refer the reader to texts on algebraic topology.

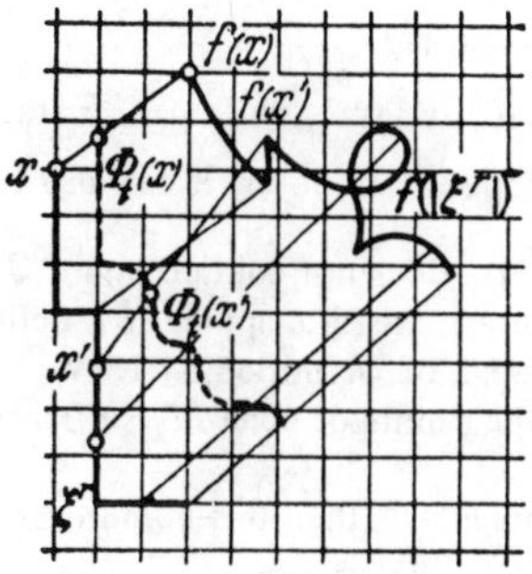

FIGURE 143

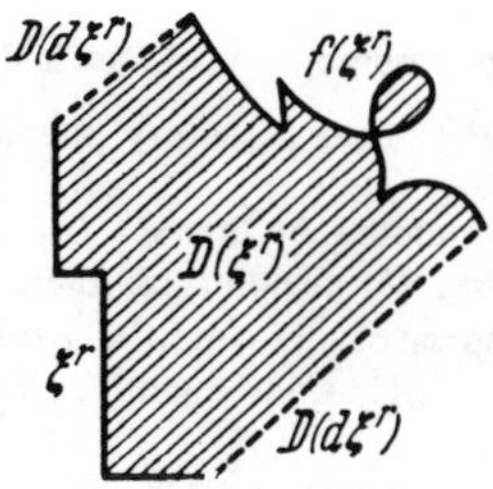

FIGURE 144

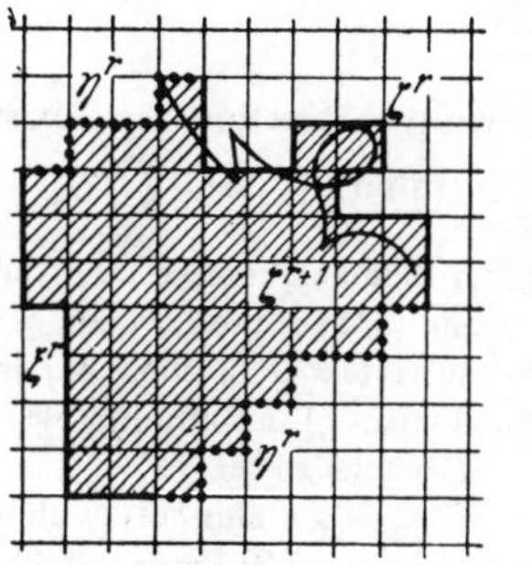

FIGURE 145

Now let us pass directly to the proof of Theorem 35.1. If $r_i = n - \dim L_i$, $i = 1, \ldots, k$, then we have

$$r_1 + \ldots + r_k - (k-1)n = n - (\dim L_1 + \ldots + \dim L_k) = 0.$$

Let us also select points $O_1, \ldots, O_k$ such that in system $(O; e_1, \ldots, e_n)$ point O_j has coordinates $(o_j^1, \ldots o_j^n)$ satisfying the following two conditions:

1) $o_j^i = 0$ for $i = q_{j-1}+1, q_{j-1}+2, \ldots, q_j$; $j = 1, \ldots, k$ (that is, point O_j has zero coordinates relative to axes situated in plane L_j);

2) numbers $o_j^i - o_l^i$ are irrational for $j \neq l$ $(i = 1, \ldots, n;\ j,\ l = 1, \ldots, k)$. For example, such points can be obtained if all the numbers o_j^i for $i = q_{j-1}+1,\ q_{j-1}+2, \ldots, q_j$; $j = 1, \ldots, k$, are set equal to zero, while as the rest of the numbers o_j^i the numbers π, π^2, $\pi^3, \ldots, \pi^{(k-1)n}$ are taken in any order.

For such a judicious choice of points $O_1, \ldots, O_k$, the number $\dfrac{o_j^i - o_l^i}{h}$ will be *nonintegral* for any $j \neq l$ $(i = 1, \ldots, n;\ j,\ l = 1, \ldots, k)$ and for any rational h, so that for any rational h we can apply formula (35.2) to cubilages $K_h(O_1), \ldots, K_h(O_k)$. Moreover, thanks to this choice of points $O_1, \ldots, O_k$, the equations

$$x^i = 0 \quad \text{for} \quad i = q_{j-1}+1,\ q_{j-1}+2, \ldots, q_j$$

(where j is one of the numbers $1, \ldots, k$) define in system $(O_j;\ e_1, \ldots, e_n)$ a plane G_j which passes through point O (that is, it is a subspace) and has a dimension $n - (q_j - q_{j-1}) = n - \dim L_j = r_j$. In addition, subspace G_j contains subspace L_i for any $i \neq j$, so that it decomposes into the direct sum of the subspaces $L_1, \ldots, L_{j-1}, L_{j+1}, \ldots, L_k$. Consequently, G_j is contained in the carrier plane of cone M_j.

Let P be some set of space E^n. In addition, let $U_\delta(P)$ be the union of all open spheres of radius δ centered on points of set P; in other words, $U_\delta(P)$ is the δ neighborhood of set P (in space E^n).

For each $j = 1, \ldots, k$ we select a coordinate parallelepiped P_j having G_j as its carrier plane and containing point O. Here we assume that $\overline{P}_j \subset M_j$ and $Q \notin \overline{P}_j$, where $\overline{P}_j = P_j \cup (\mathrm{relbd}\, P_j)$ is the corresponding closed parallelepiped. Since intersection $G_1 \cap \ldots \cap G_k$ consists only of one point O, it follows that intersection $\overline{P}_1 \cap \ldots \cap \overline{P}_k$ contains only one point O, which is an interior point of each of the polyhedrons $\overline{P}_j$ relative to its carrier plane G_j. Hence it follows that for any $j = 1, \ldots, k$ the intersection

$$(\mathrm{relbd}\, \overline{P}_j) \cap \left(\bigcap_{i \neq j} \overline{P}_i \right)$$

is empty. Therefore, there exists a number $\delta > 0$ such that for any $j = 1, \ldots, k$ the sets $U_{4\delta}(\mathrm{relbd}\, \overline{P}_j)$ and $\bigcap_{i \neq j} U_{4\delta}(\overline{P}_i)$ have an empty intersection. Moreover, it can also be assumed (after reducing, if necessary, the number δ) that set $U_\delta(\overline{P}_j)$ does not contain point Q.

Since M_j is the covering of set Ω_j at point Q, there must exist a mapping ψ_j, defined close to the vertex Q of cone M_j and satisfying conditions 1 and 2 of Definition 34.1. We designate the domain of definition of ψ_j as M_j^*. Consequently, all the points of cone M_j sufficiently close to Q belong to set M_j^*.

If g_ε is a homothety with a center Q and a coefficient $\varepsilon > 0$, then this homothety g_ε will carry set $U_{4\delta}(P)$ into the $4\delta\varepsilon$ neighborhood of set $g_\varepsilon(P)$, that is,

$$g_\varepsilon(U_{4\delta}(P)) = U_{4\delta\varepsilon}(g_\varepsilon(P)).$$

This implies that for any $j = 1, \ldots, k$ the sets $U_{4\delta\varepsilon}(\mathrm{relbd}\, g_\varepsilon(\overline{P}_j))$ and $\bigcap_{i \neq j} U_{4\delta\varepsilon}(g_\varepsilon(\overline{P}_i))$ have an empty intersection; moreover, set $U_{\delta\varepsilon}(g_\varepsilon(\overline{P}_j))$ does not contain point Q.

Since

$$\psi_j(A) = A + o_Q(A), \quad \text{that is,} \quad \rho(A, \psi_j(A)) = o_Q(A), \tag{35.7}$$

there must exist some $\varepsilon > 0$ such that for all $j = 1, \ldots, k$ the following inequality is satisfied, with $A \in g_\varepsilon(\overline{P}_j)$:

$$\rho(A, \psi_j(A)) < \delta\varepsilon \qquad (A \in g_\varepsilon(\overline{P}_j)). \tag{35.8}$$

Let us select such a number ε and leave it unchanged in the following.

The coordinate parallelepiped $g_\varepsilon(P_j)$ (where $j = 1, \ldots, k$) is defined in system $(O; e_1, \ldots, e_n)$ by relations of the type

$$\begin{aligned} x^i &= \alpha^i_j && \text{for } i = q_{j-1}+1, q_{j-1}+2, \ldots, q_j; \\ \alpha^i_j &< x^i < \beta^i_j && \text{for other values of } i. \end{aligned} \tag{35.9}$$

The intersection $g_\varepsilon(P_1) \cap \ldots \cap g_\varepsilon(P_k)$ consists of a single point having coordinates $(\alpha^1, \ldots, \alpha^n)$ in system $(O; e_1, \ldots, e_n)$. If numbers α^i_j, β^i_j in relations (35.9) are changed slightly, then we obtain a new coordinate parallelepiped P^*_j "close" to parallelepiped $g_\varepsilon(P_j)$. Let us change the numbers α^i_j, β^i_j in (35.9) in such a way that the resulting parallelepipeds $P^*_1, \ldots, P^*_k$ "close" to parallelepipeds $g_\varepsilon(P_1), \ldots, g_\varepsilon(P_k)$ possess the properties:

a) parallelepiped P^*_j is rational in system $(O_j; e_1, \ldots, e_n)$;

b) parallelepiped $\overline{P}^*_j$ is contained in the $\delta\varepsilon$ neighborhood of set $g_\varepsilon(\overline{P}_j)$, and its boundary $\operatorname{relbd} \overline{P^*_j}$ is contained in the $\delta\varepsilon$ neighborhood of set $\operatorname{relbd} g_\varepsilon(\overline{P}_j)$;

c) intersection $P^*_1 \cap \ldots \cap P^*_k$ is nonempty (and thus consists of a single point).

Obviously, all these conditions will be satisfied if the numbers α^i_j, β^i_j in (35.9) vary only slightly during the transition from parallelepiped $g_\varepsilon(P_j)$ to parallelepiped P^*_j. Since P^*_j is (for each $j = 1, \ldots, k$) a rational parallelepiped in system $(O_j; e_1, \ldots, e_n)$, there exists a rational number h^* such that P^*_j is wholly made up of basis h^* cubes of cubilage $K_{h^*}(O_j)$, and the same will be true for $\operatorname{relbd} P^*_j$. Clearly, for any natural p, the number $h = h^*/p$ possesses the same property, that is, each of sets P^*_j and $\operatorname{relbd} P^*_j$ is the union of a finite number of basis h cubes of cubilage $K_h(O_j)$, $j = 1, \ldots, k$. Thus, there exists an arbitrarily small number h possessing the property that each of sets P^*_j and $\operatorname{relbd} P^*_j$ is the union of a finite number of basis h cubes of cubilage $K_h(O_j)$, $j = 1, \ldots, k$.

Now let us show that the intersection

$$\psi_1(g_\varepsilon(\overline{P}_1)) \cap \ldots \cap \psi_k(g_\varepsilon(\overline{P}_k)) \tag{35.10}$$

is nonempty. First we assume the contrary, namely that intersection (35.10) is empty; then there must exist a number $\gamma > 0$ such that the following intersection is empty as well:

$$U_\gamma(\psi_1(g_\varepsilon(\overline{P}_1))) \cap \ldots \cap U_\gamma(\psi_k(g_\varepsilon(\overline{P}_k))). \tag{35.11}$$

Here it can also be assumed that $\gamma < \delta\varepsilon$. Next we select some $h > 0$ such that $h\sqrt{n} < \gamma$, where each of sets P^*_j and $\operatorname{relbd} P^*_j$ constitutes the intersection of a finite number of basis h cubes of cubilage $K_h(O_j)$, $j = 1, \ldots, k$. The relation $h\sqrt{n} < \gamma$ signifies that the diameter of each basis h cube of cubilage $K_h(O_j)$ is less than γ.

For any point $A \in \overline{P^*_j}$ we designate as $\varphi_j(A)$ the point of parallelepiped $g_\varepsilon(\overline{P}_j)$ closest to A. In view of Theorem 29.5, mapping $\varphi_j\colon \overline{P^*_j} \to g_\varepsilon(\overline{P}_j)$ is defined uniquely and is clearly continuous, while for any point $A \in \overline{P^*_j}$ the relation $\rho(A, \varphi_j(A)) < \delta\varepsilon$ (see condition b, above) is satisfied. Taking (35.8) into account, we thus find that for any point $A \in \overline{P^*_j}$ we have

$$\rho(A, \psi_j(\varphi_j(A))) \leqslant \rho(A, \varphi_j(A)) + \rho(\varphi_j(A), \psi_j(\varphi_j(A))) < \delta\varepsilon + \delta\varepsilon = 2\delta\varepsilon.$$

Therefore, mapping $f_j = \psi_j \circ \varphi_j$, defined and continuous on parallelepiped $\overline{P_j^*}$, possesses the property that

$$\rho(A, f_j(A)) < 2\delta\varepsilon; \qquad A \in \overline{P_j^*}, \quad j = 1, \ldots, k. \tag{35.12}$$

Now let ξ_j be the chain of dimension r_j, consisting of all the r_j-dimensional cubes of cubilage $K_h(O_j)$ contained in parallelepiped $\overline{P_j^*}$, $j = 1, \ldots, k$. Therefore, $|\xi_j| = \overline{P_j^*}$, $|d\xi_j| = \text{relbd}\, \overline{P_j^*}$. By applying to chain ξ_j and mapping f_j the approximation theorem formulated on pages 258-259 and taking into account relation (35.12) and the relation $\gamma < \delta\varepsilon$, we obtain in cubilage $K_h(O_j)$ a chain ζ_j^* of dimension $r_j + 1$ and chains ζ_j, η_j of dimension r_j, such that

$$|\zeta_j^*| \subset U_{3\delta\varepsilon}(|\xi_j|) = U_{3\delta\varepsilon}(\overline{P_j^*}) \subset U_{4\delta\varepsilon}(g_\varepsilon(\overline{P}_j)),$$
$$|\zeta_j| \subset U_\gamma(f_j(|\xi_j|)) \subset U_\gamma(\psi_j(g_\varepsilon \overline{P}_j)) \subset U_{2\delta\varepsilon}(g_\varepsilon(\overline{P}_j)),$$
$$|\eta_j| \subset U_{3\delta\varepsilon}(|d\xi_j|) = U_{3\delta\varepsilon}(\text{relbd}\, \overline{P_j^*}) \subset U_{4\delta\varepsilon}(\text{relbd}\, g_\varepsilon(\overline{P}_j)),$$
$$|d\xi_j| = \text{relbd}\, \overline{P_j^*} \subset U_{\delta\varepsilon}(\text{relbd}\, g_\varepsilon(\overline{P}_j))$$

(see condition b, above, and relation (35.8)), the following relations being valid:

$$d\zeta_j^* = \xi_j + \zeta_j + \eta_j, \quad d\zeta_j = d\xi_j + d\eta_j, \qquad j = 1, \ldots, k.$$

Now consider the chain

$$\lambda_j = \zeta_1 \times \ldots \times \zeta_{j-1} \times \zeta_j^* \times \xi_{j+1} \times \ldots \times \xi_k, \qquad j = 1, \ldots, k.$$

This chain is defined, in view of the choice of cubilage $K_h(O_j)$, $j = 1, \ldots, k$, and it has a dimension

$$\dim \lambda_j = r_1 + \ldots + r_{j-1} + (r_j + 1) + r_{j+1} + \ldots + r_k - (k-1)n = 1.$$

According to formula (35.2), the boundary of chain λ_j has the form

$$d\lambda_j = \sum_{i=1}^{j-1} (\ldots \times d\zeta_i \times \ldots \times \zeta_j^* \times \ldots) + (\ldots \times d\zeta_j^* \times \ldots) +$$
$$+ \sum_{i=j+1}^{k} (\ldots \times \zeta_j^* \times \ldots \times d\xi_i \times \ldots), \tag{35.13}$$

where the dots indicate terms not written out (which before ζ_j^* have the form ζ_l and after ζ_j^* have the form ξ_l). Consider the term $\ldots \times d\zeta_i \times \ldots \times \zeta_j^* \times \ldots$, entering into the first sum, or, in more detail, the term

$$\zeta_1 \times \ldots \times \zeta_{i-1} \times d\zeta_i \times \zeta_{i+1} \times \ldots \times \zeta_{j-1} \times \zeta_j^* \times \xi_{j+1} \times \ldots \times \xi_k. \tag{35.14}$$

Since each of chains ζ_l, ζ_l^*, ξ_l is situated in $U_{4\delta\varepsilon}(g_\varepsilon(\overline{P}_l))$, whereas chain $d\zeta_i = d\xi_i + d\eta_i$ is situated in $U_{4\delta\varepsilon}(\text{relbd}\, g_\varepsilon(\overline{P}_i))$, chain (35.14) must be in the set

$$U_{4\delta\varepsilon}(\text{relbd}\, g_\varepsilon(\overline{P}_i)) \cap \left(\bigcap_{l \neq i} U_{4\delta\varepsilon}(g_\varepsilon(\overline{P}_l)) \right).$$

The fact that this intersection is by definition empty implies that chain (35.14) is equal to zero. Similar considerations indicate that each term of the second sum on the right-hand side of (35.13) is zero as well. Consequently, formula (35.13) becomes

$$d\lambda_j = \zeta_1 \times \dots \times \zeta_{j-1} \times d\zeta_j^* \times \xi_{j+1} \times \dots \times \xi_k.$$

Now, recalling relation $d\zeta_j^* = \xi_j + \zeta_j + \eta_j$, we obtain

$$d\lambda_j = \zeta_1 \times \dots \times \zeta_{j-1} \times \xi_j \times \dots \times \xi_k + \zeta_1 \times \dots \times \zeta_j \times \xi_{j+1} \times \dots \times \xi_k + \\ + \zeta_1 \times \dots \times \zeta_{j-1} \times \eta_j \times \xi_{j+1} \times \dots \times \xi_k.$$

The last term on the right-hand side of this expression is equal to zero (for the same reasons as previously), so that

$$d\lambda_j = \zeta_1 \times \dots \times \zeta_{j-1} \times \xi_j \times \dots \times \xi_k + \zeta_1 \times \dots \times \zeta_j \times \xi_{j+1} \times \dots \times \xi_k.$$

Next we write these relations for $j = 1, \dots, k$ and we add up all the equations so obtained; this gives (taking into account that terms occurring twice can be canceled in the sum of the chains):

$$d(\lambda_1 + \dots + \lambda_k) = \xi_1 \times \dots \times \xi_k + \zeta_1 \times \dots \times \zeta_k.$$

Now recall that $|\zeta_j| \subset U_\gamma(\psi_j(g_\varepsilon(\overline{P}_j)))$, $j = 1, \dots, k$, indicating that the field of chain $\zeta_1 \times \dots \times \zeta_k$ is contained in set (35.11). Since, by assumption, set (35.11) is empty, it follows that $\zeta_1 \times \dots \times \zeta_k = 0$, and thus that

$$d(\lambda_1 + \dots + \lambda_k) = \xi_1 \times \dots \times \xi_k.$$

However, chain $\xi_1 \times \dots \times \xi_k$ is actually only *a single* point, since the intersection $P_1^* \cap \dots \cap P_k^*$ is a single point. We have thus found that the boundary of the one-dimensional chain $\lambda_1 + \dots + \lambda_k$ is a single point, whereas the boundary of any one-dimensional chain should consist of an *even* number of points. This contradiction indicates that intersection (35.10) is *nonempty*.

Let B be an arbitrary point of set (35.10). Since in view of (35.8)

$$B \in \psi_1(g_\varepsilon(\overline{P}_1)) \subset U_{\delta\varepsilon}(g_\varepsilon(\overline{P}_1)) \text{ and } Q \notin U_{\delta\varepsilon}(g_\varepsilon(\overline{P}_1)),$$

it follows that $B \neq Q$. Moreover, for any $j = 1, \dots, k$ we have

$$B \in \psi_j(g_\varepsilon(\overline{P}_j)) \subset \psi_j(M_j^*) \subset \Omega_j,$$

so that $B \in \Omega_1 \cap \dots \cap \Omega_k$. Consequently, B is the point being sought.

NOTE 35.2. Another, considerably simpler, proof of Theorem 35.1 is possible, provided that the following additional condition is satisfied: mappings $\psi_1, \dots, \psi_k$ (indicating that $M_1, \dots, M_k$ are coverings of sets $\Omega_1, \dots, \Omega_k$) are defined in a certain neighborhood around point Q and are *smooth* mappings.

As previously, let O be a common interior point of cones $M_1, \dots, M_k$. If $N_1, \dots, N_k$ are the carrier planes of cones $M_1, \dots, M_k$, and if open spheres $E_1, \dots, E_k$, situated in planes $N_1, \dots, N_k$ and centered on Q, are so selected that E_i is contained in the domain of definition of mapping ψ_i, $i = 1, \dots, k$, then $\psi_i(E_i)$ is a smooth manifold (of dimension $\dim N_i$) in space E^n, passing through point Q and having at point Q a tangent plane N_i $(i=1, \ldots, k)$.

Let us show that for any $j=1, \ldots, k$ the intersection $\psi_1(E_1) \cap \ldots \cap \psi_j(E_j)$ is, in the vicinity of point Q, a smooth manifold having at point Q a tangent plane $N_1 \cap \ldots \cap N_j$. For $j=1$ this statement is valid. Now let us assume that it is valid for some $j<k$, and let us proceed to show its validity for the number $j+1$, that is, let us demonstrate that $\psi_1(E_1) \cap \ldots \cap \psi_j(E_j) \cap \psi_{j+1}(E_{j+1})$ is a smooth manifold having at point Q a tangent plane $N_1 \cap \ldots \cap N_j \cap N_{j+1}$. Since

$$\psi_1(E_1) \cap \ldots \cap \psi_j(E_j) \cap \psi_{j+1}(E_{j+1}) = \\ = (\psi_1(E_1) \cap \ldots \cap \psi_j(E_j)) \cap \psi_{j+1}(E_{j+1}),$$

where $\psi_1(E_1) \cap \ldots \cap \psi_j(E_j)$ and $\psi_{j+1}(E_{j+1})$ are, in the vicinity of point Q, smooth manifolds having at point Q tangent planes $N_1 \cap \ldots \cap N_j$ and N_{j+1}, it is sufficient to establish that $N_1 \cap \ldots \cap N_j$ and N_{j+1} do not lie in a single hyperplane (see Theorem 33.12). However, this follows directly from the fact that $L_i \subset N_1 \cap \ldots \cap N_j$ for $i=j+1, \ldots, k$ and $L_i \subset N_{j+1}$ for $i=1, \ldots, j$ (where $L_1, \ldots, L_k$ are the subspaces constructed at the beginning of the proof of Theorem 35.1; p. 251). Thus our statement has been proven for all $j=1, \ldots, k$. In particular, for $j=k$ we find that $\psi_1(E_1) \cap \ldots \cap \psi_k(E_k)$ is, in the vicinity of point Q, a smooth manifold having at point Q a tangent plane $N_1 \cap \ldots \cap N_k$.

Since $O \in N_1 \cap \ldots \cap N_k$, there must exist in manifold $\psi_1(E_1) \cap \ldots \cap \psi_k(E_k)$ a curve Λ originating at point Q and having at point Q a tangent vector $\overrightarrow{QO}$.

The mapping ψ_j: $E_j \to \psi_j(E_j)$ is (if sphere E_j is sufficiently small) one-to-one, that is, there exists a mapping φ_j: $\psi_j(E_j) \to E_j$ *inverse* to ψ_j. This means that $\psi_j(\varphi_j(x))=x$ for any point $x \in \psi_j(E_j)$, where mapping φ_j, like ψ_j, has the identity mapping as its tangent mapping at point Q. Consequently, curve $\varphi_j(\Lambda)$ situated in sphere E_j also has $\overrightarrow{QO}$ as its tangent vector at point Q. But since O is an *interior* point of cone M_j relative to its carrier plane, a point C_j must exist on curve $\varphi_j(\Lambda)$ such that the whole arc of this curve from point Q to C_j belongs to M_j (Figure 146). Consequently, the whole arc of $\psi_j(\varphi_j(\Lambda)) = \Lambda$ from point Q to point $B_j = \psi_j(C_j)$ belongs to set $\psi_j(M_j \cap E_j) \subset \Omega_j$. Here $B_j \neq Q$, since mapping ψ_j is one-to-one. This will be the situation for any $j=1, \ldots, k$. Thus the whole arc of curve Λ from point Q to some point B different from Q is contained in the set $\Omega_1 \cap \ldots \cap \Omega_k$, implying that Theorem 35.1 is valid.

If we limit ourselves to this version of the proof (that is, if we make the additional assumption in Theorem 35.1 that $\psi_1, \ldots, \psi_k$ are *smooth* mappings), then we must also limit ourselves to sets Ω_i and coverings M_i of these such that there exists a *smooth* mapping ψ_i satisfying conditions 1 and 2 of Definition 34.1. The following example indicates that a *smooth* mapping ψ_i of the required type does not always exist, so that the proof presented in this note, although simpler, provides us with a somewhat weaker result than does Theorem 35.1.

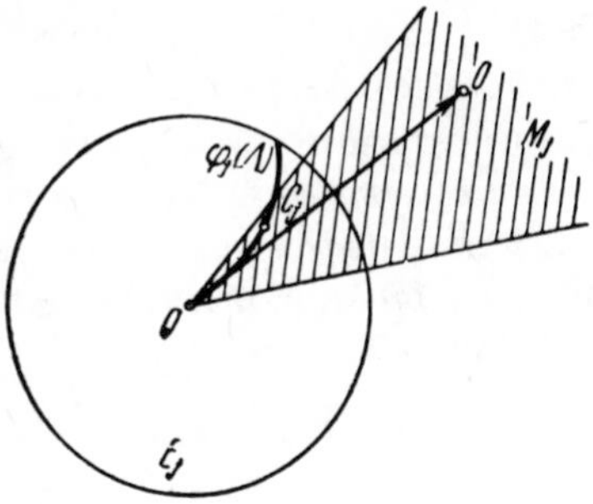

FIGURE 146

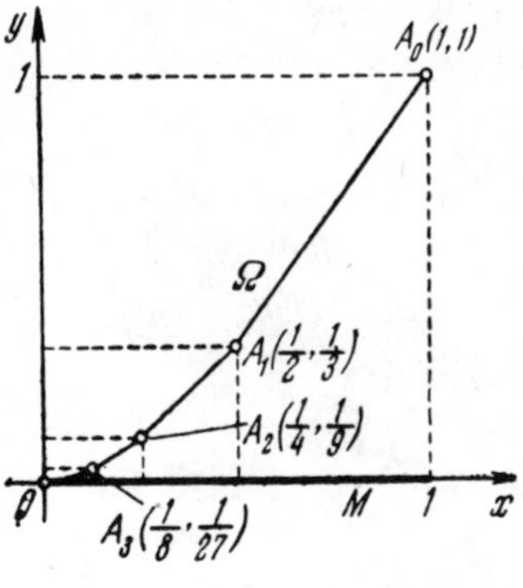

FIGURE 147

EXAMPLE 35.3. Let us construct in the $x\,y$ plane an infinite-element broken line $A_0A_1A_2A_3\,\ldots$ with vertices

$$A_i=\left(\frac{1}{2^i},\ \frac{1}{3^i}\right), \qquad i=0,\ 1,\ 2,\ \ldots$$

By adding the point $Q(0,\ 0)$ to this broken line, we obtain a closed set in the plane, which we will call Ω (Figure 147). In addition, let M be the nonnegative abscissa semiaxis. Finally, let ψ be the projection of segment $[0,\ 1]$ of the abscissa axis onto set Ω parallel to the ordinate axis. This mapping ψ satisfies conditions 1 and 2 of Definition 34.1, that is, M is the covering of set Ω at point Q. It is easy to see that no *smooth* mapping ψ will possess the required properties.

§10. Extremum Criteria

36. Necessary condition for extremum of function. In this subsection we will formulate and prove a necessary condition for the extremum of a function which is more general (that is, has a wider field of application) and more strict (that is, imposes more requirements) than the condition in Subsection 7. Moreover, the condition given in this subsection has the advantage that it is, at any rate in form, universal, being formulated *in like manner* for any point in the domain of definition of the function. The method of Subsection 7, on the other hand, calls for a separate consideration of points lying inside the "curvilinear polyhedron" or on its faces of different dimensions.

Let M be a convex cone with a vertex at point z_0, located in space E^n, and let $\boldsymbol{\delta z}=\{\delta z^1,\ \ldots,\ \delta z^n\}$ be some vector of space E^n. We will say that *the direction of vector* $\boldsymbol{\delta z}$ *lies in cone* M if there exists a point $z\in M$ such that the vector $z-z_0$ is equal to $\boldsymbol{\delta z}$ (Figure 148).

THEOREM 36.1. *Let* $\Omega_1, \ldots, \Omega_l$ *be certain sets in space* E^n. *In addition, let* $F^0(z)$ *be a smooth function defined on some open set of space* E^n, *containing the set* $\Sigma = \Omega_1 \cap \ldots \cap \Omega_l$. *The problem is: to find the point at which function* F^0, *considered only on set* Σ, *reaches its lowest value. Let* $z_0 \in \Sigma$ *and let* M_i *be a covering of set* Ω_i *at point* z_0, $i = 1, \ldots, l$. *In order for point* z_0 *to constitute a solution of the problem, that is, in order for the following relation to be valid:*

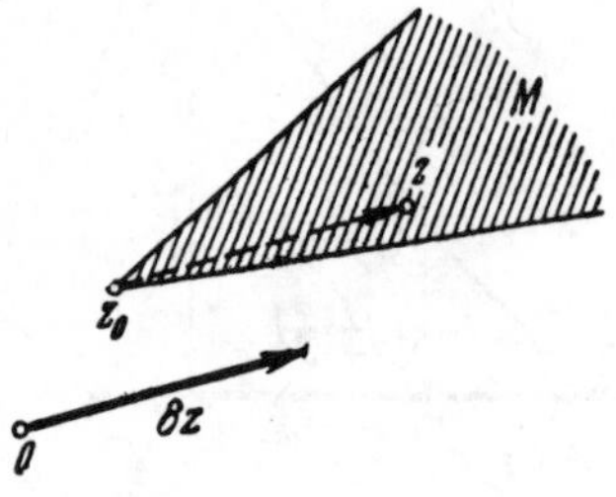

FIGURE 148

$$F^0(z_0) \leqslant F^0(z) \quad \text{for} \quad z \in \Sigma,$$

there must exist a number ψ_0 *and vectors* $\boldsymbol{a}_1, \ldots, \boldsymbol{a}_l$, *such that the direction of vector* $\boldsymbol{a}_i$ *lies in the dual cone* $D(M_i)$, $i = 1, \ldots, l$ *and the following conditions are satisfied:*

(α) $\psi_0 \leqslant 0$;

(β) $\psi_0 \operatorname{grad} F^0(z_0) = \boldsymbol{a}_1 + \boldsymbol{a}_2 + \ldots + \boldsymbol{a}_l$;

(γ) *if* $\psi_0 = 0$, *then at least one of vectors* $\boldsymbol{a}_1, \boldsymbol{a}_2, \ldots, \boldsymbol{a}_l$ *is different from zero.*

If the problem of the maximum (rather than the minimum) of function F^0 *on set* Σ *is considered, then the sign of inequality* (α) *must be changed.**

Proof. If $\operatorname{grad} F^0(z_0) = \mathbf{0}$, then by setting $\psi_0 = -1$, $\boldsymbol{a}_1 = \boldsymbol{a}_2 = \ldots = \boldsymbol{a}_k = \mathbf{0}$ we obtain values of $\psi_0, \boldsymbol{a}_1, \ldots, \boldsymbol{a}_l$ satisfying all of conditions (α) through (γ) (independently of whether or not function $F^0(z)$ is a minimum at point z_0). Thus for $\operatorname{grad} F^0(z_0) = \mathbf{0}$ the statement of the theorem is trivially valid. Hence in the following we will assume that $\operatorname{grad} F^0(z_0) \neq \mathbf{0}$.

Let Ω^* be the set containing point z_0 and all points $z \in E^n$ for which function $g(z) = F^0(z) - F^0(z_0)$ assumes negative values (that is, $F^0(z) < F^0(z_0)$). In addition, let M^* be the half-space consisting of all points $z \in E^n$ for which $(z - z_0) \operatorname{grad} g(z_0) \leqslant 0$ (or, equivalently, $(z - z_0) \operatorname{grad} F^0(z_0) \leqslant 0$). According to Note 34.5, set M^* is a covering of set Ω^* at point z_0.

Let us show that the system of cones $M^*, M_1, \ldots, M_l$ with a common vertex z_0 possesses separability in E^n. First, we assume the converse, namely that this system of cones does not possess separability. Then, according to Theorem 35.1, there exists some point $z_1 \in \Omega^* \cap \Omega_1 \cap \ldots \cap \Omega_l$ which is different from z_0 (note that M^* is a half-space, that is, M^* is not a plane). Since $z_1 \in \Omega^*$ and $z_1 \neq z_0$, it follows that $g(z_1) < 0$, that is, $F^0(z_1) < F^0(z_0)$. Consequently, we have found a point $z_1 \in \Sigma$ such that $F^0(z_1) < F^0(z_0)$. However, this contradicts the fact that function F^0, considered on Σ, reaches its lowest value at point z_0. This contradiction also indicates that the system of cones $M^*, M_1, \ldots, M_l$ possesses separability.

According to Theorem 32.2, there exist vectors $\boldsymbol{a}^*, \boldsymbol{a}_1, \ldots, \boldsymbol{a}_l$ whose directions lie, respectively, in cones $D(M^*), D(M_1), \ldots, D(M_l)$, such that at least one of

* This remark (concerning the change in sign of inequality (α) when considering the maximum of the function) also pertains to all the subsequent theorems in the present subsection.

these vectors is different from zero and, at the same time,

$$a^* + a_1 + \ldots + a_l = 0.$$

Since M^* is a half-space defined by the inequality $(z - z_0)\,\mathrm{grad}\, F^0(z_0) \leqslant 0$, it follows that $a^* = -\psi_0\,\mathrm{grad}\, F^0(z_0)$, where $\psi_0 \leqslant 0$. Obviously, the quantities $\psi_0, a_1, \ldots, a_l$ are the ones sought.

The problem of the *maximum* of function $F^0(z)$ is easily reduced to the case just analyzed: we need only consider the problem of the maximum of the function $F^0(z_0)$.

THEOREM 36.2. *Let Ω' be a set defined in the space E^n of variables $z^1, \ldots, z^n$ by the equations*

$$F^1(z) = 0, \ \ldots, \ F^k(z) = 0; \tag{36.1}$$

moreover, let Ω'' be a set defined by the inequalities

$$f^1(z) \leqslant 0, \ \ldots, \ f^q(z) \leqslant 0, \tag{36.2}$$

and let $\Omega_1, \ldots, \Omega_l$ be arbitrary sets of space E^n. Finally, let $F^0(z)$ be a smooth function defined on some open set of space E^n containing the set $\Sigma = \Omega' \cap \Omega'' \cap \cap \Omega_1 \cap \ldots \cap \Omega_l$. The problem is: to find the point at which function F^0, considered only on set Σ, reaches its lowest value. Let $z_0 \in \Sigma$ and let M_i be a covering of set Ω_i at point z_0, $i = 1, \ldots, l$. Finally, let $I(z_0)$ be the zone of activity at point z_0. In order for point z_0 to constitute a solution of the problem, that is, in order for the following relation to be valid:

$$F^0(z_0) \leqslant F^0(z) \quad \text{for} \quad z \in \Sigma,$$

there must exist numbers $\psi_0, \psi_1, \ldots, \psi_k$, vectors $a_1, \ldots, a_l$ whose directions lie, respectively, in the dual cones $D(M_1), \ldots, D(M_l)$, and nonpositive numbers λ_j, $j \in I(z_0)$, all such that the following conditions are satisfied:

(α) $\psi_0 \leqslant 0$;

(β) $\sum_{\alpha=0}^{k} \psi_\alpha\,\mathrm{grad}\, F^\alpha(z_0) + \sum_{j \in I(z_0)} \lambda_j\,\mathrm{grad}\, f^j(z_0) = a_1 + \ldots + a_l$;

(γ) *if all the numbers $\psi_0, \psi_1, \ldots, \psi_k$ and λ_j, $j \in I(z_0)$ are equal to zero, then at least one of vectors $a_1, \ldots, a_l$ is different from zero.*

Proof. Let Ω_i' be the hypersurface defined in space E^n by the equation $F^i(z) = 0$, and let Ω_j'' be the set defined in E^n by the constraint $f^j(z) \leqslant 0$ $(i = 1, \ldots, k;\ j \in I(z_0))$. Note that, if $\mathrm{grad}\, F^i(z_0) = 0$ for some $i = \alpha$, then, by assuming that $\psi_\alpha = -1$ and that all the other quantities ψ_i, λ_j, a_h are zero, we satisfy the requirements of the theorem (independently of whether or not function $F^0(z)$ reaches a minimum at point z_0). Similarly, if $\mathrm{grad}\, f^j(z_0) = 0$ for some $j = \beta \in I(z_0)$, then the requirements of the theorem are satisfied by assuming that $\lambda_\beta = -1$ and that all the other quantities ψ_i, λ_j, a_h are zero. Therefore we can assume that

$\operatorname{grad} F^i(z_0) \neq \mathbf{0}$ and that $\operatorname{grad} f^j(z_0) \neq \mathbf{0}$ for all i, j. In this case hyperplane M'_i, defined by the equation $(z-z_0)\operatorname{grad} F^i(z_0)=0$, is a covering of set Ω'_i at point z_0 $(i=1, \ldots, k)$, while the half-space defined by the inequality $(z-z_0)\operatorname{grad} f^j(z_0) \leqslant 0$ is a covering of set Ω''_j at point $z_0\,(j \in I(z_0))$.

By applying Theorem 36.1 to function $F^0(z)$, considered on set Σ, we find that there must exist a number $\psi_0 \leqslant 0$ and vectors $\boldsymbol{a}'_i$ $(i=1, \ldots, k)$, $\boldsymbol{a}''_j$ $(j \in I(z_0))$, $\boldsymbol{a}_h$ $(h=1, \ldots, l)$, such that

$$\psi_0 \operatorname{grad} F^0(z_0) = \sum_i \boldsymbol{a}'_i + \sum_j \boldsymbol{a}''_j + \sum_h \boldsymbol{a}_h,$$

where, if $\psi_0 = 0$, at least one of vectors $\boldsymbol{a}'_i$, $\boldsymbol{a}''_j$, $\boldsymbol{a}_h$ is different from zero. Note, too, that in view of the definition of sets M'_i and M''_j we can write

$$\boldsymbol{a}'_i = -\psi_i \operatorname{grad} F^i(z_0), \quad \boldsymbol{a}''_j = -\lambda_j \operatorname{grad} f^j(z_0),$$

where ψ_i $(i=1, \ldots, k)$ are real and λ_j $(j \in I(z_0))$ are nonpositive numbers.

In the theorem proven, k, q, l may be any nonnegative whole numbers. In particular, any of them (or any two of them) may be assumed to be zero, so that from Theorem 36.2 we obtain six particular cases. For instance, Theorem 36.1 is a particular case of Theorem 36.2, obtained for $k=0$, $q=0$. Of all the other particular cases, we will formulate only the one for $l=0$. In this case sets $\Omega_1, \ldots, \Omega_l$ are absent, that is, the entire space E^n is to be considered as the intersection $\Omega_1 \cap \ldots \cap \Omega_l$ (it can also be assumed that $l=1$ and $\Omega_1 = E^n$, so that $M_1 = E^n$ and $D(M_1) = \{z_0\}$).

THEOREM 36.3. *Let Σ be a set defined in space E^n by relations* (36.1), (36.2), *and let $F^0(z)$ be a smooth function defined on some open set of space E^n containing set Σ. The problem is: to find the point at which function F^0, considered only on set Σ, reaches its lowest value. In order for a point $z_0 \in \Sigma$ to constitute a solution of the problem, that is, in order for the following relation to be valid:*

$$F(z_0) \leqslant F(z) \quad \text{for} \quad z \in \Sigma,$$

there must exist numbers $\psi_0, \psi_1, \ldots, \psi_k$ and nonpositive numbers λ_j, $j \in I(z_0)$, such that at least one of the numbers ψ_i, λ_j, where $i=0, 1, \ldots, k$, $j \in I(z_0)$, is different from zero, the following conditions being satisfied:

(α) $\psi_0 \leqslant 0$;

(β) $\sum_{\alpha=0}^{k} \psi_\alpha \operatorname{grad} F^\alpha(z_0) + \sum_{j \in I(z_0)} \lambda_j \operatorname{grad} f^j(z_0) = \mathbf{0}$.

If for $j \notin I(z_0)$ we set $\lambda_j = 0$, then condition (β) becomes

$$\sum_{\alpha=0}^{k} \psi_\alpha \operatorname{grad} F^\alpha(z_0) + \sum_{j=1}^{q} \lambda_j \operatorname{grad} f^j(z_0) = \mathbf{0},$$

and thus the following equations are valid:

$$\lambda_1 f^1(z_0) = \lambda_2 f^2(z_0) = \ldots = \lambda_q f^q(z_0) = 0$$

(since $f^j(z_0) = 0$ for $j \in I(z_0)$ and $\lambda_j = 0$ for $j \notin I(z_0)$). Consequently, Theorem 36.3 directly implies Theorem 11.4, which was formulated back in Chapter I: it is enough just to replace λ_i by $-\sigma_i$.

NOTE 36.4. Let us consider the particular case of Theorem 36.3 obtained if inequalities (36.2) are not valid, that is, if set Σ is defined in space E^n by equations (36.1). In this case condition (β) in Theorem 36.3 becomes

$$\sum_{\alpha=0}^{k} \psi_\alpha \operatorname{grad} F^\alpha(z_0) = \mathbf{0}, \tag{36.3}$$

or, in coordinate form,

$$\sum_{\alpha=0}^{k} \psi_\alpha \frac{\partial F^\alpha(z_0)}{\partial z^i} = 0, \qquad i = 1, \ldots, n. \tag{36.4}$$

Therefore, *in order for function* $F^0(z)$, *considered on the set* Ω *defined by equations* (36.1), *to reach an extremum at a point* z_0, *there must exist a nonzero vector* $\psi = \{\psi_0, \psi_1, \ldots, \psi_k\}$ *satisfying condition* (36.3) *(or* (36.4)*)*. This is the classical Lagrange theorem concerning a conditional extremum. Theorem 35.3 (and all the more Theorem 35.2) is a generalization of this theorem.

NOTE 36.5. In Theorems 36.2 and 36.3 functions $F^0, F^1, \ldots, F^k$ were not required to be independent, that is, the rank of the functional matrix $\left(\frac{\partial F^i}{\partial z^j}\right)$ did not have to be $k+1$. However, if these functions $F^0, F^1, \ldots, F^k$ are dependent, the statements of these theorems become empty of all meaning.

Let us consider the more general Theorem 36.2. If at point $z_0 \in \Sigma$ the rank of functional matrix $\left(\frac{\partial F^i}{\partial z^j}\right)$ is less than $k+1$, that is, if the vectors

$$\operatorname{grad} F^0(z_0),\ \operatorname{grad} F^1(z_0),\ \ldots,\ \operatorname{grad} F^k(z_0)$$

are linearly dependent, then conditions (α), (β), (γ) are satisfied at point z_0 (irrespective of the form of set Σ or the behavior of function F^0 on this set). Actually, in the given case there must exist numbers $\psi_0, \psi_1, \ldots, \psi_k$, not all equal to zero, such that

$$\sum_{\alpha=0}^{k} \psi_\alpha \operatorname{grad} F^\alpha(z_0) = \mathbf{0}.$$

After changing, if necessary, the signs of all the numbers $\psi_0, \psi_1, \ldots, \psi_k$, we can assume that $\psi_0 \leqslant 0$, so that for $\lambda_j = 0$, $\boldsymbol{a}_h = \boldsymbol{0}$ all of conditions (α), (β), (γ) will be satisfied.

Thus, Theorem 36.2 is always formally valid, but in reality it is of interest only when the vectors $\operatorname{grad} F^i(z_0)$, $i = 0, 1, \ldots, k$ are linearly independent. This will also be the case for the following theorems.

NOTE 36.6. In Theorem 36.1 it is *not required* that the system of cones $M_1, \ldots, M_l$ possess separability. However, if this system of cones does possess separability, then the statement of Theorem 36.1 becomes empty of meaning.

If we assume, on the other hand, that system of cones $M_1, \ldots, M_l$ possesses separability in E^n, then in view of Theorem 32.2 there must exist vectors $\boldsymbol{a}_1, \ldots, \boldsymbol{a}_l$, at least one of which is different from zero, such that the direction of vector $\boldsymbol{a}_j$ lies in cone $D(M_j)$, $j = 1, \ldots, l$, and the relation $\boldsymbol{a}_1 + \ldots + \boldsymbol{a}_l = \boldsymbol{0}$ is valid. Therefore, for $\psi_0 = 0$ all the conditions (α), (β), (γ) of Theorem 36.1 will be satisfied.

Consequently, Theorem 36.1 is always formally valid, but in reality it will be of interest only if the system of cones $M_1, \ldots, M_l$ does not possess separability.

A similar statement can be made for the subsequent theorems as well (since they are derived from Theorem 36.1). For instance, let us return to Theorem 36.2, where we call Π_j $(j \in I(z_0))$ the half-space defined by the inequality $(z - z_0) \operatorname{grad} f^j(z_0) \leqslant 0$ and we call Γ_i $(i = 1, \ldots, k)$ the tangent hyperplane of hypersurface $F^i(z) = 0$ constructed at point z_0. Theorem 36.2 will be of interest only if the system of cones Γ_i, Π_j, M_h $(i = 1, \ldots, k;\ j \in I(z_0);\ h = 1, \ldots, l)$ with a common vertex z_0 does not possess separability in E^n (otherwise the statement in this theorem is empty of meaning).

Note, in particular, that Theorem 36.2 (like Theorem 36.3) becomes empty of meaning if system of constraints (36.2) is not nonsingular at point z_0. Actually, the condition of nonsingularity of this system of constraints at point z_0 signifies that the entire half-space Π_j, $j \in I(z_0)$, has a common *interior* point. If this condition is not satisfied, then in view of Theorem 32.5 the system of cones Π_j $(j \in I(z_0))$ possesses separability, and thus the system of cones Γ_i, Π_j, M_h possesses separability all the more.

Consequently, although formally this is *not a requirement* in Theorems 36.2 and 36.3 either, the indicated theorems are of interest only if system of constraints (36.2) is nonsingular at point z_0.

THEOREM 36.7. *Let $\Omega_1, \ldots, \Omega_l$ and $\Xi_1, \ldots, \Xi_m$ be certain sets in space E^n. In addition, let $F^0(z)$ be a smooth function defined on some open set of space E^n, containing the set $\Sigma = \Omega_1 \cap \ldots \cap \Omega_l \cap \Xi_1 \cap \ldots \cap \Xi_m$. The problem is: to find the point at which function F^0, considered only on set Σ, reaches its lowest value. Let $z_0 \in \Sigma$. In addition, let M_i be a covering of set Ω_i at point z_0 $(i = 1, \ldots, l)$ and let N_j be a covering of set Ξ_j at point z_0 $(j = 1, \ldots, m)$. Finally, assume that the system of cones $N_1, \ldots, N_m$ (with a common vertex z_0) does not possess separability. In order for point z_0 to constitute a solution of the problem, that is, in order* *for the following relation to be valid:*

$$F^0(z_0) \leqslant F^0(z) \quad \text{for} \quad z \in \Sigma,$$

there must exist a number ψ_0 *and vectors* $\boldsymbol{a}_1, \ldots, \boldsymbol{a}_l$*, such that the direction of vector* $\boldsymbol{a}_i$ *lies in dual cone* $D(M_i)$, $i = 1, \ldots, l$*, the following conditions being satisfied:*

(α) $\psi_0 \leqslant 0$;

(β) $(\psi_0 \operatorname{grad} F^0(z_0) - \boldsymbol{a}_1 - \ldots - \boldsymbol{a}_l)\,\boldsymbol{\delta z} \leqslant 0$ *for any vector* $\boldsymbol{\delta z}$ *the direction of which lies in the cone* $N_1 \cap \ldots \cap N_m$;

(γ) *if* $\psi_0 = 0$, *then at least one of vectors* $\boldsymbol{a}_1, \boldsymbol{a}_2, \ldots, \boldsymbol{a}_l$ *is different from zero.*

Proof. In view of Theorem 36.1, there exists a number $\psi_0 \leqslant 0$ and vectors $\boldsymbol{a}_1, \ldots, \boldsymbol{a}_l$, $\boldsymbol{b}_1, \ldots, \boldsymbol{b}_m$, such that the direction of vector $\boldsymbol{a}_i$ lies in cone $D(M_i)$, $i = 1, \ldots, l$, the direction of vector $\boldsymbol{b}_j$ lies in cone $D(N_j)$, $j = 1, \ldots, m$, and the following relation is valid:

$$\psi_0 \operatorname{grad} F(z_0) = \boldsymbol{a}_1 + \ldots + \boldsymbol{a}_l + \boldsymbol{b}_1 + \ldots + \boldsymbol{b}_m, \tag{36.5}$$

where for $\psi_0 = 0$ at least one of vectors $\boldsymbol{a}_1, \ldots, \boldsymbol{a}_l, \boldsymbol{b}_1, \ldots, \boldsymbol{b}_m$ is different from zero. If we assume that $\psi_0 = 0$ and $\boldsymbol{a}_1 = \ldots = \boldsymbol{a}_l = 0$, then we obtain $\boldsymbol{b}_1 + \ldots + \boldsymbol{b}_m = 0$, some of vectors $\boldsymbol{b}_1, \ldots, \boldsymbol{b}_m$ being different from zero. However, this contradicts the assumption that system of cones $N_1, \ldots, N_m$ does not possess separability. Consequently, some of the quantitites $\psi_0, \boldsymbol{a}_1, \ldots, \boldsymbol{a}_l$ are different from zero, that is, condition (γ) is satisfied. We have also seen that $\psi_0 \leqslant 0$, that is, condition (α) is also satisfied.

It remains to verify that condition (β) is satisfied. Let

$$\boldsymbol{n} = \psi_0 \operatorname{grad} F(z_0) - \boldsymbol{a}_1 - \ldots - \boldsymbol{a}_l. \tag{36.6}$$

Since $\boldsymbol{n} = \boldsymbol{b}_1 + \ldots + \boldsymbol{b}_m$ (see (36.5)), it follows from Lemma 32.1 that the direction of vector $\boldsymbol{n}$ lies in the cone serving as the convex hull of cones $D(N_1), \ldots, D(N_m)$. According to Corollary 31.4, this means that the direction of vector $\boldsymbol{n}$ lies in the cone $D(N_1 \cap \ldots \cap N_m)$, that is, $\boldsymbol{n}\,\boldsymbol{\delta z} \leqslant 0$ for any vector $\boldsymbol{\delta z}$, whose direction lies in the cone $N_1 \cap \ldots \cap N_m$. However, in view of (36.6), this also means that condition (β) is satisfied.

THEOREM 36.8. *Let* Ω' *be a set defined in space* E^n *by the equations*

$$F^1(z) = 0, \ \ldots, \ F^k(z) = 0.$$

and let Ξ' *be a set defined in* E^n *by the equations*

$$G^1(z) = 0, \ \ldots, \ G^r(z) = 0.$$

In addition, let Ω'' *be a set defined by the inequalities*

$$f^1(z) \leqslant 0, \ \ldots, \ f^q(z) \leqslant 0,$$

let Ξ'' *be a set defined by the inequalities*

$$g^1(z) \leqslant 0, \ \ldots, \ g^p(z) \leqslant 0, \tag{36.7}$$

and let $\Omega_1, \ldots, \Omega_l, \ \Xi_1, \ldots, \Xi_m$ *be arbitrary sets of space* E^n. *Finally, let* $F^0(z)$ *be a smooth function defined on some open set of space* E^n, *containing the set*

$$\Sigma = \Omega' \cap \Omega'' \cap \Omega_1 \cap \ldots \cap \Omega_l \cap \Xi' \cap \Xi'' \cap \Xi_1 \cap \ldots \cap \Xi_m.$$

The problem is: to find the point at which function F^0, *considered only on set* Σ, *reaches its lowest value.*

Let $z_0 \in \Sigma$, *let* M_i *be a covering of set* Ω_i *at point* z_0 $(i = 1, \ldots, l)$, *and let* N_j *be a covering of set* Ξ_j *at* z_0 $(j = 1, \ldots, m)$. *In addition, let* $I(z_0)$ *be the set of numbers* $i = 1, \ldots, q$ *for which* $f^i(z_0) = 0$, *and let* $J(z_0)$ *be the set of numbers* $j = 1, \ldots, p$ *for which* $g^j(z_0) = 0$. *We assume that all the vectors*

$$\operatorname{grad} G^i(z_0), \quad i = 1, \ldots, r; \qquad \operatorname{grad} g^j(z_0), \quad j \in J(z_0),$$

are different from zero. Finally, let L_i *be the tangent hyperplane of hypersurface* $G^i(z) = 0$ *at point* z_0 $(i = 1, \ldots, r)$ *and let* P_j *be the half-space defined by the inequality* $(z - z_0)\operatorname{grad} g^j(z_0) \leqslant 0$, *and let us assume that the system of convex cones* $L_i, \ P_j, \ N_h$ *(where* $i = 1, \ldots, r; \ \ j \in J(z_0); h = 1, \ldots, m$*) does not possess separability.*

In order for point z_0 *to constitute a solution of the problem, that is, in order for the following relation to be valid:*

$$F^0(z_0) \leqslant F^0(z) \quad \text{for} \quad z \in \Sigma,$$

there must exist numbers $\psi_0, \psi_1, \ldots, \psi_k$, *vectors* $\boldsymbol{a}_1, \ldots, \boldsymbol{a}_l$ *whose directions lie, respectively, in cones* $D(M_1), \ldots, D(M_l)$, *and nonpositive numbers* λ_j, $j \in I(z_0)$, *all such that the following conditions are satisfied:*

(α) $\psi_0 \leqslant 0$;

(β) $\left(\sum_{\alpha=0}^{k} \psi_\alpha \operatorname{grad} F^\alpha(z_0) + \sum_{j \in I(z_0)} \lambda_j \operatorname{grad} f^j(z_0) - \boldsymbol{a}_1 - \ldots - \boldsymbol{a}_l \right) \boldsymbol{\delta z} \leqslant 0$

for any vector $\boldsymbol{\delta z}$, *whose direction lies in the intersection of cones* $L_i, \ P_j, \ N_h$ $(i = 1, \ldots, r; \ j \in J(z_0); \ h = 1, \ldots, m)$;

(γ) *if all the numbers* $\psi_0, \ \psi_1, \ldots, \psi_k$ *and* λ_j, $j \in I(z_0)$, *are equal to zero, then at least one of vectors* $\boldsymbol{a}_1, \ldots, \boldsymbol{a}_l$ *is different from zero.*

This theorem follows directly from the previous one (cf. proof of Theorem 36.2). Theorem 36.8 is the most general of the theorems in this subsection. The quantities k, r, q, p, l, m entering into it may be any nonnegative whole numbers. By assuming some of them to be zero, we arrive at an entire series of particular cases of Theorems 36.8. For example, Theorem 36.2 is a particular case of Theorem 36.8, obtained for $r = p = m = 0$ (or else it can be assumed that $r = p = 0, \ m = 1$

and $\Xi_1 = N_1 = E^n$). Of all the other particular cases we will formulate only two theorems (the first is obtained for $r = q = p = l = 0$ and the second for $r = q = l = m = 0$, if we take into account that the condition for nonsingularity of system of constraints (36.7) at point z_0 is equivalent to assuming nonseparability of cones P_j, $j \in J(z_0)$).

THEOREM 36.9. *Let Ω' be a set defined in space E^n by the equations*

$$F^1(z) = 0, \ldots, F^k(z) = 0,$$

and let $\Xi_1, \ldots, \Xi_m$ be certain sets in space E^n. In addition, let $F^0(z)$ be a smooth function defined on some open set of space E^n containing set $\Sigma = \Omega' \cap \Xi_1 \cap \ldots \cap \Xi_m$. The problem is: to find the point at which function F^0, considered only on set Σ, reaches its lowest value. Let $z_0 \in \Sigma$ and let N_j be a covering of set Ξ_j at point z_0 ($j = 1, \ldots, m$). We assume that the system of cones $N_1, \ldots, N_m$ does not possess separability.

In order for point z_0 to constitute a solution of the problem, that is, for the following relation to be valid:

$$F^0(z_0) \leqslant F^0(z) \quad \text{for} \quad z \in \Sigma,$$

there must exist a nonzero vector $\psi = \{\psi_0, \psi_1, \ldots, \psi_k\}$ satisfying the following two conditions:

(α) $\psi_0 \leqslant 0$;

(β) $\left(\sum_{\alpha=0}^{k} \psi_\alpha \operatorname{grad} F^\alpha(z_0) \right) \boldsymbol{\delta z} \leqslant 0$ *for any vector* $\boldsymbol{\delta z}$ *whose direction lies in cone $N_1 \cap \ldots \cap N_m$.*

COROLLARY 36.10. *Let Ω' be a set defined in the space E^n of variables $z^1, \ldots, z^n$ by the equations*

$$F^1(z) = 0, \ldots, F^k(z) = 0,$$

and let Ξ'' be a set defined by the inequalities

$$g^1(z) \leqslant 0, \ldots, g^p(z) \leqslant 0. \tag{36.8}$$

In addition, let $z_0 \in \Omega' \cap \Xi''$ and let system of constraints (36.8) be nonsingular at point z_0, that is, there exists in E^n a vector $\boldsymbol{a}$ forming a negative scalar product with each of the vectors

$$\operatorname{grad} g^i(z_0), \qquad i \in J(z_0). \tag{36.9}$$

Finally, let $F^0(z)$ be a smooth function defined in some open set of space E^n containing set $\Omega' \cap \Xi''$. In order for function $F^0(z)$, considered in set $\Omega' \cap \Xi''$, to reach a minimum at point z_0, there must exist a nonzero vector $\psi = \{\psi_0, \psi_1, \ldots, \psi_n\}$ satisfying the following two conditions:

(α) $\psi_0 \leqslant 0$;

(β) $\left(\sum_{\alpha=0}^{k} \psi_\alpha \operatorname{grad} F^\alpha(z_0)\right) \delta z \leqslant 0$ *for any vector* δz, *forming a nonpositive scalar product with each of vectors* (36.9).

Note another particular case of Theorem 36.8, obtained for $r=p=0$ and $m=1$. This case is interesting in view of the fact that here the condition of nonseparability of the system of cones L_i, P_j, N_h is absent.

THEOREM 36.11. *Let* Ω' *be a set defined in space* E^n *by the equations*

$$F^1(z)=0, \ \ldots, \ F^k(z)=0,$$

let Ω'' *be a set defined by the inequalities*

$$f^1(z)\leqslant 0, \ \ldots, \ f^q(z)\leqslant 0,$$

and let $\Omega_1, \ldots, \Omega_l$, Ξ *be arbitrary sets of space* E^n. *In addition, let* $F^0(z)$ *be a smooth function defined on some open set of space* E^n *containing the set* $\Sigma = \Omega' \cap \Omega'' \cap \Omega_1 \cap \ldots \cap \Omega_l \cap \Xi$. *The problem is: to find the point at which function* F^0, *considered only on set* Σ, *reaches its lowest value. We assume that* $z_0 \in \Sigma$ *and that* M_i *is a covering of set* Ω_i *at point* z_0 $(i=1, \ldots, l)$, *while* N *is a covering of set* Ξ *at* z_0.

In order for point z_0 *to constitute a solution of the problem, that is, in order for the following relation to be valid:*

$$F^0(z_0)\leqslant F^0(z) \quad \text{for} \quad z\in\Sigma,$$

there must exist numbers $\psi_0, \psi_1, \ldots, \psi_k$, *vectors* $\boldsymbol{a}_1, \ldots, \boldsymbol{a}_l$ *whose directions lie, respectively, in the dual cones* $D(M_1), \ldots, D(M_l)$, *and nonpositive numbers* λ_j, $j\in I(z_0)$, *all such that the following conditions are satisfied:*

(α) $\psi_0 \leqslant 0$;

(β) $\left(\sum_{\alpha=0}^{k} \psi_\alpha \operatorname{grad} F^\alpha(z_0) + \sum_{j\in I(z_0)} \lambda_j \operatorname{grad} f^j(z_0) - \boldsymbol{a}_1 - \ldots - \boldsymbol{a}_l\right) \delta z \leqslant 0$

for any vector δz *whose direction lies in cone* N;

(γ) *if all the numbers* $\psi_0, \psi_1, \ldots, \psi_k$ *and* λ_j, $j\in I(z_0)$, *are equal to zero, then at least one of vectors* $\boldsymbol{a}_1, \ldots, \boldsymbol{a}_l$ *will be different from zero.*

As a proof, it is sufficient to note that the statements presented during the proof of Theorem 36.7 remain in force for $m=1$ as well (while in this case no conditions need be imposed on cone N_1). By the way, Theorem 36.11 can be derived from Theorem 36.8 without referring to its proof. It is sufficient to set $r=p=0$, $m=2$, $\Xi_1=\Xi$, $N_1=N$, and $\Xi_2=N_2=E^n$; then $N_1\cap N_2=N$, and, since the system of cones N_1, N_2 does not possess separability, Theorem 36.8 can be directly applied.

NOTE 36.12. If any of sets $\Omega_1, \ldots, \Omega_l, \Xi_1, \ldots, \Xi_m$ in Theorem 36.8 (or in the other theorems of this subsection) is closed and *convex*, then as the corresponding cone $M_1, \ldots, M_l, N_1, \ldots, N_m$ we can take the *supporting cone* of this convex set at point z_0. Actually, according to Theorem 34.2, this supporting cone is a *covering* of the given convex set at point z_0. Using this note, it is then possible to formulate a number of theorems pertaining to the case where sets $\Omega_1, \ldots, \Omega_l, \Xi_1, \ldots, \Xi_m$ (or some of them) are convex.

As an example, we give the following theorem, derived on the basis of Note 36.12 from Theorem 36.11 (here we limit ourselves to the case $l=0,\ q=0$).

THEOREM 36.13. *Let Ω' be a set defined in the space E^n of variables $z^1, \ldots, z^n$ by the equations*

$$F^1(z)=0, \ldots, F^k(z)=0,$$

and let Ξ be some closed convex set in E^n. In addition, let $F^0(z)$ be a smooth function defined on some open set of space E^n containing set $\Omega' \cap \Xi$. In order for function $F^0(z)$, considered on set $\Omega' \cap \Xi$, to reach a minimum at a point $z_0 \in \Omega' \cap \Xi$, there must exist a nonzero vector $\boldsymbol{\psi}=\{\psi_0, \psi_1, \ldots, \psi_k\}$ satisfying the following two conditions:

(α) $\psi_0 \leqslant 0$;

(β) $\left(\sum_{\alpha=0}^{k} \psi_\alpha \operatorname{grad} F^\alpha(z_0)\right) \boldsymbol{\delta z} \leqslant 0$ *for any vector $\boldsymbol{\delta z}$, whose direction lies in the supporting cone of set Ξ at z_0.*

EXAMPLE 36.14. (Gibbs' Lemma). *Let $\varphi_1(t), \varphi_2(t), \ldots, \varphi_n(t)$ be continuously differentiable functions. If the function*

$$F^0(z)=\varphi_1(z^1)+\varphi_2(z^2)+\ldots+\varphi_n(z^n),$$

considered on a set Σ defined by the relations

$$z^1+z^2+\ldots+z^n=a; \quad z^1 \geqslant 0, \quad z^2 \geqslant 0, \ldots, z^n \geqslant 0,$$

reaches a maximum at a point $z_0=(z_0^1, z_0^2, \ldots, z_0^n)$, then there exists a number λ, such that

$$\varphi_i'(z_0^i)=\lambda \quad \text{for} \quad z_0^i>0,$$
$$\varphi_i'(z_0^i) \leqslant \lambda \quad \text{for} \quad z_0^i=0.$$

For the proof, we apply Theorem 36.3.* Set Σ is defined in the space E^n of variables $z^1, \ldots, z^n$ by the equation

$$F^1(z)=z^1+\ldots+z^n-a=0 \tag{36.10}$$

and by the inequalities

$$f^i(z)=-z^i \leqslant 0, \qquad i=1, \ldots, n. \tag{36.11}$$

* Gibbs' Lemma can be proven simply and directly, without applying the necessary conditions for an extremum cited in the present subsection. Here it is given just for purposes of illustration.

If all of constraints (36.11) are active (that is, $z_0^1 = \ldots = z_0^n = 0$), then clearly Gibbs' Lemma doesn't state anything. Thus let us assume that at least one of the indexes is not active. According to Theorem 36.3, there exists a vector $\psi = \{\psi_0, \psi_1\}$ such that the condition $\psi_0 \geqslant 0$ is satisfied (since function F^0 reaches a *maximum* at z_0 ,rather than a minimum), as well as condition (β). Since

$$\operatorname{grad} F^0(z) = \{\varphi_1'(z^1),\ \varphi_2'(z^2),\ \ldots,\ \varphi_n'(z^n)\},$$
$$\operatorname{grad} F^1(z) = \{1,\ 1,\ \ldots,\ 1\},$$
$$\operatorname{grad} f^i(z) = \{0,\ 0,\ \ldots,\ 0,\ -1,\ 0,\ \ldots,\ 0\}$$

(where -1 is in the ith place), condition (β) becomes

$$\psi_0\{\varphi_1'(z_0^1),\ \varphi_2'(z_0^2),\ \ldots,\ \varphi_n'(z_0^n)\} + \psi_1\{1,\ 1,\ \ldots,\ 1\} = \{\lambda^1,\ \lambda^2,\ \ldots,\ \lambda^n\},$$

where $\lambda^i \leqslant 0$, with $\lambda^i = 0$ for $i \notin I(z_0)$ (that is, for $z_0^i > 0$). In other words,

$$\psi_0\varphi_i'(z_0^i) = -\psi_1 + \lambda^i, \qquad i = 1,\ \ldots,\ n.$$

In particular, for inactive indexes (at least one being such) we have $\psi_0\varphi_i'(z_0^i) = -\psi_1$, from which it follows that $\psi_0 \neq 0$, that is, $\psi_0 > 0$. Consequently,

$$\varphi_i'(z_0^i) = -\frac{\psi_1}{\psi_0} + \frac{\lambda^i}{\psi_0}, \qquad i = 1,\ \ldots,\ n,$$

that is, $\varphi_i'(z_0^i) = -\psi_1/\psi_0 = \lambda$ for the inactive indexes and $\varphi_i'(z_0^i) \leqslant \lambda$ (since $\lambda^i \leqslant 0$) for the active indexes.

EXAMPLE 36.15.* Let us apply Gibbs' Lemma to find the maximum of the function

$$F^0(z) = \sum_{i=1}^{n} a_i\left(1 - e^{-b_i z^i}\right)$$

on the set Σ defined by relations (36.10) and (36.11). The numbers a_i, b_i are assumed to be positive.

Let $z_0 = (z_0^1,\ \ldots,\ z_0^n)$ be the point at which function $F^0(z)$, considered on Σ, reaches a maximum. According to Gibbs' Lemma, there exists a number λ, such that

$$a_i b_i e^{-b_i z_0^i} = \lambda \quad \text{for} \quad z_0^i > 0 \text{ and } a_i b_i e^{-b_i z_0^i} \leqslant \lambda \quad \text{for} \quad z_0^i = 0.$$

For $z_0^i > 0$ we have $e^{-b_i z_0^i} < 1$, so that $a_i b_i > \lambda$. This means that $z_0^i = 0$ for

* See: Danskin, J.M. *The Theory of Max-Min and Its Application to Weapons Allocation Problems.* – Springer, New York. 1967, p. 11.

$a_ib_i \leqslant \lambda$. On the other hand, for $a_ib_i > \lambda$ we have $z_0^i > 0$. Consequently,

$$z_0^i = \begin{cases} \frac{1}{b_i} \ln \frac{a_ib_i}{\lambda} & \text{for} \quad a_ib_i > \lambda; \\ 0 & \text{for} \quad a_ib_i \leqslant \lambda. \end{cases} \tag{36.12}$$

It remains to determine λ. To do this, we note that in view of (36.10) and (36.12) the following relation is valid:

$$\sum_{a_ib_i > \lambda} \frac{1}{b_i} \ln \frac{a_ib_i}{\lambda} = a,$$

which is an equation involving λ. The left-hand side of this equation is clearly a continuous function of λ, defined for $0 < \lambda < \infty$ and decreasing monotonically from ∞ to 0. Consequently, for any $a > 0$ this equation has a unique solution.

EXAMPLE 36.16. *If a smooth function* $F^0(z)$, *considered on the set* Σ *defined by relations* (36.10) *and* (36.11), *reaches a maximum at a point* $z_0 = (z_0^1, z_0^2, \ldots, z_0^n)$, *then there exists a number* λ *such that*

$$\delta \boldsymbol{z} \operatorname{grad} F^0(z_0) \leqslant \lambda \sum_{i=1}^{n} \delta z^i \tag{36.13}$$

for any vector $\delta \boldsymbol{z} = \{\delta z^1, \ldots, \delta z^n\}$ *satisfying the conditions:*

$$\delta z^i \geqslant 0 \quad \text{for} \quad z_0^i = 0 \quad (\text{that is, for} \quad i \in I(z_0)). \tag{36.14}$$

For the proof, we note that in view of Corollary 36.10 there must exist a nonzero vector $\psi = \{\psi_0, \psi_1\}$ satisfying the condition

$$(\psi_0 \operatorname{grad} F^0(z_0) + \psi_1 \operatorname{grad} F^1(z_0))\, \delta \boldsymbol{z} \leqslant 0 \tag{36.15}$$

for any vector $\delta \boldsymbol{z} = \{\delta z^1, \ldots, \delta z^n\}$ satisfying condition (36.14), with $\psi_0 \geqslant 0$. This relation can be rewritten as

$$\psi_0 (\delta \boldsymbol{z} \operatorname{grad} F^0(z_0)) \leqslant -\psi_1 \sum_{i=1}^{n} \delta z^i.$$

If $\psi_0 \neq 0$ (that is, $\psi_0 > 0$), then by setting $\lambda = -\psi_1/\psi_0$ we obtain relation (36.13). Now we have only to consider the case of $\psi_0 = 0$. In this case $\psi_1 \neq 0$ and relation (36.15) becomes

$$\psi_1 (\delta z^1 + \ldots + \delta z^n) \leqslant 0$$

for any vector $\delta \boldsymbol{z}$ satisfying conditions (36.14). If there exists at least one inactive index i (for which $z_0^i \neq 0$), then by assuming that $\delta z^i = \pm 1$, $\delta z^j = 0$ for $j \neq i$ we obtain a vector $\delta \boldsymbol{z}$ satisfying condition (36.14). Consequently, $\pm \psi_1 \leqslant 0$, which implies that $\psi_i = 0$, and this contradicts what was stated above. Thus all the indexes are active, that is, $z_0 = (0, 0, \ldots, 0)$. However, in this case relation (36.13) is obvious: it is sufficient to take as λ the largest of the numbers

$$\frac{\partial F^0(z_0)}{\partial z^1}, \ldots, \frac{\partial F^0(z_0)}{\partial z^n}.$$

Before formulating the next statement, let us recall the definition of a convex function. Let $F(z)$ be a function defined on some *convex* set P of space E^n (possibly over this entire space). Function F is called *convex* if for any two points z_1, z_2 of set P and any point z of segment $[z_1, z_2]$, that is, for a point $z = (1-\lambda) z_1 + \lambda z_2$, where $0 \leqslant \lambda \leqslant 1$, the following relation is valid:

$$F(z) \leqslant (1-\lambda) F(z_1) + \lambda F(z_2). \tag{36.16}$$

Function $F(z)$ is called *concave* if the function $-F(z)$ is convex, that is, if function $F(z)$ satisfies the following condition (for $z = (1-\lambda) z_1 + \lambda z_2$, where $0 \leqslant \lambda \leqslant 1$):

$$F(z) \geqslant (1-\lambda) F(z_1) + \lambda F(z_2). \tag{36.17}$$

Obviously, an *affine* function (considered on an arbitrary convex set) is simultaneously both convex and concave. It is easy to show that the converse is true as well, namely that, if a function F (defined on a convex set P) is simultaneously convex and concave, then it has to be affine (more precisely, F is identical to some affine function considered on set P).

An analysis indicates the following condition for convexity of a function. Let $F(z)$ be a twice continuously differentiable function defined on an open convex set P of space E^n. *For function $F(z)$ to be convex, it is necessary and sufficient that for any point $z_0 \in P$ the quadratic form*

$$\sum_{i,j=1}^{n} \frac{\partial^2 F(z_0)}{\partial z^i \partial z^j} \xi^i \xi^j \tag{36.18}$$

assumes (for any $\xi^1, \ldots, \xi^n$) only nonnegative values. If, in particular, at some point $z_0 \in P$ the quadratic form (36.18) is *positive-definite* (that is, for any nonzero vector $\xi = \{\xi^1, \ldots, \xi^n\}$ it has a *positive* value), then in some sufficiently small neighborhood around point z_0 function $F(z)$ will be convex. A function possessing this property (that is, one which is convex in some neighborhood around a point z_0) is called *locally convex at point z_0*.

COROLLARY 36.17. *Let Ω' be a set defined in the space E^n of variables $z^1, \ldots, z^n$ by the equations*

$$F^1(z) = 0, \ldots, F^k(z) = 0, \tag{36.19}$$

and let Ξ'' *be a set defined by the inequalities*

$$g^1(z) \leqslant 0, \ \ldots, \ g^p(z) \leqslant 0. \tag{36.20}$$

In addition, let $F^0(z)$ *be a smooth function defined on some open set of space* E^n *containing the set* $\Omega' \cap \Xi''$. *Finally, let* $z_0 \in \Omega' \cap \Xi''$ *and let the functions* $g^i(z)$, $i \in J(z_0)$, *be locally concave at point* z_0. *In order for function* $F^0(z)$, *considered on set* $\Omega' \cap \Xi''$, *to reach a minimum at* z_0 *there must exist a nonzero vector* $\psi = \{\psi_0, \psi_1, \ldots, \psi_k\}$ *satisfying the following two conditions:*

(α) $\psi_0 \leqslant 0$;

(β) $\left(\sum_{\alpha=0}^{k} \psi_\alpha \operatorname{grad} F^\alpha(z_0)\right) \delta z \leqslant 0$ *for any vector* δz, *forming a nonpositive scalar product with each of the vectors* $\operatorname{grad} g^i(z_0)$, $i \in J(z_0)$.

This statement differs from Corollary 36.10 in that system of constraints (36.20) is not assumed to be nonsingular at point z_0. Instead the requirement is imposed that functions $g^i(z)$, $i \in J(z_0)$, be locally concave at point z_0. The validity of Corollary 36.17 is implied by the following considerations.

Let $i \in J(z_0)$, that is, $g^i(z_0) = 0$. In addition, let P_i be the half-space consisting of all points z for which $(z - z_0) \operatorname{grad} g^i(z_0) \leqslant 0$, and let z_1 be a point quite close to z_0 (more precisely, located inside the region of local concavity of function $g^i(z)$). If $g^i(z_1) > 0$, then for any point $z = (1 - \lambda) z_0 + \lambda z_1$ in the interval (z_0, z_1) (that is, for $0 < \lambda < 1$) we have, according to (36.17),

$$g^i(z) \geqslant (1 - \lambda) g^i(z_0) + \lambda g^i(z_1), \quad \text{that is,} \quad g^i(z) \geqslant \lambda g^i(z_1).$$

Consequently,

$$\frac{g^i(z) - g^i(z_0)}{|z - z_0|} = \frac{g^i(z)}{\lambda |z_1 - z_0|} \geqslant \frac{\lambda g^i(z_1)}{\lambda |z_1 - z_0|} = \frac{g^i(z_1)}{|z_1 - z_0|} = \text{const} > 0.$$

Hence it follows that the derivative of function $g^i(z)$, taken at point z_0 in the direction of the vector $z_1 - z_0$, is *positive*, that is, $(z_1 - z_0) \operatorname{grad} g^i(z_0) > 0$, so that point z_1 *does not belong to* half-space P_i. Thus, if $g^i(z_1) > 0$ (point z_1 being sufficiently close to z_0), then $z_1 \notin P_i$, that is, *all points of half-space* P_i *sufficiently close to* z_0 *satisfy the constraint* $g^i(z) \leqslant 0$.

The intersection $N = \bigcap_{i \in J(z_0)} P_i$ of all the half-spaces P_i corresponding to the constraints (36.20) active at point z_0 is a convex cone with a vertex z_0, where, according to what has been proven, all the points of this cone sufficiently close to its vertex satisfy all of constraints (36.20), that is, all points of cone N close to its vertex are contained in set Ξ''. Consequently, N is a *covering* of set Ξ'' at point z_0 (as the mapping ψ satisfying conditions 1 and 2 formulated in Definition 34.1, we can take the *identity* mapping of cone N). Here, by definition, cone N consists of all the points $z \in E^n$ satisfying the conditions

$$(z - z_0) \operatorname{grad} g^i(z_0) \leqslant 0, \qquad i \in J(z_0).$$

Corollary 36.17 now follows directly from Theorem 36.11 (for $q=l=0$).

NOTE 36.18. An analysis of the proof indicates that the requirement of concavity of functions $g^i(z)$, $i \in J(z_0)$, can be made somewhat less stringent. Actually, it is enough to require that the above consideration be valid just for points z_1 belonging to cone N (and lying sufficiently close to its vertex z_0), that is, it is sufficient to require that each of functions $g^i(z)$, $i \in J(z_0)$, be concave not in the entire *neighborhood* of z_0, but rather just on cone N (close to its vertex). In other words, the requirement of local concavity of functions $g^i(z)$, $i \in J(z_0)$, can be replaced by the following less stringent requirement: *each of functions* $g^i(z)$, $i \in J(z_0)$, *is for some* $\varepsilon > 0$ *concave on set* $N \cap \Sigma_\varepsilon$, *where* $N = \bigcap_{i \in J(z_0)} P_i$ *is the cone constructed for the proof of Corollary* 36.17, *and* Σ_ε *is a sphere of radius* ε *centered on* z_0. This note widens somewhat the region of applicability of Corollary 36.17.

EXAMPLE 36.19. Let Ω' be a set defined in the space E^3 of variables z^1, z^2, z^3 by the equation $z^3 = 0$, and let Ξ'' be a set defined by the inequalities

$$(z^1)^3 - z^2 \leqslant 0, \quad z^2 \leqslant 0. \tag{36.21}$$

In other words, Ω' and Ξ'' are defined by relations of the type indicated in Corollary 36.10, but with $n=3$, $k=1$, $p=2$, and

$$F^1(z^1, z^2, z^3) = z^3, \quad g^1(z^1, z^2, z^3) = (z^1)^3 - z^2, \quad g^2(z^1, z^2, z^3) = z^2.$$

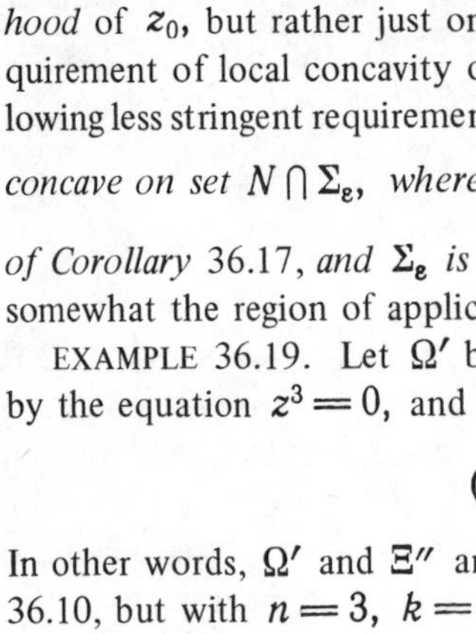

FIGURE 149

In addition, let us consider the function $F^0 = -z^1 - z^2 - z^3$ and let us try to find the minimum of function F^0 on the set $\Omega' \cap \Xi''$. It is easy to see that set $\Omega' \cap \Xi''$ lies entirely in the quadrant $z^1 \leqslant 0$, $z^2 \leqslant 0$ of the plane $z^3 = 0$ (Figure 149), so that function F^0 reaches a minimum on this set at the point $z_0 = (0, 0, 0)$.

Moreover, both of constraints (36.21) are active at point z_0. We have:

$$\left.\begin{aligned} &\operatorname{grad} F^0(z_0) = \{-1, -1, -1\};\\ &\operatorname{grad} F^1(z_0) = \{0, 0, 1\};\\ &\operatorname{grad} g^1(z_0) = \{0, -1, 0\};\\ &\operatorname{grad} g^2(z_0) = \{0, 1, 0\}. \end{aligned}\right\} \tag{36.22}$$

Condition (β), entering into the formulation of Corollaries 36.10 and 36.17, now becomes

$$(\psi_0\{-1, -1, -1\} + \psi_1\{0, 0, 1\})\,\boldsymbol{\delta z} \leqslant 0$$

for any vector $\boldsymbol{\delta z} = \{\delta z^1, \delta z^2, \delta z^3\}$ forming a nonpositive scalar product with each of vectors (36.22), that is, for any vector $\boldsymbol{\delta z} = \{\delta z^1, 0, \delta z^3\}$. This implies that: $\psi_0 = 0$,

$\psi_1 = 0$. Therefore, *there cannot exist any nonzero* vector $\psi = \{\psi_1, \psi_2\}$ satisfying condition (β), that is, although at point $z_0 = (0, 0, 0)$ a minimum is reached, still the necessary conditions specified in Corollaries 36.10 and 36.17 are not satisfied. This is because these corollaries are *not applicable* here. Corollary 36.10 is inapplicable because system of constraints (36.21) is not nonsingular at point z_0 (there does not exist a vector $\boldsymbol{a}$ forming a negative scalar product with each of the vectors $\operatorname{grad} g^1(z_0)$, $\operatorname{grad} g^2(z_0)$). Corollary 36.17 is inapplicable because function $g^1(z^1, z^2, z^3)$ is not locally concave at z_0.

This example indicates that the condition of nonsingularity of system of constraints (36.8) at point z_0 is significant in Corollary 36.10 (and thus, in Theorem 36.7, the assumption that system of cones $N_1, \ldots, N_m$ does not possess separability is significant).

NOTE 36.20. In this subsection we have presented a number of theorems giving the necessary conditions for the extremum of a function. In all cases the proof was based on Theorem 35.1 and on Theorem 36.1, proven with the aid of Theorem 35.1. Theorems of this type can also be obtained using other methods.

Let us consider briefly the method of Jordan and Polak,* using as an example the proof of Theorem 36.9 for $m = 1$ (that is, we assume only a single set $\Xi = \Xi_1$ having a covering $N = N_1$ at point z_0).

Let z_0 be a point giving a solution of the problem of minimizing function F^0 on set $\Omega' \cap \Xi$. Consider the $(k+1)$-dimensional space E^{k+1} of variables $x^0, x^1, \ldots, x^k$, where g is the mapping of space E^n into E^{k+1} defined by the formulas $x^i = F^i(z)$, $i = 0, \ldots, k$. If we set

$$a^i_j = \frac{\partial F^i(z_0)}{\partial z^j}, \qquad i = 0, 1, \ldots, k; \quad j = 1, \ldots, n,$$

then with the aid of numbers a^i_j we can define an affine mapping $f\colon E^n \to E^{k+1}$ using the formulas:

$$x^i = \sum_{j=1}^{n} a^i_j (z^j - z^j_0) + F^i(z_0), \qquad i = 0, 1, \ldots, k.$$

Since functions $F^0, F^1, \ldots, F^k$ are smooth, it follows that

$$F^i(z) = F^i(z_0) + \sum_{j=1}^{n} \frac{\partial F^i(z_0)}{\partial z^j}(z^j - z^j_0) + o_{z_0}(z), \qquad i = 0, 1, \ldots, k,$$

or, in vector form,

$$g(z) = f(z) + o_{z_0}(z).$$

This relation shows that at point z_0 f is a tangent affine mapping to the mapping g of space E^n into E^{k+1}.

* Jordan, B.W., and E. Polak. *Theory of a Class of Discrete Optimal Control Systems.* – J. Electron. and Control, 17, 6. 1964; Cannon, M., C. Cullum, and E. Polak. *Constrained Minimization Problems in Finite-Dimensional Spaces.* – J. SIAM Control, 4, No. 3. 1966.

The image of convex cone N for affine mapping f is a convex cone $f(N)$ in space E^{k+1}, with a vertex at the point $f(z_0)=g(z_0)$. Let Λ be the ray originating at point $f(z_0)=g(z_0)$ of space E^{k+1} and going in the direction of the negative x^0 semiaxis; in other words, ray Λ consists of the points $(x^0, x^1, \ldots, x^k)$ satisfying the conditions

$$x^0 \leqslant F^0(z_0), \quad x^1=F^1(z_0), \quad x^2=F^2(z_0), \ldots, x^k=F^k(z_0).$$

Assume that Λ is in space E^{k+1} an interior ray of convex cone $f(N)$. Then, in view of Theorem 34.7 (as applied to set Ξ and mapping g), there exists a cone $M \subset f(N)$ with a vertex $g(z_0)$, such that Λ is an interior ray of cone M, M being a covering of set $g(\Xi)$ at point $g(z_0)$. Therefore, cones M and Λ are nonseparable in E^{k+1}, so that in view of Theorem 35.1 there exists some point $x \in g(\Xi) \cap \Lambda$ different from $g(z_0)$, that is, there exists a point $z \in \Xi$ such that point $x=g(z)$ is different from $g(z_0)$ and lies on ray Λ. In other words,

$$F^0(z) < F^0(z_0), \quad F^1(z)=F^1(z_0), \ldots, F^k(z)=F^k(z_0),$$

that is, $z \in \Omega'$ and $F^0(z) < F^0(z_0)$. In addition, as we have seen, $z \in \Xi$. Consequently, $z \in \Omega' \cap \Xi$ and $F^0(z) < F^0(z_0)$. However, this contradicts the fact that function F^0, considered on set $\Omega' \cap \Xi$, reaches a minimum at point z_0. This contradiction indicates that Λ is not an interior ray of convex cone $f(N)$, so that in view of Theorem 30.4 there exists a hyperplane Γ separating ray Λ from cone $f(N)$.

Now let $\psi=\{\psi_0, \psi_1, \ldots, \psi_k\}$ be a normal vector of hyperplane Γ, so directed that cone $f(N)$ lies in the negative (closed) half-space (ray Λ thus being in the positive half-space), and let us show that vector ψ is the desired vector. Since the vector $l_0=\{-1, 0, \ldots, 0\}$ has the direction of ray Λ, and thus is directed into the positive half-space, we know that $l_0\psi \geqslant 0$, that is, $-1\cdot\psi_0 + 0\cdot\psi_1 + \ldots + 0\cdot\psi_n \geqslant 0$. Accordingly, condition (α) of Theorem 36.9 is satisfied.

Moreover, let δz be an arbitrary vector whose direction lies in cone N. In other words, $\delta z = z - z_0$, where z is some point of cone N. Then the direction of the vector $f(\delta z)=f(z)-f(z_0)=f(z)-g(z_0)$ lies in cone $f(N)$, so that this vector is directed into the negative half-space. Consequently, $\psi f(\delta z) \leqslant 0$ Recalling the definition of mapping f and writing this scalar product in coordinate form, we obtain the inequality given in condition $(\beta)$of Theorem 36.9.

This elegant discussion can (in a somewhat more complicated form) be applied to the proof of Theorem 36.11. On the other hand, the more general Theorem 36.8 cannot be derived using this method (to derive it, we would have to show that, if sets $\Xi_1, \ldots, \Xi_m$ have at their common point z_0 the coverings $N_1, \ldots, N_m$, and if the system of cones $N_1, \ldots, N_m$ does not possess separability, then $N_1 \cap \ldots \cap N_m$ is a covering of set $\Xi_1 \cap \ldots \cap \Xi_m$ at point z_0; however, this is simply *incorrect* in the general case).

NOTE 36.21. The widely known *Dubovitskii-Milyutin method* (as applied to problems of extrema of functions) uses Theorem 32.8 instead of the general Theorem 32.5, while instead of the more detailed Theorem 35.1 a result of the same type is used, but one that pertains only to the case in which functions ψ_i are smooth (cf. Note 35.2), all of cones $M_1, M_2, \ldots, M_k$ except one being solid. Therefore, the method applied here to obtain the necessary conditions for an extremum constitutes a *generalization* of the Dubovitskii-Milyutin method. Accordingly, the results presented in this subsection are more general than those obtained with the Dubovitskii-Milyutin method.

As an example, we present a proof of Corollary 36.10 using the Dubovitskii-Milyutin method. Let L be the tangent plane of an $(n-k)$-dimensional manifold Ω', defined in E^n by the equations $F^1(z)=0, \ldots, F^k(z)=0$ (here, in view of Note 36.5, we can assume functions $F^1(z), \ldots, F^k(z)$ to be independent at point z_0, that is, they define an $(n-k)$-dimensional manifold in the neighborhood of point z_0.) Moreover, to simplify the notation, we assume the first s constraints (36.8) to be active at point z_0 and we call M_i, $i=1, \ldots, s$, the half-space of space E^n, consisting of all points z, for which

$$(z-z_0)\,\mathrm{grad}\, g^i(z_0) \leqslant 0,$$

and we call M_0 the half-space consisting of all points z for which

$$(z - z_0) \operatorname{grad} F^0(z_0) \leqslant 0.$$

Obviously, $M_0, M_1, \ldots, M_s$ are closed solid cones in E^n with a common vertex z_0, and plane L also passes through point z_0. This makes it possible to apply Theorem 32.7.

Assume that the intersection

$$L \cap (\operatorname{int} M_0) \cap (\operatorname{int} M_1) \cap \ldots \cap (\operatorname{int} M_s) \tag{36.23}$$

is nonempty, that is, that there exists a ray l originating at point z_0 and lying in plane L which is an *interior* ray of each of cones $M_0, M_1, \ldots, M_s$. Then there must exist a smooth curve Λ, originating at point z_0, having l as its tangent ray at point z_0 and lying on manifold Ω'. Since l is an interior ray of half-space M_0, the angle between ray l and the vector $\operatorname{grad} F^0(z_0)$ is *obtuse*. Consequently, for all points of line Λ, sufficiently close to z_0 and different from z_0, the relation $F^0(z) < F^0(z_0)$ is valid. Similarly, for all points of line Λ, sufficiently close to z_0 and different from z_0, the relation $g^i(z) < g^i(z_0) = 0$, $i = 1, \ldots, s$, is also valid. Therefore, on curve Λ we can choose a point z, satisfying all the relations $F^1(z) = 0, \ldots, F^k(z) = 0$ and all the constraints (36.8) (that is, belonging to the set $\Omega' \cap \Xi''$), such that $F^0(z) < F^0(z_0)$. However, then function F^0, considered on set $\Omega' \cap \Xi''$, does not reach a minimum at z_0.

Thus, in order for function F^0, considered on set $\Omega' \cap \Xi''$, to reach a minimum at z_0, intersection (36.23) has to be empty. However, if this intersection is empty, then according to Theorem 32.8 there must exist affine functions $f^*, f^0, f^1, \ldots, f^s$, going to zero at point z_0 but not all identically zero, such that function f^* is identically zero in plane L, function f^i is nonpositive in half-space M_i, $i = 0, 1, \ldots, s$, and the following relation is valid:

$$f^* + f^0 + f^1 + \ldots + f^s \equiv 0.$$

Since $f^0(z_0) = 0$ and function f^0 is nonpositive in half-space M_0, it follows that $\operatorname{grad} f^0 = -\psi_0 \operatorname{grad} F^0(z_0)$, where $\psi_0 \leqslant 0$. Similarly, $\operatorname{grad} f^j = \lambda_j \operatorname{grad} g^j(z_0)$, where $\lambda_j \geqslant 0$ ($j = 1, \ldots, s$, that is, $j \in J(z_0)$). Moreover, since affine function f^* is identically zero in L, it follows that

$$\operatorname{grad} f^* = -\psi_1 \operatorname{grad} F^1(z_0) - \ldots - \psi_k \operatorname{grad} F^k(z_0).$$

Let us show that the vector $\psi = \{\psi_0, \psi_1, \ldots, \psi_k\}$ is the vector sought.

As we have seen, condition (α) of Corollary 36.10 is satisfied. Let us verify that condition (β) is satisfied. If $\boldsymbol{\delta z}$ is a vector forming a nonpositive scalar product with each of vectors (36.9), that is, with each of the vectors $\operatorname{grad} g^j(z_0)$, $j = 1, \ldots, s$, then

$$f^0(z_0 + \delta z) = f^0(z_0 + \delta z) - f^0(z_0) = \boldsymbol{\delta z} \operatorname{grad} f^0 = -\psi_0 \boldsymbol{\delta z} \operatorname{grad} F^0(z_0);$$

$$f^*(z_0 + \delta z) = f^*(z_0 + \delta z) - f^*(z_0) = \boldsymbol{\delta z} \operatorname{grad} f^* = -\sum_{i=1}^{k} \psi_i \boldsymbol{\delta z} \operatorname{grad} F^i(z_0);$$

$$f^j(z_0 + \delta z) = f^j(z_0 + \delta z) - f^j(z_0) = \boldsymbol{\delta z} \operatorname{grad} f^j = \\ = \lambda_j \boldsymbol{\delta z} \operatorname{grad} g^j(z_0) \leqslant 0 \quad \text{for} \quad j = 1, \ldots, s.$$

By adding all these relations together and taking into account that $f^* + f^0 + f^1 + \ldots + f^s \equiv 0$, we now obtain

$$0 = f^*(z_0 + \delta z) + f^0(z_0 + \delta z) + f^1(z_0 + \delta z) + \ldots + f^s(z_0 + \delta z) \leqslant \\ \leqslant -\sum_{\alpha=0}^{k} \psi_\alpha (\boldsymbol{\delta z} \operatorname{grad} F^\alpha(z_0)).$$

Consequently, relation (β) of Corollary 36.10 is also valid.

It remains to show that vector $\psi=\{\psi_0, \psi_1, \ldots, \psi_k\}$ is different from zero. If we assume the converse, namely that $\psi_0=\psi_1= \ldots =\psi_k=0$, then $f^0\equiv 0$, $f^*\equiv 0$, so that functions $f^1, \ldots, f^s$ are linked by the relation $f^1+ \ldots +f^s\equiv 0$. If a is a vector forming a negative scalar product with each of vectors (36.9), then

$$f^j(z_0+a)=f^j(z_0+a)-f^j(z_0)=a \operatorname{grad} f^j=\lambda_j(a \operatorname{grad} g^j(z_0))\leqslant 0$$

for any $j\in J(z_0)$, that is, for any $j=1, \ldots, s$. Hence, in view of the relation $f^1+ \ldots + + f^s\equiv 0$, we obtain $f^j(z_0+a)=0$, $j=1, \ldots, s$, that is, $\lambda_j(a \operatorname{grad} g^j(z_0))=0$, so that $\lambda_j=0$ for all $j=1, \ldots, s$. This implies that $f^1\equiv 0, \ldots, f^s\equiv 0$. Consequently, *all* the functions $f^*, f^0, f^1, \ldots, f^s$ are identically equal to zero, which contradicts the assumption. This contradiction shows that the vector $\psi=\{\psi_0, \psi_1, \ldots, \psi_s\}$ is different from zero.

As an example, let us present an application of Theorem 36.8 which does not lie within the framework of the Dubovitskii-Milyutin method.

EXAMPLE 36.22. Let Ξ_1 be the set described in space E^n by the relations $|z^2|\leqslant(z^1)^2$, $z^1\geqslant 0$, let Ξ_2 be the set described by the relations $|z^4|\leqslant(z^3)^2$, $z^3\geqslant 0$, and let Ω' be the hypersurface defined by the equation

$$F^1(z)=z^6-(z^5)^2=0.$$

It is easy to see that the half-plane N_1 defined by the relations $z^2=0$, $z^1\geqslant 0$ is the maximum covering of set Ξ_1 at all points of the $(n-2)$-dimensional plane L_1 defined by the equations $z^1=z^2=0$; similarly, the half-space N_2 defined by the relations $z^4=0$, $z^3\geqslant 0$ is the maximum covering of set Ξ_2 at all points of the $(n-2)$-dimensional plane L_2 defined by the equations $z^3=z^4=0$. Finally, the maximum covering of hypersurface Ω' (at any point of it) is the hyperplane Γ, tangent to this hypersurface.

Therefore, at points of each of the sets

$$L_1\cap\Xi_2\cap\Omega', \quad \Xi_1\cap L_2\cap\Omega'$$

two of the three coverings N_1, N_2, Γ are not solid, while at points of the set $L_1\cap L_2\cap\Omega'$ all three coverings are not solid, so that at points of these sets the Dubovitskii-Milyutin method is inapplicable (no matter what the function F^0, the minimum of which is sought on the set $\Omega'\cap\Xi_1\cap\Xi_2$). Naturally, we can consider function F^0 separately at points of each of the sets

$$L_1\cap\Xi_2\cap\Omega', \quad \Xi_1\cap L_2\cap\Omega', \quad L_1\cap L_2\cap\Omega',$$

and at points not belonging to one of these sets, but this takes us back to the method of separate study of a function at the faces of a "curvilinear polyhedron," discussed in Subsection 7.

Let us consider the application of Theorem 36.8 under these conditions, using as an example the search for the minimum on set $\Omega'\cap\Xi_1\cap\Xi_2$ of the function

$$F^0(z)=\sum_{i=1}^{n}(z^i)^2+\frac{19}{36}\sum_{i=1}^{n}z^i.$$

Assume that this function reaches a minimum at the point $z_0=\left(z_0^1, \ldots, z_0^n\right)\in\Omega'\cap\Xi_1\cap\Xi_2$. Now let N_1 and N_2 be the maximum coverings of sets Ξ_1, Ξ_2 at point z_0 (for $z_0\notin L_i$ these maximum coverings are described by Theorem 34.4). Note that cones N_1 and N_2 are clearly non-separable, so that Theorem 36.8 is applicable. Condition (β) of Theorem 36.8 now becomes

$$\left(\psi_0 \operatorname{grad} F^0(z_0)\right)\delta z+\psi_1\left(\delta z^6-2z_0^5\,\delta z^5\right)\leqslant 0 \tag{36.24}$$

for any vector $\delta z = \{\delta z^1, \ldots, \delta z^n\}$, the direction of which lies in cone $N_1 \cap N_2$. It is easy to show that for

$$\delta z^1 = -z_0^1, \quad \delta z^2 = -2z_0^2, \quad \delta z^3 = -z_0^3, \quad \delta z^4 = -2z_0^4 \tag{36.25}$$

(and for *any* $\delta z^5, \ldots, \delta z^n$) the direction of vector δz lies in cone $N_1 \cap N_2$. Thus, for $\psi_0 = 0$, $\psi_1 \neq 0$, in view of (36.24) we would have $\psi_1 \left(2\delta z^6 - 2z_0^5 \delta z^5\right) \leqslant 0$ for *any* δz^5, δz^6, which is impossible. Consequently, $\psi_0 \neq 0$. In view of condition (α), we obtain $\psi_0 < 0$, and thus it can be assumed that $\psi_0 = -1$. Condition (36.24) now becomes

$$-\sum_{i=1}^{n} \left(2z_0^i + \frac{19}{36}\right) \delta z^i + \psi_1 \left(\delta z^6 - 2z_0^5 \delta z^5\right) \leqslant 0. \tag{36.26}$$

Since this relation is valid, in particular, when δz^1, δz^2, δz^3, δz^4 assume values (36.25), and when $\delta z^5 = \ldots = \delta z^n = 0$, it follows that

$$\left(2z_0^1 + \frac{19}{36}\right) z_0^1 + 2\left(2z_0^2 + \frac{19}{36}\right) z_0^2 + \left(2z_0^3 + \frac{19}{36}\right) z_0^3 + 2\left(2z_0^4 + \frac{19}{36}\right) z_0^4 \leqslant 0.$$

By writing this relation in the form

$$\frac{19}{18}\left((z_0^1)^2 + z_0^2\right) + \frac{19}{18}\left((z_0^3)^2 + z_0^4\right) + \frac{17}{18}(z_0^1)^2 + \frac{17}{18}(z_0^3)^2 + \\ + 4(z_0^2)^2 + 4(z_0^4)^2 + \frac{19}{36}\left(z_0^1 + z_0^3\right) \leqslant 0$$

and noting that all the terms on the left except the last are obviously nonnegative (the first two terms are nonnegative because of the definition of sets Ξ_1 and Ξ_2), we get $z_0^1 + z_0^3 \leqslant 0$. Hence, once again in view of the definition of sets Ξ_1, Ξ_2, we obtain $z_0^1 = 0$, $z_0^3 = 0$, and thus $z_0^2 = 0$ and $z_0^4 = 0$. Now relations (36.25) become $\delta z^1 = \delta z^2 = \delta z^3 = \delta z^4 = 0$, while for these values relation (36.26) can be rewritten as

$$-\sum_{i=5}^{n} \left(2z_0^i + \frac{19}{36}\right) \delta z^i + \psi_1 \left(\delta z^6 - 2z_0^5 \delta z^5\right) \leqslant 0,$$

this relation being valid for *any* $\delta z^5, \ldots, \delta z^n$. Consequently,

$$\begin{aligned} &-\left(2z_0^5 + \frac{19}{36}\right) - 2\psi_1 z_0^5 = 0, \\ &-\left(2z_0^6 + \frac{19}{36}\right) + \psi_1 = 0, \\ &-\left(2z_0^7 + \frac{19}{36}\right) = 0, \\ &\ldots\ldots\ldots\ldots\ldots \\ &-\left(2z_0^n + \frac{19}{36}\right) = 0. \end{aligned} \tag{36.27}$$

The first two of these relations, together with the relation $z_0^6 - (z_0^5)^2 = 0$ (which is valid, since $z_0 \in \Omega'$), constitute a system of three equations in three unknowns: z_0^5, z_0^6, ψ_1. By eliminating z_0^6 and ψ_1 from this system, we get

$$2z_0^5 + \frac{19}{36} + 2z_0^5\left(2(z_0^5)^2 + \frac{19}{36}\right) = 0,$$

or

$$\left(z_0^5 + \frac{1}{6}\right)\left(4(z_0^5)^2 - \frac{2}{3}z_0^5 + \frac{19}{6}\right) = 0.$$

This gives $z_0^5 = -1/6$ and thus $z_0^6 = 1/36$. Finally, the subsequent relations of (36.27) give the values $z_0^7 = \ldots = z_0^n = -19/72$.

Consequently, the minimum of function $F^0(z)$ on set $\Omega' \cap \Xi_1 \cap \Xi_2$ can be reached *only* at a point z_0 having the coordinates

$$z_0^1 = z_0^2 = z_0^3 = z_0^4 = 0, \quad z_0^5 = -\frac{1}{6}, \quad z_0^6 = \frac{1}{36}, \quad z_0^7 = \ldots = z_0^n = -\frac{19}{72}.$$

Hence at this point function $F^0(z)$ actually *attains* a minimum, since set $\Omega' \cap \Xi_1 \cap \Xi_2$ is closed, while function F^0 is continuous and increases without limit as a point z moves off to infinity (in any direction).

NOTE 36.23. In conclusion, we note that the necessary conditions derived in this subsection impose *more* requirements than do the conditions arrived at using the method of Subsection 7.

For example, let z_0 be a point of "curvilinear polyhedron" M lying, say, on an $(n-1)$-dimensional "face" of it, the necessary conditions of Subsection 7 being satisfied at this point, that is, the derivatives of function $F^0(z)$ taken at point z_0 in the directions lying in this face are zero. From the point of view of the methods in this subsection, on the other hand, the situation can be interpreted as follows. The "curvilinear polyhedron" is an n-dimensional set Ω'' described by a system of constraints of type (36.2), while its $(n-1)$-dimensional "face" is a set of points in which *one* of the constraints (the first, say) is active, while the rest are inactive. Therefore, the "face" in question is described by the relations

$$f^1(z) = 0, \quad f^2(z) < 0, \ldots, f^q(z) < 0. \tag{36.28}$$

Moreover, relations (36.1) do not apply (since $k = 0$), that is, set Ω' coincides with the entire space E^n, so that $\Omega' \cap \Omega'' = \Omega''$ is the "curvilinear polyhedron" being considered.

Assume that function $F^0(z)$, considered on Ω'' reaches a minimum at some point belonging to the $(n-1)$-dimensional "face" (36.28), so that at this point just the first constraint (36.2) is active. Then, conditions (α), (β) of Theorem 36.3 become

$$\psi_0 \operatorname{grad} F^0(z_0) = -\lambda_1 \operatorname{grad} f^1(z_0), \qquad \psi_0 < 0, \quad \lambda_1 \leqslant 0 \tag{36.29}$$

(if it were true that $\psi_0 = 0$, then in view of the relation $\operatorname{grad} f^1(z_0) \neq 0$ we would also have $\lambda_1 = 0$, which is impossible). If a is an arbitrary vector *tangent* to "face" (36.28) at point z_0, that is, a vector satisfying the relation $a \operatorname{grad} f^1(z_0) = 0$, then from (36.29) we have

$$a \operatorname{grad} F^0(z_0) = 0.$$

In other words, the derivative of function $F^0(z)$ at point z_0, in any direction tangent to "face" (36.28), is zero. However, this is the necessary condition for an extremum which we used in Subsection 7, that is, the necessary conditions of Subsection 7 *follow from* Theorem 36.3.

On the other hand, Theorem 36.3 (that is, in the given case, relation (36.29)) implies somewhat more than this. Actually, so far we have made use only of the *collinearity* of the vectors $\text{grad}\, F^0(z_0)$ and $\text{grad}\, f^1(z_0)$, while relation (36.29) states that, in addition, the *directions* of these vectors must be *opposite*. In the given case (that is, for points belonging to the $(n-1)$-dimensional "faces") this, then, represents the additional information which is contained in Theorem 36.3, in comparison with the methods of Subsection 7. The geometrical meaning of this additional condition is obvious: if vectors $\text{grad}\, F^0(z_0)$ and $\text{grad}\, f^1(z_0)$ are *oppositely* directed, then *within* domain Ω'' (that is, in the direction of *decrease* of function $f^1(z)$) function $F^0(z)$ will *increase*, and it is precisely this which corresponds to attaining a *minimum* of function $F^0(z)$ at point z_0.

Similarly, we could discern the additional information which Theorem 36.3 contains in the case where a minimum is reached on "faces" of lower dimension. Consequently, Theorem 36.3 (being only a particular case of the general results arrived at in this subsection) constitutes a refinement and reinforcement of the methods discussed in Subsection 7.

37. Sufficient condition for extremum of function. In this subsection we will derive, with certain constraints imposed on the functions and sets involved, the necessary and sufficient conditions for an extremum. The necessity of these conditions follows from the results of the foregoing subsection, and the sufficiency will be established here.

As a basis for our discussion, we take the following theorem on the minimum of an affine function* at an intersection of convex cones. As in the previous subsection, if the problem involves a maximum of a function, then the sign of the inequality in condition (α), below, is simply reversed.

THEOREM 37.1. *Let f^0 be an affine function in E^n and let $M_1, \ldots, M_l, N_1, \ldots, N_m$ be a system of convex cones, with a common vertex z_0 in E^n, that does not possess separability. In order for function f^0, considered on the cone*

$$K = M_1 \cap \ldots \cap M_l \cap N_1 \cap \ldots \cap N_m,$$

to reach a minimum at point z_0, it is necessary and sufficient that there exist a number ψ_0 and vectors $\boldsymbol{a}_1, \ldots, \boldsymbol{a}_l$, with directions, respectively, in the dual cones $D(M_1), \ldots, D(M_l)$, such that the following conditions are satisfied:

(α) $\psi_0 < 0$;

(β) $(\psi_0 \,\text{grad}\, f^0 - \boldsymbol{a}_1 - \ldots - \boldsymbol{a}_l)\, \boldsymbol{\delta z} \leqslant 0$ *for any vector $\boldsymbol{\delta z}$, whose direction lies in the cone $N_1 \cap \ldots \cap N_m$.*

Proof. Assume that function f^0, considered on cone K, reaches a minimum at point z_0. Since each cone M_i is a covering of set $\Omega_i = M_i$ at point z_0 $(i = 1, \ldots, l)$

* Note that the term "linear" does not have just one meaning in mathematics. For example, in functional analysis homogeneity has to be assumed when considering *linear* operators and *linear* functionals. On the other hand, in algebra and geometry (*linear* equations, *linear* transformations) homogeneity is not assumed. To avoid confusion, in this book we refer to *homomorphisms* of vector spaces (instead of linear operators, p. 95) and *affine* mappings (p. 129). The term "linear" will everywhere be used just in its algebraic sense, that is, an equation of the form $F(x) = 0$, where $F(x)$ is an affine function (p. 136), is called a *linear equation.*

and since each cone N_j is a covering of $\Xi_j = N_j$ at z_0 $(j=1, \ldots, m)$, it follows from Theorem 36.7 (applied to the function $F^0 = f^0$) that there must exist a number ψ_0 and vectors $\boldsymbol{a}_1, \ldots, \boldsymbol{a}_l$ satisfying conditions (α), (β), (γ) of Theorem 36.7. Condition (β) of Theorem 36.7 is identical to condition (β) of Theorem 37.1. Let us show that $\psi_0 < 0$.

First we assume the converse, namely that $\psi_0 = 0$. Then condition (β) (its validity was already shown) becomes: $(-\boldsymbol{a}_1 - \ldots - \boldsymbol{a}_l)\,\delta \boldsymbol{z} \leqslant 0$ for any vector $\delta\boldsymbol{z}$ whose direction lies in cone $N_1 \cap \ldots \cap N_m$. This means that the direction of vector $-\boldsymbol{a}_1 - \ldots - \boldsymbol{a}_l$ lies in cone $D(N_1 \cap \ldots \cap N_m)$, and thus from Corollary 31.4 and Lemma 32.1 we know that there exist vectors $\boldsymbol{b}_1, \ldots, \boldsymbol{b}_m$ such that the direction of vector $\boldsymbol{b}_j$ lies in the cone $D(N_j)$ $(j=1, \ldots, m)$, where

$$-\boldsymbol{a}_1 - \ldots - \boldsymbol{a}_l = \boldsymbol{b}_1 + \ldots + \boldsymbol{b}_m.$$

But, in view of Theorem 32.2, this contradicts the fact that the system of convex cones $M_1, \ldots, M_l, N_1, \ldots, N_m$ does not possess separability (note that, according to condition (γ) of Theorem 36.7 and the assumption $\psi_0 = 0$, some of vectors $\boldsymbol{a}_1, \ldots, \boldsymbol{a}_l$ are different from zero). Thus the assumption that $\psi_0 = 0$ leads to a contradiction: consequently, condition (α) of Theorem 36.7 implies that $\psi_0 < 0$. Consequently, the *necessity* of the conditions formulated has been established.

Now let us prove *sufficiency*. Let there be a number ψ_0 and vectors $\boldsymbol{a}_1, \ldots, \boldsymbol{a}_l$ satisfying conditions (α) and (β) of Theorem 37.1. Consider an arbitrary point $\boldsymbol{z}$ of cone K. Since the direction of vector $\delta\boldsymbol{z} = z - z_0$ lies in cone $N_1 \cap \ldots \cap N_m$, it follows from condition (β) that

$$(\psi_0 \operatorname{grad} f^0 - \boldsymbol{a}_1 - \ldots - \boldsymbol{a}_l)(z - z_0) \leqslant 0.$$

Moreover, we know that the direction of vector $\boldsymbol{a}_i$ lies in cone $D(M_i)$, and that the direction of vector $z - z_0$ lies in cone K, and even more so in cone M_i $(i = 1, \ldots, l)$. Consequently, $\boldsymbol{a}_i(z - z_0) \leqslant 0$, so that

$$(\psi_0 \operatorname{grad} f^0)(z - z_0) \leqslant (\boldsymbol{a}_1 + \ldots + \boldsymbol{a}_l)(z - z_0) \leqslant 0.$$

Hence, in view of Theorem 21.8, $\psi_0 (f^0(z) - f^0(z_0)) \leqslant 0$. Now, recalling that $\psi_0 < 0$, we obtain: $f^0(z) - f^0(z_0) \geqslant 0$, that is, $f^0(z_0) \leqslant f^0(z)$. Therefore, function f^0, considered on cone K, reaches a minimum at point z_0.

NOTE 37.2. To prove sufficiency, we did not make use of the assumption that the system of convex cones $M_1, \ldots, M_l, N_1, \ldots, N_m$ does not possess separability. Thus *if there exist a number* ψ_0 *and vectors* $\boldsymbol{a}_1, \ldots, \boldsymbol{a}_l$, *satisfying conditions* (α) *and* (β) *of Theorem* 37.1, *then function* f^0, *considered on cone* K, *reaches a minimum at point* z_0 (this being valid without imposing any demands whatsoever on the system of cones $M_1, \ldots, M_l, N_1, \ldots, N_m$).

NOTE 37.3. Condition (α) in Theorem 37.1 can be made still less stringent. For example, we can replace it with conditions (α) and (γ) of Theorem 36.7 (leaving condition (β) in Theorem 37.1 unchanged). Then, assuming conditions (α) and (γ) of

Theorem 36.7 and condition (β) of Theorem 37.1 to be satisfied, it follows, as we saw during the proof of necessity, that the condition $\psi_0 < 0$ is satisfied as well (assuming that the system of convex cones $M_1, \ldots, M_l, N_1, \ldots, N_m$ does not possess separability).

NOTE 37.4. In the formulation of Theorem 37.1, cones $N_1, \ldots, N_m$ were not considered separately, and only their intersection $N_1 \cap \ldots \cap N_m$ was dealt with in all cases. Therefore, we could have limited ourselves to a *single* cone $N = N_1 \cap \ldots$ $\ldots \cap N_m$, that is, we could have considered only the case $m = 1$. However, the theorem was presented in the above formulation, so that it would have a form somewhat more convenient for the following discussion.

Next let us consider the problem of the minimum of function F^0 on the set $\Sigma =$ $= \Omega_1 \cap \ldots \cap \Omega_l \cap \Xi_1 \cap \ldots \cap \Xi_m$ (as in Theorem 36.7). Here we impose certain constraints on sets Ω_i, Ξ_j and on function F^0. First of all, the following demand is imposed, in the present subsection on function $F^0(z)$:

(a) *function* $F^0(z)$, *defined on an open set containing* Σ, *is smooth, and there exists an affine function* $f^0(z)$ *satisfying the condition:*

$$f^0(z) \leqslant F^0(z) \text{ in the domain of definition of function } F^0(z) \text{ and } f^0(z_0) = F^0(z_0). \quad (37.1)$$

It would be enough to assume, as in the preceding subsection, that $F^0(z)$ is not smooth, but has continuous first derivatives only *close to* the considered point z_0.

Note that satisfaction of condition (a) implies validity of the following relation:

$$\operatorname{grad} F^0(z_0) = \operatorname{grad} f^0. \quad (37.2)$$

Actually, since $F^0(z) \geqslant f^0(z)$, it follows that for any point z different from z_0 (where z is in the domain of definition of function $F^0(z)$)

$$\frac{F^0(z) - F^0(z_0)}{|z - z_0|} \geqslant \frac{f^0(z) - F^0(z_0)}{|z - z_0|} = \frac{f^0(z) - f^0(z_0)}{|z - z_0|},$$

and thus that the derivative of function $F^0(z)$ at point z_0 in *any* direction will not be less than the derivative of function $f^0(z)$ in this same direction. In other words, $\boldsymbol{a} \operatorname{grad} F^0(z_0) \geqslant \boldsymbol{a} \operatorname{grad} f^0$, or

$$\boldsymbol{a}(\operatorname{grad} F^0(z_0) - \operatorname{grad} f^0) \geqslant 0$$

for any vector $\boldsymbol{a}$. Now, replacing $\boldsymbol{a}$ by $-\boldsymbol{a}$, we get

$$\boldsymbol{a}(\operatorname{grad} F^0(z_0) - \operatorname{grad} f^0) = 0$$

for any vector $\boldsymbol{a}$. In particular, for $\boldsymbol{a} = \operatorname{grad} F^0(z_0) - \operatorname{grad} f^0$ we obtain $\operatorname{grad} F^0(z_0) - \operatorname{grad} f^0 = 0$, which also proves equation (37.2).

Note that condition (a) is obviously satisfied if the following is true of function $F^0(z)$:

(a′) *function* $F^0(z)$ *is defined on a convex open set containing* Σ, *and is smooth and convex.*

Let us assume condition (a$'$) to be satisfied, and let us define affine function $f^0(z)$ by the equation

$$f^0(z) = F^0(z_0) + (z - z_0)\,\mathrm{grad}\,F^0(z_0).$$

Clearly, $f^0(z_0) = F^0(z_0)$. Moreover, let z_1 be some arbitrary point different from z_0 lying in the domain of definition of function $F^0(z)$. Then, for any λ satisfying the conditions $0 < \lambda < 1$, a point $z = (1-\lambda) z_0 + \lambda z_1$ lies in the interval (z_0, z_1), so that function F^0 is defined at point z. In view of the convexity of this function, we have (see (36.16))

$$F^0(z) \leqslant (1-\lambda) F^0(z_0) + \lambda F^0(z_1),$$

that is,

$$\frac{F^0(z) - F^0(z_0)}{\lambda} \leqslant F^0(z_1) - F^0(z_0),$$

so that

$$\frac{F^0(z) - F^0(z_0)}{|z - z_0|} = \frac{F^0(z) - F^0(z_0)}{\lambda\,|z_1 - z_0|} \leqslant \frac{F^0(z_1) - F^0(z_0)}{|z_1 - z_0|}.$$

The left-hand side of this relation has, as $\lambda \to 0$ $(\lambda > 0)$, a limit equal to the derivative of function $F^0(z)$ at point z_0 in the direction of the vector $z_1 - z_0$. In other words, this limit is equal to $\boldsymbol{a}\,\mathrm{grad}\,F^0(z_0)$, where $\boldsymbol{a}$ is the unit vector in the direction of vector $z_1 - z_0$. Accordingly,

$$\boldsymbol{a}\,\mathrm{grad}\,F^0(z_0) \leqslant \frac{F^0(z_1) - F^0(z_0)}{|z_1 - z_0|}. \tag{37.3}$$

Since, in addition, $z_1 - z_0 = |z_1 - z_0|\,\boldsymbol{a}$ (in view of the fact that vector $z_1 - z_0$ has the same direction as $\boldsymbol{a}$, while it is $|z_1 - z_0|$ times as long), it follows that, after multiplying relation (37.3) by the positive number $|z_1 - z_0|$, we obtain

$$(z_1 - z_0)\,\mathrm{grad}\,F^0(z_0) = |z_1 - z_0|\,\boldsymbol{a}\,\mathrm{grad}\,F^0(z_0) \leqslant F^0(z_1) - F^0(z_0),$$

or

$$F^0(z_1) \geqslant F^0(z_0) + (z_1 - z_0)\,\mathrm{grad}\,F^0(z_0) = f^0(z_1).$$

Thus, for $z_1 \in \Sigma$ the inequality $f^0(z_1) \leqslant F^0(z_1)$ is valid, that is, function $F^0(z)$ satisfies condition (a).

Note that conditions (a) and (a$'$) were formulated as applied to the problem of the *minimum* of function $F^0(z)$ on some set, to be considered further. For the problem of the *maximum* of the function, the sign of the inequality in condition (a) is changed, and in condition (a$'$) the function is assumed to be *concave* rather than convex. The problem of the maximum of function $F^0(z)$ can always be replaced by the problem of the minimum of function $-F^0(z)$.

Now let us state the restrictions which will be imposed on set Σ. Assume that for each $i=1, \ldots, l$ there exists a cone M_i, satisfying the following condition:

(b) *cone* M_i *is a covering of set* Ω_i *at point* z_0 *and, in addition,* $\Omega_i \subset M_i$, $i=1, \ldots, l$.

Moreover, assume that for each $j=1, \ldots, m$ there exists a cone N_j, satisfying the following condition:

(c) *cone* N_j *is a covering of set* Ξ_j *at point* z_0 *and, in addition,* $\Xi_j \subset N_j$, $j=1, \ldots, m$.

Finally, we impose an additional condition, which will be called the *condition of generality of position:*

(d) *the system of convex cones* $M_1, \ldots, M_l, N_1, \ldots, N_m$ *does not in* E^n *possess separability.*

First of all, let us explain the meaning of the condition of generality of position. If this condition were not satisfied, then in view of Theorem 32.2 there would exist vectors $\boldsymbol{a}_1, \ldots, \boldsymbol{a}_l, \boldsymbol{b}_1, \ldots, \boldsymbol{b}_m$, at least one of which is nonzero, such that the direction of vector $\boldsymbol{a}_i$ lies in cone $D(M_i)$, $i=1, \ldots, l$, and the direction of vector $\boldsymbol{b}_j$ lies in cone $D(N_j)$, $j=1, \ldots, m$, the following relation being valid:

$$\boldsymbol{a}_1+\ldots+\boldsymbol{a}_l+\boldsymbol{b}_1+\ldots+\boldsymbol{b}_m=\boldsymbol{0}. \quad (37.4)$$

There are two possible alternatives: vector $\boldsymbol{a}_1+\ldots+\boldsymbol{a}_l$ equal to zero or not equal to zero.

First, if $\boldsymbol{a}_1+\ldots+\boldsymbol{a}_l=\boldsymbol{0}$, then in view of (37.4) the equation $\boldsymbol{b}_1+\ldots+\boldsymbol{b}_m=\boldsymbol{0}$ will also be valid. We consider the case in which some of vectors $\boldsymbol{a}_1, \ldots, \boldsymbol{a}_l$ are different from zero (if $\boldsymbol{a}_1=\ldots=\boldsymbol{a}_l=\boldsymbol{0}$, then the ensuing discussion should be referred to $M_1, \ldots, M_l$ rather than $N_1, \ldots, N_m$). Hence, according to Theorem 32.2, the system of convex cones $M_1, \ldots, M_l$ possesses separability, that is, some hyperplane Γ separates one of these from the intersection of the others. Consequently, $M_1 \cap \ldots \cap M_l \subset \Gamma$ and thus, all the more so,

$$M_1 \cap \ldots \cap M_l \cap N_1 \cap \ldots \cap N_m \subset \Gamma.$$

Conditions (b) and (c) thus imply that the set

$$\Sigma=\Omega_1 \cap \ldots \cap \Omega_l \cap \Xi_1 \cap \ldots \cap \Xi_m$$

lies wholly in hyperplane Γ. However, then function $F^0(z)$ can also be considered just in hyperplane Γ, that is, instead of E^n we consider the Euclidean space $\Gamma \subset E^n$, of *lower* dimension.

Thus if, instead of (37.4), the relation $\boldsymbol{a}_1+\ldots+\boldsymbol{a}_l=\boldsymbol{0}$ is valid, then we can *reduce* by one the dimension of the space E^n, containing the set Σ, on which the minimum of function F^0 is being sought. If, on the other hand, in space Γ the system of cones $M_1, \ldots, M_l, N_1, \ldots, N_m$ does not after all satisfy the condition of generality of position, then it is possible to apply these same considerations again.

Now let $\boldsymbol{a}_1 + \ldots + \boldsymbol{a}_l \neq \boldsymbol{0}$ and assume that $\psi_0 = 0$. Then the number ψ_0 obviously satisfies condition (α) of Theorem 36.7. It is also obvious that some of vectors $\boldsymbol{a}_1, \ldots, \boldsymbol{a}_l$ are different from zero, that is, condition (γ) of Theorem 36.7 is also satisfied. Finally, from (37.4) we have

$$-\boldsymbol{a}_1 - \ldots - \boldsymbol{a}_l = \boldsymbol{b}_1 + \ldots + \boldsymbol{b}_m,$$

and thus, according to Corollary 31.4 and Lemma 32.1, the direction of the vector $-\boldsymbol{a}_1 - \ldots - \boldsymbol{a}_l$ lies in the cone $D(N_1 \cap \ldots \cap N_m)$, that is,

$$(-\boldsymbol{a}_1 - \ldots - \boldsymbol{a}_l)\,\boldsymbol{\delta z} \leqslant 0$$

for any vector $\boldsymbol{\delta z}$, whose direction lies in cone $N_1 \cap \ldots \cap N_m$. However, this means (in view of the equation $\psi_0 = 0$) that condition (β) of Theorem 36.7 is also satisfied. Therefore, for $\boldsymbol{a}_1 + \ldots + \boldsymbol{a}_l \neq \boldsymbol{0}$, the necessary conditions for a minimum specified in Theorem 36.7 will clearly be satisfied *independently* of which function $F^0(z)$ was selected (since $\psi_0 = 0$, the role of function $F^0(z)$ in these necessary conditions is eliminated).

In other words, if the condition of generality of position is not satisfied, then either the dimension of the space E^n, containing set Σ, can be reduced or else the necessary conditions for a minimum derived in the preceding subsection must be satisfied, independently of whether function $F^0(z)$ reaches a minimum at point z_0 or not, that is, these necessary conditions become meaningless and there is no sense in discussing the sufficiency of these conditions. All this shows that the condition of generality of position is completely natural.

THEOREM 37.5. *In* E^n, *let sets* $\Omega_1, \ldots, \Omega_l, \Xi_1, \ldots, \Xi_m$ *and cones* $M_1, \ldots, M_l, N_1, \ldots, N_m$ *with a common vertex* z_0, *be defined, this vertex being in the set*

$$\Sigma = \Omega_1 \cap \ldots \cap \Omega_l \cap \Xi_1 \cap \ldots \cap \Xi_m,$$

and conditions (b), (c), *and* (d) *being satisfied. Moreover, let a function* $F^0(\boldsymbol{z})$ *satisfying condition* (a) *be defined as well. In order for function* $F^0(z)$, *considered on set* Σ, *to reach a minimum at point* $\boldsymbol{z}_0$, *it is necessary and sufficient that there exist a number* ψ_0 *and vectors* $\boldsymbol{a}_1, \ldots, \boldsymbol{a}_l$, *such that the direction of vector* $\boldsymbol{a}_i$ *lies in the dual cone* $D(M_i)$, $i = 1, \ldots, l$, *the following conditions being satisfied:*

(α) $\psi_0 < 0$;

(β) $(\psi_0 \operatorname{grad} F^0(z_0) - \boldsymbol{a}_1 - \ldots - \boldsymbol{a}_l)\,\boldsymbol{\delta z} \leqslant 0$ *for any vector* $\boldsymbol{\delta z}$ *whose direction lies in cone* $N_1 \cap \ldots \cap N_m$.

Proof. Let function $F^0(\boldsymbol{z})$, considered on set Σ, reach a minimum at point $\boldsymbol{z}_0 \in \Sigma$. Then, according to Theorem 36.7, there must exist a number ψ_0 and vectors $\boldsymbol{a}_1, \ldots, \boldsymbol{a}_l$, such that the direction of vector $\boldsymbol{a}_i$ lies in the cone $D(M_i)$, $i = 1, \ldots, l$, and conditions (α), (β), (γ) of Theorem 36.7 are satisfied. Since condition (β) is formulated in the same way in Theorems 36.7 and 37.5, we just have to verify that $\psi_0 < 0$. This proof is carried out exactly as in Theorem 37.1. Consequently, the *necessity* of the conditions formulated is proven.

Let us now prove *sufficiency*. If conditions (a) through (d) above, as well as conditions (α) and (β) of Theorem 37.5, are satisfied, then the affine function $f^0(z)$ whose existence was assumed in condition (a) satisfies the relation

$$\operatorname{grad} F^0(z_0) = \operatorname{grad} f^0.$$

Thus condition (β) becomes

$$(\psi_0 \operatorname{grad} f^0 - \boldsymbol{a}_1 - \ldots - \boldsymbol{a}_l)\, \boldsymbol{\delta z} \leqslant 0$$

for any vector $\boldsymbol{\delta z}$ whose direction lies in cone $N_1 \cap \ldots \cap N_m$. In view of Theorem (37.1), this implies that function $f^0(z)$, considered on the cone

$$K = M_1 \cap \ldots \cap M_l \cap N_1 \cap \ldots \cap N_m,$$

reaches a minimum at point z_0. Since according to conditions (b) and (c) we have $\Sigma \subset K$, it follows that $f^0(z)$, considered on set Σ, reaches a minimum at z_0, that is,

$$f^0(z_0) \leqslant f^0(z) \quad \text{for} \quad z \in \Sigma.$$

From condition (a) it is now easy to show that

$$F^0(z_0) = f^0(z_0) \leqslant f^0(z) \leqslant F^0(z) \quad \text{for} \quad z \in \Sigma,$$

indicating that function $F^0(z)$, considered in set Σ, reaches a minimum at z_0.

NOTE 37.6. To prove sufficiency, condition (d) was not used (cf. Note 37.2). Therefore, *if there exist a number* ψ_0 *and vectors* $\boldsymbol{a}_1, \ldots, \boldsymbol{a}_l$ *(whose directions lie in the cones* $D(M_1), \ldots, D(M_l)$*), such that conditions* (a), (b), (c) *are satisfied, as as well as conditions* (α) *and* (β) *of Theorem* 37.5, *then function* $F^0(z)$, *considered on set* Σ, *reaches a minimum at point* z_0 (this is true even if no demands are imposed on the system of cones $M_1, \ldots, M_l, N_1, \ldots, N_m$).

In order to obtain a number of corollaries from Theorem 37.5, we will mention some special situations in which conditions (b) and (c) are definitely satisfied.

Let F be some nonconstant *affine* function defined in space E^n and let $\Omega = \operatorname{Ker} F$ be its kernel. Obviously, the set $M = \Omega$ is a covering of set Ω (at any point $z_0 \in \Omega$), with $\Omega \subset M$. Consequently, condition (b) is obviously satisfied if for set Ω_i and cone M_i the following is true:

(b′) *set* Ω_i *is the kernel of some nonconstant affine function (that is, it is a hypersurface), and* $M_i = \Omega$

Similarly, condition (c) is obviously satisfied if for Ξ_j and N_j the following is true:

(c′) *set* Ξ_j *is the kernel of some nonconstant affine function (that is, it is a hypersurface) and* $N_j = \Xi_j$.

Another situation in which conditions (b) and (c) are satisfied is the case of a convex set and its supporting cone. If Ω if a convex set and M is its supporting cone at point $z_0 \in \Omega$, then in view of Theorem 34.2 M is a covering of set Ω at z_0, where obviously $\Omega \subset M$. Thus condition (b) will be satisfied if the following is true:

(b″) *set* Ω_i *is convex and* M_i *is the supporting cone of set* Ω_i *at point* $z_0 \in \Omega_i$.

Similarly, condition (c) is obviously satisfied if the following is true:

(c″) *set* Ξ_j *is convex and* N_j *is the supporting cone of set* Ξ_j *at point* $z_0 \in \Xi_j$.

A particular case of the given situation is the case of *convex constraints.* Let $f(z)$ be a smooth convex function and let Ω be the set of all points satisfying the condition $f(z) \leqslant 0$. Moreover, let z_0 be an arbitrary point at which $f(z_0)=0$ and let M be a half-space defined by the inequality $(z-z_0)\,\mathrm{grad}\, f(z_0) \leqslant 0$. It follows directly from the definition of a convex function (see (36.16)) that set Ω is convex, and the fact that function f is smooth implies that half-space M is the supporting cone of convex set Ω at point z_0.

Therefore, condition (b″), and thus (b) as well, are clearly satisfied if the following is true:

(b‴) *set* Ω_i *is defined by constraint* $f^i(z) \leqslant 0$, *where* f^i *is a smooth convex function that goes to zero at point* z_0, *and* M_i *is the half-space* $(z-z_0)\,\mathrm{grad}\, f^i(z_0) \leqslant 0$.

Similarly, condition (c) is satisfied if the following is true:

(c‴) *set* Ξ_j *is defined by constraint* $g^j(z) \leqslant 0$, *where* g^j *is a smooth convex function that goes to zero at point* z_0 *and* N_j *is the half-space* $(z-z_0)\,\mathrm{grad}\, g^j(z_0) \leqslant 0$.

Assuming that in Theorem 37.5 some of sets Ω_i and cones M_i satisfy condition (b'), while some satisfy condition (b‴) (the same being true of sets Ξ_j and cones N_j) we obtain from Theorem 37.5 the following statement:

THEOREM 37.7. *Let* Ω' *be a set defined in space* E^n *by the equations*

$$F^1(z)=0, \ldots, F^k(z)=0,$$

and let Ξ' *be a set defined in* E^n *by the equations*

$$G^1(z)=0, \ldots, G^r(z)=0,$$

where $F^1(z), \ldots, F^k(z), G^1(z), \ldots, G^r(z)$ *are nonconstant affine functions. In addition, let* Ω'' *be a set defined by the inequalities*

$$f^1(z) \leqslant 0, \ldots, f^q(z) \leqslant 0,$$

and let Ξ'' *be a set defined by the inequalities*

$$g^1(z) \leqslant 0, \ldots, g^p(z) \leqslant 0,$$

where $f^1(z), \ldots, f^q(z), g^1(z), \ldots, g^p(z)$ *are smooth convex functions. Finally, let* $\Omega_1, \ldots, \Omega_l, \Xi_1, \ldots, \Xi_m$ *be arbitrary sets of space* E^n *and let* $F^0(z)$ *be a function defined on some open set of space* E^n, *containing the set*

$$\Sigma=\Omega' \cap \Omega'' \cap \Omega_1 \cap \ldots \cap \Omega_l \cap \Xi' \cap \Xi'' \cap \Xi_1 \cap \ldots \cap \Xi_m.$$

Assume that at point $z_0 \in \Sigma$ *function* $F^0(z)$ *satisfies condition* (a) *(or condition* (a') *and that cones* $M_1, \ldots, M_l, N_1, \ldots, N_m$ *with a vertex* z_0 *are defined, these cones satisfying conditions* (b) *and* (c) *(for example, the latter are satisfied if sets* $\Omega_1, \ldots, \Omega_l, \Xi_1, \ldots, \Xi_m$ *are convex and* $M_1, \ldots, M_l, N_1, \ldots, N_m$ *are the supporting cones of these sets at point* z_0). *Let* $I(z_0)$ *be the set of numbers* $i=1, \ldots, q$ *for which* $f^i(z_0)=0$, *and let* $J(z_0)$ *be the set of numbers* $j=1, \ldots, p$ *for which* $g^j(z_0)=0$. *Assume that all the vectors* $\operatorname{grad} f^i(z_0)$, $i \in I(z_0)$, *and also* $\operatorname{grad} g^j(z_0)$, $j \in J(z_0)$, *are different from zero.*

We set $K_i = \operatorname{Ker} F^i$, $i=1, \ldots, k$; $L_j = \operatorname{Ker} G^j$, $j=1, \ldots, r$, *and we call* Π_i *the half-space defined by the inequality* $(z-z_0)\operatorname{grad} f^i(z_0) \leqslant 0$, $i \in I(z_0)$ *and* P_j *the half-space defined by the inequality* $(z-z_0)\operatorname{grad} g^j(z_0) \leqslant 0$, $j \in J(z_0)$, *it being assumed that the system of convex cones*

$$K_i\,(i=1, \ldots, k),\quad L_j\,(j=1, \ldots, r),\quad \Pi_i\,(i \in I(z_0)),$$
$$P_j\,(j \in J(z_0)),\quad M_i\,(i=1, \ldots, l),\quad N_j\,(j=1, \ldots, m) \tag{37.5}$$

does not possess separability in E^n.

In order for function $F^0(z)$, *considered on set* Σ, *to reach a minimum at point* z_0, *it is necessary and sufficient that there exist numbers* $\psi_0, \psi_1, \ldots, \psi_k$, *nonpositive numbers* λ_i, $i \in I(z_0)$, *and vectors* $a_1, \ldots, a_l$, *whose directions lie, respectively, in dual cones* $D(M_1), \ldots, D(M_l)$, *all such that the following conditions are satisfied:*

(α) $\psi_0 < 0$;

(β) $$\Big(\sum_{\alpha=0}^{k} \psi_\alpha \operatorname{grad} F^\alpha(z_0) + \sum_{j \in I(z_0)} \lambda_j \operatorname{grad} f^j(z_0) - a_1 - \ldots - a_l\Big)\,\delta z \leqslant 0$$

for any vector δz *whose direction lies in the intersection of cones* L_i, P_j, N_h $(i=1, \ldots, r;\ j \in J(z_0);\ h=1, \ldots, m)$.

For the proof, it is sufficient to apply Theorem 37.5 and to note that set W, defined by the relations

$$f^i(z) \leqslant 0,\ i \notin I(z_0);\quad g^j(z) \leqslant 0,\ j \notin J(z_0),$$

has z_0 as an *interior* point (in space E^n) and thus that set W has the entire space E^n as its covering at point z_0. Therefore, we can add set W to the system of sets

$$K_i\,(i=1, \ldots, k),\quad L_j\,(j=1, \ldots, r),\quad f^i(z) \leqslant 0\,(i \in I(z_0)),$$
$$g^j(z) \leqslant 0\,(j \in J(z_0)),\quad \Omega_i\,(i=1, \ldots, l),\quad \Xi_j\,(j=1, \ldots, m)$$

(and cone E^n to the system of cones (37.5)), without violating properties (b), (c), and (d).

NOTE 37.8. As in the previous theorems, the assumption that the system of all cones (37.5) does not possess separability was used only to prove necessity. However, sufficiency of the formulated conditions exists, even without this assumption. Note also that in this theorem it is enough to consider as smooth and convex only those of functions f^i, g^j which correspond to the constraints $f^i(z) \leqslant 0$, $g^j(z) \leqslant 0$ that are *active* at point z_0.

Theorem 37.7 is the most general of all the theorems presented in this subsection. The numbers k, r, q, p, l, m in it can be any nonnegative whole numbers. By setting some of these numbers equal to zero, we obtain from Theorem 37.7 a number of particular cases.

For instance, Theorem 37.5 is a particular case of Theorem 37.7, obtained by setting $k=r=p=q=0$. Of all the other particular cases, we will formulate only three theorems (the first for $k=r=p=q=l=0$, the second for $r=p=l==m=0$, and the third for $r=q=l=m=0$).

THEOREM 37.9. *Let* $\Sigma = \Xi_1 \cap \ldots \cap \Xi_m$, *where* $\Xi_1, \ldots, \Xi_m$ *are certain sets of space* E^n, *and let* $N_1, \ldots, N_m$ *be convex cones with a vertex at point* $z_0 \in \Sigma$, *such that* N_i *is a covering of set* Ξ_i *at point* z_0, *with* $N_i \supset \Xi_i$, $i=1, \ldots, m$. *Assume, in addition, that system of cones* $N_1, \ldots, N_m$ *does not possess separability in* E^n. *Finally, let* $F^0(z)$ *be a function satisfying condition* (a) *(or* (a')*). In order for function* $F^0(z)$, *considered on set* Σ, *to reach a minimum at point* z_0, *it is necessary and sufficient that the vector* $\operatorname{grad} F^0(z_0)$ *form a nonnegative scalar product with each vector whose direction lies in cone* $N_1 \cap \ldots \cap N_m$ *(or, in other words, that there exist vectors* $\boldsymbol{a}_1, \ldots, \boldsymbol{a}_m$, *whose directions lie, respectively, in dual cones* $D(N_1), \ldots, D(N_m)$, *such that* $\operatorname{grad} F^0(z_0) = -\boldsymbol{a}_1 - \ldots - \boldsymbol{a}_m$).

The last remark (in parentheses) entering into the formulation of this theorem is implied by Corollary 31.4 and Lemma 32.1. It should be noted that the condition imposed in this theorem on sets Ξ_j and cones N_j is satisfied, in particular, if set Ξ_j is convex, while N_j is the supporting cone of this set at point $z_0 \in \Sigma$, $j=1, \ldots, m$.

THEOREM 37.10. *Let plane* E^n *be defined in* Ω' *by the system of linear equations*

$$F^1(z)=0, \ \ldots, \ F^k(z)=0$$

(that is, functions $F^1, \ldots, F^k$ *are affine*), the vectors*

$$\operatorname{grad} F^1, \ \ldots, \ \operatorname{grad} F^k$$

being linearly independent, and let Ω'' *be a set defined by the inequalities*

$$f^1(z) \leqslant 0, \ \ldots, \ f^q(z) \leqslant 0. \tag{37.6}$$

In addition, let z_0 *be some point of set* $\Omega' \cap \Omega''$ *and let* $F^0(z)$ *be a function satisfying condition* (a) *(or* (a')*). Assume that all functions* $f^i(z)$, $i \in I(z_0)$, *are convex and smooth. Finally, let us suppose that there exists a point* $z_1 \in \Omega'$ *for which all the scalar products*

$$(z_1 - z_0) \operatorname{grad} f^i(z_0), \quad i \in I(z_0), \tag{37.7}$$

are negative.

* See footnote on p. 287.

In order for function $F^0(z)$, *considered on set* $\Omega' \cap \Omega''$, *to reach a minimum at point* z_0, *it is necessary and sufficient that the vector* $\operatorname{grad} F^0(z_0)$ *be representable in the form*

$$\operatorname{grad} F^0(z_0) = \sum_{i=1}^{k} \psi_i \operatorname{grad} F^i(z_0) + \sum_{j \in I(z_0)} \lambda_j \operatorname{grad} f^j(z_0), \qquad (37.8)$$

where all the numbers λ_j, $j \in I(z_0)$, *are nonpositive.*

Proof. Let us set $K_i = \operatorname{Ker} F^i$, $i = 1, \ldots, k$, and let us call Π_j the half-space defined by the inequality $(z - z_0) \operatorname{grad} f^j(z_0) \leqslant 0$, $j \in I(z_0)$. We will show that the system of convex cones K_i, Π_j $(i = 1, \ldots, k;\ j \in I(z_0))$ does not possess separability.

First we assume the converse. Then, in view of Theorem 32.2, there will exist vectors $\boldsymbol{a}_i$, $\boldsymbol{b}_j$ (where $i = 1, \ldots, k;\ j \in I(z_0)$), not all equal to zero, such that the directions of vectors $\boldsymbol{a}_i$ and $\boldsymbol{b}_j$ lie, respectively, in cones $D(K_i)$ and $D(\Pi_j)$, with

$$\sum \boldsymbol{a}_i + \sum \boldsymbol{b}_j = 0.$$

Taking the form of cones K_i and Π_j into account, we find that there must exist numbers α_i, β_j $(i = 1, \ldots, k;\ j \in I(z_0))$, not all equal to zero, such that the numbers β_j are nonnegative and the following relation is valid:

$$\sum_{i=1}^{k} \alpha_i \operatorname{grad} F^i + \sum_{j \in I(z_0)} \beta_j \operatorname{grad} f^j(z_0) = 0. \qquad (37.9)$$

Note that in this relation *not all* the numbers β_j are equal to zero (that is, at least one of the β_j is *positive*), since otherwise the vectors $\operatorname{grad} F^i$, $i = 1, \ldots, k$, would have to be linearly dependent. Since $z_0 \in K_i$ and $z_1 \in K_i$, it follows that $(z_1 - z_0) \operatorname{grad} F^i = 0$, $i = 1, \ldots, k$. Therefore, after multiplying relation (37.9) scalarly by vector $z_1 - z_0$, we obtain

$$\sum_{j \in I(z_0)} \beta_j (z_1 - z_0) \operatorname{grad} f^j(z_0) = 0.$$

However, this contradicts the fact that all the numbers (37.7) are negative, while of the nonnegative numbers β_j, $j \in I(z_0)$, some are different from zero. This contradiction indicates that the system of cones K_i, Π_j $(i = 1, \ldots, k;\ j \in I(z_0))$ does not possess separability.

Now let us set $\Xi_1 = N_1 = E^n$ and let us apply Theorem 37.7 for $r = q = l = 0$, $m = 1$. Then there must exist numbers ψ_i, λ_j $(i = 0, 1, \ldots, k;\ j \in I(z_0))$ satisfying the relation

$$\left(\sum_{i=0}^{k} \psi_i \operatorname{grad} F^i(z_0) + \sum_{j \in I(z_0)} \lambda_j \operatorname{grad} f^j(z_0) \right) \delta z \leqslant 0$$

for *any* vector δz (that is, a vector δz whose direction lies in cone $N_1 = E^n$), where $\psi_0 < 0$ and all the numbers λ_j, $j \in I(z_0)$, are nonpositive. This means that

$$\sum_{i=0}^{k} \psi_i \operatorname{grad} F^i(z_0) + \sum_{j \in I(z_0)} \lambda_j \operatorname{grad} f^j(z_0) = 0,$$

where it can be assumed that $\psi_0 = -1$, which also implies relation (37.8).

THEOREM 37.11. *Let Ω' be a plane defined in E^n by the system of linear equations $F^i(z) = 0$, $i = 1, \ldots, k$, the vectors* $\operatorname{grad} F^i$, $i = 1, \ldots, k$, *being linearly independent, and let Ξ'' be a set defined by the constraints $g^j(z) \leqslant 0$, $j = 1, \ldots, p$. In addition, let z_0 be some point of set $\Omega' \cap \Xi''$ and let $F^0(z)$ be a function satisfying condition* (a) *(or* (a')*). Assume all the functions $g^j(z)$, $j \in J(z_0)$, to be convex and smooth. Finally, assume that a point $z_1 \in \Omega'$ exists for which all the scalar products*

$$(z_1 - z_0) \operatorname{grad} g^j(z_0), \qquad j \in J(z_0),$$

are negative.

In order for function $F^0(z)$, considered on set $\Omega' \cap \Xi''$, to reach a minimum at point z_0, it is necessary and sufficient that there exist a vector $\psi = \{\psi_0, \psi_1, \ldots, \psi_k\}$ such that the following two conditions are satisfied:

(α) $\psi_0 < 0$;

(β) $\left(\sum_{\alpha=0}^{k} \psi_\alpha \operatorname{grad} F^\alpha(z_0)\right) \delta z \leqslant 0$ *for any vector δz forming a nonpositive scalar product with each of the vectors* $\operatorname{grad} g^j(z_0)$, $j \in J(z_0)$.

This theorem is proven in the same way as the previous one.

NOTE 37.12. Let us consider the particular case of Theorem 37.11 corresponding to $k = 0$, that is, the case where we seek the minimum of a function $F^0(z)$ defined on set Ξ'' by the system of inequalities $g^j(z) \leqslant 0$, $j = 1, \ldots, p$, this system being nonsingular at point z_0.

If K is a cone defined by the system of inequalities

$$(z - z_0) \operatorname{grad} g^j(z_0) \leqslant 0, \qquad j \in J(z_0),$$

and if l is a ray originating at z_0 and lying *inside* cone K, then there must exist a point z on ray l such that the entire segment $[z_0, z]$ is contained in set Ξ''. In other words, each such ray *enters into* set Ξ'' (Figure 150). Thus the condition specified in Theorem 37.11 (namely that $\delta z \operatorname{grad} F^0(z_0) \geqslant 0$ for any vector δz whose direction lies in cone K) implies that *the derivative of function $F^0(z)$ at point z_0 in any direction entering into set Ξ'' is nonnegative.* It is easy to show that this derivative is *positive.* In other words, along any ray originating at point z_0 and entering into set Ξ'', function $F^0(z)$ *increases* close to z_0. This circumstance agrees well with the fact that at point z_0 function $F^0(z)$, considered on set Ξ'', reaches a *minimum.*

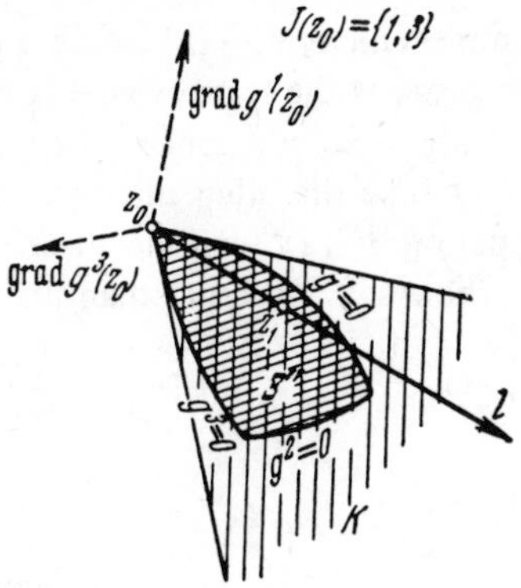

FIGURE 150

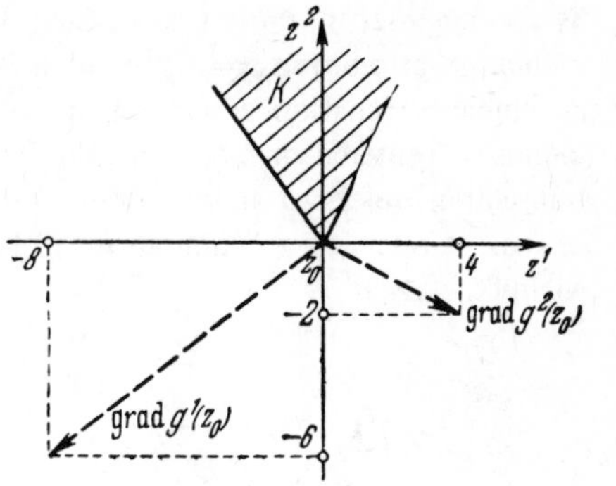

FIGURE 151

It should be noted, however, that just this circumstance alone, without satisfaction of the other conditions of Theorem 37.11 (in particular, without assuming convexity of functions $g^j(z)$, $j \in J(z_0)$), does not guarantee that a minimum is reached at z_0. The following example provides an illustration.

EXAMPLE 37.13. Let us consider the set Ξ'' described in the plane of variables z^1, z^2 by the relations

$$(z^1)^2 + (z^2)^2 - 8z^1 - 6z^2 \leqslant 0, \quad 4z^1 - (z^1)^2 - (z^2)^2 - 2z^2 \leqslant 0, \qquad (37.10)$$

and let us try to find the minimum of function $F^0(z^1, z^2) = z^2$ on this set. Here, set Ξ'' is defined by inequalities of the form $g^1(z^1, z^2) \leqslant 0$, $g^2(z^1, z^2) \leqslant 0$, where g^1, g^2 are the functions on the left-hand sides of relations (37.10). Point $z_0 = (0, 0)$ obviously belongs to set Ξ'' and at this point both of the constraints (37.10) are active.

Next we consider the behavior of function F^0 on set Ξ'' close to point z_0. Since

$$\operatorname{grad} g^1(z_0) = \{-8, -6\}, \quad \operatorname{grad} g^2(z_0) = \{4, -2\},$$

it follows that the cone K considered in Note 37.12 here has the form of an angle lying wholly in the upper half-plane (Figure 151). Therefore, the vector $\operatorname{grad} F^0(z_0) = \{0, 1\}$ forms a *positive* scalar product with each vector $\boldsymbol{\delta z}$ whose direction lies in cone K, that is, along each ray originating at point z_0 and entering into set Ξ'', function $F^0(z)$ *increases* close to point z_0. In other words, at point z_0 the condition specified in Theorem 37.11 is satisfied.

However, Theorem 37.11 is inapplicable here, since function $g^2(z^1, z^2)$ is not convex. Actually, at the point (4, –2), which also belongs to set Ξ'' (Figure 152), function $F^0(z)$ has a *lower* value than at z_0:

$$F^0(0, 0) = 0, \qquad F^0(4, -2) = -2.$$

This example shows that the requirement of convexity of functions $g^i(z)$, $i \in J(z_0)$, imposed in Theorem 37.11 (or in Theorem 37.10) is necessary (it is not difficult to present an example showing that the requirement of convexity of function $F^0(z)$

is also necessary). Note that in the selected example function $F^0(z)$ reaches a *local minimum* at point $z_0 = (0, 0)$, that is, its value at z_0 is the least of all the values assumed by function $F^0(z)$ on the set $\Xi'' \cap \Sigma_\varepsilon$, where Σ_ε is a sphere of quite small radius ε centered on z_0. However, Theorem 37.11 (like the other theorems of this subsection) refers to an *absolute minimum* of function $F^0(z)$, considered on set Ξ'', that is, it refers to a minimum of $F^0(z)$ over a whole set Ξ'' rather than just in the vicinity of z_0.

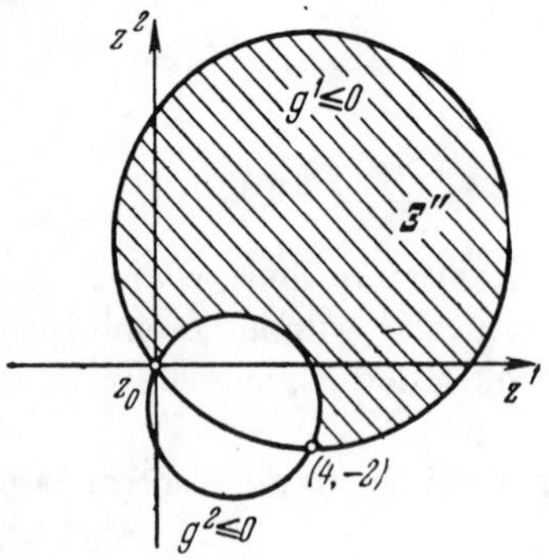

FIGURE 152

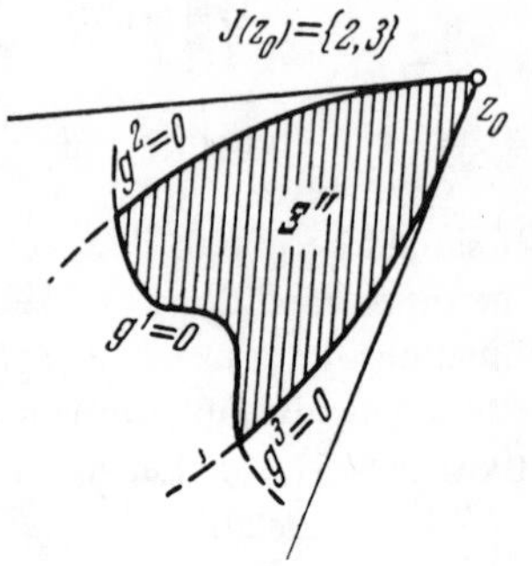

FIGURE 153

NOTE 37.14. If *all* the functions on the left-hand sides of the constraints $g^j(z) \leqslant 0$ are convex (rather than just those corresponding to the constraints active at point z_0), then it is easy to show that set Ξ'' is *convex*. If, on the other hand, only those functions corresponding to the constraints active at z_0 are convex, then set Ξ'' in Theorem 37.11 may also be nonconvex (Figure 153).

NOTE 37.15. The necessary and sufficient conditions for a minimum presented in this subsection constitute a generalization of the classical methods for studying extrema according to the first and second derivatives. Let us consider, for example, Theorem 37.10 in the particular case where $k = q = 0$, that is, the case where the domain of definition of function $F^0(z)$ coincides with the entire space E^n. This leads us to the following statement. *Let $F^0(z)$ be a convex function defined in E^n; in order for $F^0(z)$ to reach a minimum at point z_0, it is necessary and sufficient that the equation* $\operatorname{grad} F^0(z_0) = 0$ *be valid.* This is the classical condition for a minimum; the equation $\operatorname{grad} F^0(z_0) = 0$, that is, the going to zero of the partial derivatives $\dfrac{\partial F^0(z_0)}{\partial z^i}$, $i = 1, \ldots, n$, is identical to the classical *necessary* condition for an extremum, while convexity of the function (certain demands being imposed upon the *second* derivatives; see pp. 278–279) guarantees that an extremum (here a minimum) is reached at point z_0.

38. Maximum principle. In this subsection we will formulate a number of theorems which can be united under the general name of a "maximum principle." These theorems give a necessary and sufficient condition for an extremum which is equivalent to the theorems of the preceding subsection, but which differs from them in form.

Once again, as in Theorem 37.11, we consider the minimum of a function $F^0(z)$ on a set $\Omega' \cap \Xi''$, where Ω' is a plane defined by the equations $F^i(z)=0$, $i=1, \ldots, k$, the functions $F^1, \ldots, F^k$ being affine; the assumptions relating to function $F^0(z)$ and set Ξ'' will be given below. Now let us introduce an auxiliary vector $\psi=\{\psi_0, \psi_1, \ldots, \psi_k\}$ analogous to the one in Theorem 37.11, and let us consider the function

$$H=H(z)=\psi_0 F^0(z)+\psi_1 F^1(z)+\ldots+\psi_k F^k(z). \qquad (38.1)$$

Introduction of this function enables us to write condition (β) of Theorem 37.11 in a simpler form:

$\delta z \operatorname{grad} H(z_0) \leqslant 0$ *for any vector* δz *whose direction lies in the cone* K *defined by the inequalities* $(z-z_0)\operatorname{grad} g^j(z_0) \leqslant 0$, $j \in J(z_0)$.

In other words, the derivative of function H at point z_0, in any direction lying in cone K, is *nonpositive.* It looks as if (for the values of $\psi_0, \psi_1, \ldots, \psi_k$ whose existence was attested to in Theorem 37.11) function H *decreases* (more precisely, does not increase) when we move from point z_0 into cone K, that is, function H, considered on cone K (and thus on set $\Xi'' \subset K$ as well), appears to reach a *local maximum* at point z_0. And what is more, it is to be supposed that function H, considered on Ξ'', reaches an *absolute maximum* at point z_0; since function $F^0(z)$ is convex, it follows that function $\psi_0 F^0(z)$ is concave (since $\psi_0 < 0$), so that in view of the affineness of functions $F^1(z), \ldots, F^k(z)$ function H is concave as well (Figure 154).

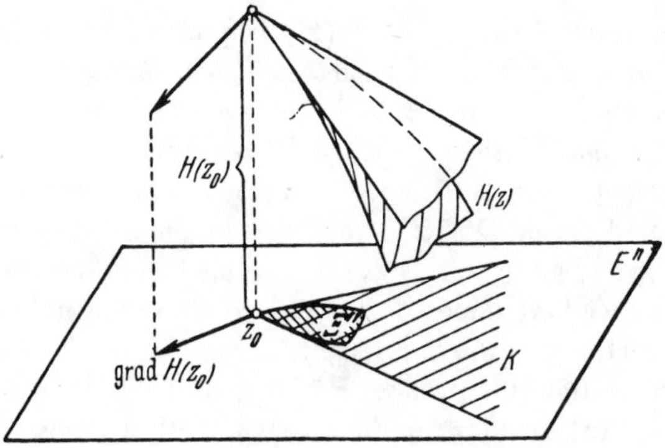

FIGURE 154

Consequently, we arrive at the following hypothesis: *in order for (with the conditions of Theorem* 37.11 *satisfied) function* $F^0(z)$, *considered on set* $\Omega' \cap \Xi''$, *to reach a minimum at point* z_0, *it is necessary and sufficient that there exist a vector* $\psi=\{\psi_0, \psi_1, \ldots, \psi_k\}$ *such that* $\psi_0 < 0$ *and such that the function* H *of variable* z, *defined by equation* (38.1), *reaches a maximum on set* Ξ'' *at point* z_0.

Note that *sufficiency* of this condition can be proven trivially (even without placing any constraints on set Ξ'' or on the functions $F^1(z), \ldots, F^k(z)$ defining set Ω').

Actually, on set Ω' (and thus on $\Omega' \cap \Xi''$ as well) all the functions $F^1(z), \ldots, F^k(z)$ go to zero, so that on set $\Omega' \cap \Xi''$ function H reduces to the expression $\psi_0 F^0(z)$. But then it is clear that, if function H, considered on Ξ'', reaches a maximum at point z_0, then on set $\Omega' \cap \Xi''$ this function, that is, $\psi_0 F^0(z)$, all the more so reaches a maximum at z_0, meaning that $F^0(z)$ on $\Omega' \cap \Xi''$ reaches a minimum at z_0 (since $\psi_0 < 0$).

On the other hand, the *necessity* of the condition formulated above is not evident: let it be said that function H reaches a maximum at point z_0 not only on set $\Omega' \cap \Xi''$, where the relation $H = \psi_0 F^0(z)$ is valid, but rather on the *entire* set Ξ''. Let us show that, given the constraints imposed in Theorem 37.11, necessity exists as well, that is, the following throrem verifying the correctness of the stated hypothesis is valid.

THEOREM 38.1 *(maximum principle). In E^n let Ω' be a plane defined by the system of linear equations $F^i(z) = 0$, $i = 1, \ldots, k$, the vectors* $\operatorname{grad} F^i$, $i = 1, \ldots, k$, *being linearly independent, and let Ξ'' be a set defined by the inequalities $g^j(z) \leqslant 0$, $j = 1, \ldots, p$. Moreover, let z_0 be some point of set $\Omega' \cap \Xi''$ and let $F^0(z)$ be a function defined on a set containing Ξ'' and satisfying condition* (a) *(or* (a')*) (see p. 289). All the functions $g^j(z)$, $j \in J(z_0)$, are assumed to be convex and smooth. Finally, assume also that there exists a point $z_1 \in \Omega'$ for which all the scalar products*

$$(z_1 - z_0) \operatorname{grad} g^j(z_0), \quad j \in J(z_0),$$

are negative. In order for function $F^0(z)$, considered on set $\Omega' \cap \Xi''$, to reach a minimum at point z_0, it is necessary and sufficient that there exist a vector $\psi = \{\psi_0, \psi_1, \ldots, \psi_k\}$ such that $\psi_0 < 0$ and such that the function $H(z)$, defined by formula (38.1) *and considered on set Ξ'', reaches a maximum at z_0.*

Proof. As noted above, sufficiency is obvious. Let us prove necessity. If function $F^0(z)$, considered on set $\Omega' \cap \Xi''$, reaches a minimum at point z_0, then there must exist a vector $\psi = \{\psi_0, \psi_1, \ldots, \psi_k\}$ satisfying conditions (α) and (β) of Theorem 37.11 (since here we have imposed the same conditions as in Theorem 37.11). Let us show that this vector ψ is the one sought.

Let z be an arbitrary point of set Ξ'', so that at this point function F^0 is defined. Since function $F^0(z)$ satisfies condition (a) (p. 289), it follows that $f^0(z) \leqslant F^0(z)$, where f^0 is an affine function the existence of which was attested to in condition (a). Consequently, taking into account that $\psi_0 < 0$, we obtain (see (37.1) and (37.2)):

$$\begin{aligned}
H(z) - H(z_0) &= \psi_0 F^0(z) + \psi_1 F^1(z) + \ldots + \psi_k F^k(z) - \psi_0 F^0(z_0) - \\
&- \psi_1 F^1(z_0) - \ldots - \psi_k F^k(z_0) \leqslant \psi_0 f^0(z) + \psi_1 F^1(z) + \ldots + \psi_k F^k(z) - \\
&- \psi_0 f^0(z_0) - \psi_1 F^1(z_0) - \ldots - \psi_k F^k(z_0) = \psi_0 (f^0(z) - f^0(z_0)) + \\
&+ \psi_1 (F^1(z) - F^1(z_0)) + \ldots + \psi_k (F^k(z) - F^k(z_0)) = \\
&= \psi_0 (z - z_0) \operatorname{grad} f^0 + \psi_1 (z - z_0) \operatorname{grad} F^1 + \ldots + \psi_k (z - z_0) \operatorname{grad} F^k = \\
&= (z - z_0)(\psi_0 \operatorname{grad} f^0 + \psi_1 \operatorname{grad} F^1 + \ldots + \psi_k \operatorname{grad} F^k) = \\
&= (z - z_0)(\psi_0 \operatorname{grad} F^0(z_0) + \psi_1 \operatorname{grad} F^1 + \ldots + \psi_k \operatorname{grad} F^k).
\end{aligned}$$

However, the latter expression is *nonpositive* in view of condition (β) of Theorem 37.11, since $z \in \Xi'' \subset K$, where K is a cone defined by the inequalities $(z - z_0)\,\mathrm{grad}\, g^j(z_0) \leqslant 0$, $j \in J(z_0)$, and thus the direction of vector $\delta z = z - z_0$ lies in cone K. Accordingly, $H(z) - H(z_0) \leqslant 0$, that is, $H(z) \leqslant H(z_0)$. Since in this discussion point $z \in \Xi''$ was assumed to be arbitrary, it follows that function $H(z)$, considered on set Ξ'', reaches a maximum at point z_0.

The following statement is proven in exactly the same way (except that Theorem 37.11 is used, and Theorem 37.7 with $q = l = 0$):

THEOREM 38.2 *(maximum principle). Let Ω' be a set defined in space E^n by the equations*

$$F^1(z) = 0, \ \ldots, \ F^k(z) = 0,$$

and let Ξ' be a set defined in E^n by the equations

$$G^1(z) = 0, \ \ldots, \ G^r(z) = 0,$$

where $F^1(z), \ldots, F^k(z), G^1(z), \ldots, G^r(z)$ are nonconstant affine functions. In addition, let Ξ' be a set defined by the inequalities

$$g^1(z) \leqslant 0, \ \ldots, \ g^p(z) \leqslant 0,$$

where $g^1(z), \ldots, g^p(z)$ are smooth convex functions. Finally, let $\Xi_1, \ldots, \Xi_m$ be arbitrary sets of space E^n and let $F^0(z)$ be a function whose domain of definition contains the set

$$\Xi = \Xi' \cap \Xi'' \cap \Xi_1 \cap \ldots \cap \Xi_m.$$

Assume that at point $z_0 \in \Omega' \cap \Xi$ function $F^0(z)$ satisfies condition (a) *(or* (a')*) (p. 289) and that cones $N_1, \ldots, N_m$ with vertex z_0, satisfying condition* (c) (p. 290), *are defined (for instance, these conditions are satisfied if sets $\Xi_1, \ldots, \Xi_m$ are convex and $N_1, \ldots, N_m$ are the supporting cones of these sets at point z_0). Assume that all the vectors $\mathrm{grad}\, g^j(z_0)$, $j \in J(z_0)$, are different from zero. Let us set $K_i = \mathrm{Ker}\, F^i$, $i = 1, \ldots, k$; $L_j = \mathrm{Ker}\, G^j$, $j = 1, \ldots, r$, and let us call P_j the half-space defined by the inequality $(z - z_0)\,\mathrm{grad}\, g^j(z_0) \leqslant 0$, it being assumed that the system of convex cones*

$$K_i \ (i = 1, \ldots, k), \quad L_i \ (i = 1, \ldots, r),$$
$$P_j \ (j \in J(z_0)), \quad N_j \ (j = 1, \ldots, m)$$

does not possess separability in E^n.

In order for function $F^0(z)$, considered on set $\Omega' \cap \Xi$, to reach a minimum at point z_0, it is necessary and sufficient that there exist a vector $\psi = \{\psi_0, \psi_1, \ldots, \psi_k\}$ such that $\psi_0 < 0$ and such that function $H(z)$, defined by formula (38.1) *and considered on set Ξ, reaches a maximum at z_0.*

Finally, let us formulate the maximum principle for the case where set Ω is defined by linear equations and convex constraints.

THEOREM 38.3. *Let* Ω' *be a set defined in space* E^n *by the equations* $F^i(z)=0$, $i=1, \ldots, k$, *where* $F^1(z), \ldots, F^k(z)$ *are nonconstant affine functions. In addition, let* $\Xi_1, \ldots, \Xi_m$ *be arbitrary sets of space* E^n *and let* $F^0(z)$ *be a function whose domain of definition contains the set* $\Xi=\Xi_1 \cap \ldots \cap \Xi_m$. *Finally, let* Ω'' *be a set defined by the inequalities* $f^j(z) \leqslant 0$, $j=1, \ldots, q$, *where* $f^1(z), \ldots, f^q(z)$ *are smooth convex functions, whose domains of definition all contain set* Ξ. *Assume that at point* $z_0 \in \Omega' \cap \Omega'' \cap \Xi$ *function* $F^0(z)$ *satisfies condition* (a) *(or* (a')*)* (p.289) *and that cones* $N_1, \ldots, N_m$ *with vertex* z_0, *satisfying condition* (c) (p.290), *are defined. All the vectors* $\operatorname{grad} f^j(z_0)$, $j \in I(z_0)$, *are assumed to be nonzero. Now let us set* $K_i = \operatorname{Ker} F^i$, $i=1, \ldots, k$, *and let us call* Π_j *the half-space defined by the inequality* $(z-z_0)\operatorname{grad} f^j(z_0) \leqslant 0$, *it being assumed that the system of convex cones*

$$K_i\,(i=1, \ldots, k), \quad \Pi_j\,(j \in I(z_0)), \quad N_h\,(h=1, \ldots, m)$$

does not possess separability in E^n.

In order for function $F^0(z)$, *considered on set* $\Omega' \cap \Omega'' \cap \Xi$, *to reach a minimum at point* z_0, *it is necessary and sufficient that there exist numbers* $\psi_0, \psi_1, \ldots, \psi_k$ *and nonpositive numbers* $\lambda_1, \ldots, \lambda_q$ *such that the following conditions are satisfied:*

(α) $\psi_0 < 0$;

(β) *function* $H(z)=\sum_{\alpha=0}^{k} \psi_\alpha F^\alpha(z)+\sum_{j=1}^{q} \lambda_j f^j(z)$, *considered on set* Ξ, *reaches a maximum at point* z_0;

(γ) $\lambda_1 f^1(z_0)= \ldots =\lambda_q f^q(z_0)=0$.

Proof. Assume that conditions (α), (β), (γ) are satisfied. If we take some arbitrary point $z \in \Omega' \cap \Omega'' \cap \Xi$, then $H(z) \leqslant H(z_0)$ (in view of condition (β)). Since on set Ω' (and, in particular, at point z) all the functions $F^1(z), \ldots, F^k(z)$ go to zero, it follows from condition (γ) that this inequality becomes

$$\psi_0 F^0(z)+\sum \lambda_j f^j(z) \leqslant \psi_0 F^0(z_0).$$

Therefore,

$$\psi_0(F^0(z)-F^0(z_0)) \leqslant -\sum \lambda_j f^j(z) \leqslant 0,$$

that is, according to (α) we have $F^0(z)-F^0(z_0) \geqslant 0$. But this also means that function $F^0(z)$, considered on set $\Omega' \cap \Omega'' \cap \Xi$, reaches a minimum at z_0. This establishes *sufficiency* of the conditions formulated.

Now let us prove *necessity*. If function $F^0(z)$, considered on set $\Omega' \cap \Omega'' \cap \Xi$, reaches a minimum at point z_0, then there will exist numbers $\psi_0, \psi_1, \ldots, \psi_k$ and nonpositive numbers λ_j, $j \in I(z_0)$, satisfying conditions (α) and (β) of Theorem 37.7 (for $r=p=l=0$). Let us assume that $\lambda_j=0$ for $j \notin I(z_0)$ and let us show that the numbers $\psi_0, \psi_1, \ldots, \psi_k, \lambda_1, \ldots, \lambda_q$ so obtained satisfy conditions (α), (β), (γ) of Theorem 38.3. Condition (α) is identical to condition (α) of Theorem 37.7, and thus it is satisfied. Condition (γ) is also satisfied, since for $j \in I(z_0)$ we have $f^j(z_0) = 0$, while for $j \notin I(z_0)$ we have $\lambda_j = 0$. It only remains to show that condition (β) of Theorem 38.3 is satisfied.

Let z be an arbitrary point of set Ξ, so that at this point all the functions $F^0(z)$, $f^1(z), \ldots, f^q(z)$ are defined. Since function $F^0(z)$ satisfies condition (a) (p. 289), it follows that $f^0(z) \leqslant F^0(z)$, where f^0 is an affine function whose existence is attested to in condition (a). Then, taking into account that $\psi_0 < 0$, we get (see (37.1) and (37.2)):

$$H(z) - H(z_0) = \psi_0 F^0(z) + \psi_1 F^1(z) + \ldots + \psi_k F^k(z) + \\ + \lambda_1 f^1(z) + \ldots + \lambda_q f^q(z) - \psi_0 F^0(z_0) - \psi_1 F^1(z_0) - \ldots - \psi_k F^k(z_0) - \\ - \lambda_1 f^1(z_0) - \ldots - \lambda_q f^q(z_0) \leqslant \psi_0 (f^0(z) - f^0(z_0)) + \\ + \psi_1 (F^1(z) - F^1(z_0)) + \ldots + \psi_k (F^k(z) - F^k(z_0)) + \\ + \sum_{j=1}^{q} \lambda_j (f^j(z) - f^j(z_0)) = \psi_0 (z - z_0) \operatorname{grad} f^0 + \\ + \sum_{i=1}^{k} \psi_i (z - z_0) \operatorname{grad} F^i + \sum_{j=1}^{q} \lambda_j f^j(z) = \\ = \sum_{\alpha=0}^{k} \psi_\alpha (z - z_0) \operatorname{grad} F^\alpha(z_0) + \sum_{j \in I(z_0)} \lambda_j f^j(z).$$

Function $f^j(z)$ is convex, so that for $j \in I(z_0)$ we have

$$f^j(z) = f^j(z) - f^j(z_0) \geqslant (z - z_0) \operatorname{grad} f^j(z_0)$$

(cf. pp. 289–290), and thus, taking the relation $\lambda_j \leqslant 0$ into account, we get

$$H(z) - H(z_0) \leqslant \sum_{\alpha=0}^{k} \psi_\alpha (z - z_0) \operatorname{grad} F^\alpha(z_0) + \\ + \sum_{j \in I(z_0)} \lambda_j (z - z_0) \operatorname{grad} f^j(z_0) = \\ = \left(\sum_{\alpha=0}^{k} \psi_\alpha \operatorname{grad} F^\alpha(z_0) + \sum_{j \in I(z_0)} \lambda_j \operatorname{grad} f^j(z_0) \right) (z - z_0).$$

However, the right-hand side of this equation is *nonpositive* in view of condition (β) of Theorem 37.7, since $z \in \Xi \subset N_1 \cap \ldots \cap N_m$, so that the direction of the vector $z - z_0$ lies on cone $N_1 \cap \ldots \cap N_m$. Consequently, $H(z) - H(z_0) \leqslant 0$, that is, $H(z) \leqslant H(z_0)$.

In the theorem proven it can be assumed (as in Theorem 38.2) that some of sets $\Xi_1, \ldots, \Xi_m$ and some of cones $N_1, \ldots, N_m$ satisfy condition (c′), some satisfy condition (c‴), and the rest satisfy condition (c). In this way a more general theorem can be obtained directly from Theorem 38.3 (cf. Theorem 38.2), which, however, will not be formulated here.

NOTE 38.4. As the proof shows, in Theorem 38.3 it is enough to require that functions $f^j(z)$ be convex and smooth only for $j \in I(z_0)$.

NOTE 38.5. If in Theorem 38.3 only *local* convexity of functions $F^0(z)$ and $f^i(z)$, $i \in I(z_0)$, is required at point z_0, then the necessary and sufficient condition for a *local minimum* is obtained, that is, a minimum of function $F^0(z)$, considered on the set $\Sigma_r \cap \Omega' \cap \Omega'' \cap \Xi$, where Σ_r is a sphere with some sufficiently small radius r, centered on z_0 (here the maximum of function $H(z)$ also has to be considered on set $\Sigma_r \cap \Xi$). On the other hand, it is clear that the necessary condition for a local minimum also constitutes the necessary condition for an absolute minimum (that is, a minimum of function $F^0(z)$, considered on the entire set $\Omega' \cap \Omega'' \cap \Xi$).

Consequently, if only local convexity of functions $F^0(z)$ and $f^i(z)$, $i \in I(z_0)$, is required at point z_0, and if the final requirement is replaced by the condition that function $H(z)$, considered on set Ξ, reaches a local maximum at z_0, then the *necessary* condition for an (absolute) minimum is obtained from Theorem 38.3. Similarly, we can arrive at the necessary condition for an (absolute) minimum in the form of a maximum principle from Theorems 38.1 and 38.2 as well.

EXAMPLE 38.6. Let $F^0(z)$ be a smooth convex function defined in an open set of space E^n, and let Ξ be a convex set contained in the domain of definition of function $F^0(z)$. Finally, let Ω' be some plane and let $z_0 \in \Omega' \cap \Xi$. We call N the supporting cone of convex set Ξ at point z_0 and we assume that Ω' and N are nonseparable in E^n. Now, if we select affine functions $F^1(z), \ldots, F^k(z)$, the gradients of which are linearly independent, such that

$$\Omega' = (\mathrm{Ker}\, F^1) \cap \cdots \cap (\mathrm{Ker}\, F^k)$$

then it is easy to see that the system of convex cones

$$\mathrm{Ker}\, F^1, \ldots, \mathrm{Ker}\, F^k, N$$

does not possess separability in E^n, that is, all the conditions of Theorem 38.2 (in which we now set $r = p = 0$, $m = 1$) are satisfied.

Consequently, in order for function $F^0(z)$, considered on set $\Omega' \cap \Xi$, to reach a minimum at point z_0, it is necessary and sufficient that there exist a vector $\psi = \{\psi_0, \psi_1, \ldots, \psi_k\}$, such that $\psi_0 < 0$, and such that the function $H(z)$, defined by formula (38.1) and considered on set Ξ, reaches a maximum at z_0. For this, it can be assumed that $\psi_0 = -1$.

We now give a geometrical interpretation of the proposition just formulated (Figure 155). To do this, it will be convenient to designate as E^{n+1} the space of dimension $n+1$ containing E^n and to introduce in E^{n+1} an (orthonormal) system of coordinates $z^1, \ldots, z^n, z^{n+1}$ such that the hyperplane $z^{n+1} = 0$ coincides with E^n

Let W be the *graph* of function $z^{n+1}=F^0(z)$, considered on set Ξ, that is, the set of all points $(z^1, \ldots, z^n, z^{n+1})$ for which point $z=(z^1, \ldots, z^n)$ belongs to Ξ and $F^0(z)=z^{n+1}$. In addition, we let Ω^* be the plane in E^{n+1} parallel to Ω' and passing through point $z^*=(z_0^1, \ldots, z_0^n, F^0(z_0))\in W$ (where $z_0^1, \ldots, z_0^n$ are the coordinates of point z_0 in E^n). Finally, let Γ be the hyperplane of space E^{n+1} serving as a graph of function $z^{n+1}=\Phi(z)$, where

$$\Phi(z)=\psi_1 F^1(z)+\ldots+\psi_k F^k(z)+F^0(z_0). \tag{38.2}$$

Since $F^1(z)=\ldots=F^k(z)=0$ for $z\in\Omega'$ (and, in particular, for $z=z_0$), it follows that $\Phi(z)\equiv F^0(z_0)$ for $z\in\Omega'$. Consequently, hyperplane Γ contains plane Ω^*.

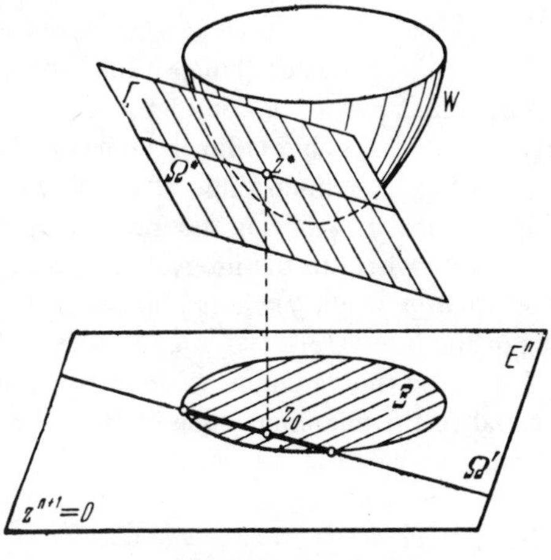

FIGURE 155

Now note that $H(z_0)=-F^0(z_0)$, so that

$$F^0(z)-\Phi(z)=H(z_0)-H(z).$$

Accordingly, the fact that function $H(z)$, considered on set Ξ, reaches a maximum at z_0 implies that $F^0(z)\geqslant\Phi(z)$ on set Ξ and also that in E^{n+1} set W is situated *above* hyperplane Γ (that is, in the half-space $z^{n+1}\geqslant\Phi(z)$).

Conversely, if there exists a hyperplane Γ, passing through plane Ω^*, such that set W is situated above hyperplane Γ, then Γ will be defined by the equation $z^{n+1}=\Phi(z)$, where function $\Phi(z)$ has form (38.2), so that in Ξ the inequality $H(z)\leqslant H(z_0)$ is satisfied.

Therefore, *in order for a smooth convex function* $F^0(z)$, *considered on set* $\Omega'\cap\Xi$ *(where* Ξ *is a convex set contained in the domain of definition of function* $F^0(z)$ *and nonseparable from plane* $\Omega'\subset E^n$), *to reach a minimum at point* $z_0\in\Omega'\cap\Xi$, *it is necessary and sufficient that there exist in space* $E^{n+1}\supset E^n$ *a hyperplane* Γ *such that the graph of function* $z^{n+1}=F^0(z)$, *considered on set* Ξ, *is situated above* Γ, *where* Γ *contains the plane* Ω^* *parallel to* Ω' *and passing through point* $z^*=(z_0, F^0(z_0))$.

This proposition (which is geometrically quite obvious; see Figure 155) also serves as an illustration of the maximum principle in the given case.

39. Method of dynamic programming. In this subsection we will describe a method for finding the minimum which is completely different from the methods considered in the preceding subsections. This method, which is known as the method of *dynamic programming,* was introduced by R. Bellman.

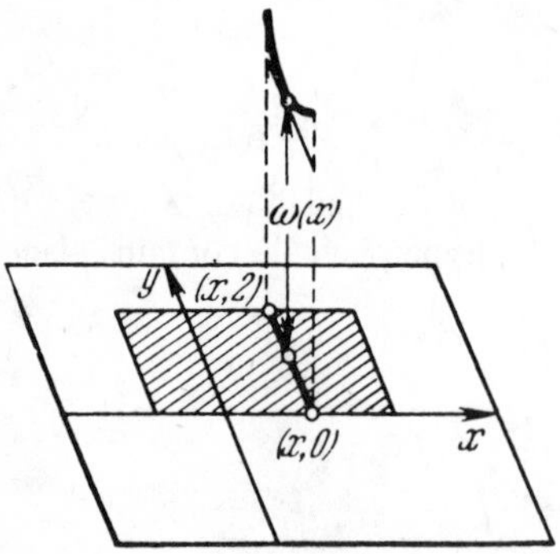

FIGURE 156

In order to grasp the principle of this method, we consider another solution of the problem posed in Example 7.1 (p. 33).

EXAMPLE 39.1. Find the minimum of function (7.4), considered in the rectangle $-1 \leqslant x \leqslant 3$, $0 \leqslant y \leqslant 2$.

Solution. Let us assign to variable x a *fixed* value (lying in the interval $-1 \leqslant x \leqslant 3$) and let us consider F to be a function just of variable y (in the interval $0 \leqslant y \leqslant 2$). The minimum value of this function is designated as $\omega(x)$ (Figure 156). Therefore, for any y, satisfying the inequalities $0 \leqslant y \leqslant 2$, the relation $F(x, y) \geqslant \omega(x)$ will be valid, where in the interval $0 \leqslant y \leqslant 2$ there exists at least one point for which the *equation* $F(x, y) = \omega(x)$ is reached. The number $\omega(x)$ is defined for any x lying in the interval $-1 \leqslant x \leqslant 3$, that is, $\omega(x)$ is a function defined in this interval. Let m be the minimum value of this function.

Clearly, m will be equal to the minimum value of function $F(x, y)$ in the rectangle being considered. Actually, since

$$m = \min_{-1 \leqslant x \leqslant 3} \omega(x),$$

there must exist in the interval $-1 \leqslant x \leqslant 3$ a point x_0 such that $\omega(x_0) = m$. Moreover, since

$$\omega(x_0) = \min_{0 \leqslant y \leqslant 2} F(x_0, y),$$

it follows that in the interval $0 \leqslant y \leqslant 2$ there is some point y_0, such that $F(x_0, y_0) = \omega(x_0)$, that is, $F(x_0, y_0) = m$. On the other hand, for any point (x, y) of the given rectangle, we have

$$F(x, y) \geqslant \min_{0 \leqslant y \leqslant 2} F(x, y) = \omega(x) \geqslant \min_{-1 \leqslant x \leqslant 3} \omega(x) = m.$$

Thus, $F(x, y) \geqslant m$ for any point of the rectangle, and there exists (at least one) point (x_0, y_0) for which $F(x_0, y_0) = m$. However, this also means that m is the minimum value of function F.

From the foregoing it follows that the minimum value of function $F(x, y)$ can be found in two stages: first function $\omega(x)$ is found, and then its minimum value is determined. We will now carry out this two-stage process.

To find the function

$$\omega(x) = \min_{0 \leqslant y \leqslant 2} F(x, y)$$

we set the partial derivative F_y of function F equal to zero (see the second relation (7.6)). Obviously, if $x \neq 0$, derivative F_y will go to zero at one point only, $y=1$. Consequently, for a fixed $x \neq 0$, function $F(x, y)$ may reach its minimum in the interval $0 \leqslant y \leqslant 2$ either at point $y=1$ or at the end points of the segment (at $y=0$ or $y=2$). In other words, $\omega(x)$ is the least of three numbers: $F(x, 0)$, $F(x, 1)$, $F(x, 2)$ (this statement also being true for $x=0$, since $F(0, y) \equiv 3 = \text{const}$). According to (7.4),

$$\left.\begin{aligned} F(x, 0) = F(x, 2) &= \frac{2}{3}x^3 + x^2 - 4x + 3; \\ F(x, 1) &= \frac{2}{3}x^3 + 3. \end{aligned}\right\} \tag{39.1}$$

Since the function $x^2 - 4x = x(x-4)$ is positive for $-1 \leqslant x < 0$ and negative for $0 < x \leqslant 3$, we obtain the following expression for function $\omega(x)$:

$$\omega(x) = \begin{cases} \frac{2}{3}x^3 + 3 & \text{for} \quad -1 \leqslant x \leqslant 0, \\ \frac{2}{3}x^3 + x^2 - 4x + 3 & \text{for} \quad 0 \leqslant x \leqslant 3. \end{cases}$$

Now it is clear that function $\omega(x)$ may assume its lowest value m either at one of the points $x=-1$, $x=0$, $x=3$, or else at the points where its derivative goes to zero, that is, at the points $x=-2$, $x=1$. But since the point $x=-2$ does not lie in the segment $-1 \leqslant x \leqslant 3$, we need only compare the numbers $\omega(-1)$, $\omega(0)$, $\omega(1)$, $\omega(3)$. A direct calculation shows the least of these to be the value $\omega(1) = 2/3$. Thus the minimum of function $F(x, y)$ is $m = 2/3$; it follows from (39.1) that this minimum of F is reached at two points: (1,0) and (1,2).

The solution method employed in this example reduces to a successive calculation of the minima of functions of a *single* variable. First we seek a minimum of the function $F(x, y)$ of one variable y for a fixed (but arbitrary) value of x, this minimum being designated as $\omega(x)$, and then we seek a minimum of the obtained function $\omega(x)$ of variable x.

This method can also be used for find the minimum of a function of any number of variables. For instance, let $F(z^1, \ldots, z^n)$ be a function defined on some set $\Omega \subset E^n$. We fix arbitrarily the numbers $z^1 = z_0^1, \ldots, z^{n-1} = z_0^{n-1}$ and we consider the function $F(z_0^1, \ldots, z_0^{n-1}, z^n)$ of *one* variable z^n (it is defined for those values of z^n for which the point $(z_0^1, \ldots, z_0^{n-1}, z^n)$ belongs to set Ω). The least value of this function (depending, naturally, on the choice of the numbers $z_0^1, \ldots, z_0^{n-1}$) will be designated as $\omega_{n-1}(z_0^1, \ldots, z_0^{n-1})$. Therefore, we obtain a function $\omega_{n-1}(z^1, \ldots, z^{n-1})$ of the $n-1$ variables $z^1, \ldots, z^{n-1}$:

$$\omega_{n-1}(z^1, \ldots, z^{n-1}) = \min_{z^n} F(z^1, \ldots, z^{n-1}, z^n). \tag{39.2}$$

Now let us fix arbitrarily the numbers $z_0^1, \ldots, z_0^{n-2}$ and let us consider the function $\omega_{n-1}(z_0^1, \ldots, z_0^{n-2}, z^{n-1})$ of *one* variable z^{n-1}. We designate the least value of this function as $\omega_{n-2}(z_0^1, \ldots, z_0^{n-2})$, and we obtain a function $\omega_{n-2}(z^1, \ldots, z^{n-2})$ of the $n-2$ variables $z^1, \ldots, z^{n-2}$:

$$\omega_{n-2}(z^1, \ldots, z^{n-2}) = \min_{z^{n-1}} \omega_{n-1}(z^1, \ldots, z^{n-2}, z^{n-1}).$$

In general, if a function $\omega_k(z^1, \ldots, z^k)$ of k variables has already been found, then we determine $\omega_{k-1}(z^1, \ldots, z^{k-1})$ as the minimum of this function with respect to variable z^k (for fixed $z^1, \ldots, z^{k-1}$):

$$\omega_{k-1}(z^1, \ldots, z^{k-1}) = \min_{z^k} \omega_k(z^1, \ldots, z^{k-1}, z^k); \quad k = 2, \ldots, n-1. \tag{39.3}$$

Therefore, the following functions are determined successively:

$$\omega_{n-1}(z^1, \ldots, z^{n-2}, z^{n-1}), \quad \omega_{n-2}(z^1, \ldots, z^{n-2}), \ldots \\ \ldots, \omega_2(z^1, z^2), \quad \omega_1(z^1). \tag{39.4}$$

Let m be the least value of the last of these functions:

$$m = \min_{z^1} \omega_1(z^1). \tag{39.5}$$

It is easy to see that *m is the least value of the initial function $F(z^1, \ldots, z^n)$*. A more precise formulation and a proof of this result will be given below, in Theorem 39.2. Consequently, after considering the sequence of functions (39.4) defined by recurrence relations (39.2) and (39.3), we can use formula (39.5) to find the least value m of function $F(z^1, \ldots, z^n)$, where the application of this method involves finding each time the minimum of a function depending just on one variable. This, then, is the method of dynamic programming (as applied to finding the minimum of a function).

A definite advantage of this method is its universality (that is, its wide range of applicability). From the description given in general terms above, it is clear that this method can be applied without making *any* assumptions whatsoever about the nature of function $F(z^1, \ldots, z^n)$ or its domain of definition. Therefore, of all the considered (and not considered) methods for finding extrema, dynamic programming is the most general.

Another advantage of this method is that it makes possible a *reduction of the dimension* of the problem, that is (in the application considered above), the calculation of the minimum of a function of many variables can be reduced to a successive calculation of the minima of functions of a *single* variable. However, actually this advantage is, as a rule, only apparent and in an overwhelming majority of cases becomes a disadvantage of the method.

This is because, in sequence (39.4), the functions become more and more complex, even for a suitable analytical notation of the initial function $F(z^1, \ldots, z^n)$. This was already evident in Example 39.1, where in "sequence" (39.4) there was only a single function $\omega(x)(=\omega_1(z^1))$: in contrast to the initial function $F(x, y)$, function $\omega(x)$ no longer had a *single* analytical representation, being defined differently in the two segments making up its domain of definition. Thus it is easy to imagine what would happen if the method were applied in the general case: the number of "pieces" on which function $\omega_k(z^1, \ldots, z^k)$ will be defined differently will be all the more numerous as k is reduced, making the number of calculations required for the sequential execution of the method practically unlimited.

Moreover, although formally it is always the minimum of a function of a *single* variable which is involved, the problem is actually complicated by the presence of several parameters (for instance, in order to find function $\omega_{n-1}(z^1, \ldots, z^{n-1})$ using formula (39.2), we indeed seek the minimum of a function depending on only one variable z^n, but at the same time all possible values of parameters $z_0^1, \ldots, z_0^{n-1})$ are involved). This complication is inevitable, since during the recurrent calculation of functions (39.4) we do not know until the last moment (that is, until the number m has been found) what the values of variables $z^1, \ldots, z^n$ will be for which the initial function $F(z^1, \ldots, z^n)$ reaches a minimum. The values of functions (39.4) have to be found (and remembered) for *all* the values of the variables.

Therefore, in the general case, for any high value of n, the method of dynamic programming cannot be carried out completely even using electronic computers, since it involves an excessive number of calculations and demands a memory capacity which is far too great. This is not surprising, for the method of dynamic programming essentially requires a consideration of *all* the values of function $F(z^1, \ldots, z^n)$. For instance, in Example 39.1 we first calculated the values of function $F(x, y)$ in *each* segment $x = \text{const}$ (that is, we considered *all* the values of this function), after which we chose the least of the minima of the functions in these segments.

These things make it difficult to apply the method of dynamic programming. However, sometimes certain additional considerations can be invoked which make it possible to complete the solution of the problem with the aid of this method. This is illustrated by Example 39.3, given below.

Now let us formulate and prove a theorem which gives an account of the method of dynamic programming (as applied to finding the minimum of a function). First, however, it will be convenient to introduce the following notation. Let $F(z^1, \ldots, z^n)$ be a function defined on some set $\Omega \subset E^n$. In addition, let E^{n-1} be the space of variables $z^1, \ldots, z^{n-1}$ and let Ω_{n-1} be the set of all points $(z^1, \ldots, z^{n-1}) \in E^{n-1}$

for each of which it is possible to choose a number z^n such that $(z^1, \ldots, z^{n-1}, z^n) \in \Omega$. In other words, the point $(z^1, \ldots, z^{n-1})$ belongs to set Ω_{n-1} if, and only if, a line passing through this point and lying parallel to the z^n axis intersects set Ω. On set Ω_{n-1} we define a function $\omega_{n-1}(z^1, \ldots, z^{n-1})$ using formula (39.2), where the minimum is taken over all values of z^n for which $(z^1, \ldots, z^{n-1}, z^n) \in \Omega$. It is assumed that function $\omega_{n-1}(z^1, \ldots, z^{n-1})$ is defined on the *entire* set Ω_{n-1}, that is, minimum (39.2) exists for any point $(z^1, \ldots, z^{n-1}) \in \Omega_{n-1}$.

We assume that the function $\omega_k(z^1, \ldots, z^k)$, defined on some set Ω_k of the space E^k of variables $z^1, \ldots, z^k$, has already been constructed (for some $k = 2, \ldots, n-1$). Then, let E^{k-1} be the space of variables $z^1, \ldots, z^{k-1}$, and let Ω_{k-1} be the set of all points $(z^1, \ldots, z^{k-1}) \in E^{k-1}$, for each of which a number z^k can be selected such that $(z^1, \ldots, z^{k-1}, z^k) \in \Omega_k$. On set Ω_{k-1} we define a function $\omega_{k-1}(z^1, \ldots, z^{k-1})$ using formula (39.3), where the minimum is taken over all the values of z^k for which $(z^1, \ldots, z^{k-1}, z^k) \in \Omega_k$. It will be assumed that the function $\omega_{k-1}(z^1, \ldots, z^{k-1})$ is defined on the *entire* set Ω_{k-1}, that is, that for any point $(z^1, \ldots, z^{k-1}) \in \Omega_{k-1}$ minimum (39.3) exists.

The assumption that minima (39.2), (39.3), and (39.5) exist is the only limitation imposed during the application of the method of dynamic programming. Note that this assumption is definitely valid if Ω is a closed bounded set of space E^n, while function $F(z^1, \ldots, z^n)$ is continuous on Ω.

THEOREM 39.2. *Let $F(z^1, \ldots, z^n)$ be some function defined on set $\Omega \subset E^n$. Minimum (39.2) is assumed to exist for any point $(z^1, \ldots, z^{n-1}) \in \Omega_{n-1}$. Moreover, it is assumed that for any point $(z^1, \ldots, z^{k-1}) \in \Omega_{k-1}$ minimum (39.3) exists, with $k = 2, \ldots, n-1$. Finally, we also assume that minimum (39.5) exists (over set Ω_1). Then the number m is the minimum of function $F(z^1, \ldots, z^n)$ on set Ω. In addition, point $z_* = (z_*^1, \ldots, z_*^n) \in \Omega$ is a point of minimum of function $F(z^1, \ldots, z^n)$ on set Ω if, and only if, the following relations are valid:*

$$\begin{aligned} F(z_*^1, \ldots, z_*^n) &= \omega_{n-1}(z_*^1, \ldots, z_*^{n-1}) = \\ &= \omega_{n-2}(z_*^1, \ldots, z_*^{n-2}) = \ldots = \omega_2(z_*^1, z_*^2) = \omega_1(z_*^1) = m. \end{aligned} \tag{39.6}$$

Proof. For any point $z = (z^1, \ldots, z^n) \in \Omega$ we have

$$F(z^1, \ldots, z^n) \geqslant \min_{z^n} F(z^1, \ldots, z^n) = \omega_{n-1}(z^1, \ldots, z^{n-1}),$$

$$\omega_{n-1}(z^1, \ldots, z^{n-1}) \geqslant \min_{z^{n-1}} \omega_{n-1}(z^1, \ldots, z^{n-1}) = \omega_{n-2}(z^1, \ldots, z^{n-2}),$$

$$\ldots\ldots\ldots\ldots\ldots\ldots\ldots\ldots\ldots\ldots\ldots\ldots$$

$$\omega_2(z^1, z^2) \geqslant \min_{z^2} \omega_2(z^1, z^2) = \omega_1(z^1),$$

$$\omega_1(z^1) \geqslant \min_{z^1} \omega_1(z^1) = m.$$

Consequently,

$$\begin{aligned} F(z^1, \ldots, z^n) &\geqslant \omega_{n-1}(z^1, \ldots, z^{n-1}) \geqslant \\ &\geqslant \omega_{n-2}(z^1, \ldots, z^{n-2}) \geqslant \ldots \geqslant \omega_2(z^1, z^2) \geqslant \omega_1(z^1) \geqslant m. \end{aligned} \tag{39.7}$$

In particular, $F(z^1, \ldots, z^n) \geqslant m$ for any point $(z^1, \ldots, z^n) \in \Omega$.

Let us show that a point $(z_0^1, \ldots, z_0^n) \in \Omega$ exists for which $F(z_0^1, \ldots, z_0^n) = m$. This will indicate that m is the least value of function $F(z^1, \ldots, z^n)$ on set Ω. From relation (39.5) we know that there exists a number $z_0^1 \in \Omega_1$, for which $\omega_1(z_0^1) = = m$. Moreover, the relation

$$\omega_1(z_0^1) = \min_{z^2} \omega_2(z_0^1, z^2)$$

(see (39.3)) implies the existence of a number z_0^2 such that $(z_0^1, z_0^2) \in \Omega_2$ and $\omega_2(z_0^1, z_0^2) = \omega_1(z_0^1)$, that is, $\omega_2(z_0^1, z_0^2) = m$. Similarly, we can prove the existence of a number z_0^3 such that $(z_0^1, z_0^2, z_0^3) \in \Omega_3$ and $\omega_3(z_0^1, z_0^2, z_0^3) = m$, etc. By continuing this process, we find ultimately numbers $z_0^1, \ldots, z_0^{n-1}$ such that $(z_0^1, \ldots, z_0^{n-1}) \in \Omega_{n-1}$ and $\omega_{n-1}(z_0^1, \ldots, z_0^{n-1}) = m$. Now the relations

$$\omega_{n-1}(z_0^1, \ldots, z_0^{n-1}) = \min_{z^n} F(z_0^1, \ldots, z_0^{n-1}, z^n)$$

imply the existence of a number z_0^n such that $(z_0^1, \ldots, z_0^n) \in \Omega$ and

$$\omega_{n-1}(z_0^1, \ldots, z_0^{n-1}) = F(z_0^1, \ldots, z_0^{n-1}, z_0^n),$$

that is, $F(z_0^1, \ldots, z_0^n) = m$. Consequently, m is the minimum of function $F(z^1, \ldots, z^n)$ on set Ω.

Let us next prove the concluding part of the theorem. If $z_* = (z_*^1, \ldots, z_*^n)$ is the point of minimum of function $F(z^1, \ldots, z^n)$, that is, if $F(z_*^1, \ldots, z_*^n) = m$, then *equals* signs have to appear everywhere in relations (39.7) (otherwise, we would have $F(z_*^1, \ldots, z_*^n) > m$), meaning that relations (39.6) are valid. Conversely, if relations (39.6) are valid, then, in particular, $F(z_*^1, \ldots, z_*^n) = m$, that is, $(z_*^1, \ldots, z_*^n)$ is the point of minimum.

To illustrate the application of the method of dynamic programming, we once again consider the problem posed in Example 9.2 (p. 46).

EXAMPLE 39.3. *Find n nonnegative numbers $z^1, \ldots, z^n$ whose sum does not exceed a and which when multiplied together have the maximum possible product.*

Solution. We wish to find the minimum value of the function

$$F(z^1, \ldots, z^n) = -z^1 z^2 \cdot \ldots \cdot z^n, \tag{39.8}$$

considered on a set Ω, defined by the inequalities

$$z^1 \geqslant 0,\ z^2 \geqslant 0,\ \ldots,\ z^n \geqslant 0,\quad z^1 + \ldots + z^n \leqslant a. \tag{39.9}$$

Let us calculate successively functions (39.4). Clearly, the set Ω_{n-1} is defined by the inequalities

$$z^1 \geqslant 0,\ \ldots,\ z^{n-1} \geqslant 0,\quad z^1 + \ldots + z^{n-1} \leqslant a. \tag{39.10}$$

Let $(z^1, \ldots, z^{n-1})$ be an arbitrary point of set Ω_{n-1}. In order for the point $(z^1, \ldots, z^{n-1}, z^n)$ to belong to set Ω, that is, to satisfy inequalities (39.9), it is necessary and sufficient that the number z^n be nonnegative and satisfy the last relation (39.9), the following relations being then satisfied:

$$0 \leqslant z^n \leqslant a - (z^1 + \ldots + z^{n-1}). \tag{39.11}$$

Now we find (the minimum is selected over segment (39.11)):

$$\begin{aligned}
\omega_{n-1}(z^1, \ldots, z^{n-1}) &= \min_{z^n} F(z^1, \ldots, z^n) = \\
&= \min_{z^n}(-z^1z^2 \cdot \ldots \cdot z^n) = -\max_{z^n}(z^1z^2 \cdot \ldots \cdot z^{n-1}z^n) = \\
&= -z^1z^2 \cdot \ldots \cdot z^{n-1} \cdot \max_{z^n} z^n = \\
&= -z^1z^2 \cdot \ldots \cdot z^{n-1}(a - (z^1 + \ldots + z^{n-1})).
\end{aligned} \tag{39.12}$$

Next let us find function $\omega_{n-2}(z^1, \ldots, z^{n-2})$. It is clear from (39.10) that the set Ω_{n-2} is defined by the inequalities

$$z^1 \geqslant 0, \ldots, z^{n-2} \geqslant 0, \quad z^1 + \ldots + z^{n-2} \leqslant a. \tag{39.13}$$

Let $(z^1, \ldots, z^{n-2}) \in \Omega_{n-2}$ and $z^{n-1} \geqslant 0$. In order for point $(z^1, \ldots, z^{n-2}, z^{n-1})$ to belong to set Ω_{n-1}, that is, to satisfy inequalities (39.10), it is necessary and sufficient that the number z^{n-1} satisfy the inequalities

$$0 \leqslant z^{n-1} \leqslant a - (z^1 + \ldots + z^{n-2}).$$

If we designate the number $a - (z^1 + \ldots + z^{n-2})$ as c, then the number z^{n-1} can be replaced in the segment $0 \leqslant z^{n-1} \leqslant c$, so that

$$\begin{aligned}
\omega_{n-2}(z^1, \ldots, z^{n-2}) &= \min_{0 \leqslant z^{n-1} \leqslant c} \omega_{n-1}(z^1, \ldots, z^{n-2}, z^{n-1}) = \\
&= \min_{0 \leqslant z^{n-1} \leqslant c} [-z^1z^2 \cdot \ldots \cdot z^{n-1}(a - (z^1 + \ldots + z^{n-2} + z^{n-1}))] = \\
&= \min_{0 \leqslant z^{n-1} \leqslant c} [-z^1 \ldots z^{n-2} \cdot (z^{n-1}(c - z^{n-1}))] = \\
&= \min_{0 \leqslant x \leqslant c} \left[-z^1 \ldots z^{n-2}\left(x\left(\frac{c-x}{1}\right)^1\right)\right] = \\
&= -z^1 \ldots z^{n-2} \cdot \max_{0 \leqslant x \leqslant c} x\left(\frac{c-x}{1}\right)^1 = -z^1 \ldots z^{n-2} \cdot \left(\frac{c}{2}\right)^2 = \\
&= -z^1z^2 \cdot \ldots \cdot z^{n-2} \cdot \left(\frac{a - (z^1 + \ldots + z^{n-2})}{2}\right)^2
\end{aligned}$$

(here we used Lemma 9.1, p. 46, for $k=1$). Thus,

$$\omega_{n-2}(z^1, \ldots, z^{n-2}) = -z^1 \cdot \ldots \cdot z^{n-2}\left(\frac{a-(z^1+\ldots+z^{n-2})}{2}\right)^2. \quad (39.14)$$

In addition, let us calculate the function $\omega_{n-3}(z^1, \ldots, z^{n-3})$. Set Ω_{n-3} is described by the inequalities

$$z^1 \geqslant 0, \ldots, z^{n-3} \geqslant 0, \quad z^1+\ldots+z^{n-3} \leqslant a.$$

Let $(z^1, \ldots, z^{n-3}) \in \Omega_{n-3}$ and $z^{n-2} \geqslant 0$. In order for point $(z^1, \ldots, z^{n-3}, z^{n-2})$ to belong to set Ω_{n-2} (see (39.13)), it is necessary and sufficient that the number z^{n-2} satisfy the inequalities

$$0 \leqslant z^{n-2} \leqslant a-(z^1+\ldots+z^{n-3}).$$

If we designate the number $a-(z^1+\ldots+z^{n-3})$ as d, then the number z^{n-2} can be replaced in the segment $0 \leqslant z^{n-2} \leqslant d$, so that

$$\begin{aligned}\omega_{n-3}(z^1, \ldots, z^{n-3}) &= \min_{0\leqslant z^{n-2}\leqslant d} \omega_{n-2}(z^1, \ldots, z^{n-3}, z^{n-2}) = \\ &= \min_{0\leqslant z^{n-2}\leqslant d}\left[-z^1 \ldots z^{n-2}\left(\frac{a-(z^1+\ldots+z^{n-2})}{2}\right)^2\right] =\end{aligned}$$

$$\begin{aligned}&= \min_{0\leqslant z^{n-2}\leqslant d}\left[-z^1 \ldots z^{n-3}\left(z^{n-2}\left(\frac{d-z^{n-2}}{2}\right)^2\right)\right] = \\ &= \min_{0\leqslant x\leqslant d}\left[-z^1 \ldots z^{n-3}\left(x\left(\frac{d-x}{2}\right)^2\right)\right] = \\ &= -z^1 \ldots z^{n-3} \cdot \max_{0\leqslant x\leqslant d} x\left(\frac{d-x}{2}\right)^2 = -z^1 \ldots z^{n-3} \cdot \left(\frac{d}{3}\right)^3 = \\ &= -z^1 \cdot \ldots \cdot z^{n-3}\left(\frac{a-(z^1+\ldots+z^{n-3})}{3}\right)^3\end{aligned}$$

(here Lemma 9.1 was used, for $k=2$). Thus,

$$\omega_{n-3}(z^1, \ldots, z^{n-3}) = -z^1 \ldots z^{n-3} \cdot \left(\frac{a-(z^1+\ldots+z^{n-3})}{3}\right)^3.$$

Continuing similarly, we find by induction:

$$\omega_k(z^1, \ldots, z^k) = -z^1 \cdot \ldots \cdot z^k \cdot \left(\frac{a-(z^1+\ldots+z^k)}{n-k}\right)^{n-k},$$
$$k=1, \ldots, n-1.$$

In particular, for $k=1$ we have

$$\omega_1(z^1) = -z^1\left(\frac{a-z^1}{n-1}\right)^{n-1}.$$

Consequently, since set Ω_1 is defined by the inequality $0 \leqslant z^1 \leqslant a$, we get

$$m = \min_{0\leqslant z^1\leqslant a} \omega_1(z^1) = \min_{0\leqslant z^1\leqslant a}\left[-z^1\left(\frac{a-z^1}{n-1}\right)^{n-1}\right] =$$
$$= -\max_{0\leqslant z^1\leqslant a}\left(z^1\left(\frac{a-z^1}{n-1}\right)^{n-1}\right) = -\left(\frac{a}{n}\right)^n$$

(here Lemma 9.1 was used, for $k=n-1$). Thus,

$$\min_{\Omega} F(z^1, \ldots, z^n) = -\left(\frac{a}{n}\right)^n,$$

that is, the product $z^1z^2 \ldots z^n$ attains (with constraints (39.9)) its largest value, equal to $(a/n)^n$. With the aid of Lemma 9.1, it is easy to see that this largest value is reached only for $z^1 = z^2 = \ldots = z^n = a/n$, which is in complete agreement with the result of Example 9.2.

Chapter V. OPTIMALITY CRITERIA FOR DISCRETE PROCESSES

§11. Dynamic Programming

40. Description of method. For simplicity let us first consider the *fundamental problem* (see Subsection 3). In other words, we consider the controlled object

$$x(t) = f_t(x(t-1),\ u(t)), \qquad t = 1, \ldots, N, \tag{40.1}$$
$$u(t) \in U_t(x(t-1)), \qquad t = 1, \ldots, N, \tag{40.2}$$

for which the initial state $x(0) = x_0$ is fixed and no constraints are placed on the subsequent states $x(1), \ldots, x(N)$. The problem is to find a process

$$u(1),\ u(2),\ \ldots,\ u(N), \tag{40.3}$$
$$x(0),\ x(1),\ \ldots,\ x(N), \tag{40.4}$$

satisfying conditions (40.1), (40.2), for which the initial condition $x(0) = x_0$ is satisfied, with the functional

$$J = \sum_{t=1}^{N} f_t^0(x(t-1),\ u(t)) \tag{40.5}$$

assuming its least possible value (if the problem involves finding the *maximum* of functional (40.5) rather than the minimum, then in all the formulas presented below, and in particular (40.6), (40.7), the min symbol should be replaced by the max symbol).

We define as follows the sets $\Omega_0,\ \Omega_1,\ \ldots,\ \Omega_N$, located in phase space E^n The set Ω_0 consists of a single point x_0. Further, if a set Ω_{t-1} is already defined (where $1 \leqslant t \leqslant N$), then Ω_t will denote the set of all points of form $f_t(x,\ u)$, where $x,\ u$ satisfy the conditions $x \in \Omega_{t-1},\ u \in U_t(x)$ (function f_t and set U_t were specified in relations (40.1), (40.2)). This definition directly implies that for all $t = 0, 1, \ldots, N$ the relations $x(t) \in \Omega_t$ are satisfied, provided process (40.3), (40.4) satisfies relations (40.1).

Now let us define inductively certain functions $\omega_N(x), \omega_{N-1}(x), \ldots, \omega_1(x), \omega_0(x)$. First of all, we set $\omega_N(x) \equiv 0$. The function $\omega_N(x)$ is defined throughout the entire phase space E^n (and, in particular, on set Ω_N). Assume that a function $\omega_t(x)$ (where $1 \leqslant t \leqslant N$), defined on a set Ω_t (or on some set clearly containing

Ω_t), has already been constructed. Then the function $\omega_{t-1}(x)$ is defined by the relation*

$$\omega_{t-1}(x) = \min_{u \in U_t(x)} \left(\omega_t(f_t(x, u)) + f_t^0(x, u)\right), \qquad (40.6)$$

where f_t^0 are the functions used to formulate functional (40.5). Note that for any $x \in \Omega_{t-1}$ the point $f_t(x, u)$ belongs (for $u \in U_t(x)$) to set Ω_t, so that the expression following the min symbol in (40.6) *has meaning.* Now let us assume (this being the only restriction imposed during the application of the dynamic-programming method) that function ω_{t-1} is defined on the whole set Ω_{t-1}, that is, *for any* $x \in \Omega_{t-1}$ *the minimum on the right-hand side of* (40.6) *exists.* The method of finding optimal processes (in discrete systems) with the aid of the following theorem is known as dynamic programming.

THEOREM 40.1. *Let* (40.3), (40.4) *be some process (with an initial state* $x(0)$*) in the discrete controlled object* (40.1), (40.2). *For optimality of this process (in the sense of the minimum of functional* (40.5) *for a fixed initial state* $x(0)$ *and without placing any restrictions on states* $x(1), \ldots, x(N)$*), it is necessary and sufficient that the following relation be satisfied for each* $t = 1, \ldots, N$:

$$f_t^0(x(t-1), u(t)) + \omega_t(f_t(x(t-1), u(t))) = \\ = \min_{u \in U_t(x(t-1))} \left(f_t^0(x(t-1), u) + \omega_t(f_t(x(t-1), u))\right). \qquad (40.7)$$

Proof. First we show the *necessity* of condition (40.7). Let process (40.3), (40.4) be optimal. We select an arbitrary τ, $1 \leqslant \tau \leqslant N$, and an arbitrary point $v \in U_\tau(x(\tau-1))$. Using these quantities we now construct some process

$$\tilde{x}(0), \tilde{x}(1), \ldots, \tilde{x}(N); \quad \tilde{u}(1), \ldots, \tilde{u}(N)$$

with an initial state $\tilde{x}(0) = x(0)$ satisfying relations (40.1), (40.2). Let us set

$$\tilde{x}(t) = x(t), \quad \tilde{u}(t) = u(t) \quad \text{for} \quad t = 0, 1, \ldots, \tau - 1, \qquad (40.8)$$

and then also

$$\tilde{u}(\tau) = v, \quad \tilde{x}(\tau) = f_\tau(\tilde{x}(\tau-1), \tilde{u}(\tau)). \qquad (40.9)$$

Finally, for $t = \tau + 1, \ldots, N$ we select points $\tilde{x}(t)$, $\tilde{u}(t)$ in such a way that the following relations are satisfied:

$$\omega_{t-1}(\tilde{x}(t-1)) = \omega_t(f_t(\tilde{x}(t-1), \tilde{u}(t))) + \\ + f_t^0(\tilde{x}(t-1), \tilde{u}(t)), \qquad t = \tau + 1, \ldots, N. \qquad (40.10)$$

* Known as *Bellman's equation*; in our interpretation this "equation" is just the *definition* of the functions ω_t.

It is easy to see that this choice of points $\tilde{x}(t)$, $\tilde{u}(t)$ (for $t=\tau+1, \ldots, N$) is possible. We substitute $t=\tau+1$, $x=\tilde{x}(\tau)$ into (40.6). Since $\tilde{x}(\tau)\in\Omega_\tau$, this means that relation (40.6) is valid for this value of x. Consequently, there exists a point

$$u=\tilde{u}(\tau+1)\in U_{\tau+1}(\tilde{x}(\tau)),$$

for which in (40.6) a minimum is reached, that is, relation (40.10) is valid for $t=\tau+$ $+1$. Using this value $\tilde{u}(\tau+1)$, we get $\tilde{x}(\tau+1)$ from (40.1):

$$\tilde{x}(\tau+1)=f_{\tau+1}(\tilde{x}(\tau),\ \tilde{u}(\tau+1)).$$

The required points $\tilde{u}(\tau+1)$, $\tilde{x}(\tau+1)$ have thus been found. By substituting into (40.6) the values $t=\tau+2$, $x=\tilde{x}(\tau+1)$, we determine $\tilde{u}(\tau+2)$ similarly, and then $\tilde{x}(\tau+2)$, etc.

Hence the process $\tilde{x}(t)$, $\tilde{u}(t)$ has been constructed. In view of (40.1), we can rewrite relation (40.10) as

$$\begin{aligned}&\omega_{t-1}(f_{t-1}(\tilde{x}(t-2),\ \tilde{u}(t-1)))=\\&\quad=\omega_t(f_t(\tilde{x}(t-1),\ \tilde{u}(t)))+f_t^0(\tilde{x}(t-1),\ \tilde{u}(t)),\quad t=\tau+1,\ \ldots,\ N.\end{aligned}$$

By summing these relations over all $t=\tau+1, \ldots, N$ and striking out like terms on the left-hand and right-hand sides, we obtain (taking into account (40.8), (40.9), and the relation $\omega_N\equiv 0$):

$$\begin{aligned}\omega_\tau(f_\tau(x(\tau-1),\ v))=f_{\tau+1}^0(\tilde{x}(\tau),\ \tilde{u}(\tau+1))+\ \ldots\\ \ldots+f_N^0(\tilde{x}(N-1),\ \tilde{u}(N)).\end{aligned}\qquad(40.11)$$

Moreover, since $x(t)$, $u(t)$ is an *optimal* process, we know by definition that

$$\begin{aligned}&f_1^0(x(0),\ u(1))+\ \ldots\ +f_N^0(x(N-1),\ u(N))\leqslant\\&\qquad\leqslant f_1^0(\tilde{x}(0),\ \tilde{u}(1))+\ \ldots\ +f_N^0(\tilde{x}(N-1),\ \tilde{u}(N)).\end{aligned}$$

In view of (40.8), the first $\tau-1$ terms on the left-hand and right-hand sides cancel out, so that we have

$$\begin{aligned}&f_\tau^0(x(\tau-1),\ u(\tau))+\ \ldots\ +f_N^0(x(N-1),\ u(N))\leqslant\\&\qquad\leqslant f_\tau^0(\tilde{x}(\tau-1),\ \tilde{u}(\tau))+\ \ldots\ +f_N^0(\tilde{x}(N-1),\ \tilde{u}(N)),\end{aligned}$$

and this relation can in turn be rewritten as (see (40.9), (40.11))

$$\begin{aligned}&f_\tau^0(x(\tau-1),\ u(\tau))+\ \ldots\ +f_N^0(x(N-1),\ u(N))\leqslant\\&\qquad\leqslant f_\tau^0(x(\tau-1),\ v)+\omega_\tau(f_\tau(x(\tau-1),\ v)).\end{aligned}\qquad(40.12)$$

Recall that this relation is valid for any $\tau=1, \ldots, N$ and $v\in U_\tau(x(\tau-1))$.

If we set $\tau = N$ in (40.12), then we obtain (since $\omega_N \equiv 0$)

$$f_N^0(x(N-1),\ u(N)) \leqslant f_N^0(x(N-1),\ v), \quad v \in U_N(x(N-1)),$$

and this implies that for $t = N$ relation (40.7) is valid.

Assume that relation (40.7) has already been proven valid for $t = N, \ldots, \tau + 1$. In view of (40.6) this can be rewritten as

$$f_t^0(x(t-1),\ u(t)) + \omega_t(f_t(x(t-1),\ u(t))) = \omega_{t-1}(x(t-1)),$$

or (see (40.1))

$$f_t^0(x(t-1),\ u(t)) + \\ + \omega_t(f_t(x(t-1),\ u(t))) = \omega_{t-1}(f_{t-1}(x(t-2),\ u(t-1))).$$

By summing these relations over $t = N, \ldots, \tau + 1$ and canceling like terms on both sides, we get

$$f_{\tau+1}^0(x(\tau),\ u(\tau+1)) + \ldots + f_N^0(x(N-1),\ u(N)) = \\ = \omega_\tau(f_\tau(x(\tau-1),\ u(\tau))).$$

Therefore, relation (40.12) becomes

$$f_\tau^0(x(\tau-1),\ u(\tau)) + \omega_\tau(f_\tau(x(\tau-1),\ u(\tau))) \leqslant \\ \leqslant f_\tau^0(x(\tau-1),\ v) + \omega_\tau(f_\tau(x(\tau-1),\ v)),$$

which means that relation (40.7) is valid for $t = \tau$. It can then be shown by induction that equation (40.7) is valid for all $t = N, \ldots, 1$. The necessity of condition (40.7) is thereby proven.

Now let us prove *sufficiency*. Assume that process $x(t),\ u(t)$ satisfies conditions (40.7) and that $x^*(t),\ u^*(t)$ is any other process (satisfying conditions (40.1), (40.2)) having the same initial state $x^*(0) = x(0)$. Since $x^*(t-1) \in \Omega_{t-1}$, it follows that (40.6) is valid for $x = x^*(t-1)$:

$$\omega_{t-1}(x^*(t-1)) = \min_{u \in U_t(x^*(t-1))} (\omega_t(f_t(x^*(t-1),\ u)) + f_t^0(x^*(t-1),\ u))$$

so that from the inclusion $u^*(t) \in U_t(x^*(t-1))$ (see (40.2)) the following inequality is obtained:

$$\omega_{t-1}(x^*(t-1)) \leqslant \omega_t(f_t(x^*(t-1),\ u^*(t))) + f_t^0(x^*(t-1),\ u^*(t)).$$

According to (40.1), this inequality can be rewritten as

$$\omega_{t-1}(x^*(t-1)) \leqslant \omega_t(x^*(t)) + f_t^0(x^*(t-1),\ u^*(t)).$$

Then, by summing these relations over $t = 1, \ldots, N,$ we get

$$\omega_0(x^*(0)) \leqslant f_1^0(x^*(0), u^*(1)) + \ldots + f_N^0(x^*(N-1), u^*(N)).$$

On the other hand, by rewriting (40.7) as

$$f_t^0(x(t-1), u(t)) + \omega_t(x(t)) = \omega_{t-1}(x(t-1))$$

(see (40.1), (40.6)) and summing over $t = 1, \ldots, N,$ we obtain

$$\omega_0(x(0)) = f_1^0(x(0), u(1)) + \ldots + f_N^0(x(N-1), u(N)). \quad (40.13)$$

Finally, since $x^*(0) = x(0)$, the above relations imply that

$$f_1^0(x(0), u(1)) + \ldots + f_N^0(x(N-1), u(N)) \leqslant$$
$$\leqslant f_1^0(x^*(0), u^*(1)) + \ldots + f_N^0(x^*(N-1), u^*(N)),$$

which means that process $x(t)$, $u(t)$ is optimal.

COROLLARY 40.2. *The minimum value which functional* (40.5) *can assume for motion from initial state* $x(0)$ *according to law* (40.1), (40.2) *is* $\omega_0(x(0))$.

Actually, this minimum value is reached for the optimal process, that is, for the process satisfying condition (40.7). Thus (40.13) implies that it is equal to $\omega_0(x(0))$.

In Theorem 40.1 it was important to make use of the fact that function $\omega_{t-1}(x)$ is defined on the entire set Ω_{t-1}, $t = 1, \ldots, N,$ that is, that in relations (40.6) a minimum is indeed *attained* for any $x \in \Omega_{t-1}$. Theorem 40.3, below, specifies a simple, frequently applied condition, the satisfaction of which guarantees that these requirements are met (and thus that the method of dynamic programming described in Theorem 40.1 is applicable).

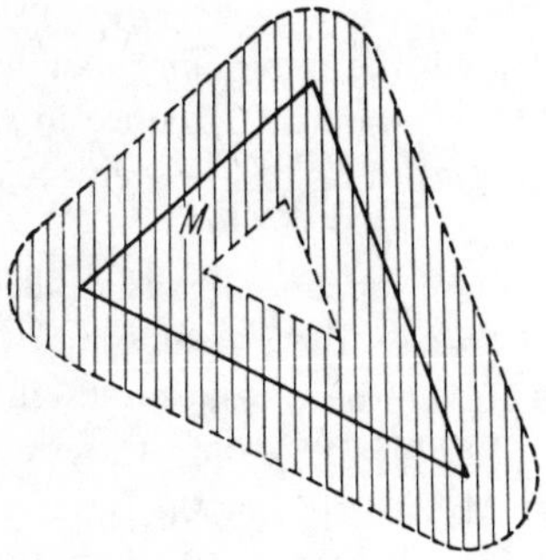

FIGURE 157

In order to formulate Theorem 40.3, we introduce some additional concepts. First of all, recall that set M (situated in the space E^r of variables $u^1, \ldots, u^r$) is called *compact* if it is both closed and bounded. Moreover, let $M \subset E^r$ be some set and let ε be a positive number; the ε *neighborhood* of set M is defined as the set consisting of all points of space E^r situated a distance less than ε from M. In other words, a point $x \in E^r$ belongs to the ε neighborhood of set M if, and only if, there exists a point $a \in M$ such that the distance between a and x is less than ε. For instance, the ε neighborhood of a *circle* (in the plane of variables u^1, u^2) is a *ring* of width 2ε; Figure 157 shows the ε neighborhood of a *triangle* (considered as a contour, that is, without interior points).

Finally, we will say that set $U(x)$ (in the space E^r of variables $u^1, \ldots, u^r$) depends *continuously* on a point $x \in E^n$ provided the following condition is satisfied: for any point $x_0 \in E^n$ and any number $\varepsilon > 0$, there exists a $\delta > 0$ such that for any two points x', x'' situated a distance less than δ from x_0 each of sets $U(x')$, $U(x'')$ will lie completely in the ε neighborhood of the other. Graphically, a continuous dependence of set $U(x)$ on point x means that set $U(x)$ varies "smoothly" as point x is moved.

THEOREM 40.3. *Assume that in phase space E^n some set G is defined and that functions $f_t(x, u)$, $f_t^0(x, u)$ are defined and continuous (over the aggregate of variables x, u) for $x \in G$ and any u. Assume in addition that for any $x \in G$ the set $U_t(x)$ is compact and continuously dependent on x. Finally, assume that for any $x \in G, u \in U(x)$ a point $f_t(x, u)$ belongs to set G. Then functions $\omega_t(x)$ are defined on the entire set G and continuous, so that for any initial state $x(0) \in G$ Theorem 40.1 is applicable.*

Proof. Function $\omega_N \equiv 0$ is obviously defined and continuous on the entire set G. Assume that function $\omega_t(x)$ (where $1 \leqslant t \leqslant N$) has been shown to be defined and continuous on set G. Then, in relation (40.6) defining the function $\omega_{t-1}(x)$, the expression following the min symbol is meaningful (since $f_t(x, u) \in G$) and it is a *continuous* function. When u runs through a *compact* set $U_t(x)$ (where $x \in G$), this function, according to a familiar theorem of analysis, does not fail to reach its least value, that is, minimum (40.6) exists for any point $x \in G$. Therefore, function $\omega_{t-1}(x)$ is defined on the entire set G. That it is continuous is easy to prove (using the fact that $U_t(x)$ depends continuously on x). This completes the induction process establishing the continuity of all functions $\omega_t(x)$ on set G.

It remains to note that for any initial state $x(0) \in G$ all the sets $\Omega_0, \ldots, \Omega_N$ are contained in G (since $f_t(x, u) \in G$ for any $x \in G$, $u \in U_t(x)$). Consequently, function $\omega_t(x)$ is defined, in particular, on the entire set Ω_t ($t = 0, 1, \ldots, N$), that is, Theorem 40.1 is applicable.

NOTE 40.4. With the conditions of Theorem 40.3 satisfied, the minimum possible value of functional (40.5) depends *continuously* on the initial state $x(0)$ (since in view of Corollary 40.2 this minimum value is $\omega_0(x(0))$, while according to Theorem 40.3 function $\omega_0(x)$ is continuous).

This concludes the description of the method of dynamic programming, as applied to the fundamental problem. For a problem with constraints on the phase coordinates (in particular, for a problem with variable endpoints), there will be certain nonessential complications. For instance, let us examine, for object (40.1), (40.2), only those processes (40.3), (40.4) which satisfy the condition $x(t) \in M_t$ for all $t = 0, 1, \ldots, N$. The states $x(t)$ entering into processes of this type fill some *subset* of set M_t. This *subset* will now be denoted by the symbol Ω_t. In relation (40.6) the minimum is selected over all $u \in U_t(x)$ satisfying the additional condition $f_t(x, u) \in \Omega_t$. Formulas (40.7) are modified similarly. Finally, if set Ω_0 contains more than one point, then the following obvious additional condition has to be added to the system of relations (40.7):

$$\omega_0(x(0)) = \min_{x \in \Omega_0} \omega_0(x)$$

(it can be assumed formally that this is relation (40.7) for $t=0$). These, then, are the modifications in the formulation of Theorem 40.1. The proof remains the same (with the obvious alterations).

EXAMPLE 40.5. Consider the controlled object

$$\begin{aligned} x^1(t) &= (1 - x^1(t-1) - x^2(t-1))\, u(t), \\ x^2(t) &= x^1(t-1) + x^2(t-1), \end{aligned} \tag{40.14}$$

where u is a scalar control parameter which assumes values in the interval $[0, 1]$. In other words, object (40.1), (40.2) is considered with a two-dimensional phase space (that is, $n=2$), where

$$f_t^1(x^1, x^2, u) = (1 - x^1 - x^2)\, u, \quad f_t^2(x^1, x^2, u) = x^1 + x^2, \quad U_t(x) = [0, 1]$$

for any $t = 1, \ldots, N$. For this object we pose the problem of the maximum of functional (40.5), where

$$f_t^0(x(t-1), u(t)) = (u(t))^2 - x^1(t-1).$$

Here the initial state x_0 is fixed, no constraints being imposed on the subsequent states; in addition, the number N is specified.

Theorem 40.1 can be applied to solve this problem (that is, to find the desired optimal process); note that the possibility of applying this theorem is guaranteed by Theorem 40.3 (where G coincides with the entire space E^2 of variables x^1, x^2).

By definition, $\omega_N \equiv 0$. Moreover, in view of (40.6) (where, because of the problem stated, a *maximum* is sought rather than a minimum), we have

$$\begin{aligned} \omega_{N-1}(x^1, x^2) &= \max_{u \in [0, 1]} (\omega_N(f_N(x^1, x^2, u)) + f_N^0(x^1, x^2, u)) = \\ &= \max_{u \in [0, 1]} ((u)^2 - x^1) = 1 - x^1, \end{aligned}$$

the maximum being reached for $u=1$ in this case. Consequently, according to (40.7), for an optimal process the relation $u(N)=1$ is valid (for any state $x^1(N-1)$, $x^2(N-1)$).

In addition,

$$\begin{aligned} \omega_{N-2}(x^1, x^2) &= \max_{u \in [0, 1]} (\omega_{N-1}(f_{N-1}(x^1, x^2, u)) + f_{N-1}^0(x^1, x^2, u)) = \\ &= \max_{u \in [0, 1]} (1 - (1 - x^1 - x^2)\, u + (u)^2 - x^1). \end{aligned}$$

Since the expression on the right-hand side of this equation (after the max symbol) is a quadratic trinomial in u with a positive coefficient for $(u)^2$, it follows that this expression, considered in the interval [0, 1], reaches a maximum at one of the *ends* of the interval. Therefore,

$$\omega_{N-2}(x^1, x^2) = \max_{u=0,1} (1 - (1 - x^1 - x^2)u + (u)^2 - x^1) = \\ = \max(1 - x^1, x^2 + 1),$$

where for $1 - x^1 \geqslant x^2 + 1$ the maximum is reached at the point $u = 0$, while for $1 - x^1 \leqslant x^2 + 1$ it is reached at the point $u = 1$. According to (40.7), this means that the following relation is valid for the optimal process:

$$u(N-1) = \begin{cases} 0 & \text{for} \quad 1 - x^1(N-2) \geqslant x^2(N-2) + 1, \\ 1 & \text{for} \quad 1 - x^1(N-2) \leqslant x^2(N-2) + 1. \end{cases}$$

Thus, for $1 - x^1(N-2) = x^2(N-2) + 1$, both of the values $u(N-1) = 0$, $u(N-1) = 1$ satisfy the optimality condition.

The function ω_{N-3} is calculated similarly:

$$\begin{aligned} \omega_{N-3}(x^1, x^2) &= \max_{u \in [0,1]} (\omega_{N-2}(f_{N-2}(x^1, x^2, u)) + f^0_{N-2}(x^1, x^2, u)) = \\ &= \max_{u \in [0,1]} ((u)^2 - x^1 + \max(1 - f^1_{N-2}(x^1, x^2, u), f^2_{N-2}(x^1, x^2, u) + 1)) = \\ &= \max_{u \in [0,1]} ((u)^2 - x^1 + \max(1 - (1 - x^1 - x^2)u, x^1 + x^2 + 1)) = \\ &= \max_{u \in [0,1]} (\max((u)^2 - x^1 + 1 - (1 - x^1 - x^2)u, (u)^2 + x^2 + 1)) = \\ &= \max_{u=0,1} (\max((u)^2 - x^1 + 1 - (1 - x^1 - x^2)u, (u)^2 + x^2 + 1)) = \\ &= \max(\max_{u=0,1}((u)^2 - x^1 + 1 - (1 - x^1 - x^2)u), \max_{u=0,1}((u)^2 + x^2 + 1)) = \\ &= \max(\max(1 - x^1, x^2 + 1), \max(x^2 + 1, x^2 + 2)) = \\ &= \max(1 - x^1, x^2 + 2). \end{aligned}$$

Here, for $1 - x^1 \geqslant x^2 + 2$ the maximum is reached at the point $u = 0$, while for $1 - x^1 \leqslant x^2 + 2$ it is reached at $u = 1$. According to (40.7), this means that the following relation is valid for the optimal process:

$$u(N-2) = \begin{cases} 0 & \text{for} \quad 1 - x^1(N-3) \geqslant x^2(N-3) + 2, \\ 1 & \text{for} \quad 1 - x^1(N-3) \leqslant x^2(N-3) + 2. \end{cases}$$

Hence it turns out that, even after calculating functions ω_{N-2}, ω_{N-3}, we are not able to find the values $u(N-1)$, $u(N-2)$ of optimal control; all we know is that they depend on $x(N-2)$, $x(N-3)$, which are as yet unknown.

Continuing in this manner, we can find the subsequent functions $\omega_t(x^1, x^2)$:

$$\omega_t(x^1, x^2) = \max(1 - x^1, x^2 + N - t - 1), \qquad t = N-2, \ldots, 1, 0,$$

where for the optimal process the following relation is valid:

$$u(t+1)=\begin{cases} 0 & \text{for} \quad 1-x^1(t) \geqslant x^2(t)+N-t-1, \\ 1 & \text{for} \quad 1-x^1(t) \leqslant x^2(t)+N-t-1. \end{cases} \tag{40.15}$$

In particular, for $t=0$ we get

$$u(1)=\begin{cases} 0 & \text{for} \quad x^1(0)+x^2(0) \leqslant -(N-2), \\ 1 & \text{for} \quad x^1(0)+x^2(0) \geqslant -(N-2). \end{cases}$$

If the point $x_0=(x^1(0),\ x^2(0))$ satisfies the condition

$$x^1(0)+x^2(0) < -(N-2),$$

then $u(1)=0$, so that from (40.14) we obtain

$$x^1(1)=0, \quad x^2(1)=x^1(0)+x^2(0).$$

However, then in (40.15) for $t=1$ the *upper* condition is valid (since $x^1(1)+x^2(1)=$ $=x^1(0)+x^2(0)<-(N-2)<-(N-3)$), that is, $u(2)=0$. Consequently, in view of (40.14), we have $x^1(2)=0$, $x^2(2)=x^2(1)$. Now, by continuing in this manner, we arrive at the optimal process:

$$\left.\begin{aligned} &u(1)=u(2)=\ \dots\ =u(N-1)=0, \quad u(N)=1; \\ &x^1(1)=x^1(2)=\ \dots\ =x^1(N-1)=0, \\ &x^1(N)=1-(x^1(0)+x^2(0)); \\ &x^2(1)=x^2(2)=\dots=x^2(N-1)=x^2(N)=x^1(0)+x^2(0). \end{aligned}\right\} \tag{40.16}$$

If, on the other hand, the point $x_0=(x^1(0),\ x^2(0))$ satisfies the condition

$$x^1(0)+x^2(0) > -(N-2),$$

then $u(1)=1$, so that from (40.14) we get

$$x^1(1)=1-x^1(0)-x^2(0), \quad x^2(1)=x^1(0)+x^2(0).$$

However, then $x^1(1)+x^2(1)=1$, that is, the *lower* condition in (40.15) is valid for $t=1$, so that $u(2)=1$. Consequently, in view of (40.14), we have $x^1(2)=0$, $=x^2(2)=1$. By continuing in this manner, we arrive at the optimal process:

$$\left.\begin{aligned} &u(1)=u(2)=\ \dots\ =u(N-1)=u(N)=1; \\ &x^1(1)=1-x^1(0)-x^2(0), \quad x^1(2)=\ \dots\ =x^1(N)=0; \\ &x^2(1)=x^1(0)+x^2(0), \quad x^2(2)=\ \dots\ =x^2(N)=1. \end{aligned}\right\} \tag{40.17}$$

Finally, for $x^1(0)+x^2(0)=-(N-2)$, both processes (40.16) and (40.17) are optimal.

41. Connection with theory of extrema of functions. In this subsection we will show how Theorem 40.1 can be derived from Theorem 39.2. First of all we note that Theorem 39.2 remains valid if $z^1, \ldots, z^n$ are assumed to be *points* of an m-dimensional space. For instance, let $z^i = (z_1^i, \ldots, z_m^i)$, $i = 1, \ldots, n$, where z_j^i are *scalar* variables, so that the space of variables $z^1, \ldots, z^n$ is actually a space of dimension mn; in spite of this, let us denote the space of variables $z^1, \ldots, z^n$ by the symbol E^n, as previously. In other words, the symbol z^i (for any $i = 1, \ldots, n$) will now designate a point (of m-dimensional space) joining the m coordinates of a point in the mn-dimensional space E^n. For such a definition of symbols $z^1, \ldots, z^n$ and E^n, Theorem 39.2 (together with the proof of this theorem presented in Subsection 39) remains valid. This fact will be made use of for the proof of Theorem 40.1.

Now let us introduce the auxiliary variables x_t^i, u_t^j, where $i = 1, \ldots, n$; $j = 1, \ldots, r$; $t = 1, \ldots, N$, and let us call z_t the point uniting those of the indicated variables which have a subscript t:

$$z_t = (x_t^1, \ldots, x_t^n, u_t^1, \ldots, u_t^r).$$

The space of variables $z_1, \ldots, z_N$ (that is, the space of all the variables x_t^i, u_t^j) will be called E^N. We also define

$$x_t = (x_t^1, \ldots, x_t^n), \quad u_t = (u_t^1, \ldots, u_t^r), \quad z_t = (x_t, u_t); \qquad t = 1, \ldots, N.$$

Next let Ω be the set of all points

$$z = (z_1, \ldots, z_N) = (x_1, u_1, x_2, u_2, \ldots, x_N, u_N)$$

of space E^N satisfying the relations

$$x_t = f_t(x_{t-1}, u_t), \qquad t = 1, \ldots, N; \tag{41.1}$$

$$u_t \in U_t(x_{t-1}), \qquad t = 1, \ldots, N, \tag{41.2}$$

where $x_0 = x(0)$, $f_t(x, u)$, $U_t(x)$ are elements entering into the definition of discrete controlled object (40.1), (40.2). On the set $\Omega \subset E^N$ we define a function $F(z)$:

$$F(z) = F(z_1, \ldots, z_N) = F(x_1, u_1, x_2, u_2, \ldots, x_N, u_N) = \sum_{t=1}^{N} f_t^0(x_{t-1}, u_t), \tag{41.3}$$

where $f_t^0(x, u)$, $t = 1, \ldots, N$, are functions entering into the definition of functional (40.5).

Now let (40.3), (40.4) be some admissible process in the discrete controlled object being considered, that is, relations (40.1), (40.2) are valid for it. We define a point $z = (x_1, u_1, x_2, u_2, \ldots, x_N, u_N)$ of space E^N, after setting

$$x_t = x(t), \quad u_t = u(t), \qquad t = 1, \ldots, N. \tag{41.4}$$

From (40.1), (40.2) it follows that point z satisfies relations (41.1), (41.2), that is, $z \in \Omega$. On the other hand, any point $z \in \Omega$ will, according to formulas (41.4), define an admissible process in the given discrete controlled object.

Thus the admissible processes are found to have a one-to-one correspondence with points of set Ω. The value of functional J for admissible process (40.3), (40.4) is here equal, in view of (40.5) and (41.3), to the value of function F at the corresponding point $z \in \Omega$. Consequently, relations (41.4) reduce the problem of optimal control under consideration to a problem of the minimum of a function $F(z)$ specified on set Ω. Theorem 39.2 can be applied to the latter problem.

We thus have a function $F(z_1, \ldots, z_N)$ (see (41.3)) specified on a set $\Omega \subset E^N$. Let us construct the sets $\Omega_{N-1}, \ldots, \Omega_1$, which were referred to in connection with Theorem 39.2 (p. 312). If E^{N-1} is the space of variables $z_1, \ldots, z_{N-1}$, then Ω_{N-1} is the set of all those points $(z_1, \ldots, z_{N-1}) \in E^{N-1}$ for each of which a point z_N can be selected such that $(z_1, \ldots, z_{N-1}, z_N) \in \Omega$. It follows from the definition of set Ω that point $(z_1, \ldots, z_{N-1})$ belongs to set Ω_{N-1} if, and only if, relations (41.1) and (41.2) are valid for $t = 1, \ldots, N-1$ (since then an arbitrary point $u_N \in U_N(x_{N-1})$ can be selected, while x_N is determined from (41.1) for $t = N$, and we thereby obtain a point z_N such that $(z_1, \ldots, z_{N-1}, z_N) \in \Omega$). Similarly, point $(z_1, \ldots, z_k)$ belongs to set Ω_k if, and only if, relations (41.1) and (41.2) are valid for $t = 1, \ldots, k$.

In addition, we formulate the functions $\omega_{N-1}, \ldots, \omega_1$, specified on sets $\Omega_{N-1}, \ldots, \Omega_1$, in accordance with formulas (39.2) and (39.3), only now these functions are called $\overset{*}{\omega}_{N-1}, \ldots, \overset{*}{\omega}_1$ (we retain the notation $\omega_{N-1}, \ldots, \omega_1$ for those functions which will be formulated in the following and which will satisfy the conditions of Theorem 40.1). According to (39.2),

$$\overset{*}{\omega}_{N-1}(z_1, \ldots, z_{N-1}) = \min_{z_N} F(z_1, \ldots, z_{N-1}, z_N) =$$

$$= \min_{z_N} \sum_{t=1}^{N} f_t^0(x_{t-1}, u_t) = \sum_{t=1}^{N-1} f_t^0(x_{t-1}, u_t) + \min_{z_N} f_N^0(x_{N-1}, u_N).$$

Here the point $(z_1, \ldots, z_{N-1}) \in \Omega_{N-1}$ is assumed to be fixed, and the minimum is taken over all z_N for which $(z_1, \ldots, z_{N-1}, z_N) \in \Omega$. However, if $(z_1, \ldots, z_{N-1}) \in \Omega_{N-1}$, that is, if relations (41.1), (41.2) are valid for $t = 1, \ldots, N-1$, then in order for point $(z_1, \ldots, z_{N-1}, z_N)$ to belong to set Ω it is necessary and sufficient that the point $z_N = (x_N, u_N)$ satisfy relations (41.1) and (41.2) for $t = N$, that is, that $u_N \in U_N(x_{N-1})$, point x_N being determined from (41.1) for $t = N$. Now, noting that point x_N does not enter into the expression $f_N^0(x_{N-1}, u_N)$, we conclude that the minimum is taken over all $u_N \in U_N(x_{N-1})$. Thus,

$$\overset{*}{\omega}_{N-1}(z_1, \ldots, z_{N-1}) = \sum_{t=1}^{N-1} f_t^0(x_{t-1}, u_t) + \min_{u_N \in U_N(x_{N-1})} f_N^0(x_{N-1}, u_N).$$

Next, taking into account that $\omega_N(x) \equiv 0$ (see p. 318), we get

$$\min_{u_N \in U_N(x_{N-1})} f_N^0(x_{N-1}, u_N) = \min_{u \in U_N(x_{N-1})} [\omega_N(f_N(x_{N-1}, u)) + f_N^0(x_{N-1}, u)].$$

According to (40.6), the right-hand side of this equation is equal to $\omega_{N-1}(x_{N-1})$, so that

$$\overset{*}{\omega}_{N-1}(z_1, \ldots, z_{N-1}) = \sum_{t=1}^{N-1} f_t^0(x_{t-1}, u_t) + \omega_{N-1}(x_{N-1}),$$

where ω_{N-1} is the function introduced prior to the formulation of Theorem 40.1.

Thus function ω^*_{N-1} has been calculated. Assume that for some $k = 2, \ldots, N-1$ the following formula has already been established:

$$\omega^*_k(z_1, \ldots, z_k) = \sum_{t=1}^{k} f^0_t(x_{t-1}, u_t) + \omega_k(x_k). \tag{41.5}$$

Let us show that this formula also remains valid when k is replaced by $k-1$. In view of relations (39.3), (41.5), (41.1), and (40.6), we have (omitting the details, which are analogous to those in the preceding discussion):

$$\omega^*_{k-1}(z_1, \ldots, z_{k-1}) = \min_{z_k} \omega^*_k(z_1, \ldots, z_{k-1}, z_k) =$$
$$= \min_{z_k=(x_k, u_k)} \left[\sum_{t=1}^{k} f^0_t(x_{t-1}, u_t) + \omega_k(x_k)\right] =$$
$$= \sum_{t=1}^{k-1} f^0_t(x_{t-1}, u_t) + \min_{(x_k, u_k)} (f^0_k(x_{k-1}, u_k) + \omega_k(x_k)) =$$
$$= \sum_{t=1}^{k-1} f^0_t(x_{t-1}, u_t) + \min_{(x_k, u_k)} [f^0_k(x_{k-1}, u_k) + \omega_k(f_k(x_{k-1}, u_k))] =$$
$$= \sum_{t=1}^{k-1} f^0_t(x_{t-1}, u_t) + \min_{u \in U_k(x_{k-1})} [f^0_k(x_{k-1}, u) + \omega_k(f_k(x_{k-1}, u))] =$$
$$= \sum_{t=1}^{k-1} f^0_t(x_{t-1}, u_t) + \omega_{k-1}(x_{k-1}).$$

Consequently, it has been shown by induction that formula (41.5) is valid for all $k = 1, \ldots, N-1$. In particular,

$$\omega^*_1(z_1) = f^0_1(x_0, u_1) + \omega_1(x_1) = f^0_1(x_0, u_1) + \omega_1(f_1(x_0, u_1)),$$

so that according to (39.5) we have

$$m = \min_{z_1} \omega^*_1(z_1) = \min_{u_1 \in U_1(x_0)} [f^0_1(x_0, u_1) + \omega_1(f_1(x_0, u_1))]. \tag{41.6}$$

Next, by applying Theorem 39.2 and replacing in relations (39.6) functions F and ω^*_k and the number m by their expressions (41.3), (41.5), and (41.6), we arrive at the following statement. A point $z = (z_1, \ldots, z_N) \in \Omega$ is a minimum point of function $F(z_1, \ldots, z_N)$ if, and only if, the following relations are valid:

$$\sum_{t=1}^{N} f^0_t(x_{t-1}, u_t) = \sum_{t=1}^{N-1} f^0_t(x_{t-1}, u_t) + \omega_{N-1}(x_{N-1}) =$$
$$= \sum_{t=1}^{N-2} f^0_t(x_{t-1}, u_t) + \omega_{N-2}(x_{N-2}) = \ldots = f^0_1(x_0, u_1) + \omega_1(x_1) =$$
$$= \min_{u_1 \in U_1(x_0)} [f^0_1(x_0, u_1) + \omega_1(f_1(x_0, u_1))].$$

However, these relations (taking into account that $\omega_N \equiv 0$) are equivalent to the following:

$$f_N^0(x_{N-1}, u_N) + \omega_N(x_N) = \omega_{N-1}(x_{N-1}),$$
$$f_{N-1}^0(x_{N-2}, u_{N-1}) + \omega_{N-1}(x_{N-1}) = \omega_{N-2}(x_{N-2}),$$
$$\cdots\cdots\cdots\cdots\cdots\cdots$$
$$f_2^0(x_1, u_2) + \omega_2(x_2) = \omega_1(x_1),$$
$$f_1^0(x_0, u_1) + \omega_1(x_1) = \min_{u_1 \in U_1(x_0)} [f_1^0(x_0, u_1) + \omega_1(f_1(x_0, u_1))].$$

Now, if the right-hand sides of all these relations (except the last) are rewritten with the aid of relations (40.6), we get

$$f_t^0(x_{t-1}, u_t) + \omega_t(x_t) = \min_{u \in U_t(x_{t-1})} [\omega_t(f_t(x_{t-1}, u)) + f_t^0(x_{t-1}, u)],$$
$$t = 1, \ldots, N,$$

which, in view of (41.1) and (41.4), is identical to relations (40.7). This completes the proof of Theorem 40.1 (note that the validity of Corollary 40.2 also follows from (41.6), in view of the first statement of Theorem 39.2).

The foregoing proof illustrates well the intimate relationship between problems of the minimum of a function and problems of optimal control of discrete objects. *Formally* Theorem 40.1, as we have verified, is derived from Theorem 39.2, so that in this respect it contains nothing new, except the form of its notation. On the other hand, the actual derivation of Theorem 40.1 from Theorem 39.2 is quite cumbersome, and it does not constitute a direct proof of the former theorem. Theorem 39.2 requires specialization (if Theorem 40.1 is to be obtained from it), $z^1, \ldots, z^n$ being considered as *points,* that is, the variables of function F being joined into groups: $z_1 = (x_1^i, u_1^j), \ldots, z_N = (x_N^i, u_N^j)$.

Consequently, to solve problems of optimal control of discrete objects, it is advisable *each time* to reduce them to problems of the minimum of a function (that is, to repeat the steps carried out in a general form on pages 324–328), it being advisable to have a separate theorem suited particularly to the case of a discrete controlled object, that is, Theorem 40.1. This applies to the following theorems as well.

§ 12. Necessary Conditions for Optimality

The methods for solving discrete problems of optimal control outlined in this section are a direct application of the methods of Subsection 36; it should be noted, by the way, that the results obtained could be verified independently, without being derived from the theorems of Subsection 36.

The theorems presented here do not possess the same latitude as the method of dynamic programming, since they place certain constraints on the discrete controlled object, thereby *narrowing* somewhat the class of objects considered (in particular, functions $f_t^i(x, u)$, see p. 317, are here assumed to be smooth). However, this class of discrete objects is nevertheless still very wide, and the results obtained are more profound (that is, they contain more information than Theorem 40.1). The sufficient conditions for optimality will be presented in §13.

42. Fundamental problem. First we consider the *fundamental* problem (p. 11). In other words, we examine a discrete controlled object (40.1) in which controls $u(t)$ are subject to the constraints (cf. (40.2))

$$u(t) \in U_t, \qquad t = 1, \dots, N, \tag{42.1}$$

where $U_1, \dots, U_N$ are certain sets. The initial state $x(0) = x_0$ is specified, and no constraints are imposed on states $x(t)$, for $t = 1, \dots, N$. For object (40.1) we wish to find a control (40.3) satisfying conditions (42.1) and imparting to functional (40.5) the least possible value (where (40.4) is a trajectory originating at state x_0 and corresponding to the control selected). All functions $f_t^i(x, u)$ $(i=0, 1, \dots, n;$ $t=1, \dots, N)$ are assumed to be continuously differentiable with respect to x and u (that is, with respect to $x^1, \dots, x^n, u^1, \dots, u^r$).

Auxiliary variables ψ_0 and $\psi_i(t)$ will be used in all the optimality conditions to be formulated below (cf. the formulation of the discrete maximum principle on p. 54), and sometimes other auxiliary variables as well. With the aid of these variables it will be possible to determine the auxiliary function $H_t(x, u)$, which will also enter into the formulations of the optimality conditions. For example, in the first of the theorems to be considered below, function $H_t(x, u)$ has the form

$$H_t(x, u) = \psi_0 f_t^0(x, u) + \sum_{i=1}^{n} \psi_i(t) f_t^i(x, u). \tag{42.2}$$

Note that on the left-hand side of (42.2) it is indicated explicitly that function H_t depends just on x and u (although in the definition of this function the variables ψ_i enter in as well). This is explained by the fact that it is precisely with respect to x and u that this function is differentiated (for fixed ψ_0 and $\psi_i(t)$). In particular, the following vectors are defined:

$$\operatorname{grad}_x H_t(x, u) = \left\{ \frac{\partial H_t(x, u)}{\partial x^1}, \dots, \frac{\partial H_t(x, u)}{\partial x^n} \right\},$$

$$\operatorname{grad}_u H_t(x, u) = \left\{ \frac{\partial H_t(x, u)}{\partial u^1}, \dots, \frac{\partial H_t(x, u)}{\partial u^r} \right\}.$$

THEOREM 42.1. *Let* (40.3) *be some control, and let* (40.4) *be the corresponding trajectory (with initial state* $x(0)$) *in discrete controlled object* (40.1), (42.1)*). In addition, let* L_t *be a covering of set* U_t *at point* $u(t)$, $t = 1, \dots, N$. *For optimality of this process (in the sense of minimizing functional* (40.5) *for a fixed initial state* $x(0)$ *and without any constraints on states* $x(t)$, $t = 1, \dots, N$) *, there must exist a number* ψ_0 *and vectors*

$$\psi(t) = \{\psi_1(t), \dots, \psi_n(t)\}, \qquad t = 1, \dots, N,$$

such that the following conditions are satisfied (function $H_t(x, u)$, defined by equation (42.2), enters into these conditions):

(A) $\psi_0 < 0$;

(B) $\psi(N)=0$; $\psi(t)=\operatorname{grad}_x H_{t+1}(x(t), u(t+1))$, $t=1, \ldots, N-1$;

(C) *for any* $t=1, \ldots, N$ *and any vector* δu *whose direction lies in cone* L_t, *the following inequality is valid:*

$$\delta u \operatorname{grad}_u H_t(x(t-1), u(t)) \leqslant 0.$$

Proof. We designate as E^* the space of variables x_t^i, $u_t^j; i=1, \ldots, n; j=1, \ldots, r; t=1, \ldots, N$ (making the dimension of this space equal to $Nn+Nr$). Into the functions f_t^i, $i=0, 1, \ldots, n; t=1, \ldots, N$, characterizing the discrete controlled object considered, we substitute x_{t-1}^i, u_t^j as the independent variables:

$$f_t^i(x_{t-1}, u_t)=f_t^i(x_{t-1}^1, \ldots, x_{t-1}^n, u_t^1, \ldots, u_t^r)$$

(where for $t=1$ we understand $x_0^1, \ldots, x_0^n$ to be the coordinates of the initial state $x(0)=x_0 \in E^n$). Now these functions can be assumed to be defined (and continuously differentiable) throughout the entire space E^*. Using them, we define a new system of functions in space E^*:

$$F^0(z)=F^0(x, u)=f_1^0(x_0, u_1)+f_2^0(x_1, u_2)+\ldots+f_N^0(x_{N-1}, u_N), \tag{42.3}$$

$$F_t^i(z)=F_t^i(x, u)=-x_t^i+f_t^i(x_{t-1}, u_t);\ i=1, \ldots, n;\ t=1, \ldots, N. \tag{42.4}$$

In addition, if Ω^* is the set of all points $z \in E^*$ satisfying the system of equations

$$F_t^i(z)=0; \qquad i=1, \ldots, n; \quad t=1, \ldots, N, \tag{42.5}$$

then this set Ω^* is directly connected with the discrete controlled object being considered. Specifically, let (40.3), (40.4) be some process (with an initial state $x(0)=x_0$) in controlled object (40.1). We set

$$x_t^i=x^i(t),\ u_\tau^j=u^j(\tau);\quad i=1, \ldots, n;\ j=1, \ldots, r;\ t, \tau=1, \ldots, N. \tag{42.6}$$

Then it follows from (40.1) that the point z with coordinates x_t^i, u_τ^j defined by formulas (42.6) satisfies relations (42.5) (see (42.4)), that is, it belongs to set Ω^*. Here the value of functional (40.5) for the process being considered is equal to the value of function $F^0(z)$ at the given point z (see (42.3)). Conversely, if point $z=(x_t^i, u_\tau^j)$ belongs to set Ω^*, then by determining $x(t)$, $u(t)$ with the aid of formulas (42.6) we obtain a process (40.3), (40.4) with an initial state x_0 satisfying relations (40.1).

Finally, let us designate as Ξ^* the set of all points $z=(x_t^i,\ u_\tau^j)$ of space E^* satisfying for all $\tau=1, \ldots, N$ the relations

$$u_\tau=(u_\tau^1, \ldots, u_\tau^r)\in U_\tau$$

(no restrictions being imposed on variables x_t^i). Then it is clear that, as a result of substitution (42.6), system of inclusions (42.1) is replaced by a single inclusion: $z\in\Xi^*$. Thus, instead of considering different processes in controlled object (40.1), (42.1), and the values of of functional (40.5), we can consider different points of the set $\Omega^*\cap\Xi^*$ and the values of function $F^0(z)$ at these points. In other words, *substitution* (42.6) *reduces the posed problem of optimal control to a problem of finding the point z at which function $F^0(z)$, considered on $\Omega^*\cap\Xi^*$, reaches its least value.*

Let (40.3), (40.4) be the desired optimal process. If z_0 is the corresponding point of space E^*, the coordinates of which are given by formulas (42.6), then $z_0\in\Omega^*\cap\Xi^*$ and at this point function $F^0(z)$, considered only on set $\Omega^*\cap\Xi^*$, reaches its least value. In addition, we designate as L^* the subset of space E^* consisting of all the points $z=(x_t^i,\ u_\tau^j)$ which for all $\tau=1, \ldots, N$ satisfy the relations

$$u_\tau=(u_\tau^1, \ldots, u_\tau^r)\in L_\tau.$$

It is easy to see that L^* *is a convex cone of space* E^* *with a vertex at point* z_0 *and that this cone is a covering of set* Ξ^* *at* z_0*; moreover, the direction of vector* $\boldsymbol{\delta z}=\{\delta x_t^i,\ \delta u_\tau^j\}$ *lies in cone* L^* *if, and only if, for each* $\tau=1, \ldots, N$ *the direction of vector* $\boldsymbol{\delta u}_\tau=\{\delta u_\tau^1, \ldots, \delta u_\tau^r\}$ *lies in cone* L_τ. According to these statements, in the proof the only requirement is that L^* be a covering of set Ξ^* at point z_0. Since L_τ is a covering of set U_τ at point $u_\tau=u(\tau)$ (see (42.6)), there must exist a mapping ψ_τ, defined close to the vertex of cone L_τ and assuming values in the space of variables $u_\tau^1, \ldots, u_\tau^r$, such that conditions 1 and 2 of Definition 34.1 are satisfied. Now, by setting

$$\psi(z)=\psi(x_t,\ u_\tau)=(x_t,\ \psi_\tau(u_\tau)), \qquad t=1, \ldots, N; \quad \tau=1, \ldots, N,$$

we obtain a mapping ψ defined in the vicinity of the vertex of cone L^* and assuming values in space E^*. It is easy to show that this mapping ψ satisfies conditions 1 and 2 of Definition 34.1, so that L^* is a covering of set Ξ^* at point z_0.

According to Theorem 36.9,† there must exist numbers

$$\psi_0\leqslant 0,\quad \psi_i^t, \qquad i=1, \ldots, n;\ t=1, \ldots, N,$$

such that at least one of these numbers is different from zero and satisfies condition (β), of Theorem 36.9. Let us introduce the notation

$$\psi_i(t)=\psi_i^t, \qquad i=1, \ldots, n; \quad t=1, \ldots, N,$$

† Note that here there is *only a single* cone L^*, so that the condition for nonseparability of cones $N_1, \ldots, N_m$ in Theorem 36.9 becomes unnecessary: cf. Theorem 36.11.

and let us show that condition (β) implies the conditions (B) and (C) specified during the formulation of Theorem 42.1.

Actually, in this case relation (β) becomes

$$\sum_{t=1}^{N}\left(\psi_0 \operatorname{grad}_{x_t} F^0(z_0) + \sum_{\alpha=1}^{n}\sum_{\gamma=1}^{N} \psi_\alpha^\gamma \operatorname{grad}_{x_t} F_\gamma^\alpha(z_0)\right)\boldsymbol{\delta x}_t +$$
$$+\sum_{\tau=1}^{N}\left(\psi_0 \operatorname{grad}_{u_\tau} F^0(z_0) + \sum_{\alpha=1}^{n}\sum_{\gamma=1}^{N} \psi_\alpha^\gamma \operatorname{grad}_{u_\tau} F_\gamma^\alpha(z_0)\right)\boldsymbol{\delta u}_\tau \leqslant 0 \qquad (42.7)$$

for any vector $\{\delta x_t^i,\ \delta u_\tau^j\} = \{\boldsymbol{\delta x}_t,\ \boldsymbol{\delta u}_\tau\}$ whose direction lies in cone L^*.

Relation (42.7) will be valid, in particular, if all the $\boldsymbol{\delta u}_\tau$ are assumed to be zero, while vectors $\boldsymbol{\delta x}_t$ are given *arbitrary* values (since no restrictions were imposed on variables x_t^i during the definition of cone L^*). Thus the multiplier of $\boldsymbol{\delta x}_t$ in (42.7) must go to zero:

$$\psi_0 \operatorname{grad}_{x_t} F^0(z_0) + \sum_{\alpha=0}^{n}\sum_{\gamma=1}^{N} \psi_\alpha^\gamma \operatorname{grad}_{x_t} F_\gamma^\alpha(z_0) = 0, \qquad t = 1, \ldots, N.$$

Taking into account the form of functions F^0 and F_t^i (see (42.3), (42.4)), we now get from this the condition (B) specified during the formulation of Theorem 42.1.

Next we assume that all the $\boldsymbol{\delta x}_t$ are zero; in addition, we select some whole number $\theta\,(1 \leqslant \theta \leqslant N)$ and we assume that all the $\boldsymbol{\delta u}_\tau$ for $\tau \neq \theta$ are also zero, while as $\boldsymbol{\delta u}_\theta$ we will take all possible vectors whose directions are in cone L_θ. Since the vector $\{\boldsymbol{\delta x}_t,\ \boldsymbol{\delta u}_\tau\}$ obtained has a direction lying in cone L^*, relation (42.7) being valid for it, this means that

$$\left(\psi_0 \operatorname{grad}_{u_\theta} F^0(z_0) + \sum_{\alpha=1}^{n}\sum_{\gamma=1}^{N} \psi_\alpha^\gamma \operatorname{grad}_{u_\theta} F_\gamma^\alpha(z_0)\right)\boldsymbol{\delta u}_\theta \leqslant 0$$

(for any vector $\boldsymbol{\delta u}_\theta$ whose direction is in cone L_θ). Taking the form of functions F^0 and F_t^i into account (see (42.3) and (42.4)), we now get from this the condition (C) specified during the formulation of Theorem 42.1.

It should also be noted that $\psi_0 \neq 0$ (and thus that $\psi_0 < 0$), since for $\psi_0 = 0$ the condition (B) already proven above would imply that $\psi_i(t) = \psi_i^t = 0$ for all i, t, which contradicts the choice of numbers $\psi_0,\ \psi_i^t$.

NOTE 42.2. If the problem of the maximum (rather than the minimum) of functional (40.5) is considered, then the theorem has the same form, the only difference being that in the relation $\psi_0 < 0$ the inequality sign is reversed. This remark pertains to all the subsequent theorems as well. Note that in Theorem 42.1 condition (A) may be replaced by the equation $\psi_0 = -1$ (or, in the case of a maximum, by the equation $\psi_0 = 1$), since the quantities $\psi_0,\ \psi_i(t)$ entering into formula (42.2) are *homogeneous*, that is, multiplying them all by the same factor $k > 0$ does not alter

Theorem 42.1. However, the inequality $\psi_0 \leqslant 0$ will enter into subsequent theorems, the case $\psi_0 = 0$ not being excluded.

NOTE 42.3. Condition (C) of Theorem 42.1 may (with conditions (A) and (B) remaining unchanged) take the following form:

(C′) *for each* $t = 1, \ldots, N$ *the direction of the vector*

$$\operatorname{grad}_u H_t(x(t-1), u(t))$$

lies in the dual cone $D(L_t)$.

The equivalence of conditions (C) and (C′) follows directly from the definition of a dual cone. Note also that in this form (that is, with condition (C′) instead of condition (C)) Theorem 42.1 can be (by the same means) derived from Theorem 36.2, in which we have to set $q = 0$ and $l = N$, while as $\Omega_1, \ldots, \Omega_N$ we take the sets defined, respectively, in E^* by the relations $u_t \in U_t$, $t = 1, \ldots, N$.

NOTE 42.4. According to Theorem 34.2, the condition imposed upon cones L_t during the formulation of Theorem 42.1 (that is, the requirement that L_t be a covering of set U_t at point $u(t)$, $t = 1, \ldots, N$) will be satisfied, provided that each of sets U_t is convex, while the supporting cone of convex set U_t at point $u(t)$, $t = 1, \ldots, N$, is taken as L_t. A similar observation can be made for the subsequent theorems of this subsection as well.

Note also that, if $H_{N+1}(x, u) \equiv 0$, then condition (B) of Theorem 42.1 is written in the form

$$\psi(t) = \operatorname{grad}_x H_{t+1}(x(t), u(t+1)), \qquad t = 1, \ldots, N.$$

Analogous stipulations during the formulation of condition (B) will also be assumed in the following.

Theorem 42.1 and its proof constitute an example of the application of the theorems of Subsection 36 to obtain the necessary conditions of optimality for discrete controlled objects. More general theorems will be formulated below (for the fundamental problem), which can thus be obtained from the results of Subsection 36. However, for this we need the following auxiliary proposition.

THEOREM 42.5. *Let vector space* R *decompose into the direct sum of its subspaces* $D_1, D_2, \ldots, D_s$. *If* $\overset{*}{D}_i$ $(i = 1, \ldots, s)$ *is the subspace of space* R *generated by the union of all the subspaces* $D_1, \ldots, D_s$ *except* D_i, *then*

$$R = D_i \oplus \overset{*}{D}_i \text{ and } D_i \subset \overset{*}{D}_j \quad \text{for} \quad i \neq j \qquad (i, j = 1, \ldots, s).$$

In addition, for each $i = 1, \ldots, s$ *let us define in the space a system of closed convex cones*

$$C_1^{(i)}, \ldots, C_{r_i}^{(i)} \tag{42.8}$$

with a common vertex at the zero element $\mathbf{0} \in R$, *this system either consisting just of a single cone (that is,* $r_i = 1$) *or else not possessing separability in* D_i. *If* $L_j^{(i)}$

(where $i = 1, \ldots, s$; $j = 1, \ldots, r_i$) is the convex hull of set $C_j^{(i)} \cup D_i^*$, then

$$L_j^{(i)}, \quad i = 1, \ldots, s; \quad j = 1, \ldots, r_i, \tag{42.9}$$

is a system of convex cones with a common vertex $\mathbf{0}$ *not possessing separability in* R.

Proof. Let us assume the converse, namely that system of cones (42.9) possesses separability in R, that is, one of cones (42.9) is separable from the intersection of the others. Without any loss of generality, it can be assumed (after changing, if necessary the numbering of subspaces D_i and cones (42.8)) that cone $L_1^{(1)}$ is separable in R from the intersection of the rest of cones (42.9). Consequently, there exists in R a hyperplane Γ passing through $\mathbf{0}$, such that $L_1^{(1)} \subset P_1$, where P_1 is one of the closed half-spaces defined by hyperplane Γ, while the intersection $\Pi_1^{(1)}$ of all the rest of cones (42.9) is contained in the other half-space P_2 defined by hyperplane Γ.

First we assume that $r_1 > 1$ and that plane D_1 does not lie wholly in hyperplane Γ. Then $\Gamma \cap D_1$ is a hyperplane of space D_1, while $P_1 \cap D_1$ and $P_2 \cap D_1$ are the closed half-spaces into which this hyperplane divides space D_1. Since

$$C_1^{(1)} \subset L_1^{(1)} \cap D_1 \subset P_1 \cap D_1,$$

$$\bigcap_{j=2}^{r_1} C_j^{(1)} \subset \left(\bigcap_{j=2}^{r_1} L_j^{(1)} \right) \cap D_1 \subset \Pi_1^{(1)} \cap D_1 \subset P_2 \cap D_1$$

(because $L_j^{(i)} \supset D_1$ for $i \neq 1$), this means that in space D_1 hyperplane $\Gamma \cap D_1$ *separates* cone $C_1^{(1)}$ from the intersection $C_2^{(1)} \cap \ldots \cap C_{r_1}^{(1)}$, that is, system of cones (42.8) possesses separability for $i = 1$, which, however, contradicts the assumption. This contradiction shows that (for $r_1 > 1$) hyperplane Γ must contain plane D_1.

If, on the other hand, $r_1 = 1$, then the intersection $\Pi_1^{(1)}$ of all of the cones (42.9) except $L_1^{(1)}$ contains plane D_1 (since for $i \neq 1$ we have $L_j^{(i)} \supset D_i^* \supset D_1$), so that $D_1 \subset \Pi_1^{(1)} \subset P_2$. Hence it follows that $D_1 \subset \Gamma$ in this case as well.

Thus we know that $D_1 \subset \Gamma$. Moreover, for any $i = 2, \ldots, s$ we have $D_i \subset D_1^* \subset L_1^{(1)} \subset P_1$, so that $D_i \subset \Gamma$. Consequently, all the planes $D_1, \ldots, D_s$ are contained in Γ, that is $R \subset \Gamma$, which is impossible. Accordingly, system of cones (42.9) does not possess separability in R.

THEOREM 42.6. *Let* (40.3) *be some control, and let* (40.4) *be the corresponding trajectory (with initial state* $x(0)$*) in discrete controlled object* (40.1), (42.1). *In addition, for each* $t = 1, \ldots, N$ *let set* U_t *have the form*

$$U_t = \Omega_t^{(1)} \cap \ldots \cap \Omega_t^{(l_t)} \cap \Omega_t' \cap \Omega_t'' \cap \Xi_t^{(1)} \cap \ldots \cap \Xi_t^{(m_t)},$$

where set Ω_t' *is defined in the space of variables* $u^1, \ldots, u^r$ *by the system of equations*

$$G_t^j(u^1, \ldots, u^r) = 0, \qquad j = 1, \ldots, k_t, \tag{42.10}$$

set Ω_t'' *is defined by the system of inequalities*

$$g_t^i(u^1, \ldots, u^r) \leqslant 0, \qquad i = 1, \ldots, q_t, \tag{42.11}$$

and $\Omega_t^{(1)}, \ldots, \Omega_t^{(l_t)}, \Xi_t^{(1)}, \ldots, \Xi_t^{(m_t)}$ *are sets of the space of variables* $u^1, \ldots, u^r$. *Finally, let* $K_t^{(i)}$ *be a covering of set* $\Omega_t^{(i)}$ *at point* $u(t)$, *and let* $L_t^{(j)}$ *be a covering of set* $\Xi_t^{(j)}$ *at* $u(t)$. *Assume that in the vicinity of point* $u(t)$ *all the functions* G_t^j, $j = 1, \ldots, k_t$; g_t^i, $i \in I_t(u(t))$, *are smooth and also that for each* $t = 1, \ldots, N$ *the system of convex cones* $L_t^{(h)}$ $(h = 1, \ldots, m_t)$ *does not possess separability in the space of variables* $u^1, \ldots, u^r$.

For optimality of process (40.3), (40.4) *(in the sense of the minimum of functional* (40.5) *for a fixed initial state* $x(0)$ *and with no restrictions placed on the states* $x(t)$, $t = 1, \ldots, N$), *there must exist a number* $\psi_0 \leqslant 0$ *and vectors*

$$\begin{aligned} &\psi(t) = \{\psi_1(t), \ldots, \psi_n(t)\}, \qquad t = 1, \ldots, N;\\ &\varphi(t) = \{\varphi_1(t), \ldots, \varphi_{k_t}(t)\}, \qquad t = 1, \ldots, N;\\ &\lambda(t) = \{\lambda_1(t), \ldots, \lambda_{q_t}(t)\}, \qquad t = 1, \ldots, N;\\ &a_t^{(j)}, \qquad j = 1, \ldots, l_t, \quad t = 1, \ldots, N, \end{aligned}$$

such that the direction of vector $a_t^{(i)}$ *lies in the dual cone* $D(K_t^{(j)})$ *and conditions* (A) *through* (E), *below, are satisfied. These conditions are formulated using the function* $H_t(x, u)$, *defined as*

$$H_t(x, u) = \psi_0 f_t^0(x, u) + \sum_{i=1}^{n} \psi_i(t) f_t^i(x, u) + \sum_{i=1}^{k_t} \varphi_i(t) G_t^i(u) + \\ + \sum_{i=1}^{q_t} \lambda_i(t) g_t^i(u):$$

(A) *if* $\psi_0 = 0$, *then for at least one* t *at least one of the vectors* $\psi(t)$, $\varphi(t)$, $\lambda(t)$, $a_t^{(j)}$ *is different from zero;*

(B) $\psi(t) = \mathrm{grad}_x H_{t+1}(x(t), u(t+1))$, $t = 1, \ldots, N$, *where it is assumed that* $H_{N+1} \equiv 0$;

(C) *for any* $t = 1, \ldots, N$ *and any vector* δu *whose direction lies in the intersection of cones* $L_t^{(j)}$ $(j = 1, \ldots, m_t)$ *the following inequality is valid:*

$$\left(\mathrm{grad}_u H_t(x(t-1), u(t)) - \sum_{i=1}^{l_t} a_t^{(i)}\right) \delta u \leqslant 0;$$

(D) $\lambda_j(t) \leqslant 0$ *for all* $j = 1, \ldots, q_t$; $t = 1, \ldots, N$;

(E) $\lambda_j(t)\, g_t^j(u(t)) = 0$ *for all* $j = 1, \ldots, q_t$; $t = 1, \ldots, N$.

Proof. Just as in the proof of Theorem 42.1, we consider the space E^* of variables x_t^i, u_τ^j and functions (42.3), (42.4), specified in this space. Sets Ω^* and Ξ^* are the same as in the proof of Theorem 42.1. Therefore, if (40.3), (40.4) is the optimal process sought, while z_0 is the corresponding point of space E^* (the coordinates of which are given by formulas (42.6)), this means that function $F^0(z)$, considered on set $\Omega^* \cap \Xi^*$, reaches its least value at point z_0

Set $\Omega^* \cap \Xi^*$ can be written as

$$\Omega^* \cap \Xi^* = \Omega^0 \cap \Omega^{00} \cap \Big(\bigcap_{i,\,t} \tilde{\Omega}_t^{(i)}\Big) \cap \Big(\bigcap_{j,\,t} \Xi_t^{(j)}\Big),$$

where Ω^0 is defined by system of equations (42.5) and the equations (see (42.10))

$$G_t^j(u_t) = 0, \qquad j = 1, \ldots, k_t; \quad t = 1, \ldots, N; \tag{42.12}$$

here Ω^{00} is defined by the system of inequalities (see (42.11))

$$g_t^i(u_t) \leqslant 0, \qquad i = 1, \ldots, q_t; \quad t = 1, \ldots, N;$$

finally, sets $\tilde{\Omega}_t^{(i)}$ and $\tilde{\Xi}_t^{(j)}$ $(i = 1, \ldots, l_t;\ j = 1, \ldots, m_t;\ = 1, \ldots, N)$ are defined, respectively, by the inclusions

$$u_t \in \Omega_t^{(i)}, \qquad u_t \in \Xi_t^{(j)}.$$

If we designate as $\tilde{K}_t^{(i)}$, $\tilde{L}_t^{(j)}$ the sets defined in E^* by, respectively, the inclusions

$$u_t \in K_t^{(i)}, \quad u_t \in L_t^{(j)},$$

then $\tilde{K}_t^{(i)}$ will be a covering of set $\tilde{\Omega}_t^{(i)}$ at point z_0 $(i = 1, \ldots, l_t;\ t = 1, \ldots, N)$, while $\tilde{L}_t^{(j)}$ will be a covering of set $\tilde{\Xi}_t^{(j)}$ at z_0 $(j = 1, \ldots, m_t;\ t = 1, \ldots, N)$.

It is easy to see that the system of convex cones

$$\tilde{L}_t^{(j)}, \qquad j = 1, \ldots, m_t; \quad t = 1, \ldots, N, \tag{42.13}$$

does not possess separability in E^*. Actually, let us suppose that z_0 is the zero point of space E^* (it being thus transformed into a vector space) and let us designate as $D_0, D_1, \ldots, D_N$ the subspaces of space E^* defined by the conditions

$$D_0\colon\ u_t = u(t) \qquad \text{for all} \quad t = 1, \ldots, N;$$

D_τ: $x_t = x(t)$ for all $t = 1, \ldots, N$, and $u_t = u(t)$ for all $t \neq \tau$. Then E^* decomposes into the direct sum of subspaces $D_0, D_1, \ldots, D_N$. Moreover, we designate as $D_0^*, D_1^*, \ldots, D_N^*$ the subspaces such that

$$D_i \oplus D_i^* = E^*, i = 0, 1, \ldots, N, \text{ and } D_i \subset D_j^* \quad \text{for} \quad i \neq j,$$

$$i, j = 0, 1, \ldots, N.$$

The relation $u_t \in L_t^{(j)}$ defines in subspace D_t some convex cone $\overline{L}_t^{(j)}$ with a vertex z_0 $(j = 1, \ldots, m_t)$, $L_t^{(j)}$ being the convex hull of set $\overline{L}_t^{(j)} \cup D_t^*$ $(j = 1, \ldots, m_t;$

$t=1, \ldots, N$). Now, by setting $L_0 = z_0$ and $\tilde{L}_0 = D_0^*$, we find in view of Theorem 42.5 that the system of convex cones

$$\tilde{L}_0, \ \tilde{L}_t^{(j)}, \qquad j=1, \ldots, m_t; \ t=1, \ldots, N,$$

does not possess separability in E^*. Thus, all the more so, system of cones (42.13) will not possess separability either.

Now it is clear that Theorem 36.8 is applicable (with $r=p=0$ and with sets Ω^0 and Ω^{00} used in place of Ω' and Ω''). If $I_t(z_0)$ is the set of all numbers $i=1, \ldots, q_t$ for which $g_t^i(u_t)=0$, then in view of Theorem 36.8 there must exist numbers

$$\psi_0, \ \overset{t}{\psi}_i, \ \overset{t}{\varphi}_j, \qquad i=1, \ldots, n; \ j=1, \ldots, k_t; \ t=1, \ldots, N$$

(where $\overset{t}{\psi}_i$ correspond to equations (42.5) and $\overset{t}{\varphi}_j$ to equations (42.12)), vectors $a_t^{(i)}$ ($i=1, \ldots, l_t$; $t=1, \ldots, N$), whose directions lie, respectively, in dual cones $D(\tilde{K}_t^{(i)})$, and nonpositive numbers $\overset{t}{\lambda}_j$ ($j \in I_t(z_0)$, $t=1, \ldots, N$), all such that conditions (α), (β), (γ) of Theorem 36.8 are satisfied. Moreover, let us set $\overset{t}{\lambda}_j=0$ for $j \notin I_t(z_0)$, so that the nonpositive numbers $\overset{t}{\lambda}_j$ will be defined for all $j=1, \ldots, q_t$; $t=1, \ldots, N$, where for all these values of j, t it will be true that

$$\overset{t}{\lambda}_j g_t^j(u_t)=0. \tag{42.14}$$

Now note that in this case condition (β) of Theorem 36.8 becomes

$$\begin{aligned}
&\sum_{t=1}^{N}\left(\psi_0 \operatorname{grad}_{x_t} F^0(z_0) + \sum_{\alpha=1}^{n}\sum_{\gamma=1}^{N} \overset{\gamma}{\psi}_\alpha \operatorname{grad}_{x_t} F_\gamma^\alpha(z_0)\right)\delta x_t + \\
&\quad + \sum_{\tau=1}^{N}\left(\psi_0 \operatorname{grad}_{u_\tau} F^0(z_0) + \sum_{\alpha=1}^{n}\sum_{\gamma=1}^{N} \overset{\gamma}{\psi}_\alpha \operatorname{grad}_{u_\tau} F_\gamma^\alpha(z_0)\right)\delta u_\tau + \\
&\quad + \sum_{\tau=1}^{N}\left(\sum_{\alpha=1}^{k_\tau} \overset{\tau}{\varphi}_\alpha \operatorname{grad}_{u_\tau} G_\tau^\alpha(z_0) + \sum_{\alpha=1}^{q_\tau} \overset{\tau}{\lambda}_\alpha \operatorname{grad}_{u_\tau} g_\tau^\alpha(z_0)\right)\delta u_\tau - \\
&\qquad\qquad - \sum_{t=1}^{N}\sum_{i=1}^{l_t} a_t^{(i)}\,\delta z \leqslant 0
\end{aligned} \tag{42.15}$$

for any vector $\delta z = \{\delta x_t, \delta u_\tau\}$, whose direction lies in the intersection of the cones $\tilde{L}_t^{(j)}$ ($j=1, \ldots, m_t$; $t=1, \ldots, N$).

Relation (42.15), in particular, will be valid if all the δu_τ are assumed to be zero, while vectors δx_t take on arbitrary values (since $\tilde{L}_t^{(j)} \supset D_t^* \supset D_0$). Hence, just as in the proof of Theorem 42.1, we get condition (B) of Theorem 42.6 (condition (A) is implied directly by condition (γ) of Theorem 36.8). Note that, according to the relations $\tilde{K}_t^{(i)} \supset D_0$, $\tilde{K}_t^{(i)} \supset D_\tau$ for $t \neq \tau$ (that is, $\tilde{K}_t^{(i)} \supset D_t^*$), the vector $a_t^{(i)}$ can be considered as a vector whose direction lies in cone $D(K_t^{(i)})$, where $a_t^{(i)}\,\delta z = a_t^{(i)}\,\delta u_t$; in the given case $a_t^{(i)}\,\delta z = 0$, since $\delta u_t = 0$.

Now let us set all δx_t equal to zero. In addition, we select some whole number $\theta\,(1 \leqslant \theta \leqslant N)$ and we assume that all δu_τ for $\tau \neq \theta$ are also zero, while as δu_θ we will take all possible vectors whose directions lie in the intersection of cones $L_\theta^{(1)}, \ldots, L_\theta^{(m_\theta)}$. The obtained vector $\delta z = \{\delta x_t, \delta u_\tau\}$ has a direction lying in the intersection of *all* the cones $\tilde{L}_t^{(j)}\,(j = 1, \ldots, m_t;\ t = 1, \ldots, N)$, so that relation (42.15) is valid for it. Therefore

$$\left(\psi_0 \operatorname{grad}_{u_\theta} F^0(z_0) + \sum_{\alpha=1}^{n} \sum_{\gamma=1}^{N} \psi_\alpha^\gamma \operatorname{grad}_{u_\theta} F_\gamma^\alpha(z_0)\right) \delta u_\theta + \\ + \left(\sum_{\alpha=1}^{k_\theta} \varphi_\alpha^\theta \operatorname{grad}_{u_\theta} G_\theta^\alpha(z_0) + \sum_{\alpha=1}^{q_\theta} \lambda_\alpha^\theta \operatorname{grad}_{u_\theta} g_\theta^\alpha(z_0)\right) \delta u_\theta - \\ - \sum_{i=1}^{l_\theta} a_\theta^{(i)} \delta u_\theta \leqslant 0$$

for any vector δu_θ whose direction lies in the intersection of cones $L_\theta^{(1)}, \ldots, L_\theta^{(m_\theta)}$. Taking into account the form of functions F^0 and F_t^l (see (42.3), (42.4)), we now get the condition (C) specified during the formulation of Theorem 42.6.

Finally, condition (D) is satisfied, since all the numbers λ_j^t were assumed to be non-negative, while condition (E) is implied by (42.14).

NOTE 42.7. Let us designate as $\Gamma_t^{(i)}$ the tangent hyperplane (in the space of variables $u^1, \ldots, u^r$) to hypersurface $G_t^i(u^1, \ldots, u^r) = 0$ at point $u(t)$, and as $P_t^{(j)}$ the half-space defined by the inequality

$$(u - u(t)) \operatorname{grad} g_t^j(u(t)) \leqslant 0, \quad j \in I_t(u(t)).$$

If for each $t = 1, \ldots, N$ *the system of convex cones*

$$K_t^{(i)}\,(i = 1, \ldots, l_t),\ L_t^{(i)}\,(i = 1, \ldots, m_t),\ \Gamma_t^{(i)}\,(i = 1, \ldots, k_t), \\ P_t^{(i)}\,(i \in I_t(u(t))) \tag{42.16}$$

does not possess separability in the space of variables $u^1, \ldots, u^r$, *then condition* (A) *in Theorem* 42.6 *can be replaced by the inequality* $\psi_0 < 0$.

If we assume that $\psi_0 = 0$, then condition (B) of Theorem 42.6 implies that $\psi(t) = 0,\ t = 1, \ldots, N$. Consequently, in view of condition (A) of this same theorem, for at least one t at least one of the vectors $\varphi(t), \lambda(t), a_t^{(j)}$ must be different from zero. For this t, condition (C) of Theorem 42.6 implies that the direction of the vector

$$\operatorname{grad}_u H_t(x(t-1), u(t)) - \sum_{i=1}^{l_t} a_t^{(i)}$$

lies in the cone $D\left(L_t^{(1)} \cap \ldots \cap L_t^{(m_t)}\right)$, so that (in view of Corollary 31.4 and

Lemma 32.1) this vector is equal to $b_t^{(1)} + \ldots + b_t^{(m_t)}$, where the direction of vector $b_t^{(i)}$ lies in cone $D(L_t^{(i)})$. Therefore, taking into account that $\psi_0 = 0$, $\psi(t) = 0$, we get

$$-\sum_{i=1}^{k_t} \varphi_i(t) \operatorname{grad} G_t^i(u(t)) - \sum_{i=1}^{q_t} \lambda_i(t) \operatorname{grad} g_t^i(u(t)) + \\ + \sum_{i=1}^{l_t} a_t^{(i)} + \sum_{i=1}^{m_t} b_t^{(i)} = 0.$$

In view of the selection of the number t on the left-hand side of this equation, nonzero vectors must exist. Thus from Theorem 32.2 it follows that system of cones (42.16) possesses separability. This contradiction indicates that $\psi_0 \neq 0$.

Note also that, if for at least one t system of cones (42.16) possesses separability, it follows that the statement of Theorem 42.6 becomes meaningless (cf. Note 36.6).

In the formulation of Theorem 42.6, l_t, m_t, k_t, q_t may be any nonnegative whole numbers. Assuming some of these to be zero, we can obtain a number of particular cases from this theorem. For instance, Theorem 42.1 can be derived from Theorem 42.6 for $l_t = k_t = q_t = 0$, $m_t = 1$ (for all $t = 1, \ldots, N$). Moreover, in the formulation of Theorem 42.6 it can be assumed that for each $t = 1, \ldots, N$ some of the sets $\Xi_t^{(i)}$ are hypersurfaces (in the space of variables $u^1, \ldots, u^r$), some are defined by inequalities of the type $f_t^{(i)}(u) \leqslant 0$, and some are arbitrary sets having a covering $L_t^{(i)}$. This makes it possible to diversify the various corollaries of Theorem 42.6 even more.

From all the results so obtained we can formulate two theorems, the first corresponding to $l_t = m_t = 0$ (for all t) and the second to $l_t = k_t = q_t = 0$, with the additional proviso that some of sets $\Xi_t^{(i)}$ are hypersurfaces, while the others are defined by inequalities of the type $h_t^{(i)}(u) \leqslant 0$.

THEOREM 42.8. *Let* (40.3) *be some control, and let* (40.4) *be the corresponding trajectory (with initial state* $x(0)$*) in discrete controlled object* (40.1), (42.1). *In addition, for each* $t = 1, \ldots, N$ *let set* U_t *have the form* $U_t = \Omega_t' \cap \Omega_t''$, *where set* Ω_t' *is defined in the space of variables* $u^1, \ldots, u^r$ *by system of equations* (42.10) *and set* Ω_t'' *is defined by system of inequalities* (42.11). *We assume that in the vicinity of point* $u(t)$ *all the functions* G_t^j, $j = 1, \ldots, k_t$; g_t^i, $i \in I_t(u(t))$, *are smooth.*

For optimality of process (40.3), (40.4) *(in the sense of a minimum of functional* (40.5) *for a fixed initial state* $x(0)$ *and without placing any restrictions on state* $x(t)$, $t = 1, \ldots, N$*), there must exist a number* $\psi_0 \leqslant 0$ *and vectors*

$$\psi(t) = \{\psi_1(t), \ldots, \psi_n(t)\}, \qquad t = 1, \ldots, N;$$
$$\varphi(t) = \{\varphi_1(t), \ldots, \varphi_{k_t}(t)\}, \qquad t = 1, \ldots, N;$$
$$\lambda(t) = \{\lambda_1(t), \ldots, \lambda_{q_t}(t)\}, \qquad t = 1, \ldots, N,$$

such that conditions (A) *through* (E), *below, are satisfied. These conditions are formulated using the function* $H_t(x, u)$ *defined as*

$$H_t(x, u) = \psi_0 f_t^0(x, u) + \sum_{i=1}^{n} \psi_i(t) f_t^i(x, u) + \\ + \sum_{i=1}^{k_t} \varphi_i(t) G_t^i(u) + \sum_{j=1}^{q_t} \lambda_j(t) g_t^j(u):$$

(A) *if* $\psi_0 = 0$, *then for at least one* t *at least one of the vectors* $\psi(t)$, $\varphi(t)$, $\lambda(t)$ *is different from zero;*

(B) $\psi(t) = \text{grad}_x H_{t+1}(x(t), u(t+1))$, $t = 1, \ldots, N$, *where it is assumed that* $H_{N+1} \equiv 0$;

(C) $\text{grad}_u H_t(x(t-1), u(t)) = \mathbf{0}$, $t = 1, \ldots, N$;

(D) $\lambda_j(t) \leqslant 0$ *for all* $j = 1, \ldots, q_t$; $t = 1, \ldots, N$;

(E) $\lambda_j(t) g_t^j(u(t)) = 0$ *for all* $j = 1, \ldots, q_t$; $t = 1, \ldots, N$.

THEOREM 42.9. *Let* (40.3) *be some control, and let* (40.4) *be the corresponding trajectory (with initial state* $x(0)$ *) in discrete controlled object* (40.1), (42.1). *In addition, for each* $t = 1, \ldots, N$ *let set* U_t *have the form* $U_t = \Xi_t' \cap \Xi_t''$, *where set* Ξ_t' *is defined in the space of variables* $u^1, \ldots, u^r$ *by the system of equations*

$$\Phi_t^j(u^1, \ldots, u^r) = 0, \qquad j = 1, \ldots, s_t,$$

and set Ξ_t'' *is defined by the system of equations*

$$h_t^i(u^1, \ldots, u^r) \leqslant 0, \qquad i = 1, \ldots, p_t.$$

We designate as $I_t(u(t))$ *the set of numbers* $i = 1, \ldots, p_t$ *for which* $h_t^i(u(t)) = 0$. *All the vectors*

$$\text{grad}\, \Phi_t^j(u(t)), \quad j = 1, \ldots, s_t;$$
$$\text{grad}\, h_t^i(u(t)), \quad i \in I_t(u(t)), \qquad t = 1, \ldots, N,$$

are assumed to be different from zero. L_t^j *is the tangent hyperplane of hypersurface* $\Phi_t^j(u) = 0$ *at point* $u(t)$ ($j = 1, \ldots, s_t$; $t = 1, \ldots, N$), *and* P_t^i *is a half-plane defined by the inequality*

$$(u - u(t))\, \text{grad}\, h_t^i(u(t)) \leqslant 0, \quad i \in I_t(u(t)), \qquad t = 1, \ldots, N,$$

it being assumed that for each $t = 1, \ldots, N$ *the system of convex cones*

$$L_t^j, \; P_t^i, \qquad j = 1, \ldots, s_t; \quad i \in I_t(u(t)), \tag{42.16}$$

does not possess separability in the space of variables $u^1, \ldots, u^r$.

For optimality of process (40.3), (40.4) *(in the sense of the minimum of functional* (40.5) *for a fixed initial state* $x(0)$ *and with no restrictions placed on states* $x(t)$, $t = 1, \ldots, N$), *there must exist a number* ψ_0 *and vectors*

$$\psi(t) = \{\psi_1(t), \ldots, \psi_n(t)\}, \qquad t = 1, \ldots, N,$$

such that conditions (A) *through* (C), *below, are satisfied. These conditions are formulated using the function* $H_t(x, u)$ *defined by equation* (42.2):

(A) $\psi_0 < 0$;

(B) $\psi(t) = \mathrm{grad}_x H_{t+1}(x(t), u(t+1))$, $t = 1, \ldots, N$, *where it is assumed that* $H_{N+1} \equiv 0$;

(C) *for any vector* δu, *whose direction lies in the intersection of cones* (42.16), *the following inequality is valid:*

$$\delta u \, \mathrm{grad}_u H_t(x(t-1), u(t)) \leqslant 0, \qquad t = 1, \ldots, N$$

(here the inequality $\psi_0 < 0$ is proven just as in Theorem 42.1).

43. Problem with phase constraints. In this subsection we will consider the optimal control of a discrete object with a constant control domain and with phase constraints imposed. In other words, we study controlled object (40.1), (42.1), with the following constraints on the phase states:

$$x(t) \in M_t, \qquad t = 0, 1, \ldots, N.$$

For this object we wish to find a process (40.3), (40.4) satisfying the imposed constraints and yielding a minimum of functional (40.5).

First let us look at the case in which each of sets M_t, $t = 0, 1, \ldots, N$, is represented in the form $M_t = \Pi_t' \cap \Sigma_t$, where Π_t' is a set defined in E^n by the system of relations

$$W_t^i(x) = W_t^i(x^1, \ldots, x^n) = 0, \qquad i = 1, \ldots, r_t, \tag{43.1}$$

and Σ_t is some set of space E^n (below, a covering of this set at point $x(t)$ is assumed known). Analogously, each of sets U_τ, $\tau = 1, \ldots, N$, is assumed to be represented as $U_\tau = \Omega_\tau' \cap \Xi_\tau$, where Ω_τ' is a set defined in the space of variables $u^1, \ldots, u^r$ by the system of relations

$$G_\tau^j(u^1, \ldots, u^r) = 0, \qquad j = 1, \ldots, k_\tau, \tag{43.2}$$

and Ξ_τ is some set in this same space (it being assumed that a covering of set Ξ_τ at point $u(\tau)$ is known). Functions W_t^i and G_τ^j are assumed to be smooth. Under these conditions the following theorem is valid.

THEOREM 43.1. *Let* (40.3) *be some control, and let* (40.4) *be the corresponding trajectory (with initial state* $x(0) \in M_0$ *) in discrete controlled object* (40.1), (42.1), *the inclusions* $x(t) \in M_t$, $t = 1, \ldots, N$, *being satisfied. In addition, let set* M_t *be represented in the form* $M_t = \Pi_t' \cap \Sigma_t$, $t = 0, 1, \ldots, N$, *where set* Π_t' *is* *defined in* E^n *by the system of relations* (43.1), *and set* U_τ, $\tau=1, \ldots, N$, *is represented in the form* $U_\tau=\Omega'_\tau \cap \Xi_\tau$, *where set* Ω'_τ *is defined in the space of variables* $u^1, \ldots, u^r$ *by the system of relations* (43.2). *Finally, let* L_τ *be a covering of set* Ξ_τ *at point* $u(\tau)$, $\tau=1, \ldots, N$, *and let* Q_t *be a covering of set* Σ_t *at point* $x(t)$, $t=0, 1, \ldots, N$.

For optimality of the process considered (in the sense of a minimum of functional (40.5) *with constraints* $x(t) \in M_t$, $t=0, 1, \ldots, N$, *imposed), there must exist a number* $\psi_0 \leqslant 0$ *and vectors*

$$\begin{aligned} \psi(t) &= \{\psi_1(t), \ldots, \psi_n(t)\}, & t&=1, \ldots, N; \\ \varphi(t) &= \{\varphi_1(t), \ldots, \varphi_{k_t}(t)\}, & t&=1, \ldots, N; \\ \mu(t) &= \{\mu_1(t), \ldots, \mu_{r_t}(t)\}, & t&=0, 1, \ldots, N, \end{aligned}$$

such that conditions (A) *through* (C), *below, are satisfied. These conditions are formulated using the function* $H_t(x, u)$, *defined as*

$$H_t(x, u)=\psi_0 f_t^0(x, u)+\sum_{i=1}^{n} \psi_i(t) f_t^i(x, u)+\sum_{\alpha=1}^{k_t} \varphi_\alpha(t) G_t^\alpha(u):$$

(A) *if* $\psi_0=0$, *then for at least one* t *at least one of the vectors* $\psi(t)$, $\varphi(t)$, $\mu(t)$ *is different from zero;*

(B) *for any* $t=0, 1, \ldots, N$ *and any vector* $\boldsymbol{\delta x}$ *whose direction lies in cone* Q_t *the following inequality is valid:*

$$\left(-\psi(t)+\operatorname{grad}_x H_{t+1}(x(t), u(t+1))+\sum_{\alpha=1}^{r_t} \mu_\alpha(t) \operatorname{grad} W_t^\alpha(x(t))\right) \boldsymbol{\delta x} \leqslant 0,$$

where it is assumed that $\psi(0)=\mathbf{0}$ *and* $H_{N+1} \equiv 0$;

(C) *for any* $t=1, \ldots, N$ *and any vector* $\boldsymbol{\delta u}$ *whose direction lies in cone* L_t *the following inequality is valid:*

$$\boldsymbol{\delta u} \operatorname{grad}_u H_t(x(t-1), u(t)) \leqslant 0.$$

The *proof* of this theorem is carried out in the same way as the proof of Theorem 42.1. Let us call E^* the space of variables x_t^i, u_τ^j, where $i=1, \ldots, n$; $j=1, \ldots, r$; $t=0, 1, \ldots, N$; $\tau=1, \ldots, N$. Thus, in contrast to the proof of Theorem 42.1, the quantities x_0^i, $i=1, \ldots, n$, are here also assumed to be variables. Functions (42.2) are defined in space E^*, are continuous, and are continuously differentiable. Functions F^0 and F_t^i (see (42.3), (42.4)) are also defined and continuously differentiable in space E^*. We now add to (42.3), (42.4) the following functions:

$$\begin{gathered} A_t^i(z)=A_t^i(x, u)=W_t^i(x_t^1, \ldots, x_t^n); \\ i=1, \ldots, r_t; \quad t=0, 1, \ldots, N, \end{gathered}$$

$$\Phi^j_\tau(z) = \Phi^j_\tau(x, u) = G^j_\tau(u^1_\tau, \ldots, u^r_\tau);$$
$$j = 1, \ldots, k_\tau; \quad \tau = 1, \ldots, N,$$

where W^i_t and G^j_τ are the functions entering into the definitions of sets M_t and U_τ (see (43.1), (43.2)). Next we designate as Ω^* the set of all points $z = (x^i_t, u^j_\tau)$ of space E^* satisfying the system of equations

$$F^i_t(z) = 0; \quad i = 1, \ldots, n; \quad t = 1, \ldots, N;$$
$$A^i_t(z) = 0; \quad i = 1, \ldots, r_t; \quad t = 0, 1, \ldots, N;$$
$$\Phi^j_\tau(z) = 0; \quad j = 1, \ldots, k_\tau; \quad \tau = 1, \ldots, N.$$

In addition, we designate as Ξ^* the set of all points $z = (x^i_t, u^j_\tau)$ of space E^* satisfying the conditions

$$x_t = (x^1_t, \ldots, x^n_t) \in \Sigma_t, \quad t = 0, 1, \ldots, N;$$
$$u_\tau = (u^1_\tau, \ldots, u^r_\tau) \in \Xi_\tau, \quad \tau = 1, \ldots, N.$$

Just as in the proof of Theorem 42.1, substitution (42.6) (whereby t is changed from 0 to N) reduces the given problem of optimal control to a problem of finding the point z at which function $F^0(z)$, considered on the set $\Omega^* \cap \Xi^*$, reaches its least value.

Let (40.3), (40.4) be the optimal process sought. If the corresponding point of space E^*, the coordinates of which are defined by formulas (42.6), is called z_0, then we know that $z_0 \in \Omega^* \cap \Xi^*$ and that at this point function $F^0(z)$, considered only on set $\Omega^* \cap \Xi^*$, reaches its least value. Moreover, let us designate as L^* the subset of space E^* consisting of all points $z = (x^i_t, u^j_\tau)$ satisfying the relations

$$x_t = (x^1_t, \ldots, x^n_t) \in Q_t, \quad t = 0, 1, \ldots, N;$$
$$u_\tau = (u^1_\tau, \ldots, u^r_\tau) \in L_\tau, \quad \tau = 1, \ldots, N.$$

Clearly, L^* is a convex cone of space E^* with a vertex at point z_0, and this cone is a covering of set Ξ^* at point z_0 (cf. the corresponding discussion during the proof of Theorem 42.1). It is also obvious that the direction of some vector $\boldsymbol{\delta z} = \{\boldsymbol{\delta x}_t, \boldsymbol{\delta u}_\tau\}$ lies in cone L^* if and only if, for each $t = 0, 1, \ldots, N$ the direction of vector $\boldsymbol{\delta x}_t$ lies in cone Q_t, while also for any $\tau = 1, \ldots, N$ the direction of vector $\boldsymbol{\delta u}_\tau$ lies in cone L_τ.

According to Theorem 36.9, there must exist numbers

$$\psi_0 \leqslant 0, \quad \psi^t_i, \quad i = 1, \ldots, n; \quad t = 1, \ldots, N;$$

$$\mu_i^t, \quad i=1, \ldots, r_t; \quad t=0, 1, \ldots, N,$$
$$\varphi_j^\tau, \quad j=1, \ldots, k_\tau; \quad \tau=1, \ldots, N,$$

such that at least one of these is different from zero and condition (β) of Theorem 36.9 is satisfied. In this case the latter condition becomes

$$\sum_{t=0}^{N}\left(\psi_0 \operatorname{grad}_{x_t} F^0(z_0) + \sum_{\alpha=1}^{n}\sum_{\gamma=1}^{N} \psi_\alpha^\gamma \operatorname{grad}_{x_t} F_\gamma^\alpha(z_0)\right)\delta x_t + \\ + \sum_{\tau=1}^{N}\left(\psi_0 \operatorname{grad}_{u_\tau} F^0(z_0) + \sum_{\alpha=1}^{n}\sum_{\gamma=1}^{N} \psi_\alpha^\gamma \operatorname{grad}_{u_\tau} F_\gamma^\alpha(z_0)\right)\delta u_\tau + \\ + \sum_{t=0}^{N}\sum_{\alpha=1}^{r_t} \mu_\alpha^t (\delta x_t \operatorname{grad}_{x_t} A_t^\alpha(z_0)) + \\ + \sum_{\tau=1}^{N}\sum_{\alpha=1}^{k_\tau} \varphi_\alpha^\tau (\delta u_\tau \operatorname{grad}_{u_\tau} \Phi_\tau^\alpha(z_0)) \leqslant 0. \quad (43.3)$$

This relation should be valid for any vector $\delta z = \{\delta x_t, \delta u_\tau\}$ whose direction lies in cone L^*.

Let us now set all the δu_τ equal to zero. In addition, we select some whole number $\theta\,(0 \leqslant \theta \leqslant N)$ and we set all the δx_t for $t \neq \theta$ equal to zero as well, while as δx_θ we take all possible vectors whose directions lie in cone Q_θ. The resulting vector $\delta z = \{\delta x_t, \delta u_\tau\}$ has a direction lying in cone L^*, so that relation (43.3) is valid for it. Consequently,

$$\left(\psi_0 \operatorname{grad}_{x_\theta} F^0(z_0) + \sum_{\alpha=1}^{n}\sum_{\gamma=1}^{N} \psi_\alpha^\gamma \operatorname{grad}_{x_\theta} F_\gamma^\alpha(z_0) + \\ + \sum_{\alpha=1}^{r_\theta} \mu_\alpha^\theta \operatorname{grad}_{x_\theta} A_\theta^\alpha(z_0)\right)\delta x_\theta \leqslant 0$$

for any $\theta = 0, 1, \ldots, N$ and any vector δx_θ whose direction lies in cone Q_θ. Taking into account the form of functions F^0 and F_γ^α (see (42.3), (42.4)), we can rewrite this condition as follows:

$$\left(\psi_0 \operatorname{grad}_{x_\theta} f_{\theta+1}^0(x_\theta, u_{\theta+1}) - \psi^\theta + \sum_{\alpha=1}^{n} \psi_\alpha^{\theta+1} \operatorname{grad}_{x_\theta} f_{\theta+1}^\alpha(x_\theta, u_{\theta+1}) + \\ + \sum_{\alpha=1}^{r_\theta} \mu_\alpha^\theta \operatorname{grad}_{x_\theta} W_\theta^\alpha(x_\theta)\right)\delta x_\theta \leqslant 0,$$

this relation being valid for any $\theta = 0, 1, \ldots, N$ and any vector δx_θ whose direction lies in cone Q_θ. Here ψ^θ denotes the vector $\{\psi_1^\theta, \ldots, \psi_n^\theta\}$, it being assumed that $\psi^0 = 0$ and $f_{N+1}^i \equiv 0$ $(i = 0, 1, \ldots, n)$. If we next apply the change (42.6), whereby t is varied from 0 to N, and if we also set $\psi_i(t) = \psi_i^t$, $u_i(t) = u_i^t$, $\varphi_j(t) = \varphi_j^t$, then these relations lead directly to the condition (B) specified during the formulation of Theorem 43.1.

Now let us set all the δx_t equal to zero; in addition, we choose some whole number θ $(1 \leqslant \theta \leqslant N)$ and we set all the δu_τ with $\tau \neq \theta$ equal to zero as well, while as δu_θ we take all possible vectors whose directions lie in cone L_θ. The resulting vector $\delta z = \{\delta x_t, \delta u_\tau\}$ has a direction lying in cone L^*, so that relation (43.3) is valid for it. Consequently,

$$\left(\psi_0 \operatorname{grad}_{u_\theta} F^0(z_0) + \sum_{\alpha=1}^{n} \sum_{\gamma=1}^{N} \psi_\alpha^\gamma \operatorname{grad}_{u_\theta} F_\gamma^\alpha(z_0) + \right.$$
$$\left. + \sum_{\alpha=1}^{k_\theta} \varphi_\alpha^\theta \operatorname{grad}_{u_\theta} \Phi_\theta^\alpha(z_0)\right) \delta u_\theta \leqslant 0$$

for any $\theta = 1, \ldots, N$ and any vector δu_θ whose direction lies in cone L_θ. Taking into account the form of functions F^0 and F_γ^α (see (42.3), 42.4)), and also applying change (42.6), whereby t is varied from 0 to N, we obtain directly from this relation the condition (C) specified during the formulation of Theorem 43.1.

NOTE 43.2. It may happen that in Theorem 43.1, for any given value of $t = 0, 1, \ldots, N$ (or even for all t), relations (43.1) do not apply, that is, Π_t' coincides with the entire space E^n, so that $M_t = \Pi_t' \cap \Sigma_t = E^n \cap \Sigma_t = \Sigma_t$ (and then Q_t is a covering of set M_t at point $x(t)$). In this case the corresponding vector $\mu(t)$ will be *absent* in the formulation of Theorem 43.1, that is, in conditions (A) and (B) of Theorem 43.1 all the terms containing $\mu_i(t)$, $i = 1, \ldots, r_t$, can simply be crossed out.

Similarly, if for any $\tau = 1, \ldots, N$ (or for all τ) relations (43.2) do not apply, that is, Ω_τ' coincides with the entire space of variables $u^1, \ldots, u^r$, and thus $U_\tau = \Xi_\tau$ (so that L_τ is a covering of set U_τ at point $u(\tau)$), it follows that vector $\varphi(\tau)$ will be absent in the formulation of Theorem 43.1, that is, in condition (A) of Theorem 43.1 and in the expression for function $H_t(x, u)$ all the terms containing $\varphi_i(\tau)$, $i = 1, \ldots, k_\tau$, can simply be crossed out.

In particular, if neither relations (43.1) nor relations (43.2) exist, and if set M_0 consists of a single point $x(0)$, while all the other sets M_t, $t = 1, \ldots, N$, coincide with the entire space E^n (so that $Q_t = E^n$, $t = 1, \ldots, N$), it follows that Theorem 43.1 is transformed into Theorem 42.1.

NOTE 43.3. Assume that relations (43.1) are independent at point $x(t)$ (for each $t = 0, 1, \ldots, N$), that is, that Π_t' constitutes a *manifold* close to point $x(t)$. If Π_t^* is the tangent plane to this manifold at point $x(t)$, then we can assume that cones Π_t^* and Q_t (with a common vertex $x(t)$) are nonseparable in E^n, that is, there is no hyperplane Γ such that $\Pi_t^* \subset \Gamma$ and the entire cone Q_t is situated in one closed half-space defined by Γ.

In exactly the same way, we assume that relations (43.2) are independent at point $u(\tau)$ (for any $\tau = 1, \ldots, N$), that is, we assume that Ω_τ' constitutes a manifold close to point $u(\tau)$. The tangent plane to this manifold at $u(\tau)$ is called Ω_τ^*. In addition, let us suppose that cones Ω_τ^* and L_τ (with a common vertex $u(\tau)$) are nonseparable in the space of variables $u^1, \ldots, u^r$.

With these assumptions, condition (A) *in Theorem* 43.1 *can be replaced by the following condition:*

(A′) *if* $\psi_0 = 0$, *then for at least one* t *the vector* $\psi(t)$ *is different from zero.*

Actually, let us assume that the number ψ_0 and the vectors $\psi(t)$, $\varphi(t)$, $\mu(t)$ are so selected that they satisfy conditions (A), (B), (C) of Theorem 43.1, and let us also assume that $\psi_0 = 0$ and that all the vectors $\psi(t)$ are zero. Then, in view of condition (A), at least one of the vectors $\varphi(t)$, $\mu(t)$ will be different from zero for at least one t To be more definite, let us say that for some $t(=0, 1, \ldots, N)$ the vector $\mu(t) = \{\mu_1(t), \ldots, \mu_{r_t}(t)\}$ is different from zero (if one of the vectors $\varphi(t)$ is nonzero instead, then the argument is similar). Now, in view of condition (B), we have

$$\delta x \left(\sum_{\alpha=1}^{r_t} \mu_\alpha(t) \operatorname{grad} W_t^\alpha(x(t)) \right) \leqslant 0 \tag{43.4}$$

for any vector δx whose direction lies in cone Q_t. Since vector $\mu(t) = \{\mu_1(t), \ldots, \mu_{r_t}(t)\}$ is different from zero and since relations (43.1) are independent at point $x(t)$, that is, the vectors $\operatorname{grad} W_t^\alpha(x(t))$, $\alpha = 1, \ldots, r_t$, are linearly independent, it follows that the vector

$$n = \sum_{\alpha=1}^{r_t} \mu_\alpha(t) \operatorname{grad} W_t^\alpha(x(t))$$

is different from zero.

Now let Γ be a hyperplane of plane E^n which is orthogonal to vector n and passes through point $x(t)$. The expression for vector n indicates that it is orthogonal to plane Π_t^*, so that $\Pi_t^* \subset \Gamma$. Moreover, (43.4) implies that $n\,\delta x \leqslant 0$ for any vector δx whose direction lies in cone Q_t. Consequently, cone Q_t lies completely in one of the closed half-spaces defined by hyperplane Γ, so that Γ *separates* cones Π_t^* and Q_t, which contradicts the assumptions. This contradiction indicates that condition (A′) is satisfied.

THEOREM 43.4. *With the assumptions of Theorem* 43.1, *let set* Σ_t *be defined in space* E^n *by the inequalities*

$$w_t^j(x) = w_t^j(x^1, \ldots, x^n) \leqslant 0; \qquad j = 1, \ldots, p_t, \tag{43.5}$$

where w_t^j *are smooth functions, and system of constraints* (43.5) *is nonsingular at point* $x(t)$, $t = 0, 1, \ldots, N$. *Then condition* (B) *of Theorem* 43.1 *can be replaced by the following condition:*

(B′) *there must exist vectors* $\nu(t) = \{\nu_1(t), \ldots, \nu_{p_t}(t)\}$, $t = 0, 1, \ldots, N$, *such that the following relations are valid:*

$$\nu_j(t) \leqslant 0, \quad \nu_j(t)\, w_t^j(x(t)) = 0; \qquad j = 1, \ldots, p_t; \quad t = 0, 1, \ldots; N;$$

$$\psi(t) = \operatorname{grad}_x H_{t+1}(x(t), u(t+1)) + \sum_{\alpha=1}^{r_t} \mu_\alpha(t) \operatorname{grad} W_t^\alpha(x(t)) +$$

$$+\sum_{j=1}^{p_t} \nu_j(t) \operatorname{grad} w_t^j(x(t)); \qquad t=0, 1, \ldots, N,$$

where it is assumed that $\psi(0)=0$ *and* $H_{N+1}\equiv 0$.

Proof. From Theorem 34.6 we know that the set of all points $x \in E^n$ for which the vector $x - x(t)$ forms a nonpositive scalar product with each vector

$$\operatorname{grad} w_t^j(x(t)), \qquad j \in I_t(x(t)), \tag{43.6}$$

is a covering of set Σ_t at point $x(t)$ (here t is a fixed number which can assume any value $0, 1, \ldots, N$). In Theorem 43.1 Q_t can be taken to be precisely this covering of set Σ_t.

Condition (B) of Theorem 43.1 implies that the vector

$$\operatorname{grad}_x H_{t+1}(x(t), u(t+1)) - \psi(t) + \sum_{\alpha=1}^{r_1} \mu_\alpha(t) \operatorname{grad} \dot{W}_t^\alpha(x(t)) \tag{43.7}$$

forms a nonpositive scalar product with each vector whose direction lies in cone Q_t.

Hence the direction of vector (43.7) lies in the dual cone $D(Q_t)$, so that vector (43.7) can be expressed as a linear combination of vectors (43.6) with nonnegative coefficients. Those vectors $\operatorname{grad} w_t^j(x(t))$ for which $j \notin I_t(x(t))$ can also be added to this linear combination, but with zero coefficients. Then vector (43.7) becomes

$$-\sum_{j=1}^{p_t} \nu_j(t) \operatorname{grad} w_t^j(x(t)), \tag{43.8}$$

where all the coefficients $\nu_j(t)$ are nonpositive and $\nu_j(t)=0$ for $j \notin I_t(x(t))$. By equating vectors (43.7) and (43.8), we get the second relation in condition (B′). Since, in addition, $\nu_j(t)=0$ for $j \notin I_t(x(t))$ and $w_t^j(x(t))=0$ for $j \in I_t(x(t))$, it follows that $\nu_j(t)\, w_t^j(x(t))=0$ for all $j=1, \ldots, p_t$, that is, the first relation in condition (B′) is valid as well.

Now let us formulate a more general theorem (for the problem with phase constraints), the foregoing theorems being particular cases of this general theorem. The proof of this theorem, which is obtained on the basis of Theorem 36.8, is exactly analogous to the proofs of Theorems 42.1, 42.6, and 43.1.

THEOREM 43.5. *Let* (40.3) *be some control and let* (40.4) *be the corresponding trajectory with an initial state* $x(0) \in M_0$ *in discrete controlled object* (40.1), (42.1), *the inclusions* $x(t) \in M_t$, $t=1, \ldots, N$ *being satisfied. In addition, for each* $t=1, \ldots, N$ *let set* U_t *have the form*

$$U_t = \Omega_t' \cap \Omega_t'' \cap \Omega_t^{(1)} \cap \ldots \cap \Omega_t^{(l_t)} \cap \Xi_t^{(1)} \cap \ldots \cap \Xi_t^{(m_t)},$$

and let set M_τ *for each* $\tau=0, 1, \ldots, N$ *have the form*

$$M_\tau = \Pi_\tau' \cap \Pi_\tau'' \cap \Pi_\tau^{(1)} \cap \ldots \cap \Pi_\tau^{(n_\tau)} \cap \Sigma_\tau^{(1)} \cap \ldots \cap \Sigma_\tau^{(s_\tau)},$$

where set Ω'_t *is defined in the space of variables* $u^1, \ldots, u^r$ *by the system of equations*

$$G_t^j(u^1, \ldots, u^r) = 0, \qquad j = 1, \ldots, k_t,$$

set Ω''_t *is defined by the system of inequalities*

$$g_t^i(u^1, \ldots, u^r) \leqslant 0, \qquad i = 1, \ldots, q_t,$$

set Π'_τ *is defined in the space of variables* $x^1, \ldots, x^n$ *by the system of equations*

$$W_\tau^j(x^1, \ldots, x^n) = 0, \qquad j = 1, \ldots, r_\tau,$$

set Π''_τ *is defined by the system of inequalities*

$$w_\tau^i(x^1, \ldots, x^n) \leqslant 0, \qquad i = 1, \ldots, p_\tau;$$

$\Omega_t^{(1)}, \ldots, \Omega_t^{(l_t)}, \Xi_t^{(1)}, \ldots, \Xi_t^{(m_t)}$ *are certain sets of the space of variables* $u^1, \ldots, u^r$, *and* $\Pi_\tau^{(1)}, \ldots, \Pi_\tau^{(n_\tau)}, \Sigma_\tau^{(1)}, \ldots, \Sigma_\tau^{(s_\tau)}$ *are certain sets of the space of variables* $x^1, \ldots, x^n$. *Finally, let* $K_t^{(i)}$ *be a covering of set* $\Omega_t^{(i)}$ *at point* $u(t)$, *let* $L_t^{(i)}$ *be a covering of set* $\Xi_t^{(i)}$ *at* $u(t)$; *let* $P_\tau^{(j)}$ *be a covering of set* $\Pi_\tau^{(j)}$ *at* $x(\tau)$; *and let* $Q_\tau^{(j)}$ *be a covering of set* $\Sigma_\tau^{(j)}$ *at* $x(\tau)$. *All the functions* $G_t^i, g_t^i, W_\tau^i, w_\tau^i$ *are assumed to be smooth, and for each* $t = 1, \ldots, N$ *the system of convex cones* $L_t^{(i)}$ $(i = 1, \ldots, m_t)$ *does not possess separability in the space of variables* $u^1, \ldots, u^r$, *while for each* $\tau = 0, 1, \ldots, N$ *the system of convex cones* $Q_\tau^{(j)}$ $(j = 1, \ldots, s_\tau)$ *does not possess separability in the space of variables* $x^1, \ldots, x^n$.

For optimality of process (40.3), (40.4) *in the sense of minimizing functional* (40.5) *with constraints* $x(t) \in M_t$, $t = 0, 1, \ldots, N$), *there must exist a number* $\psi_0 \leqslant 0$ *and vectors*

$$\begin{aligned}
&\psi(t) = \{\psi_1(t), \ldots, \psi_n(t)\}, && t = 1, \ldots, N;\\
&\varphi(t) = \{\varphi_1(t), \ldots, \varphi_{k_t}(t)\}, && t = 1, \ldots, N;\\
&\lambda(t) = \{\lambda_1(t), \ldots, \lambda_{q_t}(t)\}, && t = 1, \ldots, N;\\
&\mu(t) = \{\mu_1(t), \ldots, \mu_{r_t}(t)\}, && t = 0, 1, \ldots, N;\\
&\nu(t) = \{\nu_1(t), \ldots, \nu_{p_t}(t)\}, && t = 0, 1, \ldots, N;\\
&a_t^{(i)}, \quad i = 1, \ldots, l_t; && t = 1, \ldots, N;\\
&b_t^{(i)}, \quad i = 1, \ldots, n_t; && t = 0, 1, \ldots, N,
\end{aligned}$$

such that the direction of vector $a_t^{(i)}$ *lies in the dual cone* $D(K_t^{(i)})$, *the direction of vector* $b_t^{(i)}$ *lies in the dual cone* $D(P_t^{(i)})$, *and conditions* (A) *through* (E), *below, are satisfied. These conditions are formulated using the function* $H_t(x, u)$, *defined as*

$$H_t(x, u) = \psi_0 f_t^0(x, u) + \sum_{i=1}^{n} \psi_i(t) f_t^i(x, u) + \sum_{i=1}^{k_t} \varphi_i(t) G_t^i(u) + \sum_{\alpha=1}^{q_t} \lambda_\alpha(t) g_t^\alpha(u):$$

(A) *if* $\psi_0 = 0$, *then for at least one* t *at least one of the vectors* $\psi(t)$, $\varphi(t)$, $\lambda(t)$, $\mu(t)$, $\nu(t)$, $a_t^{(i)}$, $b_t^{(i)}$ *is different from zero;*

(B) *for any* $t = 0, 1, \ldots, N$ *and any vector* δx *whose direction lies in the intersection of cones* $Q_t^{(i)}$ $(i = 1, \ldots, s_t)$, *the following inequality is valid:*

$$\Big(-\psi(t) + \operatorname{grad}_x H_{t+1}(x(t), u(t+1)) + \sum_{\alpha=1}^{r_t} \mu_\alpha(t) \operatorname{grad} W_t^\alpha(x(t)) + \\ + \sum_{\alpha=1}^{p_t} \nu_\alpha(t) \operatorname{grad} w_t^\alpha(x(t)) - \sum_{i=1}^{n_t} b_t^{(i)}\Big)\delta x \leqslant 0,$$

where it is assumed that $\psi(0) = 0$ *and* $H_{N+1} \equiv 0$;

(C) *for any* $t = 1, \ldots, N$ *and any vector* δu *whose direction lies in the intersection of cones* $L_t^{(i)}$ $(i = 1, \ldots, m_t)$, *the following inequality is valid:*

$$\Big(\operatorname{grad}_u H_t(x(t-1), u(t)) - \sum_{i=1}^{l_t} a_t^{(i)}\Big)\delta u \leqslant 0;$$

(D) $\lambda_j(t) \leqslant 0$, $\nu_i(\tau) \leqslant 0$ *for all* $j = 1, \ldots, q_t$; $t = 1, \ldots, N$; $i = 1, \ldots, p_\tau$; $\tau = 0, 1, \ldots, N$;

(E) $\lambda_j(t) g_t^j(u(t)) = 0$, $\nu_i(\tau) w_\tau^i(x(\tau)) = 0$ *for all* $j = 1, \ldots, q_t$; $t = 1, \ldots, N$; $i = 1, \ldots, p_\tau$; $\tau = 0, 1, \ldots, N$.

In the formulation of Theorem 43.5 the quantities l_t, m_t, n_t, s_t, k_t, q_t, r_t, p_t can be any nonnegative numbers. By assuming various of these to be zero, it is possible to obtain from Theorem 43.5 a great number of particular cases. In addition, when formulating Theorem 43.5, it can be assumed that for each $t = 1, \ldots, N$ some of sets $\Xi_t^{(j)}$ are hypersurfaces, some are defined by inequalities of the type $h_t^{(j)}(u) \leqslant 0$, and some are arbitrary sets having a covering $L_t^{(j)}$. The same applies to sets $\Sigma_t^{(i)}$. Thus the various corollaries of Theorem 43.5 can be diversified even more. For example, Theorem 43.1 is obtained from Theorem 43.5 for

$$q_t = l_t = 0, \quad p_\tau = n_\tau = 0, \quad m_t = 1, \quad s_\tau = 1 \\ (t = 1, \ldots, N; \ \tau = 0, 1, \ldots, N).$$

Similarly, Theorem 43.4 is obtained from Theorem 43.5 for $q_t = l_t = 0$, $n_\tau = s_\tau = 0$, $m_t = 1$ (note that formally we can derive a stronger result by the same means, since when applying Theorem 43.5 it is not obligatory that system of constraints (43.5) be nonsingular at point $x(t)$; however, if this system of constraints is not nonsingular, then the statement in Theorem 43.4 will be meaningless; cf. Note 36.6).

NOTE 43.6. According to Theorem 34.2, the condition imposed on sets $\Omega_t^{(i)}$, $\Xi_t^{(i)}$, $\Pi_\tau^{(j)}$, $\Sigma_\tau^{(j)}$ during the formulation of Theorem 43.5 will be satisfied if these sets are convex, while the supporting cones of these sets at the corresponding points enter in as $K_t^{(i)}$, $L_t^{(i)}$, $P_\tau^{(j)}$, $Q_\tau^{(j)}$. In the same way, these conditions will be satisfied if sets

$\Omega_t^{(i)}$, $\Xi_t^{(i)}$ are defined by a system of *locally* concave constraints (cf. Corollary 36.17) of the type

$$\xi^1(u) \leqslant 0, \ \ldots, \ \xi^\alpha(u) \leqslant 0,$$

while for $K_t^{(i)}$, $L_t^{(i)}$ the intersection of half-spaces

$$(u - u(t)) \operatorname{grad} \xi^i(u(t)) \leqslant 0, \qquad i = 1, \ldots, \alpha,$$

is selected (the same being done for $\Pi_\tau^{(j)}$, $\Sigma_\tau^{(j)}$). It may also be that some of sets $\Omega_t^{(i)}$, $\Xi_t^{(i)}$, $\Pi_\tau^{(j)}$, $\Sigma_\tau^{(j)}$ are convex, while the others are defined by systems of locally concave constraints.

44. Existence theorem. The theorems presented in the foregoing subsections (and in Subsections 45 and 46, below) contain only the *necessary* condition for optimality. Thus a solution of the optimal-control problem arrived at on the basis of these theorems is always necessary in order to put the introduction of any additional considerations on a sound basis. Actually, we can even assume the application of the necessary conditions to the given optimal-control problem to have been particularly successful, and we can suppose it to have been established that there exists *only one* process satisfying the necessary conditions formulated in Theorem 43.5 (or 42.6). This just means that the following fact has been established: *if an optimal process exists in the problem being considered, then it can be only the process found.* On the other hand, this by no means signifies that the process found is actually optimal (since it may well be that no optimal process exists, that is, no process other than the one found is optimal, but the process found is not optimal either).

If, however, there are indications that an optimal process definitely *exists,* then it can only be the process found, since no other processes satisfy the necessary conditions, that is, they cannot be optimal. The following theorem states the suitable conditions which must be fulfilled in order for an optimal process to necessarily exist.*

THEOREM 44.1. *Let functions $f_t^i(x, u)$ (see* (40.1), (40.5)*) be continuous over the aggregate of variables x, u, let sets M_0, $U_1, \ldots, U_N$ be compact, and let sets $M_1, \ldots, M_N$ be closed. Assume, in addition, that there exists in object* (40.1), (42.1) *at least one process satisfying the phase constraints $x(t) \in M_t$, $t = 0, 1, \ldots, N$. Then the problem of optimal control* (40.1), (42.1), (40.5) *with constraints M_t on the phase coordinates has a solution (that is, an optimal process exists). In particular, if functions f_t^i are continuous, while sets $U_1, \ldots, U_N$ are compact (and nonempty), it follows that the fundamental problem* (see p. 11) *has a solution.*

Proof. Let W_1 be the set of all points of form $f_1(x, u)$, where $x \in M_0$, $u \in U_1$. Since M_0 and U_1 are compact, while function f_1 is continuous, we know that W_1 is a compact set. Obviously, for any process in the object being considered (with an

* The conditions for the existence of optimal control for discrete systems of a particular form were derived by Propoi (Propoi, A.I. *O printsipe maksimuma dlya diskretnykh sistem upravleniya (The Maximum Principle for Discrete Systems of Control).* – Avtomatika i Telemekhanika, **26**, No.7. 1965).

initial state $x(0) \in M_0$), the inclusion $x(1) \in W_1$ will be satisfied. Taking into account the phase constraint $x(1) \in M_1$, we arrive at the inclusion $x(1) \in M_1 \cap W_1$. Here, since set M_1 is closed, it follows that set $M_1 \cap W_1$ is also closed, and thus compact (since W_1 is compact). In addition, let W_2 be the set of all points of form $f_2(x, u)$, where $x \in M_1 \cap W_1$, $u \in U_2$. Set W_2 is compact, and for any process in the given object (with phase constraints $x(0) \in M_0$, $x(1) \in M_1$) the inclusion $x(2) \in W_2$ is satisfied. Next, taking into account phase constraint $x(2) \in M_2$, we obtain the inclusion $x(2) \in M_2 \cap W_2$, the set $M_2 \cap W_2$ being compact.

Continuing in this manner, we can construct for each $t = 1, \ldots, N$ a compact set $M_t \cap W_t$ such that for any process satisfying phase constraints $x(t) \in M_t$ $(t = 0, 1, \ldots, N)$ the inclusion $x(t) \in M_t \cap W_t$ is satisfied. For uniformity, let us call W_0 the entire space E^n of variables $x^1, \ldots, x^n$; then the inclusion $x(0) \in M_0$ can also be written in the form $x(0) \in M_0 \cap W_0$, so that the relation $x(t) \in M_t \cap W_t$ will hold true for all $t = 0, 1, \ldots, N$. Accordingly, any process in object (40.1), (42.1) satisfying the phase constraints $x(t) \in M_t$ will also satisfy the inclusions $x(t) \in M_t \cap W_t$, $M_t \cap W_t$ being a *compact* set.

Now let us apply change (42.6), whereby t is varied from 0 to N. Then relations (40.1) are transformed into equations (42.5) (see (42.4)), while inclusions (42.1) take the form $u_t \in U_t$, $t = 1, \ldots, N$. Thus change (42.6) transforms the set of all processes in object (40.1), (42.1) satisfying the phase constraints $x(t) \in M_t$ into the set of all points of the space E^* of variable $z = (x_t^i, u_\tau^j)$ satisfying equations (42.5) and inclusions $x_t \in M_t \cap W_t$, $u_\tau \in U_\tau$ $(t = 0, 1, \ldots, N; \tau = 1, \ldots, N)$. Let us call this set Ω.

Note that inclusions $x_t \in M_t \cap W_t$, $u_\tau \in U_\tau$ (where $t = 0, 1, \ldots, N$; $\tau = 1, \ldots, N$) define in space E^* some *compact* set (since sets $M_t \cap W_t$ and U_τ are compact); let us call this set P. On the other hand, equations (42.5) define in space E^* some *closed* set Q (in view of the continuity of functions $f_t^i(x, u)$). Consequently, set $\Omega = P \cap Q$ is also *compact.* In addition, this set is *nonempty,* since by assumption there exists in object (40.1), (42.1) at least one process satisfying the phase constraints $x(t) \in M_t$.

Thus change (42.6) reduces the given discrete problem of optimal control with phase constraints to a problem of the minimum of a continuous function $F^0(z)$ (see (42.3)), considered in the nonempty compact set Ω. We know that a minimum is reached in this case, that is, there exists a point $z_0 \in \Omega$ at which function $F^0(z)$, considered on set Ω, reaches a minimum. However, then the process corresponding to point z_0 in view of change (42.6) is optimal.

Note that, by making the argument only slightly more complicated, we can arrive at an analogous theorem not for process (40.1), (42.1), but rather for process (40.1), (40.2), on condition that sets $U_t(x)$ are compact, nonempty, and continuously dependent on x for all $t = 1, \ldots, N$.

As an example, let us consider the application of Theorem 43.4 to the problem of the maximum of the product examined in Example 9.2 and solved in Subsection 39 using the method of dynamic programming. First of all, we note that in the problem as posed the control domain $U_t(x)$ actually *depends* on x, which prevents application of the methods discussed in Subsections 42 and 43. However, if we put aside the restriction $u \leqslant a - x^1$, imposing instead the condition that the sum of all the numbers $u(1), \ldots, u(N)$ does not exceed a (that is, $x^1(N) \leqslant a$), then we can state the problem somewhat differently (albeit equivalently), the control domain U being in this case constant.

Hence we consider object (40.1), where

$$f_t^1(x^1, x^2, u) = x^1 + u, \quad f_t^2(x^1, x^2, u) = x^2 u; \qquad t = 1, \ldots, N, \qquad (44.1)$$

with a constant control domain U, defined by the inequality

$$u \geqslant 0, \qquad (44.2)$$

while on the final phase state we impose the constraint $x^1(N) \leqslant a$. Next, if we introduce the function

$$w_N^1 = x^1(N) - a,$$

then this constraint can be written as

$$w_N^1 \leqslant 0. \qquad (44.3)$$

This gives a problem of optimal control for object (40.1), (42.1) (see (44.1)) with a constant control domain (44.2), the initial phase state $x^1(0) = 0$, $x^2(0) = 1$ being specified, while the constraint $x(N) \in M_N$ is placed on the final state, set M_N being defined by inequality (44.3). For this object the problem is to find the maximum of functional 40.5, where $f_1^0 = \ldots = f_{N-1}^0 \equiv 0$, $f_N^0(x^1, x^2, u) = x^2 u$.

Let $u(t), x(t)$ be an optimal process. According to Theorem 43.4, there exist numbers $\psi_0 \geqslant 0$, $\nu_1(N) \leqslant 0$ (in the following we will write just ν, rather than $\nu_1(N)$), and vectors

$$\psi(t) = \{\psi_1(t), \psi_2(t)\}, \qquad t = 1, \ldots, N,$$

such that conditions (A′), (B′), (C′) are satisfied. In this case condition (B′) becomes*

$$\begin{aligned}
&\psi_1(N) = \nu; \qquad \psi_1(t) = \psi_1(t+1), \qquad t = 1, \ldots, N-1;\\
&\psi_2(N) = 0, \quad \psi_2(N-1) = \psi_0 u(N) + \psi_2(N) u(N);\\
&\psi_2(t) = \psi_2(t+1) u(t+1), \qquad t = 1, \ldots, N-2.
\end{aligned}$$

* Condition (B′) is written out only for $t = 1, \ldots, N$. Relation (B′) for $t = 0$ includes the quantities $\mu_i(0)$ (since set M_0 is defined by the equations $x^i(0) - x_0^i = 0$, $i = 1, 2$). Therefore, condition (B′) for $t = 0$ enables us to determine numbers $\mu_1(0)$, $\mu_2(0)$ in the given case, but it provides no new information about the quantities ψ_0, $\psi_i(t)$. This situation is typical for problems with a fixed left-hand endpoint.

These relations imply directly that

$$\psi_1(t) = \nu, \qquad t = 1, \ldots, N;$$
$$\psi_2(t) = \psi_0 u(t+1) \cdot \ldots \cdot u(N), \qquad t = 1, \ldots, N-1; \qquad \psi_2(N) = 0.$$

Now note that, if $u(t) = 0$ for at least one t, then it follows from (44.1) that $x^2(t) = \ldots = x^2(N) = 0$, giving functional (40.5) a value of 0. Let us ignore this case (which obviously does not give a solution of the problem), that is, let us consider only those processes for which $u(t) \neq 0$, $t = 1, \ldots, N$. In this case δu in condition (C) (see Theorem 43.1) can assume both positive and negative values, so that condition (C) becomes

$$\psi_0 x^2(N-1) + \psi_1(N) + \psi_2(N) x^2(N-1) = 0,$$
$$\psi_1(t) + \psi_2(t) x^2(t-1) = 0, \qquad t = 1, \ldots, N-1.$$

By substituting the values found above for $\psi_1(t)$, $\psi_2(t)$, we now obtain

$$\psi_0 x^2(N-1) = -\nu, \quad \psi_0 x^2(t-1) u(t+1) \cdot \ldots \cdot u(N) = -\nu, \qquad (44.4)$$
$$t = 1, \ldots, N-1.$$

From this it is clear that $\psi_0 \neq 0$ (since otherwise we would have $\nu = 0$, giving $\psi_1(t) = \psi_2(t) = 0$ for all t, which contradicts condition (A)). Since, in addition, $x^2(N-1) \neq 0$ (in view of the previous assumption that all the numbers $u(t)$ are different from zero), it follows that $\nu \neq 0$. Thus, $\psi_0 > 0$, $\nu < 0$. The relation $\nu w_N^1(x(N)) = 0$ in condition (B′) now implies that $w_N^1(x(N)) = 0$, that is, $x^1(N) = a$.

Next, by taking into account that $x^2(t) = u(1) \cdot \ldots \cdot u(t)$ (see (40.1), (44.1)) and by multiplying the first of relations (44.4) by $u(N)$ and the second by $u(t)$, we get

$$\nu u(N) = \nu u(t), \qquad t = 1, \ldots, N-1,$$

that is, $u(1) = \ldots = u(N)$. Finally, since $u(1) + \ldots + u(N) = a$, we see that ultimately

$$u(1) = \ldots = u(N) = \frac{a}{N}.$$

Consequently, there exists a *single* process satisfying the necessary conditions specified in Theorem 43.4. From Theorem 44.1 (which is applicable, since in this case the relations $u(1) + \ldots + u(N) = x^1(N) \leqslant a$ imply that $0 \leqslant u(t) \leqslant a$, $t = 1, \ldots, N$, that is, each of the quantities $u(t)$ can vary only in the *compact* set $U_t^* = [0, a]$) we see that this process is actually optimal. Hence we are again led to the result obtained on pp. 313–315 using the method of dynamic programming.

45. Discrete objects with variable control domain. So far we have considered objects having a control domain $U_t(x) = U_t$ which for each t does not depend on x.

Now let us remove this constraint, that is, we take an object (40.1), (40.2) having a general form. However, we restrict ourselves to the case where set $U_t(x)$ for each $x=(x^1, \ldots, x^n) \in E^n$ is specified by a system of equations and inequalities of the type

$$G_t^j(x^1, \ldots, x^n, u^1, \ldots, u^r)=0, \qquad j=1, \ldots, k_t; \tag{45.1}$$
$$g_t^i(x^1, \ldots, x^n, u^1, \ldots, u^r) \leqslant 0, \qquad i=1, \ldots, q_t. \tag{45.2}$$

With regard to sets M_τ, we retain the assumptions made in Theorem 43.5, that is, we assume that

$$M_\tau = \Pi'_\tau \cap \Pi''_\tau \cap \Pi_\tau^{(1)} \cap \ldots \cap \Pi_\tau^{(n_\tau)} \cap \Sigma_\tau^{(1)} \cap \ldots \cap \Sigma_\tau^{(s_\tau)}, \tag{45.3}$$

where set Π'_τ is defined in the space of variables $x^1, \ldots, x^n$ by the system of equations

$$W_\tau^j(x^1, \ldots, x^n)=0, \qquad j=1, \ldots, r_\tau, \tag{45.4}$$

set Π''_τ is defined by the system of inequalities

$$w_\tau^i(x^1, \ldots, x^n) \leqslant 0, \qquad i=1, \ldots, p_\tau, \tag{45.5}$$

and $\Pi_\tau^{(1)}, \ldots, \Pi_\tau^{(n_\tau)}, \Sigma_\tau^{(1)}, \ldots, \Sigma_\tau^{(s_\tau)}$ are certain sets of the space of variables $x^1, \ldots, x^n$. All the functions G_t^i, g_t^i, W_t^i, w_t^i are assumed to be smooth.

THEOREM 45.1. *Consider controlled object* (40.1), (40.2), *where set* $U_t(x)$ *for each* $x=(x^1, \ldots, x^n) \in E^n$ *is defined by the system of relations* (45.1), (45.2). *In addition, we assume that for each* $t=0, 1, \ldots, N$ *a set* M_t *of form* (45.3) *is specified (see* (45.4), (45.5)*). Let* (40.3), (40.4) *be some process in the given object satisfying the phase constraints* $x(t) \in M_t$, $t=0, 1, \ldots, N$. *Moreover, for each* $\tau=0, 1, \ldots, N$ *the convex cones* $P_\tau^{(i)}$, $Q_\tau^{(j)}$ $(i=1, \ldots, n_\tau;\ j=1, \ldots, s_\tau)$ *are assumed to be defined in* E^n*; these cones, which have a vertex* $x(\tau)$, *are the coverings, respectively, of sets* $\Pi_\tau^{(i)}$, $\Sigma_\tau^{(j)}$ *at point* $x(\tau)$, *where for each* $\tau=0, 1, \ldots, N$ *the system of convex cones* $Q_\tau^{(j)}$, $j=1, \ldots, s_\tau$, *does not possess separability in* E^n.

For optimality of the process (40.3), (40.4) *under consideration (in the sense of the minimum of functional* (40.5) *with the phase constraints* $x(t) \in M_t$, $t=0, 1, \ldots, N$), *there must exist a number* $\psi_0 \leqslant 0$ *and vectors*

$$\psi(t)=\{\psi_1(t), \ldots, \psi_n(t)\}, \qquad t=1, \ldots, N;$$
$$\varphi(t)=\{\varphi_1(t), \ldots, \varphi_{k_t}(t)\}, \qquad t=1, \ldots, N;$$
$$\lambda(t)=\{\lambda_1(t), \ldots, \lambda_{q_t}(t)\}, \qquad t=1, \ldots, N;$$
$$\mu(t)=\{\mu_1(t), \ldots, \mu_{r_t}(t)\}, \qquad t=0, 1, \ldots, N;$$
$$\nu(t)=\{\nu_1(t), \ldots, \nu_{p_t}(t)\}, \qquad t=0, 1, \ldots, N;$$
$$b_t^{(i)}; \qquad i=1, \ldots, n_t; \quad t=0, 1, \ldots, N,$$

such that the direction of vector $\mathbf{b}_t^{(i)}$ *lies in the dual cone* $D(P_t^{(i)})$ *and conditions* (A) *through* (E), *below, are satisfied. These conditions are formulated using the function* $H_t(x, u)$, *defined as*

$$H_t(x, u) = \psi_0 f_t^0(x, u) + \sum_{i=1}^{n} \psi_i(t) f_t^i(x, u) + \\ + \sum_{\alpha=1}^{k_t} \varphi_\alpha(t) G_t^\alpha(x, u) + \sum_{\alpha=1}^{q_t} \lambda_\alpha(t) g_t^\alpha(x, u):$$

(A) *if* $\psi_0 = 0$, *then for at least one* t *at least one of the vectors* $\boldsymbol{\psi}(t)$, $\boldsymbol{\varphi}(t)$, $\boldsymbol{\lambda}(t)$, $\boldsymbol{\mu}(t)$, $\boldsymbol{\nu}(t)$, $\boldsymbol{b}_t^{(i)}$ *is different from zero;*

(B) *for any* $t = 0, 1, \ldots, N$ *and any vector* $\boldsymbol{\delta x}$ *whose direction lies in the intersection of cones* $Q_t^{(j)}$ $(j = 1, \ldots, s_t)$, *the following inequality is valid:*

$$\Big(-\boldsymbol{\psi}(t) + \operatorname{grad}_x H_{t+1}(x(t), u(t+1)) + \sum_{\alpha=1}^{r_t} \mu_\alpha(t) \operatorname{grad} W_t^\alpha(x(t)) + \\ + \sum_{\alpha=1}^{p_t} \nu_\alpha(t) \operatorname{grad} w_t^\alpha(x(t)) - \sum_{j=1}^{n_t} \boldsymbol{b}_t^{(j)}\Big) \boldsymbol{\delta x} \leqslant 0,$$

where it is assumed that $\boldsymbol{\psi}(0) = \mathbf{0}$ *and* $H_{N+1} \equiv 0$;

(C) $\operatorname{grad}_u H_t(x(t-1), u(t)) = \mathbf{0}$, $t = 1, \ldots, N$;

(D) $\lambda_\alpha(t) \leqslant 0$, $\lambda_\alpha(t) g_t^\alpha(x(t-1), u(t)) = 0$ *for all* $\alpha = 1, \ldots, q_t$; $t = 1, \ldots, N$;

(E) $\nu_\alpha(t) \leqslant 0$, $\nu_\alpha(t) w_t^\alpha(x(t)) = 0$ *for all* $\alpha = 1, \ldots, p_t$; $t = 0, 1, \ldots, N$.

Proof. Change (42.6) reduces the given discrete problem of optimal control to a problem of finding the minimum of function $F^0(z)$ (see (42.3)), considered on set Σ^*. The latter set is described by system of equations (42.5), (45.1), 45.4), system of inequalities (45.2), (45.5), and system of inclusions

$$x_t \in \Pi_t^{(i)}, \qquad i = 1, \ldots, n_t; \quad t = 0, 1, \ldots, N;$$
$$x_t \in \Sigma_t^{(j)}, \qquad j = 1, \ldots, s_t; \quad t = 0, 1, \ldots, N.$$

Hence Theorem 36.8 can be applied. According to this theorem, there must exist numbers

$$\psi_0 \leqslant 0, \quad \psi_i^t, \qquad i = 1, \ldots, n; \quad t = 1, \ldots, N;$$
$$\varphi_j^t, \qquad j = 1, \ldots, k_t; \quad t = 1, \ldots, N;$$
$$\lambda_i^t, \qquad i \in I_t(x(t-1), u(t)), \quad t = 1, \ldots, N;$$
$$\mu_k^t, \qquad k = 1, \ldots, r_t; \quad t = 0, 1, \ldots, N;$$
$$\nu_j^t, \qquad j \in J_t(x(t)), \quad t = 0, 1, \ldots, N,$$

and vectors $b_t^{(i)}$, $i = 1, \ldots, n_t$; $t = 0, 1, \ldots, N$, such that the conditions of Theorem 36.8 are satisfied. Here $I_t(x(t-1), u(t))$ is the set of all numbers $i = 1, \ldots, q_t$, for which the equation $g_t^i(x(t-1), u(t)) = 0$ is valid, while $J_t(x(t))$ is the set of all numbers $j = 1, \ldots, p_t$ for which the equation $w_t^j(x(t)) = 0$ is valid. Moreover, vector $b_t^{(i)}$ has a direction lying in cone $D(P_t^{(i)})$. Finally, condition (β) of Theorem 36.8 has in this case the following form (where z_0 denotes the point corresponding to process (40.3), (40.4), in view of change (42.6)):

$$\sum_{t=0}^{N}\Big(\psi_0 \operatorname{grad}_{x_t} F^0(z_0) + \sum_{\alpha=1}^{n}\sum_{\gamma=1}^{N} \psi_\alpha^\gamma \operatorname{grad}_{x_t} F_\gamma^\alpha(z_0)\Big)\,\delta x_t + \sum_{\tau=1}^{N}\Big(\psi_0 \operatorname{grad}_{u_\tau} F^0(z_0) +$$
$$+ \sum_{\alpha=1}^{n}\sum_{\gamma=1}^{N} \psi_\alpha^\gamma \operatorname{grad}_{u_\tau} F_\gamma^\alpha(z_0)\Big)\,\delta u_\tau + \sum_{t=0}^{N}\sum_{\gamma=1}^{N}\sum_{\alpha=1}^{k_\gamma} \varphi_\alpha^\gamma\big(\delta x_t \operatorname{grad}_{x_t} G_\gamma^\alpha(z_0)\big) +$$
$$+ \sum_{\tau=1}^{N}\sum_{\gamma=1}^{N}\sum_{\alpha=1}^{k_\gamma} \varphi_\alpha^\gamma\big(\delta u_\tau \operatorname{grad}_{u_\tau} G_\gamma^\alpha(z_0)\big) +$$
$$+ \sum_{t=0}^{N}\sum_{\gamma=0}^{N}\sum_{\alpha=1}^{r_\gamma} \mu_\alpha^\gamma\big(\delta x_t \operatorname{grad}_{x_t} W_\gamma^\alpha(z_0)\big) +$$
$$+ \sum_{t=0}^{N}\sum_{\gamma=1}^{N}\sum_{\alpha} \lambda_\alpha^\gamma\big(\delta x_t \operatorname{grad}_{x_t} g_\gamma^\alpha(z_0)\big) +$$
$$+ \sum_{\tau=1}^{N}\sum_{\gamma=1}^{N}\sum_{\alpha} \lambda_\alpha^\gamma\big(\delta u_\tau \operatorname{grad}_{u_\tau} g_\gamma^\alpha(z_0)\big) +$$
$$+ \sum_{t=0}^{N}\sum_{\gamma=0}^{N}\sum_{\beta} \nu_\beta^\gamma\big(\delta x_t \operatorname{grad}_{x_t} w_\gamma^\beta(z_0)\big) - \sum_{\tau=0}^{N}\sum_{i=1}^{n_\tau} b_\tau^{(i)}\,\delta z \leqslant 0$$

for any vector $\delta z = \{\delta x_t, \delta u_\tau\}$ possessing the property that for each $t = 0, 1, \ldots, N$ the direction of vector δx_t lies in the intersection of cones $Q_t^{(i)}$ $(i = 1, \ldots, s_t)$. Here all the quantities λ_α^γ, ν_β^γ are nonpositive, while in those sums for which no ranges of indexes α and β are given, it is assumed that the summation is over $\alpha \in I_\gamma(x(\gamma-1), u(\gamma))$, $\beta \in J_\gamma(x(\gamma))$. However, by setting

$$\lambda_\alpha^\gamma = 0 \quad \text{for} \quad \alpha \notin I_\gamma(x(\gamma-1), u(\gamma)) \ \text{and} \ \nu_\beta^\gamma = 0 \quad \text{for} \quad \beta \notin J_\gamma(x(\gamma)),$$

we can assume that the summation is over all $\alpha = 1, \ldots, q_\gamma$ and $\beta = 1, \ldots, p_\gamma$, all the values of λ_α^γ, ν_β^γ involved being nonpositive.

Now let θ be one of the numbers $0, 1, \ldots, N$ We set $\delta u_\tau = 0$ for all τ, and we set $\delta x_t = 0$ for $t \neq \theta$, while as δx_θ we take all possible vectors whose directions lie in the intersection of cones $Q_\theta^{(i)}$, $i = 1, \ldots, s_\theta$. The vector $\delta z = \{\delta x_t, \delta u_\tau\}$ so obtained possesses the property that for any $t = 0, 1, \ldots, N$ the direction of vector $\boldsymbol{\delta x}_t$ lies in each of cones $Q_t^{(i)}$, $i=1, \ldots, s_t$, so that for this vector the condition (β) given previously must be satisfied. Taking into account the form of functions F^0, F_t^i, G_t^i, W_t^i, g_t^i, w_t^i, we thus get condition (B) of Theorem 45.1; it is enough to set

$$\psi_\alpha(t)=\psi_\alpha^t, \quad \varphi_\alpha(t)=\varphi_\alpha^t, \quad \mu_\alpha(t)=\mu_\alpha^t, \quad \lambda_\alpha(t)=\lambda_\alpha^t, \quad \nu_\alpha(t)=\nu_\alpha^t.$$

Moreover, let us select an arbitrary θ $(1 \leqslant \theta \leqslant N)$ and let us assume that $\boldsymbol{\delta u}_\tau=\mathbf{0}$ for $\tau \neq \theta$. In addition, let $\boldsymbol{\delta x}_t=\mathbf{0}$ for all t, and finally let $\boldsymbol{\delta u}_\theta$ be *arbitrary* vectors. For the obtained vector $\boldsymbol{\delta z}=\{\boldsymbol{\delta x}_t, \boldsymbol{\delta u}_\tau\}$ the condition (β) given previously must also be satisfied, and this gives condition (C) of Theorem 45.1. Finally, it is easy to show that conditions (A), (D), (E) are satisfied.

Note that, if functions (45.1), (45.2) do not depend on x, that is, if they define a constant control domain U_t, then Theorem 45.1 is transformed into a theorem of the type considered in the preceding subsection, or, more precisely, into Theorem 43.5 with $l_t=m_t=0$.

46. Discrete maximum principle (method of local sections). In this subsection we will give another solution of the optimal-control problem for a discrete object with a variable control domain, a solution characterized by great generality. We will say that object (40.1), (40.2) *possesses local sections** if for any t, x_0, u_0 satisfying the condition $u_0 \in U_t(x_0)$ it is possible to find a smooth function $\chi_t(x)$, defined in some neighborhood of point x_0 and assuming values in the space of variables $u^1, \ldots, u^r$, such that

$$\chi_t(x) \in U_t(x) \quad \text{and} \quad \chi_t(x_0)=u_0.$$

Function $\chi_t(x)$ will be called the *local section* corresponding to the selected values t, x_0, u_0. Note that function $\chi_t(x)$ is assumed to be smooth, that is, it has continuous partial derivatives

$$\frac{\partial \chi_t^i(x)}{\partial x^j}, \qquad i=1, \ldots, r; \quad j=1, \ldots, n;$$

on the other hand, *nothing whatsoever* is assumed about the nature of the dependence of function $\chi_t(x)$ on x_0, u_0 (not even continuity).

In addition, if for object (40.1), (40.2) the control domain does not depend on x, that is, has the form (42.1), while functions $f_t^i(x, u)$ are continuously differentiable with respect to x, then this object obviously possesses local sections. Actually, let $u_0 \in U_t$. Now if we set $\chi_t(x) \equiv u_0$, then $\chi_t(x)$ is a local section (in the given case defined not locally, that is, not close to some specified point x_0, but rather throughout the entire space E^n).

* This term was borrowed from the *theory of fiber bundles,* one of the most important branches of modern topology.

THEOREM 46.1 *(maximum principle). Consider a controlled object* (40.1), (40.2) *possessing local sections. Let* $V_t(x)$ *be the subset of* $(n+1)$*-dimensional space* E^{n+1} *filled by all the points*

$$\tilde{f}_t(x, u) = (f_t^0(x, u), f_t^1(x, u), \ldots, f_t^n(x, u)),$$

where set $U_t(x)$ *is also run through, and let set* $V_t(x)$ *be compact, convex, and continuously dependent on* $x \in E^n$ *for any point* $x \in E^n$ *and any* $t = 1, \ldots, N$. *Functions* $f_t^i(x, u)$ *are assumed to be continuously differentiable with respect to* x *and* u. *Moreover, let us specify sets* M_t, $t = 0, 1, \ldots, N$, *having the form*

$$M_t = \Pi_t^{(1)} \cap \ldots \cap \Pi_t^{(n_t)},$$

where $\Pi_t^{(i)}$ *are certain sets of space* E^n. *Let* (40.3), (40.4) *be some process in the object being considered which satisfies the phase constraints* $x(t) \in M_t$, $t = 0, 1, \ldots, N$, *and let* $P_t^{(i)}$ *be a covering of set* $\Pi_t^{(i)}$ *at point* $x(t)$. *For each* $t = 1, \ldots, N$ *we select some local section* $\chi_t(x)$ *defined close to the point* $x(t-1)$:

$$\chi_t(x) \in U_t(x), \quad \chi_t(x(t-1)) = u(t), \qquad t = 1, \ldots, N.$$

For optimality of the given process (40.3), (40.4) *(in the sense of a minimum of functional* (40.5) *with constraints* $x(t) \in M_t$*), there must exist a number* $\psi_0 \leqslant 0$ *and vectors*

$$\psi(t) = \{\psi_1(t), \ldots, \psi_n(t)\}, \qquad t = 1, \ldots, N;$$
$$b_t^{(i)}, \qquad i = 1, \ldots, n_t; \quad t = 0, 1, \ldots, N,$$

such that the direction of vector $b_t^{(i)}$ *lies (in space* E^n*) in the dual cone* $D(P_t^{(i)})$ *and conditions* (A) *through* (C), *below, are satisfied. These conditions are formulated using functions* $H_t(x, u)$, $H_t^*(x)$, *defined as*

$$H_t(x, u) = \psi_0 f_t^0(x, u) + \sum_{i=1}^{n} \psi_i(t) f_t^i(x, u); \quad H_t^*(x) = H_t(x, \chi_t(x)):$$

(A) *if* $\psi_0 = 0$, *then for at least one* t *at least one of the vectors* $\psi(t)$, $b_t^{(i)}$ *is different from zero;*

(B) $\psi(t) = \operatorname{grad} H_{t+1}^*(x(t)) - \sum_{j=1}^{n_t} b_t^{(j)}$, $t = 0, 1, \ldots, N$, *where it is assumed that* $\psi(0) = 0$, $H_{N+1}^* \equiv 0$;

(C) $H_t(x(t-1), u(t)) = \max\limits_{u \in U_t(x(t-1))} H_t(x(t-1), u)$, $t = 1, \ldots, N$.

Note that, to avoid misunderstandings, during the calculation of the vector $\operatorname{grad} H_{t+1}^*(x(t))$ the formula of the total derivative has to be applied:

$$\frac{\partial H_{t+1}^*(x(t))}{\partial x^i} = \frac{\partial H_{t+1}(x(t), u(t+1))}{\partial x^i} + \sum_{j=1}^{r} \frac{\partial H_{t+1}(x(t), u(t+1))}{\partial u^j} \frac{\partial \chi_{t+1}^j(x(t))}{\partial x^i}.$$

Proof. First let us reformulate somewhat the optimal-control problem stated above. We introduce the new variables $v^0, v^1, \ldots, v^n$ and we call E^{n+1} the space of these variables. For any given x, u, the point $\tilde{f}_t(x, u)$, having coordinates

$$v^0 = f_t^0(x, u), \quad v^1 = f_t^1(x, u), \ldots, v^n = f_t^n(x, u), \tag{46.1}$$

belongs to this space. As previously, let the *trajectory* be an arbitrary sequence of points

$$x(0), \quad x(1), \ldots, x(N) \tag{46.2}$$

of space E^n, but now let the *control* be some sequence of points

$$v(1), \ldots, v(N) \tag{46.3}$$

of space E^{n+1}. According to the condition of the theorem, for any point $x \in E^n$ and any $t = 1, \ldots, N$, some set $V_t(x) \subset E^{n+1}$ is specified. We will say that trajectory (46.2) *corresponds to* control (46.3) if the following relations are valid:

$$v(t) \in V_t(x(t-1)), \qquad t = 1, \ldots, N; \tag{46.4}$$

$$x^i(t) = v^i(t), \qquad i = 1, \ldots, n; \quad t = 1, \ldots, N. \tag{46.5}$$

As a minimized functional for object (46.4), (46.5), we take

$$J = v^0(1) + v^0(2) + \ldots + v^0(N). \tag{46.6}$$

Relations (46.4) through (46.6) allow us to formulate a discrete problem of optimal control. This problem involves finding a process (that is, the control (46.3) and the trajectory (46.2) corresponding to it) which satisfies the conditions

$$x(t) \in M_t, \qquad t = 1, \ldots, N, \tag{46.7}$$

and for these conditions effects a minimum of functional (46.6).

Clearly, the problem (40.1), (40.2), (40.5), (46.7) reduces to a problem of optimal control in the form (46.4)–(46.7). This reduction is carried out precisely by formulas (46.1).

Let (40.3), (40.4) be some process for object (40.1), (40.2) that satisfies the constraints $x(t) \in M_t$, $t = 0, 1, \ldots, N$. If we set

$$v^i(t) = f_t^i(x(t-1), u(t)); \qquad i = 0, 1, \ldots, n; \ t = 1, \ldots, N, \tag{46.8}$$

then in view of (40.1), (40.2) the relations (46.4), (46.7) will be valid, while functional (40.5) has the form (46.6). In other words, change (46.1) transforms each process in object (40.1), (40.2) into some process in object (46.4), (46.5), but the value of functional (40.5) turns out to be *equal to* the corresponding value of functional (46.6). The constraints $x(t) \in M_t$ remain in force (since change (46.8) introduces just new *controls*, the trajectory remaining the same).

Conversely, for any process in object (46.4)–(46.6) (satisfying the constraints $x(t) \in M_t$), we can find the corresponding process in object (40.1), (40.2), (40.5) which is converted into it by change (46.8). To find such a process, it is sufficient to *solve* relations (46.8) for $u(t)$, $t=1, \ldots, N$, which is always possible in view of the definition of sets $V_t(x)$. Note that this backward conversion is in general not single-valued (that is, *many* processes $x(t)$, $u(t)$ may correspond to a single process $x(t)$, $v(t)$, in view of the lack of uniqueness of the solution of equation (46.8)). However, all of the processes $x(t)$, $u(t)$ corresponding via (46.8) to a single process $x(t)$, $v(t)$ impart the *very same* value to functional (40.5) (this value being equal to the value of functional (46.6) for process $x(t)$, $v(t)$).

It should also be noted that, if process $x(t)$, $u(t)$ is optimal for object (40.1), (40.2), (40.5) with the constraints $x(t) \in M_t$, then the process $x(t)$, $v(t)$ corresponding to it in view of (46.8) will be optimal for object (46.4)–(46.6) for these same constraints, and vice versa. Thus the problem of finding the optimal processes for object (40.1), (40.2), (40.5) is equivalent to the problem of finding the optimal processes for object (46.4)–(46.6). Let us now consider the latter problem, and after solving it we will return to object (40.1), (40.2), (40.5).

Let E^* be the space of variables x_t^i, v_τ^j, where $i=1, \ldots, n$; $j=0, 1, \ldots, n$; $t=0, 1, \ldots, N$; $\tau=1, \ldots, N$. In space E^* we consider the functions

$$F^0(z) = v_1^0 + v_2^0 + \ldots + v_N^0; \tag{46.9}$$

$$F_t^i(z) = -x_t^i + v_t^i, \qquad i=1, \ldots, n;\ t=1, \ldots, N. \tag{46.10}$$

In addition, let Ω' be the set of all points of space E^* whose coordinates satisfy the system of equations

$$F_t^i(z) = 0, \qquad i=1, \ldots, n; \quad t=1, \ldots, N; \tag{46.11}$$

finally, let $\tilde{\Pi}_t^{(i)}$ be the set of all points $z=(x_t^i, v_\tau^j)$ of space E^* satisfying the condition $x_t \in \Pi_t^{(i)}$ and let Ξ_τ be the set of all points z satisfying the condition

$$v_\tau = (v_\tau^0, v_\tau^1, \ldots, v_\tau^n) \in V_\tau(x_{\tau-1}) = V_\tau(x_{\tau-1}^1, \ldots, x_{\tau-1}^n).$$

The change

$$x_t^i = x^i(t); \qquad t=0, 1, \ldots, N; \quad i=1, \ldots, n;$$

$$v_\tau^j = v^j(\tau); \qquad j=0, 1, \ldots, n; \quad \tau=1, \ldots, N, \tag{46.12}$$

reduces the given optimal-control problem to a problem of finding the point at which function $F^0(z)$, considered on the set

$$\Sigma^* = \Omega' \cap \left(\bigcap_{t=0}^{N} \bigcap_{i=1}^{n_t} \tilde{\Pi}_t^{(i)} \right) \cap \left(\bigcap_{\tau=1}^{N} \Xi_\tau \right),$$

reaches its least value.

Let (46.2), (46.3) be the optimal process in object (46.4)–(46.7) corresponding, in view of (46.8), to the given optimal process (40.3), (40.4) in object (40.1), (40.2), (40.5) (with constraints $x(t) \in M_t$). If we set

$$\hat{x}^i_t = x^i(t), \quad \hat{v}^j_\tau = v^j(\tau) = f^j_\tau(x(\tau-1),\ u(\tau)),$$

then we obtain the point $\hat{z} = (\hat{x}^i_t,\ \hat{v}^j_\tau)$ at which function $F^0(z)$, considered on set Σ^*, reaches a minimum. Thus the necessary condition for an extremum stated in Theorem 36.8 is applicable here.

In order to use this theorem, let us call $\widetilde{P}^{(i)}_t$ the set of all points $z = (x^i_t,\ v^j_\tau)$ of space E^* satisfying the condition $x_t \in P^{(i)}_t$. Then $\widetilde{P}^{(i)}_t$ is a covering of set $\widetilde{\Pi}^{(i)}_t$ at point $\hat{z}$ $(i = 1, \ldots, n_t;\ t = 0, 1, \ldots, N)$. In addition, if we set

$$\tilde{\chi}_t(x) = \tilde{f}_t(x,\ \chi_t(x)),$$

then $\tilde{\chi}_t(x)$ is a smooth function defined close to point $\hat{x}_{t-1}$, where

$$\tilde{\chi}_t(x) \in V_t(x) \text{ and } \tilde{\chi}_t(\hat{x}_{t-1}) = \hat{v}_t.$$

Finally, for any $\theta = 1, \ldots, N$, let $\widetilde{L}_\theta$ be the set of all points of form $\hat{z} + \boldsymbol{\delta z}$, where $\boldsymbol{\delta z} = \{\delta x^i_t,\ \delta v^j_\tau\}$ satisfies the following condition:

(α) *the direction of vector* $\boldsymbol{\delta v}_\theta - \sum_{i=1}^{n} \frac{\partial \tilde{\chi}_\theta(\hat{x}_{\theta-1})}{\partial x^i_{\theta-1}} \delta x^i_{\theta-1}$ *lies in cone* L_θ, *where* L_θ *is the supporting cone of convex field* $V_\theta(\hat{x}_{\theta-1})$ *at point* $\hat{v}_\theta$.

Set $\widetilde{L}_\theta$ is obviously a closed convex cone of space E^* with a vertex at point $\hat{z}$. We will show that this cone is a covering of set Ξ_θ at point $\hat{z}$ and also that the system of convex cones $\widetilde{L}_\theta$, $\theta = 1, \ldots, N$, does not possess separability in E^*. Let us assume this to have been established already and let us show that then the theorem is proven.

In view of the assumption made, Theorem 36.8 (with $r = q = p = 0$) can be applied to function $F^0(z)$, considered on set Σ^*. Actually, set Σ^* is represented as the intersection of the set Ω', defined by system of equations (46.11), and the two systems of sets $\widetilde{\Pi}^{(i)}_t$ and Ξ_τ; here the coverings $\widetilde{P}^{(i)}_t$, $\widetilde{L}_\tau$ of these sets at point $\hat{z}$ are known, and in addition the system of cones $\widetilde{L}_\tau$, $\tau = 1, \ldots, N$, does not possess separability in E^*. According to Theorem 36.8, there must exist numbers $\psi_0 \leqslant 0$, ψ^t_i $(i = 1, \ldots, n, \ldots, N)$ and vectors $\tilde{\boldsymbol{b}}^{(i)}_t$ $(i = 1, \ldots, n_t;\ t = 0, 1, \ldots, N)$, all such that the direction of vector $\tilde{\boldsymbol{b}}^{(i)}_t$ lies in the dual cone $D(\widetilde{P}^{(i)}_t)$ and conditions (β) and (γ) of Theorem 36.8 are satisfied, these conditions being in the given case (see (46.9), (46.10)):

$$(\beta)\quad \sum_{t=0}^{N}\left(\psi_0 \operatorname{grad}_{x_t} F^0(\hat{z}) + \sum_{\alpha=1}^{n}\sum_{\gamma=1}^{N} \psi^\gamma_\alpha \operatorname{grad}_{x_t} F^\alpha_\gamma(\hat{z})\right)\boldsymbol{\delta x}_t +$$

$$+ \sum_{\tau=1}^{N}\left(\psi_0 \operatorname{grad}_{v_\tau} F^0(\hat{z}) + \sum_{\alpha=1}^{n}\sum_{\gamma=1}^{N} \psi^\gamma_\alpha \operatorname{grad}_{v_\tau} F^\alpha_\gamma(\hat{z})\right)\boldsymbol{\delta v}_\tau - \sum_{t=0}^{N}\sum_{i=1}^{n_t} \tilde{\boldsymbol{b}}^{(i)}_t \boldsymbol{\delta z} \leqslant 0$$

for any vector $\boldsymbol{\delta z}=\{\boldsymbol{\delta x}_t, \boldsymbol{\delta v}_\tau\}$ *whose direction lies in the intersection of cones* $\widetilde{L}_\tau$ $(\tau=1, \ldots, N)$;

(γ) *if all of the numbers* ψ_0, ψ_i^t *are zero, then at least one of the vectors* $\widetilde{\boldsymbol{b}}_t^{(i)}$ *is different from zero.*

Taking into account the form of functions F^0 and F_γ^α, we can rewrite condition (β) as

$$\psi_0 \sum_{\tau=1}^{N} \delta v_\tau^0 - \sum_{t=1}^{N} \sum_{i=1}^{n} \psi_i^t \delta x_t^i + \sum_{\tau=1}^{N} \sum_{j=1}^{n} \psi_j^\tau \delta v_\tau^j - \sum_{t=0}^{N} \sum_{i=1}^{n_t} \widetilde{\boldsymbol{b}}_t^{(i)} \boldsymbol{\delta z} \leqslant 0. \qquad (46.13)$$

Let us select some whole number $\theta\,(0 \leqslant \theta \leqslant N)$ and let us set all the $\boldsymbol{\delta x}_t$ with $t \neq \theta$ equal to zero, while as $\boldsymbol{\delta x}_\theta=\{\delta x_\theta^1, \ldots, \delta x_\theta^n\}$ we take arbitrary vectors. Moreover, vector $\boldsymbol{\delta v}_t$ is defined with the aid of the equations

$$\boldsymbol{\delta v}_t - \sum_{i=1}^{n} \frac{\partial \tilde{\chi}_t(\hat{x}_{t-1})}{\partial x^i} \delta x_{t-1}^i = \mathbf{0}, \qquad t=1, \ldots, N,$$

so that condition (α) will be satisfied. In other words,

$$\boldsymbol{\delta v}_{\theta+1} = \sum_{i=1}^{n} \frac{\partial \tilde{\chi}_{\theta+1}(\hat{x}_\theta)}{\partial x_\theta^i} \delta x_\theta^i, \quad \boldsymbol{\delta v}_t = \mathbf{0} \quad \text{for} \quad t \neq \theta+1.$$

The vector $\boldsymbol{\delta z}=\{\delta x_t^i, \delta v_\tau^j\}$ obtained has a direction lying in each of the cones $\widetilde{L}_\tau$, $\tau=1, \ldots, N$, so that relation (46.13) is valid for it. Therefore, assuming for convenience that $\psi_i^0=0$ $(i=1, \ldots, n)$, $\tilde{\chi}_{N+1}^\alpha=0$ $(\alpha=0, 1, \ldots, n)$, we get

$$\psi_0 \sum_{i=1}^{n} \frac{\partial \tilde{\chi}_{\theta+1}^0(\hat{x}_\theta)}{\partial x_\theta^i} \delta x_\theta^i - \sum_{i=1}^{n} \psi_i^\theta \delta x_\theta^i + \sum_{i=1}^{n} \sum_{j=1}^{n} \psi_j^{\theta+1} \frac{\partial \tilde{\chi}_{\theta+1}^j(\hat{x}_\theta)}{\partial x_\theta^i} \delta x_\theta^i -$$

$$- \sum_{i=1}^{n_\theta} \boldsymbol{b}_\theta^{(i)} \boldsymbol{\delta x}_\theta \leqslant 0, \qquad \theta=0, 1, \ldots, N, \qquad (46.14)$$

where $\boldsymbol{b}_\theta^{(i)}$ is a vector of space E^n having (on axes $\overset{1}{x}, \ldots, \overset{n}{x}$) the same coordinates as the vector $\widetilde{\boldsymbol{b}}_\theta^{(i)}$ has on axes $\overset{1}{x}_\theta, \ldots, \overset{n}{x}_\theta$ (the remaining coordinates of vector $\widetilde{\boldsymbol{b}}_\theta^{(i)}$ are zero, in view of the choice of cone $\widetilde{P}_\theta^{(i)}$). Relation (46.14) is valid for any vector $\boldsymbol{\delta x}_\theta=\{\delta x_\theta^1, \ldots, \delta x_\theta^n\}$, which implies condition (B) of Theorem 46.1 (because of the change $\psi_\alpha(t)=\psi_\alpha^t$ and the relation $\tilde{\chi}_t(x)=\tilde{f}_t(x, \chi_t(x))$).

Finally, we select an arbitrary number $\theta\,(=1, \ldots, N)$ and we set all the $\boldsymbol{\delta v}_\tau$ with $\tau \neq \theta$ equal to zero, while as $\boldsymbol{\delta v}_\theta=\{\delta v_\theta^0, \ldots, \delta v_\theta^n\}$ we will take arbitrary vectors whose directions lie in cone L_θ. In addition, we set $\boldsymbol{\delta x}_t=\mathbf{0}$ for all t. The vector $\boldsymbol{\delta z}=\{\boldsymbol{\delta x}_t, \boldsymbol{\delta v}_\tau\}$ obtained has a direction lying in each of the cones

$\tilde{L}_\tau$, $\tau = 1, \ldots, N$, so that relation (46.13) is valid for it. Therefore,

$$\psi_0 \delta v_\theta^0 + \sum_{j=1}^{n} \psi_j^\theta \delta v_\theta^j \leqslant 0, \qquad \theta = 1, \ldots, N,$$

for any vector $\boldsymbol{\delta v}_\theta = \{\delta v_\theta^0, \ldots, \delta v_\theta^n\}$ whose direction lies in cone L_θ.

This obviously means that function $\psi_0 v_\theta^0 + \sum_{\alpha=1}^{n} \psi_\alpha^\theta v_\theta^\alpha$, considered on cone L_θ, reaches a maximum at the vertex of this cone, that is, at point $\hat{v}_\theta$. But the convex set $V_\theta(\hat{x}_{\theta-1})$ *is contained* in its supporting cone L_θ (as is implied directly by the definition of the supporting cone). Thus the function $\psi_0 v_\theta^0 + \sum_{\alpha=1}^{n} \psi_\alpha^\theta v_\theta^\alpha$, considered on set $V_\theta(\hat{x}_{\theta-1})$, reaches a maximum at $\hat{v}_\theta$:

$$\psi_0 \hat{v}_\theta^0 + \sum_{\alpha=1}^{n} \psi_\alpha^\theta \hat{v}_\theta^\alpha = \max_{v \in V_\theta(\hat{x}_{\theta-1})} \left(\psi_0 v^0 + \sum_{\alpha=1}^{n} \psi_\alpha^\theta v^\alpha \right), \qquad \theta = 1, \ldots, N.$$

Hence we also obtain (in view of the change $\psi_i^t = \psi_i(t)$ and relation (46.8)) condition (C) of Theorem 46.1. Finally, condition (A) of Theorem 46.1 is implied directly by condition (γ) formulated above.

Accordingly, to complete the proof of Theorem 46.1, it remains to establish that cone $\tilde{L}_\tau$ is a covering of set Ξ_τ at point $\hat{z}$ and that the system of convex cones $\tilde{L}_\tau$, $\tau = 1, \ldots, N$, does not possess separability in E^*. We select an arbitrary $\theta\,(= 1, \ldots, N)$, and we let $z = (x_t, v_\tau) = (x_t^i, v_\tau^j)$ be an arbitrary point of cone $\tilde{L}_\theta$. In addition, let $\psi_\theta(z)$ be a point of convex set $V_\theta(x_{\theta-1})$ close to v_θ, and let $\Psi_\theta(z)$ be a point $\bar{z} = (\bar{x}_t, \bar{v}_\tau)$ satisfying the conditions $\bar{x}_t = x_t$, $t = 0, 1, \ldots, N$; $\bar{v}_\theta = \psi_\theta(z)$, $\bar{v}_\tau = v_\tau$ for $\tau \neq \theta$. This defines the mapping $\Psi_\theta \colon \tilde{L}_\theta \to E^*$. This mapping can easily be shown to be continuous (recall that the set $V_\theta(x)$ depends continuously on $x \in E^n$).

The inclusion $\psi_\theta(z) \in V_\theta(x_{\theta-1})$ (which follows from the definition of point $\psi_\theta(z)$) implies directly (in view of the definition of mapping Ψ_θ and set Ξ_θ) that $\Psi_\theta(z) \in \Xi_\theta$ for any point $z \in \tilde{L}_\theta$. Consequently, mapping Ψ_θ satisfies condition 2 of Definition 34.1.

Let us show that mapping Ψ_θ also satisfies condition 1, that is, that the relation $\Psi_\theta(z) = z + o_{\hat{z}}(z)$ is valid. To do this, we select an arbitrary positive number ε (not exceeding unity), which will remain the same throughout the discussion to come. In space E^{n+1} we draw a sphere with a center at point $\hat{v}_\theta$ and we call Σ_0 the intersection of this sphere with cone L_0 (that is, with the supporting cone of field $V_\theta(\hat{x}_{\theta-1})$ at point $\hat{v}_\theta$). We assume that $y_0 \in \Sigma_0$, and we call $l_\theta(y_\theta)$ the ray originating at $\hat{v}_\theta$ and passing through point y_θ. In addition, let $Z_{\theta-1}$ be a closed sphere of space E^n with a center $\hat{x}_{\theta-1}$ and a radius $\delta > 0$. For a sufficiently small δ, function $\tilde{\chi}_\theta$ will be defined throughout the entire sphere $Z_{\theta-1}$.

We take an arbitrary point $x_{\theta-1} \in Z_{\theta-1}$ and we consider in space E^{n+1} all the rays originating at point $\tilde{\chi}_\theta(x_{\theta-1})$ and forming angles not exceeding ε with the ray $l_\theta(y_\theta)$. In space E^{n+1} these rays fill a closed convex cone which we will call $K_{\varepsilon,\theta}(x_{\theta-1}, y_\theta)$. Since $y_\theta \in \Sigma_\theta$, that is, since the ray $l_\theta(y_\theta)$ is contained in cone L_θ, it follows that arbitrarily close to this ray there will be rays originating at $\hat{v}_0$ and passing through points of set $V_\theta(\hat{x}_{\theta-1})$ other than $\hat{v}_0$. Consequently, *inside* cone $K_{\varepsilon,\theta}(\hat{x}_{\theta-1}, y_\theta)$ there exists a ray originating at $\hat{v}_0$ and passing through a point of set $V_\theta(\hat{x}_{\theta-1})$ other than $\hat{v}_0$. According to continuity (taking into account the compactness of set Σ_θ), the radius δ of the sphere $Z_{\theta-1}$ can be chosen so small that for any points $y_\theta \in \Sigma_\theta$, $x_{\theta-1} \in Z_{\theta-1}$ there exists inside cone $K_{\varepsilon,\theta}(x_{\theta-1}, y_\theta)$ a ray originating at the vertex $\tilde{\chi}_\theta(x_{\theta-1})$ of this cone and containing a point of convex set $V_\theta(x_{\theta-1})$ other than $\tilde{\chi}_\theta(x_{\theta-1})$.

Let us take a number δ satisfying this condition and let us leave it unchanged in the ensuing discussion. Thus, the intersection

$$K_{\varepsilon,\theta}(x_{\theta-1}, y_\theta) \cap V_\theta(x_{\theta-1}) \tag{46.15}$$

contains points situated inside cone $K_{\varepsilon,\theta}(x_{\theta-1}, y_\theta)$ $(y_\theta \in \Sigma_\theta,\ x_{\theta-1} \in Z_{\theta-1})$ other than $\tilde{\chi}_\theta(x_{\theta-1})$; point $\tilde{\chi}_\theta(x_{\theta-1})$ obviously lies in this intersection as well. Intersection (46.15) is a convex compact set. It is easy to prove that this set depends *continuously* on the pair of variables $y_\theta \in \Sigma_\theta$, $x_{\theta-1} \in Z_{\theta-1}$.

If set (46.15) has points lying outside a sphere of radius 1 with a center at point $\tilde{\chi}_\theta(x_{\theta-1})$, then we set $d(x_{\theta-1}, y_\theta) = 1$. However, if set (46.15) lies wholly inside this sphere, then we let $d(x_{\theta-1}, y_\theta)$ denote the greatest distance from $\tilde{\chi}_\theta(x_{\theta-1})$ to the points of set (46.15). Clearly, $d(x_{\theta-1}, y_\theta) > 0$, since set (46.15) contains points different from $\tilde{\chi}_\theta(x_{\theta-1})$. Moreover, function $d(x_{\theta-1}, y_\theta)$ is *continuous*, since set (46.15) depends continuously on the pair of variables $x_{\theta-1}$, y_θ. Therefore, $d(x_{\theta-1}, y_\theta)$ is a *positive* continuous function whose independent variables run through the closed bounded sets: $x_{\theta-1} \in Z_{\theta-1}$, $y_\theta \in \Sigma_\theta$. Consequently, the minimum value of the function $d(x_{\theta-1}, y_\theta)$ is positive, that is, there exists an $h > 0$ such that

$$d(x_{\theta-1}, y_\theta) \geq h \quad \text{for any} \quad x_{\theta-1} \in Z_{\theta-1},\quad y_\theta \in \Sigma_\theta.$$

Now let us select a sufficiently small positive number h_0, smaller than the radius of sphere $Z_{\theta-1}$, such that for a vector $z - \hat{z} = \delta z = \{\delta x_t^i, \delta v_\tau^j\}$ with a length $|z - \hat{z}| < h_0$ the vector

$$\delta v_\theta - \sum_{i=1}^{n} \frac{\partial \tilde{\chi}_\theta(\hat{x}_{\theta-1})}{\partial x_{\theta-1}^i} \delta x_{\theta-1}^i \tag{46.16}$$

has a length less than h. Now, if $z \in \tilde{L}_\theta$, that is, if the vector $\delta z = z - \hat{z}$ satisfies condition (α) for $\tau = \theta$, then the direction of vector (46.16) lies in cone L_θ. Thus there exists a point $y_\theta \in \Sigma_0$ such that vector (46.16) is directed along the ray $l_\theta(y_\theta)$.

Thus there exists in set (46.15) a point w_θ lying a distance not less than h from $\tilde{\chi}_\theta(x_{\theta-1})$. The point

$$s_\theta = \tilde{\chi}_\theta(x_{\theta-1}) + \left(\delta v_\theta - \sum_{i=1}^{n} \frac{\partial \tilde{\chi}_\theta(\hat{x}_{\theta-1})}{\partial x_{\theta-1}^i} \delta x_{\theta-1}^i\right) \tag{46.17}$$

lies on the axis of cone $K_{\varepsilon,\theta}(x_{\theta-1}, y_\theta)$. Since point w_θ also lies in this cone, it follows that the angle between the vectors

$$s_\theta - \tilde{\chi}_\theta(x_{\theta-1}) \text{ and } w_\theta - \tilde{\chi}_\theta(x_{\theta-1})$$

does not exceed ε. In addition, the first of these vectors (that is, vector (46.16)) has a length less than h, while the second has a length which is not less than h. Consequently, point s'_θ, which is the projection of point s_θ onto the ray originating at $\tilde{\chi}_\theta(x_{\theta-1})$ and passing through point w_θ, belongs to set $V_\theta(x_{\theta-1})$ (since it lies on the segment joining points $\tilde{\chi}_\theta(x_{\theta-1})$ and w_θ, which belong to convex set $V_\theta(x_{\theta-1})$; Figure 158). On the other hand, the distance between points s_θ and s'_θ does not exceed $|s_\theta - \tilde{\chi}_\theta(x_{\theta-1})|\sin\varepsilon$, that is, it is less than

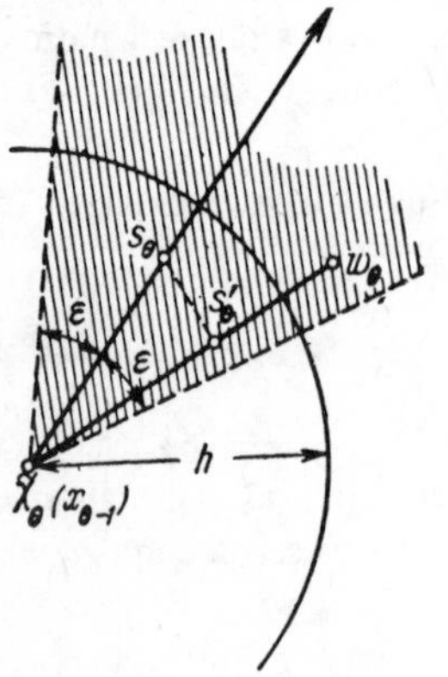

FIGURE 158

$$\varepsilon\left|\delta v_\theta - \sum_{i=1}^{n} \frac{\partial \tilde{\chi}_\theta(\hat{x}_{\theta-1})}{\partial x_{\theta-1}^i} \delta x_{\theta-1}^i\right|$$

(see (46.17)). Moreover, in view of (46.17),

$$\begin{aligned} s_\theta &= \tilde{\chi}_\theta(x_{\theta-1}) + \delta v_\theta - \sum_{i=1}^{n} \frac{\partial \tilde{\chi}_\theta(\hat{x}_{\theta-1})}{\partial x_{\theta-1}^i} \delta x_{\theta-1}^i = \\ &= \tilde{\chi}_\theta(\hat{x}_{\theta-1} + \delta x_{\theta-1}) + \delta v_\theta - \sum_{i=1}^{n} \frac{\partial \tilde{\chi}_\theta(\hat{x}_{\theta-1})}{\partial x_{\theta-1}^i} \delta x_{\theta-1}^i = \\ &= \tilde{\chi}_\theta(\hat{x}_{\theta-1}) + o(\delta x_{\theta-1}) + \delta v_\theta = \hat{v}_\theta + \delta v_\theta + o(\delta x_{\theta-1}) = v_\theta + o(\delta x_{\theta-1}), \end{aligned}$$

that is, the distance between points s_θ and v_θ is $o(\delta x_{\theta-1})$.

Therefore, at some distance from point v_θ not exceeding

$$\varepsilon\left|\delta v_\theta - \sum_{i=1}^{n} \frac{\partial \tilde{\chi}_\theta(\hat{x}_{\theta-1})}{\partial x_{\theta-1}^i} \delta x_{\theta-1}^i\right| + o(\delta x_{\theta-1}), \tag{46.18}$$

there exists a point $s'_\theta \in V_\theta(x_{\theta-1})$. Consequently, the distance from v_θ to the point $\psi_\theta(z)$ of set $V_\theta(x_{\theta-1})$ *closest* to it will, even more so, not exceed the value (46.18) (for $|\delta z| < h_0$). Since in this discussion the number $\varepsilon > 0$ was assumed to be arbitrary, this means that

$$\lim_{|\delta z| \to 0} \frac{|\psi_\theta(z) - v_\theta|}{|\delta z|} = 0,$$

that is,

$$\psi_\theta(z) = v_\theta + o_{\hat{z}}(z),$$

indicating that mapping Ψ_θ satisfies condition 1 of Definition 34.1. Thus $\tilde{L}_\tau$ is a covering of set Ξ_τ at point $\hat{z}$, $\tau = 1, \ldots, N$.

It remains to show that the system of convex cones $\tilde{L}_\tau$, $\tau = 1, \ldots, N$, does not possess separability in E^*. Let D_θ (where $\theta = 1, \ldots, N$) be the set of all points $z = (x_t, v_\tau)$ satisfying the conditions

$$x_t = \hat{x}_t \quad \text{for} \quad t = 0, 1, \ldots, N, \quad v_\tau = \hat{v}_\tau \quad \text{for} \quad \tau \neq \theta.$$

In addition, let D_0 be the set of all points $z = (x_t, v_\tau)$ satisfying the conditions

$$v_t - \hat{v}_t = \sum_{i=1}^{n} \frac{\partial \tilde{\chi}_t(\hat{x}_{t-1})}{\partial x_{t-1}^i} (x_{t-1}^i - \hat{x}_{t-1}^i), \qquad t = 1, \ldots, N.$$

Obviously, E^*, considered as a vector space with a zero point $\hat{z}$, decomposes into the direct sum of its subspaces $D_0, D_1, \ldots, D_N$. If L_θ^* (where θ is one of the numbers $1, \ldots, N$) is the set of all points $z = (x_t^i, v_\tau^i) \in D_\theta$ satisfying the condition $v_\theta \in L_\theta$, then clearly $\tilde{L}_\theta$ is the convex hull of the set constituting the union of cone $L_\theta^* \subset D_\theta$ and all the subspaces $D_0, D_1, \ldots, D_N$ except D_θ. Therefore, Theorem 42.5 implies directly (cf. p. 337) that the system of convex cones $\tilde{L}_\tau$, $\tau = 1, \ldots, N$, does not possess separability in E^*.

In Theorem 46.1 it can be assumed that for each $t = 0, 1, \ldots, N$ some of the seta $\Pi_t^{(i)}$ are hypersurfaces (in the space E^n of variables $x^1, \ldots, x^n$), some are defined by inequalities of the type $w_t^{(i)}(x) \leqslant 0$, and some are arbitrary sets having coverings $P_t^{(i)}$. As a result, from Theorem 46.1 we obtain the following more general statement.

THEOREM 46.2 *(maximum principle). We consider a controlled object* (40.1), (40.2), (40.5) *in which the control domain $U_t(x)$ is compact and depends continuously on $x \in E^n$, while the set $V_t(x)$ of all points of form $\mathfrak{f}_t(x, u)$, $u \in U_t(x)$* (see (46.1)), *is convex for any $x \in E^n$. Moreover, let us assume that the considered object possesses local sections and that functions $f_t^i(x, u)$ are continuously differentiable with respect to x and u. Finally, assume that sets M_t, $t = 0, 1, \ldots, N$, have the form*

$$M_t = \Pi_t' \cap \Pi_t'' \cap \Pi_t^{(1)} \cap \ldots \cap \Pi_t^{(n_t)},$$

where set Π_t' *is defined in space* E^n *by the system of equations*

$$W_t^i(x^1, \ldots, x^n) = 0, \quad i = 1, \ldots, r_t, \tag{46.19}$$

set Π_t'' *is defined by the system of inequalities*

$$w_t^i(x^1, \ldots, x^n) \leqslant 0, \quad i = 1, \ldots, p_t, \tag{46.20}$$

and $\Pi_t^{(i)}$, $i = 1, \ldots, n_t$, *are certain sets of space* E^n. *Let* (40.3), (40.4) *be some process in the object considered satisfying the phase constraints* $x(t) \in M_t$, $t = 0, 1, \ldots, N$, *and let* $P_t^{(i)}$ *be a covering of set* $\Pi_t^{(i)}$ *at point* $x(t)$. *In the vicinity of point* $x(t)$ *all the functions* W_t^i, $i = 1, \ldots, r_t$; w_t^i, $i = 1, \ldots, p_t$; *are assumed to be smooth. For each* $t = 1, \ldots, N$ *we select a local section* χ_t *defined close to point* $x(t-1)$:

$$\chi_t(x(t-1)) = u(t), \quad \chi_t(x) \in U_t(x).$$

For optimality of the given process (40.3), (40.4) *(in the sense of a minimum of functional* (40.5) *with constraints* $x(t) \in M_t$), *there must exist a number* $\psi_0 \leqslant 0$ *and vectors*

$$\begin{aligned} &\psi(t) = \{\psi_1(t), \ldots, \psi_n(t)\}, && t = 1, \ldots, N; \\ &\mu(t) = \{\mu_1(t), \ldots, \mu_{r_t}(t)\}, && t = 0, 1, \ldots, N; \\ &\nu(t) = \{\nu_1(t), \ldots, \nu_{p_t}(t)\}, && t = 0, 1, \ldots, N; \\ &b_t^{(i)}, \quad i = 1, \ldots, n_t; && t = 0, 1, \ldots, N, \end{aligned}$$

all such that the direction of vector $b_t^{(i)}$ *lies (in space* E^n*) in the dual cone* $D(P_t^{(i)})$ *and conditions* (A) *through* (D), *below, are satisfied. These conditions are formulated using the functions* $H_t(x, u)$, $\overset{*}{H}_t(x)$, *defined as*

$$H_t(x, u) = \psi_0 f_t^0(x, u) + \sum_{i=1}^{n} \psi_i(t) f_t^i(x, u); \quad \overset{*}{H}_t(x) = H_t(x, \chi_t(x)):$$

(A) *if* $\psi_0 = 0$, *then for at least one* t *at least one of the vectors* $\psi(t)$, $\mu(t)$, $\nu(t)$, $b_t^{(i)}$ *is different from zero;*

(B) $$\psi(t) = \operatorname{grad} \overset{*}{H}_{t+1}(x(t)) + \sum_{j=1}^{r_t} \mu_j(t) \operatorname{grad} W_t^j(x(t)) + \sum_{j=1}^{p_t} \nu_j(t) \operatorname{grad} w_t^j(x(t)) - \sum_{j=1}^{n_t} b_t^{(j)}, \quad t = 0, 1, \ldots, N,$$

where it is assumed that $\psi(0) = 0$ *and* $\overset{*}{H}_{N+1} \equiv 0$;

(C) $$H_t(x(t-1), u(t)) = \max_{u \in U_t(x(t-1))} H_t(x(t-1), u), \quad t = 1, \ldots, N;$$

(D) $\nu_j(t) \leqslant 0$, $\nu_j(t) w_t^j(x(t)) = 0$ *for all* $j = 1, \ldots, p_t$; $t = 0, 1, \ldots, N$.

Proof. Let $V_t(x)$ be the set of all points

$$\tilde{f}_t(x, u) = (f_t^0(x, u), f_t^1(x, u), \ldots, f_t^n(x, u)),$$

obtained when u runs through set $U_t(x)$. Since functions $f_t^\alpha(x, u)$ are continuous, while set $U_t(x)$ is compact, it follows that set $V_t(x)$ is compact. Moreover, by assumption, set $V_t(x)$ is convex. Finally, $V_t(x)$ depends continuously on x, since $U_t(x)$ depends continuously on x, while functions $f_t^\alpha(x, u)$ are continuous. Consequently, the demands imposed on sets $V_t(x)$ in Theorem 46.1 are met in this case.

Theorem 46.2 leads to certain results which were established earlier by various authors. As an example, let us show how a theorem obtained by Halkin can be derived (see first footnote on p. 63). To do this, we assume that we have object (40.1), (42.1), that is, that the control domains $U_t(x) = U_t$ do not depend on x. Then, as the functions $\chi_t(x)$ we can take $\chi_t(x) = u(t) = \text{const}$. Assuming as well that $n_t = 0$, we get the following statement from Theorem 46.2.

THEOREM 46.3. *We consider controlled object* (40.1), (40.2), (40.5), *in which the control domains do not depend on* x (see (42.1)) *and are compact, while the set* $V_t(x)$ *of all points of form* $\tilde{f}_t(x, u)$, $u \in U_t$, *is convex for any* $x \in E^n$. *Moreover, let us assume that functions* $f_t^i(x, u)$ *are continuous over the pair of variables* x, u *and are continuously differentiable with respect to* x, *and that sets* M_t, $t = 0, 1, \ldots, N$, *are specified by relations* (46.19), (46.20). *Finally, let* (40.3), (40.4) *be some process in the object considered that satisfies the constraints* $x(t) \in M_t$, $t = 0, 1, \ldots, N$.

For optimality of this process (in the sense of a minimum of functional (40.5) *with constraints* $x(t) \in M_t$), *there must exist a number* $\psi_0 \leqslant 0$ *and vectors*

$$\psi(t) = \{\psi_1(t), \ldots, \psi_n(t)\}, \quad t = 1, \ldots, N;$$
$$\mu(t) = \{\mu_1(t), \ldots, \mu_{r_t}(t)\}, \quad t = 0, 1, \ldots, N;$$
$$\nu(t) = \{\nu_1(t), \ldots, \nu_{p_t}(t)\}, \quad t = 0, 1, \ldots, N,$$

such that the following conditions are satisfied (function $H_t(x, u)$ *being defined by formula* (42.2)*):*

(A) *if* $\psi_0 = 0$, *then for at least one* t *at least one of the vectors* $\psi(t)$, $\mu(t)$, $\nu(t)$ *will be different from zero;*

(B) $$\psi(t) = \operatorname{grad}_x H_{t+1}(x(t), u(t+1)) + $$
$$+ \sum_{j=1}^{r_t} \mu_j(t) \operatorname{grad} W_t^j(x(t)) + \sum_{j=1}^{p_t} \nu_j(t) \operatorname{grad} w_t^j(x(t)), \quad t = 0, 1, \ldots, N,$$

where it is assumed that $\psi(0) = 0$ *and* $H_{N+1} \equiv 0$;

(C) $$H_t(x(t-1), u(t)) = \max_{u \in U_t} H_t(x(t-1), u), \quad t = 1, \ldots, N;$$

(D) $$\nu_j(t) \leqslant 0, \quad \nu_j(t)\, w_t^j(x(t)) = 0 \quad \textit{for all} \quad j = 1, \ldots, p_t; \ t = 0, 1, \ldots, N.$$

NOTE 46.4. Assume that all the vectors $\operatorname{grad} W_t^i(x(t))$, $\operatorname{grad} w_t^j(x(t))$ are different from zero, and let L_t^i be the hyperplane

$$(x - x(t)) \operatorname{grad} W_t^i(x(t)) = 0,$$

while Q_t^j is the half-space

$$(x - x(t)) \operatorname{grad} w_t^j(x(t)) \leqslant 0.$$

In addition, we assume that for each $t = 0, \ldots, N$ the system of convex cones

$$L_t^i,\ Q_t^j,\ P_t^{(h)},\ i = 1, \ldots, r_t;\ h = 1, \ldots, n_t;\ j \in J_t(x(t)), \tag{46.21}$$

does not possess separability in space E^n. Then condition (A) in Theorem 46.2 can be replaced by the following condition:

(A′) *if* $\psi_0 = 0$, *then for at least one* t *at least one of the vectors* $\psi(t)$ *is different from zero.*

Actually, if the number ψ_0 and the vectors $\psi(t)$, $\mu(t)$, $\nu(t)$, $b_t^{(i)}$ are selected so as to satisfy the conditions of Theorem 46.2, and if we assume as well that all the numbers ψ_0, $\psi_\alpha(t)$ $(\alpha = 1, \ldots, n;\ t = 1, \ldots, N)$ are zero, then in view of condition (A) at least one of the vectors $\mu(t)$, $\nu(t)$, $b_t^{(i)}$ will be different from zero for at least one t. In view of condition (B), we have for this t:

$$-\sum_{i=1}^{r_t} \mu_i(t) \operatorname{grad} W_t^i(x(t)) - \sum_{j=1}^{p_t} \nu_j(t) \operatorname{grad} w_t^j(x(t)) + \sum_{i=1}^{n_t} b_t^{(i)} = 0$$

(where not all of the vectors on the left-hand side are zero), which from Theorem 32.2 is seen to contradict the assumption that system of convex cones (46.21) does not possess separability in E^n.

NOTE 46.5. If we consider the *fundamental problem*, that is, if M_0 consists of just a single point, while sets $M_1, \ldots, M_N$ coincide with the whole space E^n, then in Theorem 46.3 relation (B) becomes unnecessary for $t = 0$ (see footnote on p. 353), while for $t = 1, \ldots, N$ functions W_t^i and w_t^j do not appear in this relation. In other words, for the fundamental problem the relations (B) and (C) written above turn exactly into relations (10.10), (10.12), (10.13). Thus Theorem 46.3 (with no phase constraints) becomes the *discrete maximum principle* considered back in Subsection 10.

Note with regard to this that in Theorems 46.1 through 46.3 the additional requirement that sets $V_t(x)$ be *convex* is imposed. Without this additional requirement, the discrete maximum principle formulated in Subsection 10 is *invalid* (we saw this in Example 10.3). Consequently, the results presented in the present subsection constitute a correctly proven (albeit considerably more general) version of the discrete maximum principle formulated in Subsection 10 in the form of a hypothesis.

The following theorem indicates how the local sections can be selected in the case where the control domain $U_t(x)$ is defined with the aid of a system of equations and inequalities.

THEOREM 46.6. *Consider controlled object* (40.1), (40.2), *in which the control domain* $U_t(x)$ *is defined by relations* (45.1), (45.2), *and let* (40.3), (40.4) *be some process in this object. Finally, let* $I_t(x(t-1), u(t))$ *be the set of all numbers* $i = 1, \ldots, q_t$, *for which the equation* $g_t^i(x(t-1), u(t)) = 0$ *is valid. If functions* G_t^j, g_t^i *are smooth, while the system of vectors*

$$\left.\begin{array}{ll} \operatorname{grad}_u G_t^j(x(t-1), u(t)), & j = 1, \ldots, k_t, \\ \operatorname{grad}_u g_t^i(x(t-1), u(t)), & i \in I_t(x(t-1), u(t)), \end{array}\right\} \tag{46.22}$$

is linearly independent, then in the object being considered there must exist a local section χ_t *corresponding to the values* t, $x(t-1)$ *and* $u(t)$:

$$\chi_t(x(t-1)) = u(t), \quad \chi_t(x) \in U_t(x).$$

Proof. Let us call l_t the number of elements of set $I_t(x(t-1), u(t))$ and let us consider the functional matrix

$$\begin{pmatrix} \dfrac{\partial G_t^j(x(t-1), u(t))}{\partial u^1} & \ldots & \dfrac{\partial G_t^j(x(t-1), u(t))}{\partial u^r} \\ \dfrac{\partial g_t^i(x(t-1), u(t))}{\partial u^1} & \ldots & \dfrac{\partial g_t^i(x(t-1), u(t))}{\partial u^r} \end{pmatrix}$$
$$j = 1, \ldots, k_t; \quad i \in I_t(x(t-1), u(t)),$$

having $k_t + l_t$ rows. Since vectors (46.22) (that is, the rows of this matrix) are linearly independent, this matrix must have a rank $k_t + l_t$. Consequently, the system of relations

$$G_t^j(x, u) = 0, \quad j = 1, \ldots, k_t; \quad g_t^i(x, u) = 0, \quad i \in I_t(x(t-1), u(t)), \tag{46.23}$$

is solvable, in the vicinity of points $x(t-1)$, $u(t)$, for *some* $k_t + l_t$ of the variables $u^1, \ldots, u^r$. Without any loss of generality, we can assume system of relations (46.23) to be solvable for the *first* m variables $u^1, \ldots, u^m$, where $m = k_t + l_t$. In other words, there exist smooth functions

$$\varphi_t^\alpha(x^1, \ldots, x^n, u^{m+1}, \ldots, u^r), \qquad \alpha = 1, \ldots, m,$$

defined in certain neighborhoods of points $x(t-1)$, $u(t)$, such that

$$\varphi_t^\alpha(x^1(t-1), \ldots, x^n(t-1), u^{m+1}(t), \ldots, u^r(t)) = u^\alpha(t), \tag{46.24}$$
$$\alpha = 1, \ldots, m,$$

where the substitution

$$u^{\alpha} = \varphi_t^{\alpha}(x^1, \ldots, x^n, u^{m+1}, \ldots, u^r), \qquad \alpha = 1, \ldots, m, \tag{46.25}$$

turns relations (46.23) into identities in $x^1, \ldots, x^n, u^{m+1}, \ldots, u^r$.

If we set

$$\chi_t^{\alpha}(x) = \varphi_t^{\alpha}(x^1, \ldots, x^n; u^{m+1}(t), \ldots, u^r(t)), \qquad \alpha = 1, \ldots, m,$$
$$\chi_t^{\beta}(x) = u^{\beta}(t), \qquad \beta = m+1, \ldots, r,$$

then we obtain the smooth function

$$\chi_t(x) = (\chi_t^1(x), \ldots, \chi_t^m(x), \chi_t^{m+1}(x), \ldots, \chi_t^r(x)),$$

defined in some neighborhood of point $x(t-1)$. In view of (46.24), we have $\chi_t(x(t-1)) = u(t)$.

Moreover, since substitution (46.25) turns relations (46.23) into identities, it follows that

$$G_t^j(x, \chi_t(x)) \equiv 0; \quad j = 1, \ldots, k_t;$$
$$g_t^i(x, \chi_t(x)) \equiv 0, \quad i \in I_t(x(t-1), u(t)).$$

Finally, for $i \notin I_t(x(t-1), u(t))$ we have

$$g_t^i(x(t-1), \chi_t(x(t-1))) = g_t^i(x(t-1), u(t)) < 0,$$

so that (after reducing, if necessary, the neighborhood of point $x(t-1)$ in which function $\chi_t(x)$ is defined) it can be assumed that $g_t^i(x, \chi_t(x)) < 0$ for $i \notin I_t(x(t-1), u(t))$ throughout the entire domain of definition of function $\chi_t(x)$. Therefore,

$$G_t^j(x, \chi_t(x)) = 0, \quad j = 1, \ldots, k_t;$$
$$g_t^i(x, \chi_t(x)) \leqslant 0, \quad i = 1, \ldots, q_t,$$

in the domain of definition of function $\chi_t(x)$, that is, $\chi_t(x) \in U_t(x)$.

THEOREM 46.7 *(maximum principle). Consider controlled object* (40.1), (40.2), (40.5), *in which the control domain* $U_t(x)$ *is defined by relations* (45.1), (45.2) *and is compact, while the set* $V_t(x)$ *of all points of form* $\tilde{f}_t(x, u)$, $u \in U_t(x)$, *is convex for any* $x \in E^n$. *Let us call* $I_t(x, u)$ *the set of all numbers* $i = 1, \ldots, q_t$ *for which the equation* $g_t^i(x, u) = 0$ *is valid, and let us assume that for* $u \in U_t(x)$ *the vectors*

$$\operatorname{grad}_u G_t^j(x, u), \; j = 1, \ldots, k_t; \quad \operatorname{grad}_u g_t^i(x, u), \; i \in I_t(x, u), \tag{46.26}$$

are linearly independent. Moreover, we assume that sets M_t, $t=0, 1, \ldots, N$, *have the same form as in Theorem* 46.2, *all the functions* f_t^i, W_t^i, w_t^i *being smooth. Let* (40.3), (40.4) *be some process in the considered object satisfying the phase constraints* $x(t) \in M_t$, $t=0, 1, \ldots, N$, *and let* $P_t^{(i)}$ *be a covering of set* $\Pi_t^{(i)}$ *at point* $x(t)$.

For optimality of this process (in the sense of a minimum of functional (40.5) *with constraints* $x(t) \in M_t$) *there must exist a number* $\psi_0 \leqslant 0$, *vectors*

$$\begin{aligned} &\psi(t) = \{\psi_1(t), \ldots, \psi_n(t)\}, \qquad t=1, \ldots, N;\\ &\mu(t) = \{\mu_1(t), \ldots, \mu_{r_t}(t)\}, \qquad t=0, 1, \ldots, N;\\ &\nu(t) = \{\nu_1(t), \ldots, \nu_{p_t}(t)\}, \qquad t=0, 1, \ldots, N;\\ &\quad b_t^{(i)}, \qquad i=1, \ldots, n_t; \quad t=0, 1, \ldots, N, \end{aligned}$$

and $r \times n$ *matrices* $\Lambda(t)$, $t=1, \ldots, N$, *all such that the direction of vector* $b_t^{(i)}$ *lies (in space* E^n) *in the dual cone* $D(P_t^{(i)})$ *and the following conditions are satisfied (function* $H_t(x, u)$ *being defined by formula* (42.2)):

(A) *if* $\psi_0 = 0$, *then for at least one* t *at least one of the vectors* $\psi(t)$, $\mu(t)$, $\nu(t)$, $b_t^{(i)}$ *is different from zero;*

(B)
$$\begin{aligned} \psi(t) = {} & \operatorname{grad}_x H_{t+1}(x(t), u(t+1)) + \\ & + (\operatorname{grad}_u H_{t+1}(x(t), u(t+1)))\Lambda(t+1) + \\ & + \sum_{j=1}^{r_t} \mu_j(t) \operatorname{grad} W_t^j(x(t)) + \sum_{j=1}^{p_t} \nu_j(t) \operatorname{grad} w_t^j(x(t)) - \sum_{j=1}^{n_t} b_t^{(j)}, \\ & t=0, 1, \ldots, N, \end{aligned}$$

where it is assumed that $\psi(0) = \mathbf{0}$, $H_{N+1} \equiv 0$;

(C) $H_t(x(t-1), u(t)) = \max\limits_{u \in U_t(x(t-1))} H_t(x(t-1), u)$, $t=1, \ldots, N$;

(D) $\nu_j(t) \leqslant 0$, $\nu_j(t) w_t^j(x(t)) = 0$; $j=1, \ldots, p_t$, $t=0, 1, \ldots, N$;

(E) $\operatorname{grad}_x g_t^i(x(t-1), u(t)) + (\operatorname{grad}_u g_t^i(x(t-1), u(t))\Lambda(t) = \mathbf{0}$ *for those* i *values for which* $g_t^i(x(t-1), u(t)) = 0$ $(t=1, \ldots, N)$;

(F) $\operatorname{grad}_x G_t^j(x(t-1), u(t)) + (\operatorname{grad}_u G_t^j(x(t-1), u(t)))\Lambda(t) = \mathbf{0}$, $j=1, \ldots, k_t$, $t=1, \ldots, N$.

Proof. Since for $u \in U_t(x)$ the system of vectors (46.26) is linearly independent, this implies that set $U_t(x)$ depends continuously on x. In addition, from Theorem 46.6 it follows that local sections exist in the given object. For each $t=1, \ldots, N$ we select the local section χ_t corresponding to the values of t, $x(t-1)$, $u(t)$. Consequently, Theorem 46.2 is applicable. If we introduce the notation

$$\frac{\partial \chi_t^i(x(t-1))}{\partial x^j} = \lambda_j^i(t), \tag{46.27}$$

$$i=1, \ldots, r; \quad j=1, \ldots, n; \quad t=1, \ldots, N,$$

and if the $r \times n$ matrix $(\lambda_j^i(t))$ is called $\Lambda(t)$, then according to Theorem 46.2 there must exist a number $\psi_0 \leqslant 0$ and vectors

$$\begin{aligned} &\psi(t) = \{\psi_1(t), \ldots, \psi_n(t)\}, \qquad t = 1, \ldots, N; \\ &\mu(t) = \{\mu_1(t), \ldots, \mu_{r_t}(t)\}, \qquad t = 0, 1, \ldots, N; \\ &\nu(t) = \{\nu_1(t), \ldots, \nu_{p_t}(t)\}, \qquad t = 0, 1, \ldots, N; \\ &\quad b_t^{(i)}, \qquad i = 1, \ldots, n_t; \quad t = 0, 1, \ldots, N, \end{aligned}$$

all such that the direction of vector $b_t^{(i)}$ lies (in space E^n) in cone $D(P_t^{(i)})$ and conditions (A), (B), (C), (D) of Theorem 46.2 are satisfied. Hence, taking (46.27) into account, we get conditions (A), (B), (C), (D) of Theorem 46.7.

Moreover, since $\chi_t(x) \in U_t(x)$, it follows that $g_t^i(x, \chi_t(x)) \leqslant 0$ for all $i = 1, \ldots, q_t$. In addition,

$$g_t^i(x(t-1), \chi_t(x(t-1))) = g_t^i(x(t-1), u(t)) = 0 \qquad \text{for } i \in I_t(x(t-1), u(t)).$$

Consequently, for $i \in I_t(x(t-1), u(t))$ the function $g_t^i(x, \chi_t(x))$ reaches a maximum at point $x = x(t-1)$, so that (see (46.27))

$$\frac{\partial}{\partial x^j} g_t^i(x, \chi_t(x)) \Big|_{x = x(t-1)} = \frac{\partial g_t^i(x(t-1), u(t))}{\partial x^j} + \sum_{\alpha=1}^{r} \frac{\partial g_t^i(x(t-1), u(t))}{\partial u^\alpha} \lambda_j^\alpha(t) = 0,$$

$$j = 1, \ldots, n; \quad i \in I_t(x(t-1), u(t)); \quad t = 1, \ldots, N.$$

This gives condition (E) of Theorem 46.7.

Finally, since $\chi_t(x) \in U_t(x)$, it follows that $G_t^j(x, \chi_t(x)) \equiv 0$ for all $j = 1, \ldots, k_t$, so that

$$\frac{\partial}{\partial x^i} G_t^j(x, \chi_t(x)) \Big|_{x = x(t-1)} = \frac{\partial G_t^j(x(t-1), u(t))}{\partial x^i} + \sum_{\alpha=1}^{r} \frac{\partial G_t^j(x(t-1), u(t))}{\partial u^\alpha} \lambda_i^\alpha(t) = 0,$$

$$i = 1, \ldots, n; \quad j = 1, \ldots, k_t; \quad t = 1, \ldots, N.$$

This gives condition (F) of Theorem 46.7.

§ 13. Sufficient Conditions for Optimality

47. Objects with constant control domain. The results of Subsection 37 will be used to obtain the sufficient conditions for optimality. In connection with this, here certain constraints corresponding to the requirements indicated in Subsection 37 will be imposed on the controlled object.

First of all, note that in Theorem 37.7 the set Ω' is defined by *linear* equations. Therefore, we can limit ourselves here to discrete controlled objects which are *linear* in x and u. In other words, equations (40.1) now have the form

$$x^i(t)=c_1^i(t)\,x^1(t-1)+\ \dots\ +c_n^i(t)\,x^n(t-1)+d_1^i(t)\,u^1(t)+\ \dots$$
$$\dots\ +d_r^i(t)\,u^r(t)+e^i(t),\qquad i=1,\ \dots,\ n,\quad t=1,\ \dots,\ N. \tag{47.1}$$

Moreover, condition (37.1), which was imposed on function $F^0(z)$ in Subsection 37, now becomes (cf. (37.2))

$$f_t^0(x,u)\geqslant f_t^0(x(t-1),u(t))+(x-x(t-1))\,\mathrm{grad}_x f_t^0(x(t-1),u(t))+$$
$$+(u-u(t))\,\mathrm{grad}_u f_t^0(x(t-1),u(t)),\qquad t=1,\ \dots,\ N. \tag{47.2}$$

In particular, condition (47.2) will be satisfied if for each $t=1,\ \dots,\ N$ the function $f_t^0(x,u)$ is *convex* (in the space of all the variables $x^1,\ \dots,\ x^n,\ u^1,\ \dots,\ u^r$).

Note that condition (47.2) was formulated for the problem of finding the *minimum* of functional (40.5). For the problem of the maximum of this functional, the inequality sign in condition (47.2) has to be reversed (in particular, this condition will be satisfied if functions $f_t^0(x,u)$, $t=1,\ \dots,\ N$, are *concave*). By the way, the problem of the maximum of functional J (see (40.5)) can always be replaced by the problem of the minimum of functional $-J$.

The conditions which we will impose on sets U_t $(t=1,\ \dots,\ N)$ and M_τ $(\tau=0,\ 1,\ \dots,\ N)$ are analogous to conditions (b), (c), (d) in Subsection 37 (pp. 289–290); they will enter into the formulation of the theorems.

THEOREM 47.1. *Let* (40.3), (40.4) *be some process in discrete controlled object* (47.1), (42.1), *the inclusions* $x(t)\in M_t$, $t=0,\ 1,\ \dots,\ N$, *being satisfied. In addition, for each* $t=1,\ \dots,\ N$, *let set* U_t *have the form*

$$U_t=\Omega_t^{(1)}\cap\ \dots\ \cap\Omega_t^{(l_t)}\cap\Xi_t^{(1)}\cap\ \dots\ \cap\Xi_t^{(m_t)},$$

and, for each $\tau=0,\ 1,\ \dots,\ N$ *let set* M_τ *have the form*

$$M_\tau=\Pi_\tau^{(1)}\cap\ \dots\ \cap\Pi_\tau^{(n_\tau)}\cap\Sigma_\tau^{(1)}\cap\ \dots\ \cap\Sigma_\tau^{(s_\tau)},$$

where $\Omega_t^{(i)}$, $\Xi_t^{(i)}$ *are certain sets of the space of variables* $u^1,\ \dots,\ u^r$, *while* $\Pi_\tau^{(i)}$, $\Sigma_\tau^{(i)}$ *are certain sets of the space of variables* $x^1,\ \dots,\ x^n$. *Finally, let* $K_t^{(i)}$ *be a covering of set* $\Omega_t^{(i)}$ *at point* $u(t)$, *let* $L_t^{(i)}$ *be a covering of set* $\Xi_t^{(i)}$ *at* $u(t)$, *let* $P_\tau^{(j)}$ *be a covering of set* $\Pi_\tau^{(j)}$ *at point* $x(\tau)$, *and let* $Q_\tau^{(j)}$ *be a covering of set* $\Sigma_\tau^{(j)}$ *at* $x(\tau)$. *The inclusions*

$$\Omega_t^{(i)} \subset K_t^{(i)}, \quad \Xi_t^{(i)} \subset L_t^{(i)}, \quad \Pi_\tau^{(j)} \subset P_\tau^{(j)}, \quad \Sigma_\tau^{(j)} \subset Q_\tau^{(j)}$$

are assumed to be satisfied (for all i, j, t, τ). *Finally, assume that the functions* $f_t^0(x, u)$ *entering into the definition of functional* (40.5) *satisfy condition* (47.2) *(in particular, this condition is satisfied if functions* $f_t^0(x, u)$ *are convex).*

For optimality of process (40.3), (40.4) *(in the sense of a minimum of functional* (40.5) *with constraints* $x(t) \in M_t$, $t = 0, 1, \ldots, N$) , *it is sufficient that there exist a number* $\psi_0 < 0$ *and vectors*

$$\begin{aligned} \psi(t) = \{\psi_1(t), \ldots, \psi_n(t)\}, \qquad & t = 1, \ldots, N; \\ a_t^{(i)}, \qquad i = 1, \ldots, l_t; \quad & t = 1, \ldots, N; \\ b_\tau^{(j)}, \qquad j = 1, \ldots, n_\tau; \quad & \tau = 0, 1, \ldots, N, \end{aligned}$$

all such that the direction of vector $a_t^{(i)}$ *lies in dual cone* $D(K_t^{(i)})$, *the direction of vector* $b_t^{(j)}$ *lies in dual cone* $D(P_t^{(j)})$, *and conditions* (A) *and* (B), *below, are satisfied. These conditions are formulated using the function* $H_t(x, u)$, *defined as*

$$H_t(x, u) = \psi_0 f_t^0(x, u) + \sum_{i=1}^{n} \psi_i(t) \left(\sum_{\alpha=1}^{n} c_\alpha^i(t)\, x^\alpha + \sum_{\alpha=1}^{r} d_\alpha^i(t)\, u^\alpha + e^i(t) \right):$$

(A) *for any* $t = 0, 1, \ldots, N$ *and any vector* δx *whose direction lies in the intersection of cones* $Q_t^{(j)}$ $(j = 1, \ldots, s_t)$, *the inequality*

$$\left(-\psi(t) + \operatorname{grad}_x H_{t+1}(x(t), u(t+1)) - \sum_{i=1}^{n_t} b_t^{(i)} \right) \delta x \leqslant 0$$

is valid, where it is assumed that $\psi(0) = 0$ *and* $H_{N+1} \equiv 0$;

(B) *for any* $t = 1, \ldots, N$ *and any vector* δu *whose direction lies in the intersection of cones* $L_t^{(i)}$ $(i = 1, \ldots, m_t)$, *the inequality*

$$\left(\operatorname{grad}_u H_t(x(t-1), u(t)) - \sum_{i=1}^{l_t} a_t^{(i)} \right) \delta u \leqslant 0$$

is valid.

Proof. Change (42.6) reduces the stated problem of optimum control to a problem of finding the minimum of the function

$$F^0(z) = f_1^0(x_0, u_1) + f_2^0(x_1, u_2) + \ldots + f_N^0(x_{N-1}, u_N)$$

on set Σ^*, which is defined by the system of equations

$$F_t^i(z) = -x_t^i + c_1^i(t)\,x_{t-1}^1 + \ldots + c_n^i(t)\,x_{t-1}^n + d_1^i(t)\,u_t^1 + \ldots \\ \ldots + d_r^i(t)\,u_t^r + e^i(t) = 0 \qquad (47.3)$$

and the system of inclusions

$$u_t \in U_t, \quad x_\tau \in M_\tau, \qquad t = 1, \ldots, N; \quad \tau = 0, 1, \ldots, N.$$

Theorem 37.7 (with $r=0$, $p=q=0$) and Note 37.8 can be applied directly to solve this problem of minimizing function $F^0(z)$ on set Σ^*. Condition (β) of Theorem 37.7 becomes in this case

$$\sum_{t=0}^{N}\left(\psi_0 \operatorname{grad}_{x_t} F^0(z_0) + \sum_{\alpha=1}^{n}\sum_{\gamma=1}^{N} \psi_\alpha^\gamma \operatorname{grad}_{x_t} F_\gamma^\alpha(z_0)\right)\boldsymbol{\delta x}_t + \\ + \sum_{\tau=1}^{N}\left(\psi_0 \operatorname{grad}_{u_\tau} F^0(z_0) + \sum_{\alpha=1}^{n}\sum_{\gamma=1}^{N} \psi_\alpha^\gamma \operatorname{grad}_{u_\tau} F_\gamma^\alpha(z_0)\right)\boldsymbol{\delta u}_\tau - \\ - \sum_{\tau=1}^{N}\sum_{i=1}^{l_\tau} \boldsymbol{a}_\tau^{(i)}\boldsymbol{\delta u}_\tau - \sum_{t=0}^{N}\sum_{j=1}^{n_t} \boldsymbol{b}_t^{(j)}\boldsymbol{\delta x}_t \leqslant 0$$

for any vector $\boldsymbol{\delta z} = \{\boldsymbol{\delta x}_t, \boldsymbol{\delta u}_\tau\}$ possessing the property that the direction of vector $\boldsymbol{\delta x}_t$ lies in the intersection of cones $Q_t^{(1)}, \ldots, Q_t^{(s_t)}$ $(t=0, 1, \ldots, N)$ and the direction of vector $\boldsymbol{\delta u}_\tau$ lies in the intersection of cones $L_\tau^{(1)}, \ldots, L_\tau^{(m_\tau)}$ $(\tau = 1, \ldots, N)$; here z_0 is the point corresponding to the given process (40.3), (40.4) in view of change (42.6).

Let $\boldsymbol{\delta z} = \{\boldsymbol{\delta x}_t, \boldsymbol{\delta u}_\tau\}$ be an arbitrary vector satisfying the specified conditions, that is, possessing the property that the direction of vector $\boldsymbol{\delta x}_t$ lies in the intersection of cones $Q_t^{(1)}, \ldots, Q_t^{(s_t)}$ $(t=0, 1, \ldots, N)$ and the direction of vector $\boldsymbol{\delta u}_\tau$ lies in the intersection of cones $L_\tau^{(1)}, \ldots, L_\tau^{(m_\tau)}$ $(\tau=1, \ldots, N)$. Then $\boldsymbol{\delta z}$ can be represented as

$$\boldsymbol{\delta z} = \sum_{t=0}^{N} \boldsymbol{\delta z}_t^{(x)} + \sum_{\tau=1}^{n} \boldsymbol{\delta z}_\tau^{(u)},$$

where vector $\boldsymbol{\delta z}_\theta^{(x)}$ has the same component $\boldsymbol{\delta x}_\theta$ as vector $\boldsymbol{\delta z}$, while the other components $\boldsymbol{\delta x}_t$, $\boldsymbol{\delta u}_\tau$ of vector $\boldsymbol{\delta z}_\theta^{(x)}$ are zero, and, analogously, vector $\boldsymbol{\delta z}_\theta^{(u)}$ has the same component $\boldsymbol{\delta u}_\theta$ as vector $\boldsymbol{\delta z}$, while the other components $\boldsymbol{\delta x}_t$, $\boldsymbol{\delta u}_\tau$ of vector $\boldsymbol{\delta z}_\theta^{(u)}$ are zero. If we show that for each vector $\boldsymbol{\delta z}_t^{(x)}$, $\boldsymbol{\delta z}_\tau^{(u)}$ $(t=0, 1, \ldots, N;$

$\tau = 1, \ldots, N)$ the condition (β) given above is satisfied, then by summing we can verify that this condition is satisfied for vector $\boldsymbol{\delta z}$ as well, thereby completing the proof.

However, it is easy to see (taking the form of functions F^0 and F_t^i into account) that for vector $\boldsymbol{\delta z}_t^{(x)}$ condition (β) *is identical to* condition (A) of Theorem 47.1 (we have only to make the change (42.6) and set $\psi_i^t = \psi_i(t)$). Therefore, if condition (A) of Theorem 47.1 is satisfied, then vector $\boldsymbol{\delta z}_t^{(x)}$ satisfies condition (β). Similarly, for vector $\boldsymbol{\delta z}_\tau^{(u)}$ condition (β) is identical to condition (B) of Theorem 47.1.

NOTE 47.2. Let $\tilde{K}_\theta^{(i)}$ be the set of all points $z = (x_t^i, u_\tau^j)$ satisfying the condition $u_\theta \in K_\theta^{(i)}$; similarly, we define the sets $\tilde{L}_\theta^{(i)}$, $\tilde{P}_\theta^{(j)}$, $\tilde{Q}_\theta^{(j)}$ by the inclusions $u_\theta \in L_\theta^{(i)}$, $x_\theta \in P_\theta^{(j)}$, $x_\theta \in Q_\theta^{(j)}$. Finally, let R_t^i be the hyperplane (of the space E^* of all the variables x_t^i, u_τ^j) defined by the equation $F_t^i = 0$ (see (47.3)). *If the system of convex cones*

$$\tilde{K}_t^{(i)},\ \tilde{L}_t^{(i)},\ \tilde{P}_t^{(i)},\ \tilde{Q}_t^{(i)},\ R_t^{(i)} \tag{47.4}$$

(for all i, t) does not possess separability in space E^*, then the conditions stated in Theorem 47.1 *are necessary and sufficient for optimality of process* (40.3), (40.4). This is implied directly by Theorem 37.7.

However, the requirement that system of cones (47.3) not possess separability is difficult to verify. If this requirement is not met, then the sufficient condition specified in Theorem 47.1 will not be necessary; a similar necessary condition is contained in Theorem 43.5.

Assuming in Theorem 47.1 that some of sets $\Omega_t^{(i)}$ are hyperplanes, some are convex (with $K_t^{(i)}$ taken to be their supporting cone), and some are defined by convex constraints (cf. conditions (b′), (b″), (b‴) on pp. 293–294), the same being the case for sets $\Xi_t^{(i)}$, $\Pi_\tau^{(j)}$, $\Sigma_\tau^{(j)}$, we can derive a number of different theorems from Theorem 47.1. Let us now formulate one of these, obtained if sets $\Pi_\tau^{(i)}$ are defined by linear equations and convex constraints, sets U_t are convex, and sets $\Sigma_t^{(i)}$ are absent.

THEOREM 47.3. *Let* (40.3), (40.4) *be some process in discrete controlled object* (47.1), (42.1), *inclusions* $x(t) \in M_t$, $t = 0, 1, \ldots, N$, *being satisfied. In addition, for each* $\tau = 1, \ldots, N$ *set* U_τ *is convex, while set* M_t *for each* $t = 0, 1, \ldots, N$ *is defined by the system of relations*

$$W_t^i(x^1, \ldots, x^n) = 0, \qquad i = 1, \ldots, r_t; \tag{47.5}$$

$$w_t^j(x^1, \ldots, x^n) \leqslant 0, \qquad j = 1, \ldots, p_t, \tag{47.6}$$

where W_t^i are nonconstant affine functions, and w_t^j are smooth convex functions. Finally, assume that the functions $f_t^0(x, u)$ entering into the definition of functional (40.5) *are convex.*

For optimality of process (40.3), (40.4) *(in the sense of a minimum of functional* (40.5) *with constraints* $x(t) \in M_t$, $t = 0, 1, \ldots, N$), *it is sufficient that there exist a number* $\psi_0 < 0$ *and vectors*

$$\begin{aligned} \psi(t) &= \{\psi_1(t), \ldots, \psi_n(t)\}, & t &= 1, \ldots, N; \\ \mu\{t\} &= \{\mu_1(t), \ldots, \mu_{r_t}(t)\}, & t &= 0, 1, \ldots, N; \\ \nu(t) &= \{\nu_1(t), \ldots, \nu_{p_t}(t)\}, & t &= 0, 1, \ldots, N, \end{aligned}$$

all such that all the numbers $\nu_i(t)$ *are nonpositive and the following conditions are satisfied (function* $H_t(x, u)$ *being the same as in Theorem* 47.1*):*

(A) $$\psi_0 \operatorname{grad}_x H_{t+1}(x(t), u(t+1)) - \psi(t) + \\ + \sum_{i=1}^{r_t} \mu_i(t) \operatorname{grad} W_t^i(x(t)) + \sum_{j=1}^{p_t} \nu_i(t) \operatorname{grad} w_t^j(x(t)) = \mathbf{0}, \\ t = 0, 1, \ldots, N,$$

where it is assumed that $\psi(0) = \mathbf{0}$ *and* $H_{N+1} \equiv 0$;

(B) *for any* $t = 1, \ldots, N$ *and any vector* δu *whose direction lies in the supporting cone of set* U_t *at point* $u(t)$, *the following inequality is valid:*

$$\delta u \operatorname{grad}_u H_t(x(t-1), u(t)) \leqslant 0;$$

(C) $\nu_\alpha(t) w_t^\alpha(x(t)) = 0, \quad \alpha = 1, \ldots, p_t; \quad t = 0, 1, \ldots, N.$

Proof. We take each of hyperplanes (47.5) and each set defined by one of constraints (47.6) as the sets $\Pi_t^{(i)}$ entering into Theorem 47.1. Moreover, for each set $\Pi_t^{(i)}$ (they are all convex) we take as $P_t^{(i)}$ the supporting cone of set $\Pi_t^{(i)}$ at point $x(t)$. Therefore, if $\Pi_t^{(i)}$ is one of hyperplanes (47.5), then the vector $b_t^{(i)}$ whose direction lies in cone $D(P_t^{(i)})$ has the form: $-\mu_i(t) \operatorname{grad} W_t^i(x(t))$, where $\mu_i(t)$ is a scalar. Moreover, if set $\Pi_t^{(j)}$ is defined by one of constraints (47.6), with $w_t^j(x(t)) = 0$, it follows that vector $b_t^{(j)}$ has the form: $-\nu_j(t) \operatorname{grad} w_t^j(x(t))$, where $\nu_j(t) \leqslant 0$. Finally, if $w_t^j(x(t)) < 0$, then $D(P_t^{(j)})$ is a point, so that $b_t^{(j)} = \mathbf{0}$, that is, $b_t^{(j)} = -\nu_j(t) \operatorname{grad} w_t^j(x(t))$, where $\nu_j(t) = 0$. It only remains to apply Theorem 47.1.

The following particular case of Theorem 47.1, relating to the *fundamental problem* (that is, the problem with a fixed initial state $x(0) = x_0$ and without any phase restraints), is of interest in that it gives the *necessary and sufficient* condition for optimality.

THEOREM 47.4. *Let* (40.3), (40.4) *be some process with an initial state* $x(0) = x_0$ *in discrete controlled object* (47.1), (42.1). *Moreover, let set* U_t *be convex for each* $t = 1, \ldots, N$. *Finally, assume that functions* $f_t^0(x, u)$ *entering into the definition of functional* (40.5) *satisfy condition* (47.2) *(in particular, this condition is satisfied if functions* $f_t^0(x, u)$ *are convex).*

For optimality of process (40.3), (40.4) *(in the sense of a minimum of functional* (40.5) *for a fixed initial state* $x(0) = x_0$ *and with no constraints on states* $x(t)$, $t = 1, \ldots, N$), *it is necessary and sufficient that there exist a number* $\psi_0 < 0$ *and vectors*

$$\psi(t) = \{\psi_1(t), \ldots, \psi_n(t)\}, \qquad t = 1, \ldots, N,$$

all such that the following conditions are satisfied (function $H_t(x, u)$ *being the same as in Theorem* 47.1*):*

(A) $\psi(t) = \text{grad}_x H_{t+1}(x(t), u(t+1)),\ t = 1, \ldots, N,$ *where it is assumed that* $H_{N+1} \equiv 0$;

(B) *for any* $t = 1, \ldots, N$ *and any vector* δu *whose direction lies in the supporting cone* L_t *of set* U *at point* $u(t)$, *the following inequality is valid:*

$$\delta u \, \text{grad}_u H_t(x(t-1), u(t)) \leqslant 0.$$

Proof. Sufficiency of the conditions formulated is implied directly by Theorem 47.1. To prove necessity according to Note 47.2, we need only establish that the system of convex cones

$$\tilde{L}_t, \ R_t^i \tag{47.7}$$

does not possess separability in the space E^* of variables x_t^i, u_τ^j $(i = 1, \ldots, n;$ $j = 1, \ldots, r; t = 0, 1, \ldots, N; \tau = 1, \ldots, N)$.

Let us call D_θ $(\theta = 1, \ldots, N)$ the set of all points $z = (x_t, u_\tau)$ of space E^* satisfying the conditions

$$x_t = x(t) \ \text{ for all } \ t = 1, \ldots, N, \quad u_\tau = u(\tau) \ \text{ for } \ \tau \neq \theta$$

(u_θ being arbitrary). In addition, if D_θ^* is the set of all points $z = (x_t, u_\tau)$ satisfying the condition $u_\theta = u(\theta)$ (x_t and u_τ being arbitrary for $\tau \neq \theta$), then $D_\theta \cap D_\theta^*$ is the point z_0 (corresponding to the considered process (40.3), (40.4) with change (42.6)). Taking z_0 to be the zero point of space E^*, we find that D_θ and D_θ^* are subspaces of the vector space E^*, where $E^* = D_\theta \oplus D_\theta^*$. In addition, note that $\tilde{L}_t \supset D_t^*$ $(t = 1, \ldots, N)$.

It is easy to see that the system of subspaces

$$R_t^i, \ D_t^*, \qquad i = 1, \ldots, n; \quad t = 1, \ldots, N, \tag{47.8}$$

does not possess separability in E^*. Actually, hyperplane R_t^i is defined by the equation $F_t^i(z) = 0$ (see (47.3)), while subspace D_t^* has the form $D_t^* = D_t^1 \cap \ldots \cap D_t^r$, where hyperplane D_t^i is defined by the equation

$$u_t^i = u^i(t). \tag{47.9}$$

Let us now number all the variables x_t^i, u_τ^j in the following order:

$$u_1^1, \ldots, u_1^r, u_2^1, \ldots, u_2^r, \ldots, u_N^1, \ldots, u_N^r,$$
$$x_1^1, \ldots, x_1^n, x_2^1, \ldots, x_2^n, \ldots, x_N^1, \ldots, x_N^n.$$

Then the matrix consisting of the coefficients of the unknowns in relations (47.9), (47.3) will be triangular, with the numbers ± 1 along its principal diagonal. Consequently, the determinant of this matrix is different from zero, so that the system of hyperplanes R_t^i, D_t^j does not possess separability in E^*.

However, then system of subspaces (47.8) does not possess separability in E^* either. Even more so, system of cones (47.7) will not possess separability (since $\tilde{L}_t \supset D_t^*$).

NOTE 47.5. It is easy to see that condition (B) of Theorem 47.4 can be written in the following equivalent form:

$$H_t(x(t-1),\ u(t)) = \max_{u \in U_t} H_t(x(t-1),\ u), \quad t = 1, \ldots, N.$$

Actually, since $\psi_0 < 0$, it follows from (47.2) that the function $H_t(x(t-1), u)$ possesses the following property:

$$H_t(x(t-1), u) \leqslant H_t(x(t-1), u(t)) + \\ + (u - u(t)) \operatorname{grad}_u H_t(x(t-1), u(t)).$$

According to condition (B) of Theorem 47.4, we now know that the function $H_t(x(t-1), u)$ of variable u, considered on the supporting cone L_t of set U_t, reaches a maximum at a point $u = u(t)$. All the more so, this function, considered on set $U_t \subset L_t$, reaches a maximum at point $u = u(t)$, that is, condition (B′) is satisfied. Conversely, condition (B′) obviously implies condition (B).

NOTE 47.6. Theorem 47.4 can also be derived from Theorem 38.2. Change (42.6) reduces the considered problem of optimal control to a problem of minimizing a function $F^0(z)$, considered on set Σ. The latter set is defined by linear equations $F_t^i(z) = 0$ (see (47.3)) and inclusions $u_t \in U_t$, where sets U_t are convex. Applying Theorem 38.2 (which is permissible, since system of cones (47.7) does not possess separability in E^*), we are led to the following statement.

In order for function $F^0(z)$. considered on set Σ, to reach a minimum at point z_0, it is necessary and sufficient that there exist numbers $\psi_0 < 0$ and ψ_i^t $(i = 1, \ldots, n;$ $t = 1, \ldots, N)$, such that the function

$$H(z) = \psi_0 F^0(z) + \sum_{t=1}^{N} \sum_{i=1}^{n} \psi_i^t F_t^i(z), \tag{47.10}$$

considered for any x_t and for $u_t \in U_t$, reaches a maximum at point z_0.

If function (47.10) reaches a maximum at z_0, then $\operatorname{grad}_{x_t} H(z_0)=0$ (this gives condition (A) of Theorem 47.4) and condition (B′) is satisfied.

Conversely, if conditions (A) and (B) of Theorem 47.4 are satisfied, then the affine function

$$D(z)=\psi_0(z-z_0)\operatorname{grad} F^0(z)+\sum_{t=1}^{N}\sum_{i=1}^{n}\psi_i^t F_t^i(z)=$$
$$=\psi_0\sum_{t=1}^{N}(x_{t-1}-x(t-1))\operatorname{grad}_x f_t^0(x(t-1),\ u(t))+$$
$$+\psi_0\sum_{t=1}^{N}(u_t-u(t))\operatorname{grad}_u f_t^0(x(t-1),u(t))+$$
$$+\sum_{t=1}^{N}\sum_{i=1}^{n}\psi_i^t F_t^i(z)-\sum_{t=1}^{N}\sum_{i=1}^{n}\psi_i^t F_t^i(z_0)$$

does not depend (in view of condition (A)) on variables x_t^i, so that (in view of condition (B)) for $u_t\in U_t$ it reaches a maximum (equal to zero) at point z_0. However, since from (47.2) we know that

$$H(z)\leqslant H(z_0)+D(z),$$

it follows that $H(z)\leqslant H(z_0)$, that is, function (47.10) reaches a maximum at z_0.

In conclusion, let us take a look at the case in which equations (40.1) defining the discrete controlled object are convex rather than linear.

THEOREM 47.7. *Let* (40.3), (40.4) *be some process in discrete controlled object* (40.1), (42.1), *the inclusions* $x(t)\in M_t$, $t=0, 1, \ldots, N$, *being satisfied. Assume that all the functions* $f_t^i(x, u)$ *(appearing on the right-hand sides of relations* (40.1)*) are convex in the aggregate of variables* $x^1, \ldots, x^n, u^1, \ldots, u^r$. *In addition, assume that sets* U_t *and* M_τ *have the same form as in Theorem* 47.1 *and that functions* $f_t^0(x, u)$ *satisfy condition* (47.2) *(in particular, this condition will be satisfied if functions* $f_t^0(x, u)$ *are convex).*

For optimality of process (40.3), (40.4) *(in the sense of a minimum of functional* (40.5) *with constraints* $x(t)\in M_t$, $t=0, 1, \ldots, N$) , *it is sufficient that there exist a number* $\psi_0<0$ *and vectors*

$$\begin{aligned}&\psi(t)=\{\psi_1(t), \ldots, \psi_n(t)\}, && t=1, \ldots, N,\\ &a_t^{(i)}, \quad i=1, \ldots, l_t; && t=1, \ldots, N;\\ &b_\tau^{(j)}, \quad j=1, \ldots, n_\tau; && \tau=0, 1, \ldots, N,\end{aligned}$$

all such that the direction of vector $a_t^{(i)}$ *lies in dual cone* $D(K_t^{(i)})$, *the direction of vector* $b_\tau^{(j)}$ *lies in dual cone* $D(P_\tau^{(j)})$, *and the following conditions are satisfied (function* $H_t(x,u)$ *being defined by formula* (42.2)*):*

(A) *for any* $t=0, 1, \ldots, N$ *and any vector* δx *whose direction lies in the intersection of cones* $Q_t^{(j)}$, *the following inequality is valid:*

$$\left(-\psi(t)+\operatorname{grad}_x H_{t+1}(x(t), u(t+1))-\sum_{i=1}^{n_t} b_t^{(i)}\right)\delta x \leqslant 0,$$

where it is assumed that $\psi(0)=0$ *and* $H_{N+1}\equiv 0$;

(B) *for any* $t=1, \ldots, N$ *and any vector* $\delta u=\{\delta u^1, \ldots, \delta u^r\}$ *whose direction lies in the intersection of cones* $L_t^{(i)}$ $(i=1, \ldots, m_t)$, *the following inequality is valid:*

$$\left(\operatorname{grad}_u H_t(x(t-1), u(t))-\sum_{i=1}^{l_t} a_t^{(i)}\right)\delta u \leqslant 0;$$

(C) $$\psi_i(t)\leqslant 0, \quad i=1, \ldots, n; \quad t=1, \ldots, N.$$

Proof. Let W_t^i be the set of all points $z=(x_t^i, u_\tau^i)$ satisfying the condition

$$F_t^i(z)=f_t^i(x_{t-1}, u_t)-x_t^i\leqslant 0.$$

Since f_t^i is a convex function, set W_t^i will be convex. If z_0 is the point corresponding to process (40.3), (40.4) in view of change (42.6), then z_0 has to lie on the *boundary* of set W_t^i (cf. (40.1)), while the half-space

$$(z-z_0)\operatorname{grad}_z F_t^i(z_0)\leqslant 0$$

serves as the supporting cone of convex set W_t^i at point z_0.

Change (42.6) reduces the problem to a search for a minimum of function $F^0(z)$ in the set defined by the system of inclusions

$$z\in W_t^i, \quad u_t\in U_t, \quad x_\tau\in M_\tau, \qquad i=1, \ldots, n; \quad t=1, \ldots, N;$$
$$\tau=0, 1, \ldots, N.$$

To solve this problem, we apply Theorem 37.5 and Note 37.6. In this case condition (β) of Theorem 37.5 has the form

$$\sum_{t=0}^{N}(\psi_0 \operatorname{grad}_{x_t} F^0(z_0))\,\delta x_t+\sum_{\tau=1}^{N}(\psi_0 \operatorname{grad}_{u_\tau} F^0(z_0))\,\delta u_\tau -$$
$$-\sum_{\tau=1}^{N}\sum_{i=1}^{l_t} a_\tau^{(i)}\delta u_\tau-\sum_{t=0}^{N}\sum_{j=1}^{n_t} b_t^{(j)}\delta x_t-\sum_{t=1}^{N}\sum_{i=1}^{n} c_t^{(i)}\delta z\leqslant 0$$

for any vector

$$\delta z = \{\delta x_t, \delta u_\tau\}$$

possessing the property that the direction of vector δx_t lies in the intersection of cones $Q_t^{(1)}, \ldots, Q_t^{(s_t)}$ $(t = 0, 1, \ldots, N)$, while the direction of vector δu_τ lies in the intersection of cones $L_\tau^{(1)}, \ldots, L_\tau^{(m_\tau)}$ $(\tau = 1, \ldots, N)$. Here $a_t^{(i)}$ is a vector whose direction lies in dual cone $D(K_t^{(i)})$, $b_\tau^{(j)}$ is a vector whose direction lies in dual cone $D(P_\tau^{(j)})$, and $c_t^{(i)}$ is a vector whose direction lies in the cone dual to the half-space

$$(z - z_0)\operatorname{grad}_z F_t^i(z_0) \leqslant 0,$$

that is,

$$c_t^{(i)} = -\psi_i(t)\operatorname{grad}_z F_t^i(z_0),$$

where $\psi_i(t) \leqslant 0$. From here on, the proof can be completed just as in the case of Theorem 47.1.

48. Objects with variable control domain. Here we will consider objects (47.1), (40.2), whose control domain $U_t(x)$ for each

$$x = (x^1, \ldots, x^n) \in E^n$$

is specified by system of equations and inequalities (45.1), (45.2); functions G_t^j are assumed to be affine, and functions g_t^i are assumed to be convex. With regard to functions f_t^0 and sets M_t, the assumptions made are the same as in the preceding subsection. As a result, we obtain the following theorem:

THEOREM 48.1. *Consider controlled object* (47.1), (40.2), *where set* $U_t(x)$ *for each given* $x = (x^1, \ldots, x^n) \in E^n$ *is defined by system of relations (45.1),* (45.2), *where functions* G_t^j *are affine and functions* g_t^i *are convex. Let* (40.3), (40.4) *be some process in the object considered, the following inclusions being satisfied:*

$$x(t) \in M_t, \quad t = 0, 1, \ldots, N,$$

where M_τ *is a set in the space* E^n *of variables* $x^1, \ldots, x^n$ *having the form*

$$M_\tau = \Pi_\tau^{(1)} \cap \ldots \cap \Pi_\tau^{(n_\tau)} \cap \Sigma_\tau^{(1)} \cap \ldots \cap \Sigma_\tau^{(s_\tau)}.$$

For each $\tau = 0, 1, \ldots, N$ *the convex cones* $P_\tau^{(i)}, Q_\tau^{(j)}$ $(i = 1, \ldots, n_\tau; j = 1, \ldots, s_\tau)$ *with a vertex* $x(\tau)$ *are defined in* E^n, *these cones being the coverings, respectively, of sets* $\Pi_\tau^{(i)}, \Sigma_\tau^{(j)}$, *where the inclusions*

$$\Pi_\tau^{(i)} \subset P_\tau^{(i)}, \quad \Sigma_\tau^{(j)} \subset Q_\tau^{(j)}$$

are satisfied. Finally, assume that functions $f_t^0(x, u)$ *entering into the definition of functional* (40.5) *satisfy condition* (47.2) *(in particular, this condition is satisfied if functions* $f_t^0(x,u)$ *are convex).*

For optimality of process (40.3), (40.4), *in the sense of a minimum of functional* (40.5) *with constraints*

$$x(t) \in M_t, \qquad t = 0, 1, \ldots, N,$$

it is sufficient that there exist a number $\psi_0 < 0$ *and vectors*

$$\begin{aligned} &\psi(t) = \{\psi_1(t), \ldots, \psi_n(t)\}, && t = 1, \ldots, N; \\ &\varphi(t) = \{\varphi_1(t), \ldots, \varphi_{k_t}(t)\}, && t = 1, \ldots, N; \\ &\lambda(t) = \{\lambda_1(t), \ldots, \lambda_{q_t}(t)\}, && t = 1, \ldots, N; \\ &b_t^{(i)}, \quad i = 1, \ldots, n_t; && t = 0, 1, \ldots, N, \end{aligned}$$

all such that the direction of vector $b_t^{(i)}$ *lies in cone* $D(P_t^{(i)})$ *and conditions* (A) *through* (C), *below, are satisfied. These conditions are formulated using the function* $H_t(x, u)$, *defined as*

$$H_t(x, u) = \psi_0 f_t^0(x, u) + \sum_{i=1}^{n} \psi_i(t) \left(\sum_{\alpha=1}^{n} c_\alpha^i(t) x^\alpha + \sum_{\alpha=1}^{r} d_\alpha^i(t) u^\alpha + e^i(t) \right) + \\ + \sum_{\alpha=1}^{k_t} \varphi_\alpha(t) G_t^\alpha(x, u) + \sum_{\alpha=1}^{q_t} \lambda_\alpha(t) g_t^\alpha(x, u):$$

(A) *for any* $t = 0, 1, \ldots, N$ *and any vector* δx *whose direction lies in the intersection of cones* $Q_t^{(j)}$ $(j = 1, \ldots, s_t)$, *the inequality*

$$\left(-\psi(t) + \operatorname{grad}_x H_{t+1}(x(t), u(t+1)) - \sum_{j=1}^{n_t} b_t^{(j)} \right) \delta x \leqslant 0$$

is valid, where it is assumed that $\psi(0) = 0$ *and* $H_{N+1} \equiv 0$;

(B) $\operatorname{grad}_u H_t(x(t-1), u(t)) = 0, \qquad t = 1, \ldots, N;$

(C) $\lambda_\alpha(t) \leqslant 0, \quad \lambda_\alpha(t) g_t^\alpha(x(t-1), u(t)) = 0, \qquad \alpha = 1, \ldots, q_t;$
$t = 1, \ldots, N.$

Proof. Change (42.6) reduces the given optimal-control problem to a problem of a minimum of function $F^0(z)$ (see (42.3)) on set Σ^*, which is defined by system of equations (47.3), (45.1), system of constraints (45.2), and system of inclusions

$$x_t \in M_t (t = 0, 1, \ldots, N).$$

This problem of minimizing function $F^0(z)$ on set Σ^* can be solved directly by applying Theorem 37.7 and Note 37.6. Condition (β) of Theorem 37.7 becomes in this case

$$\sum_{t=0}^{N}\left(\psi_0 \operatorname{grad}_{x_t} F^0(z_0)+\sum_{\alpha=1}^{n}\sum_{\gamma=1}^{N}\psi_\alpha^\gamma \operatorname{grad}_{x_t} F_\gamma^\alpha(z)\right)\delta x_t+$$
$$+\sum_{\tau=1}^{N}\left(\psi_0 \operatorname{grad}_{u_\tau} F^0(z_0)+\sum_{\alpha=1}^{n}\sum_{\gamma=1}^{N}\psi_\alpha^\gamma \operatorname{grad}_{u_\tau} F_\gamma^\alpha(z_0)\right)\delta u_\tau+$$
$$+\sum_{t=0}^{N}\sum_{\gamma=1}^{N}\sum_{\alpha=1}^{k_\gamma}\varphi_\alpha^\gamma(\delta x_t \operatorname{grad}_{x_t} G_\gamma^\alpha(z_0))+$$
$$+\sum_{\tau=1}^{N}\sum_{\gamma=1}^{N}\sum_{\alpha=1}^{k_\gamma}\varphi_\alpha^\gamma(\delta u_\tau \operatorname{grad}_{u_\tau} G_\gamma^\alpha(z_0))+\sum_{t=0}^{N}\sum_{\gamma=1}^{N}\sum_{\beta}\lambda_\beta^\gamma(\delta x_t \operatorname{grad}_{x_t} g_\gamma^\beta(z_0))+$$
$$+\sum_{\tau=1}^{N}\sum_{\gamma=1}^{N}\sum_{\beta}\lambda_\beta^\gamma(\delta u_\tau \operatorname{grad}_{u_\tau} g_\gamma^\beta(z_0))-\sum_{t=0}^{N}\sum_{j=1}^{n_t}a_t^{(j)}\delta x_t\leqslant 0$$

for any vector $\delta z=\{\delta x_t, \delta u_\tau\}$ possessing the property that the direction of vector δx_t lies in the intersection of cones

$$Q_t^{(1)}, \ldots, Q_t^{(s_t)} \qquad (t=0, 1, \ldots, N);$$

here z_0 denotes the point corresponding to the considered process (40.3), (40.4) in view of change (42.6). In addition, we note that in the above condition the summation over index β is carried out over all those numbers $j=1, \ldots, q_\gamma$ for which $g_\gamma^j(z_0)=0$; however, by setting λ_j^γ equal to zero for all the other values of j, we can assume that the summation over β is carried out from 1 to q_γ, where

$$\lambda_\beta^\gamma\leqslant 0, \quad \lambda_\beta^\gamma g_\gamma^\beta(z_0)=0$$

for all β, γ. From here on, the proof is completed just as in the case of Theorem 47.1.

NOTE 48.2. A comparison of the latter theorem with Theorem 45.1 indicates that the sufficient conditions for optimality contained in Theorem 48.1 are very close to the necessary conditions. In order to ascertain just when the conditions specified in Theorem 48.1 are both necessary and sufficient, we have to proceed exactly as in Note 47.2.

Thus, we call $\tilde{P}_\theta^{(i)}$ the set of all points $z=(x_t^i, u_\tau^i)$ satisfying the inclusion $x_\theta \in P_\theta^{(i)}$; and we define $\tilde{Q}_\theta^{(i)}$ by the inclusion $x_\theta \in Q_\theta^{(i)}$. In addition, we call R_t^j the set of all points $z=(x_t^i, u_\tau^i)$ satisfying the condition

$$(x_{t-1}-x(t-1))\operatorname{grad}_x g_t^j(x(t-1), u(t))+$$
$$+(u_t-u(t))\operatorname{grad}_u g_t^j(x(t-1), u(t))\leqslant 0.$$

Finally, let S_t^i and T_t^i be hyperplanes defined, respectively, by the equations

$$G_t^i(x_{t-1}, u_t) = 0, \; F_t^i = 0$$

(see (45.1), (47.3)).

If the system of convex cones

$$\tilde{P}_t^{(i)}, \quad \tilde{Q}_t^{(i)}, \quad R_t^i, \quad S_t^i, \quad T_t^i$$

(for all i, t*) does not possess separability in the space of all variables* x_t^i, u_τ^i, *then the conditions specified in Theorem* 48.1 *are necessary and sufficient for optimality of process* (40.3), (40.4). This follows directly from Theorem 37.7.

NOTE 48.3. In Theorem 48.1 it can be assumed that some of sets $\Pi_\tau^{(i)}$ are hyperplanes, some are defined by inequalities of the type

$$\varphi_\tau^{(i)} \leqslant 0,$$

where $\varphi_\tau^{(i)}$ is a *convex* function, and some are convex, the corresponding supporting cone being taken as $P_\tau^{(i)}$. The same can be said for sets $\Sigma_\tau^{(i)}$. As a result, it is possible to obtain from Theorem 48.1 an even more general theorem and to consider a number of particular cases of it. However, these theorems will not be presented here.

AUTHOR INDEX

Bellman, R. iii, v, 4, 43, 45, 308, 318
Boltyanskii, V.G. 43, 50, 51
Butkovskii, A.G. v

Cannon, M. 63, 281
Chang, S.S.L. v
Cullum, C. 63, 281

Danskin, J.M. 276
Dubovitskii, A.Ya. iii, vi, vii, 224, 282

Fam, H.S. v, vii
Fan, L.T. v, 6, 54

Gabasov, R. v, vii, 63
Gamkrelidze, R.V. 43
Gibbs, W. 275, 276
Glezerman, M.E. 256

Halkin, H. v, 63, 369
Holtzman, J.M. v, 63

Jordan, B.W. v, 63, 281

Katz, S. 54
Khinchin, A.Ya. 15
Kirillova, F.M. v, 63
Krasovskii, N.N. v, vii
Kuhn, H.W. vii

Lagrange, J.L. 69, 269
Lee, E.S. v
Lefshetz, S. vi, 256

Milyutin, A.A. iii, vi, vii, 224, 282
Mishchenko, E.F. vii, 43
Moiseev, N.N. v, 7
Mukurdumov, R.M. vii

Pearson, J. 63
Petrov, B.N. vii
Polak, E. v, 63, 281
Pontryagin, L.S. iii, v, vii, 43, 62, 63, 256
Propoi, A.I. v, 62, 73, 351
Pshenichnyi, B.N. v

Rozonoer, L.I. v, 62
Rozov, N.Kh. vii

Tucker, A.W. vii

Venttsel', E.S. 5

Wang, C.S. v, 6, 54
Weyl, H. vi, 144

SUBJECT INDEX

Admissible control. *See* Control
Affine function. *See* Function
Affine mapping. *See* Mapping
Affine space 112 ff, 226, 230
Approximation theorem 258–262
Axioms 143
  dimension 84–85
  scalar-product 104–105
  vector 76–80
  vector-construction 114–115

Basis 84–87, 124–126, 170–172
  orthonormal 107–108, 171
Bellman's equation 45, 50, 318
Boundary point 151, 175–181, 195–196, 204, 251, 383

Carrier plane. *See* Plane
Chain 253–262
  field of 258–260
Closure 151, 173–174, 208–209, 258
Complement, orthogonal 144, 236
Concavity, local 351
Cone
  convex vi, 189–197, 213 ff
  dual 214–218, 266–267, 271, 334, 356, 384
  separable. *See* Separability
  solid 245
  supporting 187–197, 241–242, 274, 350, 362–364, 379
  "warped" 250
Conjugate 102
Constraints 3, 54, 61, 243–247
  active 66, 296
  phase 342–352
Continued fractions 14–15
Continuous process v, 30, 36–43, 49–53
Control 2, 9, 15, 38, 44, 360
  admissible 2–3, 10, 40, 49–51
  optimal 44–47, 316 ff, 351–354
Control domain. *See* Domain
Controlled object. *See* Object
Controlling parameter 2, 9
Convex set. *See* Set
Convexity, local 306
Coordinates 154–161
  system of 154, 230
Covering. *See* Set, covering of
Cubilage 252–262
Curve, parametrized 239

Dimension 84–88
  reduction of 311
Direct complement 92
Direct sum 93, 367
Direction vector. *See* Vector
Discrete object v, 30, 38–43, 50, 63 ff, 354–358, 367
Domain
  closed 32, 65
  control 2, 10, 54, 71, 342, 353 ff
  of definition 26–27, 151–152, 235, 289
Dubovitskii-Milyutin theorem iii, vi–vii, 224–225, 282–284
Dynamic programming. *See* Programming

Effectiveness criterion 11, 18–19, 54
Endpoints 11, 42, 54, 61, 116–117, 322
Euclidean space 27–29, 70, 142 ff, 173, 182
Existence 35, 79
Existence theorems 226 ff, 351
Extrema 27, 30, 66, 226 ff
  criteria for 265 ff, 362

Feedback 12–13
Field, convex vii, 170–179, 193, 197–202
Final state 18
Function
  affine 136–142, 197–204, 278, 287, 303
  concave 278–280, 290, 375
  continuous 151, 332
  convex 278, 290, 375
  extrema of 226 ff, 326–330
  real 229
  smooth 230, 268, 329
  tangent 229
Fundamental problem 11, 30, 54–56, 317–322, 330–342, 370

Gedanken experiment 1–3
Generality of position 291–292
Geometry
  affine vi
  elementary 144
  Euclidean vi, 112 ff
  multidimensional 76 ff
Gibbs' Lemma 275–277
Gradient 63, 110–112, 229–236

Half-interval 117
Half-line. *See* Ray
Half-space 162–163, 182, 201–202
  closed 141–142, 162, 180, 193, 200, 212
  distinguished 184–187
  open 141–142, 148, 152, 163, 204, 245
  superfluous 179
Homomorphism 95–103, 287
Homothety 135–136, 260
Hull, convex 166–173, 178, 183–189, 207, 212–225, 335
Hyperplane 120 ff
  identical 158–159
  separating. *See* Separability
  supporting 195–203
  tangent 231–234, 245, 272
Hyperplanes, intersection of 180–181, 186
Hypersurface 31, 65
Hypersurfaces, intersection of 31–32

Image 95–99, 188–190
  inverse 95, 131, 142
Initial conditions 2
Initial state 10–11
Interior 174, 187, 225
Interior point 25, 34–35, 116, 150, 153, 175–186, 190–204, 247–249
Intersection of sets 150, 153, 163, 166, 253
  theorem 249–265
Intersections, theory of vi
Isomorphism 86–87, 94, 96, 100, 103
Isoperimetric problem 19–24

Kernel 95, 137–140, 198, 201, 214
Kuhn-Tucker theorem vii

Lag 13–17
Lagrange theorem 69, 269
Law of motion 10
Line segment 116, 170, 201
Line 232
  smooth 231
Linearity 287, 375
Local sections. *See* Sections

Manifold
  linear 120
  smooth 231–233, 237, 264, 282, 346
Mapping
  affine 129–136, 188–194, 226–228
  continuous 151, 160–161, 187, 239–241, 259
  identity 226, 279
  inverse 264
  smooth 263–265
  tangent 226–240
Mathematical programming. *See* Programming
Matrices 98

Maximum, local 74–75
Maximum principle v–vii, 51, 299–308
continuous 54, 62–63
discrete v, 50–63, 74–75, 358–375
Euclidean 59–61
Metric concepts 142–143
Midpoint 116
Minimum, local 306

Necessary condition 168–169, 211, 244, 265–288, 292, 318–320, 329 ff, 380
See also Optimality, conditions
Nonsingularity 244–246, 269–273, 281
Normal. *See* Vector, normal

Object
controlled 2–12, 20, 317–318, 330–340, 367–375, 384
discrete. *See* Discrete object
Optimality 43, 54, 318
conditions 30, 52, 55, 62, 71–73, 324 ff
criteria vii, 25, 30
defined 25
Orthogonal complement. *See* Complement
Orthonormal system. *See* System

Parallelepiped, coordinate 251–255, 261
Phase coordinates 2, 9–10, 322
Phase space 28, 50
Plane 120–129
carrier 173–187, 197, 211–212, 221–222, 241
complementary 127, 161, 212, 233
tangent 232–234, 346
Planes, intersection of 125
Polygon, convex 165
Polyhedron
convex 165–166, 174, 179–188, 201–203
curvilinear 33–36, 265, 284–286
faces of 182–186, 202–204
Prehistory 16–17
Process, discrete. *See* Discrete object

Programming vii
dynamic v, 43–50, 308 ff, 352
Projection 98, 233
one-to-one 233–236
orthogonal 110, 146, 187

Ray 128, 190–196 , 245
interior 283, 298, 365

Scalar 60
Scalar product 102–107, 299
Sections, local vii, 358–375
Separability vii, 204–225, 266, 270–274, 291, 335–340, 378–381
Set
boundary of 151, 173–179
closed 70, 150–152, 174, 187, 352
compact 321–322, 352–365
convex vii, 161 ff, 278, 370
covering of 240–249, 251, 274–279, 291, 364
empty 163, 243
interior of 174
open 150–152, 235
Sets
intersection of. *See* Intersection
separable. *See* Separability
union of. *See* Union
Simplex 170–173, 184–187, 207, 227–228, 248–249
Smoothness 63
Space
affine. *See* Affine space
conjugate 230
Euclidean. *See* Euclidean space
vector. *See* Vector space
Sphere, open 150, 160, 163
Subset 322, 344
Subspaces 88–95
direction 120, 142, 185
sum of 90

Sufficient condition 168, 178, 211, 287–300, 319–320, 375–387
See also Optimality, conditions
Surface, smooth 231
Symmetry 135
System
of coordinates. *See* Coordinates
orthonormal 107–109, 159–160, 234

Topology vi, 150–154, 231, 259, 358
Trajectory 2–3, 10, 36, 44, 74, 330, 335, 360
optimal 45–46

Union 90, 150, 179, 182, 258, 334
Uniqueness 79

Vector 76–78
basis 154
bound 113
collinear 287

Vector
construction of 114–118
covariant 230
direction of 124, 334
length of 104–105
normal 65, 148, 231
orthogonal 107, 144, 148, 233–236
tangent 232, 239, 286
Vector axioms. *See* Axioms
Vector parametric equation 123
Vector space vi, 76 ff, 113
Euclidean 103–112, 142
isomorphic 87
Vector sum 80–81

Weights 23

Zone of activity 243